Ecological Systems
and the Environment

HOUGHTON MIFFLIN COMPANY   BOSTON
Atlanta   Dallas   Geneva, Illinois   Hopewell, New Jersey   Palo Alto   London

Ecological Systems and the Environment

Theodore C. Foin, Jr.

Division of Environmental Studies
University of California, Davis

Copyright © 1976 by Houghton Mifflin Company.

All rights reserved. No part of this work may be reproduced or transmitted in any form or by any means, electronic or mechanical, including photocopying and recording, or by any information storage or retrieval system, without permission in writing from the publisher.

Printed in the U.S.A.
Library of Congress Catalog Card Number: 75-25010
ISBN: 0-395-20666-9

Contents

Preface / xiii

1. "Ecological Systems" and "Environmental Studies" / 6

Basic Concepts about Ecological Systems 7
Biotic Dependencies on Abiotic Factors 8
Material and Energy Transfers 10
System Structure 13
Reactions to Changing Environments 16
A Model of Ecological Systems 18
In Retrospect: Ecological Systems and Environmental Studies 20

2. Ecological Systems: Combinations and Recombinations / 23

Systems Structure 24
System Dynamics 26
Systems Evolution 34
One Viewpoint on Ecological Systems 37

3. Systems Concepts for Environmental Analysis / 40

Systems Methods and Environmental Decision-making 41
Defining a Systems Model 42
Properties Shared by All Systems 44
The Nature of System Boundaries 44
Goal Orientation 45
Hierarchical Organization 45

Section A
About Systems and Ecological Systems

Basic Components of Systems 46
System Feedback 46
Predictability of System Behavior 48
Complex Models of Particular Systems: Global Dynamics 48
The Systems Point of View 51
Appendix 55

Section B
To Cover the Earth?

4. Populations Marching Onward to an Uncertain Future / 64

The Early History and Development of Human Populations 65
Demographic Parameters: The Measure of Population Dynamics 68
Demographic Statistics: The Present 71
Demographic Projection: The Future 73
The von Foerster Growth Model and the U.N. Projections 75
Extensions and Implications 78
A Case Study: Patterns in the United States 83

5. Shutting Off the Population Valve, or Who Does What to Whom? / 88

The Biological Background 88
Social Regulation in Animals and Human Beings 89
Population Stabilization in Primitive Human Societies 90
The Tikopia of the Pacific: A Classic Case 93
Cross-cultural Comparisons among Contemporary Societies 95

The Looming Crunch   98
The Controversy over Family Planning   100
Social Factors Affecting Reproductive Attitudes   101
Zero Population Growth   103
Gazing into a Clouded Crystal Ball   105

6. Difficult Choices: Strategies for Food Production / 117

Population and Food across the Planet   119
The Population-Food System   121
Current Strategies of Food Production   122
Economic Influences in Agriculture   130
The Environmental Costs of Intensive Agriculture   137
Ecological Alternatives to High-production Agriculture   137
High Productivity, Ecological Diversity, and the World Food Problem   142

7. The Dilemmas of Water Development / 146

The Natural Hydrological Cycle   147
A Base Line for Policy Considerations: Present Supplies and Demand   148
Alternatives to Insure Continuing Water Sufficiency   151
Three Case Studies of Water Redistribution   158
Water Development, Present and Future   168

Section C
Matching Present Resources and Future Needs

8. At the Heart of the Matter / 173

De Facto Land Planning, American Style 174
Four Rationales for Land-use Control 176
Mechanisms for Land-use Planning and Control in Urban and Urbanizing Areas 178
Looking toward the Future 205

9. Sustained Yield, or What? / 210

Management Practices for Renewable Resources 211
Examples of Renewable-resource Systems Management 219
Comparisons and Conclusions 230

10. Conservation, Substitution, or Both? / 234

The Great Oil Controversy: Who Is to Be Believed? 235
A Spectrum of Potential Energy Sources 236
Energy and Environmental Considerations 245
Meeting Mineral Needs 248
Perspective on Mineral Policy 250

Section D
Environmental Quality: A Measure of Mankind

11. Ecology, Water Quality, and the Cesspool Mentality / 264

The Scope of the Problem 265
The Natural Purification Cycle: An Example 266
A Conceptual Flow Chart of Water Pollution 268
Types of Water Pollutants 269
Environmental Effects of Water Pollution: Physical and Chemical 276
Environmental Effects of Water Pollution: Biological 277
Environmental Effects of Water Pollution: Human Health 279
Environmental Effects of Water Pollution: Aesthetics 280
Approaches toward Pollution Control 281
Biological Approaches to Pollution Control 284
In Conclusion: Nature and Human Beings in Water Pollution Abatement 289

12. The Sewer in the Sky / 293

Air Pollution in the World and Nation 294
Sources of Air Pollutants 298
The Chemical Nature of Air Pollutants 299
Photochemical Smog: An Example of Interaction 300
Geophysical Influences on Air Pollution 300
Biological Influences on Air Pollution 303
Natural Pollution-destruction Mechanisms 305
Climatic Effects of Air Pollution 305
Corrosive and Soiling Effects of Air Pollution 309
Biological Effects of Air Pollution 310
Air Pollution and Human Health 311
Air Pollution and Aesthetics 314
A Case Study: Air Quality in the Los Angeles Basin 315
Air Pollution Abatement: The Special Problem of Air Quality 318
The Technological Situation 319
Social Mechanisms for Air Pollution Abatement 320
Another Case Study: The Internal Combustion Engine 323
A Final Prospectus on Air Pollution 325

13. A Question of Mounting Imbalance / 329

Solid Waste Disposal in the Here and Now 330
Transitional Solutions in Solid Waste Management 332
Innovations in Solid Waste Management 335
Building Environmental Structures from Solid Wastes 339
Energy Derived from Solid Wastes 340

14. New Problems of a Powerful Technology / 345

Waste Heat: Thermal Pollution 347
Biocides 351
Deep Disposal: Fluid Injection 359
Radioactive Wastes 365

Section E
Relicts and Islands Along the Roads of Progress

15. The Retreat of Natural Systems / 378

Human Beings and Nature: A Brief History   379
Susceptibility to Extinction   383
The Comprehensive View of Susceptibility to Extinction   386
Responses of Natural Systems   387
Do Human Beings Need Nature?   390
The Present: Contrasting Examples   392
What Might the Future Hold?   402

16. Quiet Corners and Recreation for a Diversity of Needs / 409

The Imprecise Meaning of "Open Space"   410
Open Space: Needs and Planning   411
Rural Open Spaces   415
National Parks: Their Creation and Roles   415
Preservation Objectives in National Parks   416
The Recreational Role of National Parks   423
Ecology and Recreation Planning in National Parks   424
The Wilderness Act of 1964   425
Struggles between Development and Preservation   430

17. Ecology and Urban Environments / 448

A Synopsis of the Structure and Function of Urban Areas   449
Alternative Futures for Urban Areas   454
Some Solutions for Urban Ills   454
Urban Cures and Panaceas: A Retrospective Conclusion   474

18. Environmental Perception: It's What You Think That Counts / 480

Historical Analysis of Environmental Perception   482
Environmental Perception at the National and Global Levels   484
Perceptions at the Regional Level   486
Perception of Local Environments   490
Proxemics: Perception of the Individual   499
Environmental Perception and Normative Decision-making   504
Some Questions about the Future Applications of Environmental Perception   505

19. Economic Systems, Government, and Ecological Systems: Coaction and Conflict / 510

Comparisons of Human and Natural Economies   511
Conflicts between Human and Natural Economies   511

Section F
People and their Creations

"Spaceship Earth": The Economic Solution 514
Political Influences on Economic Processes 519
Political Responses to Environmental Quality Control 520
Solutions beyond Present Mechanisms 525
A Footnote, an Opinion, and a Question or Two 531

20. Ecological Systems and Environmental Strategies / 535

The Human Being-Environment System: A Synopsis of Earlier Chapters 537
Some Problems Raised by an Analysis of the Model 538
Policy Choices 545
A Final Perspective on the Tasks that Face Us 548

Glossary / 553

Credits / 563

Index / 565

Preface

My major goal in this text is to present a unified framework for environmental analysis based on two tools: systems analysis and ecological concepts. These tools are used to generate a broad perspective on environmental problem-solving. The emphasis is on matters of strategy and tactics and the outlines of policy, rather than on the narrower (but important) achievements of engineering and technology. The strategy-tactics approach is used to reach fairly general conclusions of some lasting value—conclusions that stress the nature of the human being-environment system rather than details of highly topical events.

Within the unified framework presented here, the reader is encouraged to develop the kind of informed and independent thinking that strives for solutions to environmental problems. No pretense is made that such solutions are currently available. No attempt is made to prescribe solutions. Instead, each class of problems is carefully analyzed using systems theory and ecological concepts, and all viewpoints are considered. This will help the reader to understand major issues and their complexity, to formulate his or her own opinions, and to act as an informed citizen in influencing environmental policy.

This text is intended for the classroom students who may well be the responsible decision-makers in the next half-century. These students are enrolled in the "man-environment," "environmental awareness," and introductory environmental science/environmental studies courses in departments of biology, geography, environmental science, and planning at two-year and four-year colleges and universities. The emphasis on problem-solving, the concern for policy formulation and analysis, and the attempt to reach independent conclusions about environmental problems should provide useful foci for students in these courses. The text is also designed to be read independently by the informed citizen.

In addition, *Ecological Systems and the Environment* may be useful in survey courses in fields that do not deal exclusively with the environment. Engineers interested in the impact of engineering on environment, biologists wishing to extend their training to a form of human ecology, policy planners responsible for resource use and environmental quality control or environmental impact analyses, ecologists interested in the application of their science to human ecosystems—all these people may find some value in the broad coverage and systematic framework of this text.

No prerequisites are necessary for students to use this book, although an elementary familiarity with biology is helpful. I have attempted to minimize the use of jargon and have included a glossary of terms for the benefit of the beginning student and the interested citizen. As with any text, individual motivation for additional reading and thoughtful reflection on the problems is needed to fully accomplish the aims of this book.

Organization: The entire text is organized to promote synthesis of environmental issues. Each major section consists of a number of chapters with a short introduction. Each introduction establishes some general ideas that will be useful for all chapters in that section and provides some guidance into advanced readings and projects. In some cases, section introductions present a general conceptual model that applies to all chapters in that section.

The introduction and chapters in section A, "About Systems and Ecological Systems," develop the two analytic tools that are used throughout the text: systems concepts and ecology. In the sections that follow, these ideas are invoked to explain causation and to arrive at various possible solutions to environmental problems.

A wide range of diverse topics and environmental problems are covered thoroughly in the remaining five sections. Section B, "To Cover the Earth?" addresses population problems. Section C, "Matching Present Resources and Future Needs," discusses resources and food problems. In Section D, "Environmental Quality: A Measure of Mankind," questions of environmental quality are discussed, including air and water pollution and technology. Section E, "Relicts and Islands Along the Roads of Progress," treats natural systems, and Section F, "People and Their Creations," extends the discussion to human systems, including material on urban environments, environmental perception, politics, and economics.

The concluding chapter, "Ecological Systems and Environmental Strategies," offers a synthesis of all the foregoing chapters and suggests a highly selected set of readings, mostly philosophical in nature. It emphasizes the major problems and decisions that remain to be confronted. Although this chapter contains a personal viewpoint, it may be used in class to stimulate discussions emphasizing alternative views and to provoke the student's own development of personal viewpoints.

Study aids: Every effort has been made to design this textbook to encourage student learning and to promote the student's involvement with the material presented. The emphasis is on problem-solving and on intelligent, systematic ways of looking at the complex environmental issues and problems that face our society. Accordingly, the text contains an abundance of examples and case studies integrated into the conceptual framework of the text. To help the reader integrate the subject matter, each chapter contains a series of thought-provoking problems at varying levels of difficulty. In addition, chapters are cross-referenced, and an annotated reading list is provided at the end of each chapter. The glossary, the visually displayed self-study organizer at the beginning of each section, and the list of general sources and relevant journals are intended primarily as aids to independent readers—those students or citizens who are motivated to work through the text on their own and may wish to delve further into certain topics.

For the instructor: An instructor's manual is available to instructors using or considering this text. It contains answers to questions in the text, chapter summaries, sample test questions, suggestions for alternative sequences of material, and additional information about recent environmental developments. In addition, a complimentary set of twenty 35-mm slides related to illustrations in the text will be provided upon request to adopters of the text.

The great tide of environmental interest that rose in the late 1960s and the early 1970s seems to have receded somewhat in the face of increased costs, declining petroleum reserves, and public concern over the immediate problem of energy needs. Unfortunately, the problems and issues that first provoked this surge of environmental interest have not been solved and will not disappear. For reasons discussed throughout this text, a revival of environmental interest can be predicted sometime in the future. This time, it will occur when we realize how thoroughly humans and their environment are bound in a single global system and when we are convinced of the need for informed management to protect ourselves and our environment.

This text has been written to fill a need for texts presenting an integrated, systematic approach to environmental analysis. This need is felt today and will be even greater when public attention once again focuses on environmental questions. It is hoped that students, instructors, responsible citizens, and people seeking a professional career in environmental problem-solving will find in this text an appropriate vehicle for learning about gathering of scientific information required for making environmental management decisions, weighing evidence and viewpoints, and discovering ways to implement programs. If *Ecological Systems and the Environment* helps these people develop for themselves a way of ordering and understanding the enormous range of material that is in some sense environmentally relevant, it will accomplish its aims and prepare us for the return of greater environmental sensibility.

During its development this book has benefited from many reviewers, both colleagues and friends. There are a few who I would single out for special thanks. I wish to thank Professor Anthony Tomazinis of the University of Pennsylvania, Professor Calvin Ward of Rice University, Professor Mitchell E. Timin of San Diego State University, Professor Grahame Smith of Boston University, and Professor John Bushnell of the University of Colorado for the many structural comments and technical insights they provided in repeated reviews of the manuscript. Professors Smith and Bushnell had the fortitude to read all drafts of the manuscript; their comments alone could have formed a supplementary reader. A number of colleagues at the University of California, Davis have also read parts of the manuscript; I especially wish to thank Professors Seymour Schwartz and Thomas Dickinson for their reading of Section F. Kimi and Joel Klein have also read the entire manuscript from the viewpoint of informed citizens, and their comments were most helpful in forcing me to write clearly and eliminate the jargon. The California Department of Water Resources and Walt Disney Productions also provided useful corrections to matters in which they had part. My wife, Angela, whose patience has lasted for years of proofreading and editorial corrections, was indispensable to the final manuscript, although any errors are solely mine. Finally, the unceasing effort of Katherine Elliott and Dolores DuMont—the former a skilled typist and the latter a devoted secretary and girl Friday—were indispensable in struggling through my handwriting and helping meet deadlines. I look forward to the criticisms of readers and the development of frameworks superior to my own; when they appear, my goal will have been met.

Theodore C. Foin, Jr.

Section A
About Systems and Ecological Systems

SECTION A / SELF-STUDY ORGANIZER

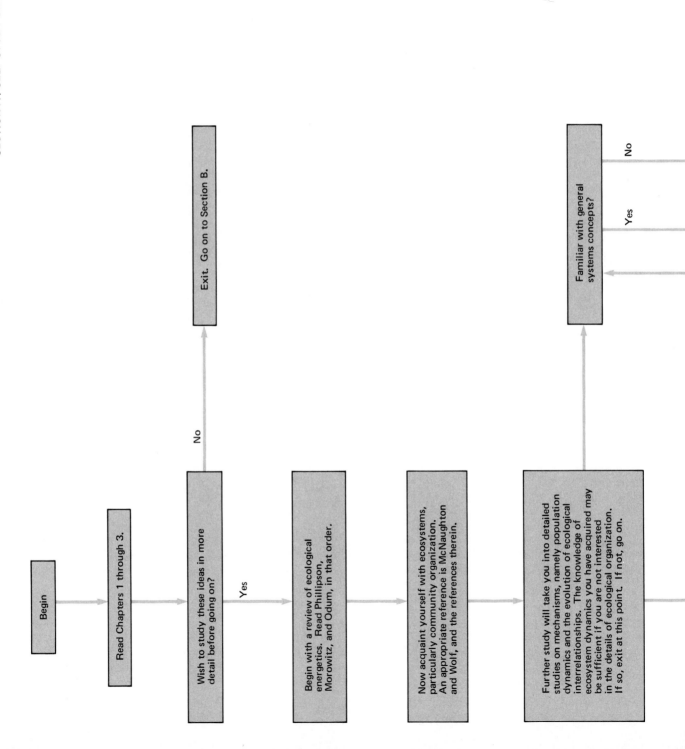

```
┌─────────────────────────────┐
│ Undertake a review of       │
│ general systems concepts.   │
│ Try a few of Churchman's    │
│ simple problems. If these   │
│ present no difficulty, read │
│ Forrester; if the ideas in  │
│ Forrester seem a bit        │
│ mysterious, return to       │
│ Churchman first and read it │
│ before going on with        │
│ Forrester.                  │
└─────────────────────────────┘
                │
                ▼
┌─────────────────────────────┐          ┌──────────────────────┐
│ For population dynamics and │          │ Exit. Go on to       │
│ two population interactions,│          │ Section B.           │
│ see Krebs, Slobodkin, and   │          └──────────────────────┘
│ MacArthur, in that order.   │                    ▲
└─────────────────────────────┘                    │ No
                │                                  │
                ▼                         ┌────────────────────┐
┌─────────────────────────────┐           │ Wish to preview the│
│ For the evolution of        │──────────▶│ application of     │
│ ecological relationships,   │           │ these ideas to     │
│ see Ricklefs for a          │           │ human systems?     │
│ comprehensive treatment.    │           └────────────────────┘
│ For a more detailed and     │                    │ Yes
│ mathematical treatment,     │                    ▼
│ consult Emlen.              │           ┌──────────────────────────┐
└─────────────────────────────┘           │ Read Meadows and Cole    │
                                          │ et al., then attempt to  │
                                          │ characterize the human   │
                                          │ ecosystem. Read Murdoch, │
                                          │ first and last chapters, │
                                          │ as a further guide. Try  │
                                          │ to keep your ideas in    │
                                          │ mind; at the end of this │
                                          │ book compare your        │
                                          │ original concept of the  │
                                          │ human ecosystem to the   │
                                          │ content of this book and │
                                          │ your own feelings then.  │
                                          └──────────────────────────┘
```

3

4

The three chapters in this section present the ideas that underlie the remainder of the book. A basic premise of this book is that ecology and systems principles are appropriate focal points around which to organize studies on the environment, studies that will help explain environmental phenomena. Ecological and systems concepts may also suggest possible approaches to the solution of environmental problems.

The rationale for this approach is simple. Human beings are an integral and dominant part of the global ecosystem, and there is scarcely a regional subsystem in which their activities are not vitally important. Since we cannot avoid this kind of interaction, we must ask ourselves two important questions: How subject are we to ecological limitations? And how should our behavior be shaped by them? These questions suggest more specific questions. For example, is the future of the human population inextricably bound with that of other populations? Can technology help us to overcome the limitations of our natural resources? Is environmental pollution a danger to the world? All these questions direct our attention toward another concept—the broad concept of the world as a system, as opposed to that brand of ecological theory derived from studies of nonhuman populations. The ideas we shall need for our studies—the study of the interaction between humans and their environment, and the study of the world, including both human and nonhuman populations, as a system—are united in Chapter 3 under "systems concepts." The underlying purpose of our studies is to help us determine what policies we should be implementing to solve environmental problems.

Ecological and systems concepts are much too extensive to be discussed fully in this section. For this reason, I urge each reader to take advantage of the additional readings listed at the end of each section introduction and of each chapter, to work the problems at the end of each chapter, and to use the self-study organizer (flow chart) that appears at the beginning of each section. The effort necessary to obtain additional background will be rewarded as you work through the remaining sections of this book.

Selected General Sources on Living and General Systems

Churchman, C. W. *The Systems Approach.* Delta Books, Dell, New York, 1968.
Cole, H. S. D., C. Freeman, M. Jahoda, and K. L. R. Pavitt (eds.). *Models of Doom.* Universe Books, New York, 1973.
Emlen, J. M. *Ecology: An Evolutionary Approach.* Addison-Wesley, Reading, Mass., 1973.
Forrester, J. W. *Principles of Systems*, preliminary edition. Wright-Allen Press, Cambridge, Mass., 1968.
Krebs, C. J. *Ecology.* Harper & Row, New York, 1972.
MacArthur, R. H. *Geographical Ecology.* Harper & Row, New York, 1972.
McNaughton, S. J., and L. L. Wolf. *General Ecology.* Holt, New York, 1973.
Meadows, D. H., D. L. Meadows, J. Randers, and W. W. Behrens. *The Limits to Growth.* Universe Books, New York, 1972.
Mesarovic, M. D. (ed.). *Views on General Systems Theory.* Wiley, New York, 1964.
Morowitz, H. J. *Energy Flow in Biology.* Academic, New York, 1968.
Murdoch, W. W. (ed.). *Environment*, 2nd edition. Sinauer Associates, Stamford, Conn., 1975.
Odum, H. T. *Environment, Power, and Society.* Wiley Interscience, New York, 1971.
Phillipson, J. *Ecological Energetics.* E. Arnold, London, 1966.
Ricklefs, R. E. *Ecology.* Chiron Press, Newton, Mass., 1973.
Roszak, T. "Technocracy: despotism of beneficent expertise," *The Nation,* (Sept. 1969), 181–188.
Slobodkin, L. B. *The Growth and Regulation of Animal Populations.* Holt, New York, 1961.

Selected Periodicals

Advances in Ecological Research
The American Naturalist
Annual Review of Ecology and Systematics
Ecology
General Systems
Human Ecology
Journal of Animal Ecology
Journal of Applied Ecology
Journal of Environmental Systems
Nature
Science
Scientific American
Simulation (Simulation Councils, Inc.)

1. "Ecological Systems" and "Environmental Studies"

The first question we must deal with is how to limit the scope of our studies. This problem is a direct consequence of what is probably the most obvious characteristic of the field of environmental studies, that it can rapidly become diffused and all-encompassing. After all, what is not in some sense "environmental"? The fact that "nothing" is the only correct answer to this question indicates the need for some program of analysis that permits orderly development of a set of general principles and insight into possible solutions for environmental problems. If the field of environmental studies is not to degenerate into a hodgepodge of unconnected case studies, the development of such a unifying device has to be our first priority.

There are two widely accepted views of what is involved in "environmental studies." The first concentrates on the living creatures that share the planet with us. The second, broader approach emphasizes that all the planet's resources, living and nonliving, are ultimately limited. In the first view, then, the emphasis is upon living systems; and in the second, it is upon the total system, both the living and the nonliving components. In both views the human species assumes a central role because human welfare and human activities command the forefront of our attention.

Further consideration of these two views of the content of environmental studies suggests that the program of analysis we need to better our understanding of environmental problems might be found in certain existing sciences. One of these is the science of ecology, described in the usual dictionary definition as "the study of the interrelationships between organisms and their environment." Since ecology is directly concerned with environments, it is a natural tool for us insofar as it can provide insight into how natural living systems operate and how they are affected by various kinds of disturbance. How much insight ecology—in the classical sense, that is, concerned mostly with nonhuman living systems—can provide for environmental studies is explored in Chapters 1 and 2.

With growing public awareness of environmental issues, ecology recently has begun to assume an additional meaning that corresponds to our second concept of environmental studies. Whether the cause of this growing awareness is an "energy crisis" or a shortage of food or space, many people identify the term "ecology" with the growing pressure upon the human population to accommodate its ambitions and activities to some set of inviolate limits. In this book we shall treat concepts such as these—that is, concepts that are based on all parts of the system (and therefore are larger than concerns about saving other creatures)—as a separate tool in our search for answers to environmental problems. Obviously this tool incorporates many separate interests, for example, policy analysis, economics, engineering, classical ecology, and so on, but the involvement occurs in a way that emphasizes how all these components fit together to generate particular consequences. The unraveling of these complex interactions provides insight into how these factors work and gives us yet another analytical tool with which to dissect environmental issues. The very breadth of the involvement of all these discrete subjects makes this tool difficult to use—but that topic is raised in Chapter 3.

Basic Concepts about Ecological Systems

We may call any collection of organisms and their environment an *ecological system*, although this term usually refers to a group of organisms that interact or that can at least potentially interact with one another. Ecological systems are organized in hierarchies, as all systems are (Chapter 3). The usual levels of organization of interest to ecologists are individual organisms, populations, and communities; each has relevant environmental applications, though certainly some levels are more important than others. For example, individual whooping cranes are of interest to people and agencies concerned with the survival of the species (Figure 1-1). For many people individual cranes are recognizable because the species is rare and there are breeding programs for particular pairs in captivity. Nevertheless, more important determinants of that species' survival ultimately are, first, the entire population of cranes (since the population is the real reproductive unit), and, second, the community within which cranes live, consisting of many species that together help determine the quality of the environment. Many of these other species may provide essential food, shelter, and nest sites for the cranes.

FIGURE 1-1 *Two whooping cranes* (Grus americana) *feeding in a marsh.* (Courtesy of Allan D. Cruickshank from National Audubon Society.)

No matter which level of organization is chosen for special study—individuals, populations, or communities—all three levels share fundamental properties of interest:

1. A living system exists only within the context of its physical and chemical environment; this environment establishes the "ground rules" for a system's continuing existence.
2. An especially important part of item 1 is the requirement that a living system capture all resources (materials and energy) needed for its continuing existence. These requirements, especially energy, impose further restrictions upon the nature of ecological systems.
3. A system will develop a characteristic structure in response to these and other influences.
4. In nearly every case the structure of a system will not be static, but rather will be dynamic and responsive to changing conditions. To promote continuing survival, there must be appropriate mechanisms that allow ecological systems to cope with change; otherwise, potential competitors would replace them.

Biotic Dependencies on Abiotic Factors

The "ideal" organism would be able to survive in all environments equally well. There is no such organism. In the real world, some organisms are better able to cope with certain conditions than others, and so different organisms tend to be distributed along an environmental gradient. For example, suppose that you were to walk a straight line from a continuously wet environment to a very dry one. In the wet environment you would encounter many trees. Where the soil is very wet, you might find many cottonwoods

BIOTIC DEPENDENCIES ON ABIOTIC FACTORS

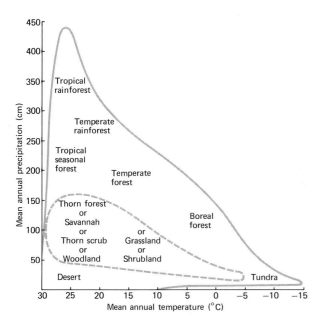

FIGURE 1-2 *The distribution of terrestrial plant ecosystems as a function of temperature and moisture. The results of community development for most combinations of temperature and moisture are easily determined except for conditions enclosed in the dashed lines. For these conditions, other factors determine exactly what kind of plant community results.* (Adapted from *Communities and Ecosystems* by R. H. Whittaker. Copyright © 1970 by The Macmillan Company, New York. Used with permission of The Macmillan Company.)

and willows, species that are adapted to saturated soils. Where the soil is less wet, you might encounter oak, aspen, or birch trees. In places where water is scarcer and trees can barely survive, you might find full-sized live oaks as well as other species stunted from the lack of water; ultimately, as soil moisture decreases, there would be no trees at all, and only grasses would be present. As the soil becomes drier and drier, the same sequence of events would occur for these grasses, starting with tall perennials where the soil moisture was adequate year round and ending with short, scrubby annuals. Finally, even grasses could not survive the lack of soil moisture, and you would find yourself in a sparse shrub- and cactus-dominated desert.

This hypothetical walk reveals the importance of water as a determinant of vegetational structure. Water, however, is not the only important environmental determinant; a walk along a gradient of temperature, altitude, or other environmental factors would produce similar results because all are important to living systems. By taking all relevant factors into account, it is possible to construct a picture of how they interact to produce patterns of recognizable ecological assemblages. For example, Whittaker has shown that the distribution of major types of plant communities is largely explicable in terms of only two environmental factors—annual mean temperature and annual mean precipitation (Figure 1-2). At the extremes of heat and moisture one will find tropical rain forests; at the extremes of cold and dryness one will find tundra. In a large range of conditions, however, other factors are important determinants of community distribution. (This range is encircled by a dashed line in Figure 1-2.) Fire and soil conditions may be important, for example, in selecting a grassland or savannah at a particular site within this range.

It is common practice in ecology to call the range of conditions in which a species can survive the *fundamental niche* of that species. This is one not very operational way to summarize the effects of physical environments upon the distribution and abundance of living systems. If we also consider the competitive interactions among living systems, we have the *realized niche*, that is, the part of the potentially habitable environment which a system actually is able to occupy. However, to remember these terms is not nearly so important as to understand that for any particular set of conditions either there will be some existing ecological system superior to all others, or, if there is not, a well-adapted system will evolve in response to the conditions at hand, given enough time.

Material and Energy Transfers

Perhaps it is surprising that all the roles of energy and materials in ecological systems are based largely upon just two laws of thermodynamics:

> The **first law** states that matter and energy are two interchangeable states. One may be converted into the other without loss.

> The **second law** states that when energy is transformed from one form to another, all energy transfers can be made only at the cost of a part of the energy involved. In other words, part of the available energy is inevitably lost, or no transfer can take place.

There are two major ecological consequences of the first law of thermodynamics. Organisms require energy for survival, and, except for green plants, they get this energy by consuming certain materials that they can convert into useful energy. In effect, this restates the fact that animals must eat to survive—

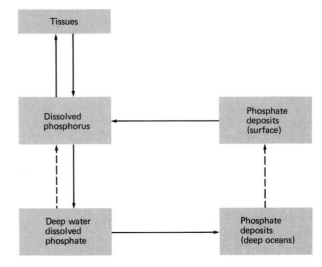

FIGURE 1-3 *The biochemical cycle for phosphorus. The solid arrows represent major pathways of flow and the dashed arrows are flows of "much less importance." The phosphorus cycle alternates between a reservoir of living biomass and dissolved phosphorus in natural waters, but the cycle is not balanced because the rate of return from the deep oceans is much less than the loss—unless parts of the ocean floor are uplifted or the sea level drops.* (Redrawn from E. P. Odum, *Fundamentals of Ecology*, 3rd edition, W. B. Saunders Co., 1971, p. 88.)

they can't eat pure energy. The second consequence is that, while materials and energy are completely interconvertible, organisms are not able to use all forms of material and energy; they are limited to a small subset of the possibilities. Organisms cannot use everything they theoretically could eat, nor can they use available energy to synthesize substances that also require materials which are not available in sufficient quantity. In other words, if a material is in critically short supply, it will limit the activity of an ecological system that needs it to synthesize an essential compound, even if the necessary energy is abundant. A mineral cycle is a network of material transfers that shows how a material is circulated through a system of users. The phosphorus cycle (Figure 1-3) is one classic example, because phosphorus is required for many biological functions. In particular, plants need it for growth. Phosphorus is limiting because much of it sinks into deep ocean waters where there are few organisms to use it; conversely, in many ocean areas, there are very few photosynthetic organisms despite the great availability of energy from the sun, because phosphorus is scarce. For this reason, places where upwelling currents move phosphorus back to the surface and where there is plenty of light are noted for enhanced biological activity.

The second law of thermodynamics sets the limitations upon the use of energy by ecological systems. Living systems are no exception to the second law; in fact, they appear to be particularly inefficient users of energy. While there is no agreement about efficiency of energy transfers, the maximum that has been observed is 30 percent, and usually it is in the range of 5 to 20 percent of the total energy available from the donor to the recipient. This means that there are limits to the number of transfers which can be made from an energy pool of given size. These limits are reflected in the structure of complex living systems quite apart from the influence of other environmental factors. Because some organisms eat certain other organisms—such patterns of feeding relationships are called *trophic webs*—it is possible to describe a network of these relationships that essentially is similar to the phosphorus cycle outlined above. Now suppose that we classify all these organisms by what they eat and we lump together all species that use sunlight, all that eat these species, and so on. Then, if we determine how much energy is contained at each level, we end up with a pyramid (Figure 1-4) that reveals the losses predicted by the second law of thermodynamics. Since the sun is the ultimate energy source for all life on earth, and since each level must be smaller than the next lower one, there can be only a few levels in the pyramid; in nature there are never more than five levels, and three or four is the usual number.

Note that material transfers are not governed by the second law of thermodynamics. In natural systems this means that trophic pyramids can be inverted, as long as they are constructed in terms of numbers of individuals or the live weight represented by these individuals—both measures of materials. In order to support the inversion, the smaller donor must produce more energy per unit time than the recipient utilizes. The laws of thermodynamics must also hold true for human beings, but the complexity of our use of materials and energy relieves us of some simple limitations imposed on other creatures. For us, energy is more than food energy: It is energy to drive machines, to produce heat, to generate electricity, to transport limited materials, and even to produce more food. At present, in fact, some of our most important energy sources (petroleum and coal) are practically worthless as food; indeed, though now used primarily as sources of energy, they actually are much more valuable as raw materials for plastics, dyes, and fibers. It is an interesting state of affairs that modern society depends highly on energy sources that have nothing to do with food and, moreover, are sources whose ultimate future uses lie in materials, not energy.

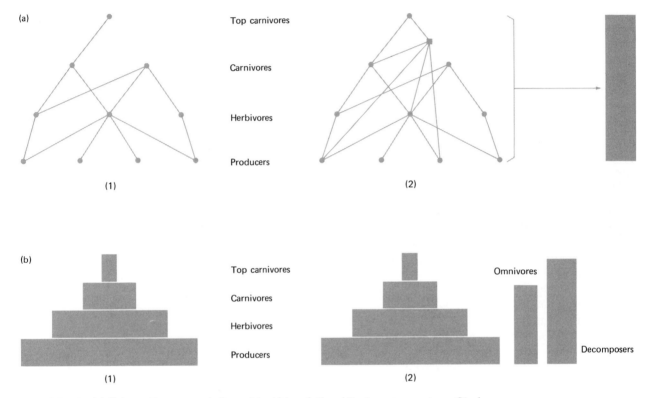

FIGURE 1-4 *(a) Schematic representation of trophic relationships in an ecosystem. Circles represent populations and lines represent known trophic relationships. (1) An incomplete system showing herbivores and carnivores of differing tastes, leading to a single top carnivore specialized to only one prey. (2) The same system with two additional elements: decomposers (large rectangle) feeding on the dead of all, and an omnivore, such as the human population (square), which feeds selectively on many levels and may be fed upon by some. (b) The trophic web can be aggregated into a trophic pyramid with units of numbers of organisms, biomass, or total energy content. The two examples in (a) might look like this.*

FIGURE 1-5 *A barn owl* (Tyto alba) *perched on a wall after having caught a small rodent.* (Courtesy of Hal Harrison from Grant Heilman Photography, Inc.)

System Structure

Because the survival of any ecological system depends upon adequate provision for energy and material needs, there inevitably will be strong selection for organisms with appropriate structural adaptations to accomplish these tasks. These adaptations can be either anatomical or behavioral: Plants have chlorophyll concentrated in efficient photosynthetic surfaces, birds have bills adapted to handling certain foods, and owls hunt at night when their prey (rodents) are active on the surface of the ground and hence vulnerable (Figure 1-5).

These specific adaptations are well known to most readers. But the same kinds of adaptations can be found at higher levels of ecological organization than the individual, and these are likely to be less familiar. Consider two important population structures, those for age and for sex. All populations have an age structure, although its nature may vary as a result of whether the population is growing, decreasing, or unchanging. If conditions are favorable long enough, the population is likely to grow and subsequently exhibit

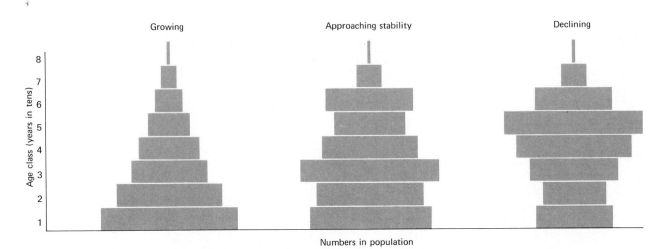

FIGURE 1-6 *An example of an age structure pyramid for a single species. Each age class is represented by a horizontal bar; the total number of bars constitutes the population. By convention, the total number in each class is distributed equally to yield a symmetrical pyramid. In some cases males are put on one side of a center point and females on the other, in which case the pyramid does not have to be symmetrical.*

a characteristic age structure (Figure 1-6). Similarly, the sex ratio may be affected greatly by various properties of the environment. If greater reproductive potential is required, the population's sex ratio may shift to favor females so that more young can be produced in a given period. Age and sex structures are only two of many that a population may possess. If we take a broader view and attempt to relate all characteristics with environmental variability, we may hypothesize that, in general, populations which occur in environments with variable supplies of materials and energy will display structural adaptations that allow for rapid and plentiful reproduction, rapid dispersal, and other characteristics that counterbalance high mortality rates. On the other hand, populations from more stable environments will have structural adaptations that provide greater individual development, more chance for specialization, and lower population growth; these result directly from restricted fluctuations in the size of the population.

Ecological communities also develop characteristic structures in response to availability and variability of energy and materials. The trophic web (food web) mentioned in the previous section is an example of one community structure; it is a detailed expansion of the trophic pyramid discussed earlier. The food web shows what each species eats as well as what eats it. While the concept requires nothing more complex than a knowledge of what eats what, additional measurements to quantify the amounts actually transferred yield a far more valuable perspective. The diagram of Silver Springs, Florida, in Figure 1-7 is one

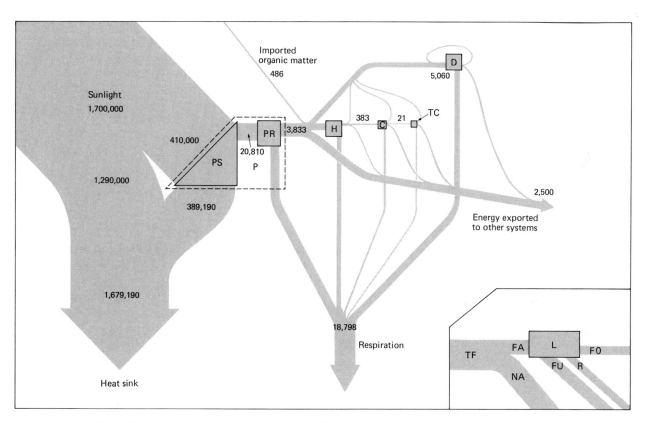

FIGURE 1-7 *Schematic representation of energy flow in Silver Springs, Florida. Numbers are representative values for energy-flow rates in kilocalories per year per meter^2 of surface. PS = producer photosynthetic tissue; PR = producer respiring tissue; P = producer trophic level; H = herbivore level; C = carnivores; TC = top carnivores; D = decomposers.* (Redrawn from H. T. Odum, 1957, *Ecological Monographs*, 27:61, Duke University Press.) *Inset: Energy flow through one level (L), showing the components of the energy budget. TF = total energy flow; NA = energy not absorbed; FA = energy absorbed (or eaten); FU = energy absorbed but not assimilated; R = respiration; FO = energy available to the next trophic level.*

example of a trophic web. By grouping all species that feed on roughly the same substances (plants are "sun-eaters," carnivores are meat-eaters, etc.), one can represent the relationships among various groups of organisms in a precise and quantitative way. Furthermore, evidence is developing that these same functions persist in natural communities even if a particular species does not. Upon analyzing data collected by others, Heatwole and Levins (1972) concluded that after the fauna of a small island was destroyed, recolonization by organisms with the same trophic classification, though not necessarily the same species, was rapid. Thus the characteristic trophic structure would be restored rapidly, even though the species involved might not be the same. This is one piece of evidence that communities can have structures other than those of their component populations. Unfortunately, communities are not yet understood sufficiently to allow us to develop more general conclusions about patterns in community structural development. This topic should occupy ecologists for a long time.

Reactions to Changing Environments

To know that some sort of pattern will result from the requirements of resource acquisition is not enough; these requirements are not the sole determinant of a system's structure. To understand how a particular structure develops and, more important, how it changes, we must incorporate the concepts of adaptation and natural selection.

Individuals may be considered "trial-and-error" assemblages of characteristics. From the array of possible characteristics some process must be able to select the "best" adaptations for any particular situation. However, since environments differ in time and in space, any one set of adaptations cannot possibly remain the "best" forever. Consequently, the selection process must operate continuously, and the "best" structure can differ from one time to another or from one place to another. Ecological systems are *dynamic* assemblages.

The selection process is one of the most basic biological tenets. When Charles Darwin proposed the concept in 1859, he conceived of natural selection in terms of individuals with different characteristics that competed, each set of characteristics being an alternative solution to a problem. But Darwin did not know the means by which successful solutions were perpetuated. When Mendel subsequently put forth genetics as the mechanism that accomplishes this, Darwin's basic ideas were not substantially altered. To put these ideas in concrete terms, consider a hypothetical example of natural selection. Suppose a pair of giraffes and a pair of gazelles suddenly were released into an area free of other competitors. Giraffes are adapted to browse from shrubs and trees, while gazelles graze on grasses (Figure 1-8). If the new habitat consisted entirely of grass-covered plains, the gazelles would be well adapted, and their population would grow until there were as many gazelles as the grass supply would support. But the pair of giraffes, whose long necks are a severe disadvantage for grazing, would eventually be eliminated from the scene if grass were the only available food. They would have difficulty finding food enough to sustain themselves, let alone to allow for the reproduction and sustenance of additional individuals.

Now suppose a second species of gazelle enters the area and is restricted to a species of short grass that thrives in bare soil or weakened turf. If the first species

FIGURE 1-8 *(a) A giraffe* (Giraffa camelopardalis) *feeding in acacia trees in the East African savannah. (b) Grants gazelles* (Gazella granti) *in the plains of Meru Park, Kenya.* (Courtesy of (a) Ylla from Photo Researchers, Inc., (b) Mary Thatcher from Photo Researchers, Inc.)

of gazelle has overgrazed its preferred food, the species of short grass could then undergo a population explosion. Under these circumstances, many outcomes are possible, including:

1. The first species could switch to the shorter grass and exclude the second species of gazelle.
2. The second species of gazelle could replace the first, if the first was not able to utilize short grass.
3. The second species of gazelle could also overgraze its food and promote still another species of grass unsuitable for either gazelle.
4. The second species of gazelle could reduce its food enough to permit the return of taller grass and its gazelle species. Eventually equilibrium could result between the two grass species and the two gazelle species that feed on them.

This is just a sample of the many alternatives that could result depending upon the exact combination of circumstances. Now suppose that a change in climate increases the mean annual precipitation. If precipitation was variable and if the two grass species had different water requirements, taller grass and one species of gazelle would be favored in some years, and shorter grass and the other species of gazelle would be favored in others. Finally, given sufficient rain, enough trees might become established to support a giraffe population.

This brief and rather artificial example of natural selection in operation provides a glimpse into the complexity of dynamic ecological systems and the

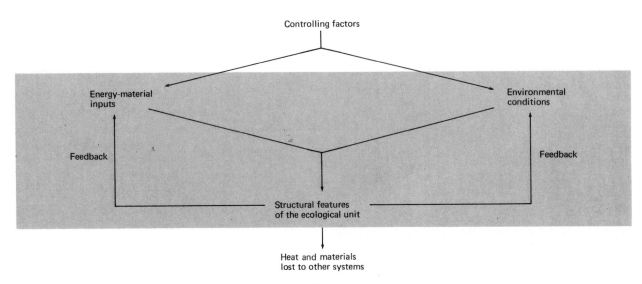

FIGURE 1-9 *Generalized model of an ecological system "frozen in time."*

bewildering number of outcomes that are possible. For our purposes, an appreciation of the influence of natural selection upon ecological systems can provide valuable insight into the types of changes to be expected when we, advertently or inadvertently, impose changing conditions upon the living systems of this planet. Fortunately, living systems are not generally so sensitive to disturbance that they collapse at slight provocation. Most commonly an ecological system will react to a changing environment by making adjustments that do not even show as gross changes; this is called *homeostasis*. Such changes involve quantitative adjustments in the system. They can occur at any level of ecological organization and can include such adjustments as an increase in an animal's metabolism and heat production when it is being chilled, a decline in the size of a population starving from a food shortage, or a readjustment in a community's structure that provides it with continuing stability despite the loss or reduction of several component populations. However, there is a point beyond which any system can no longer cope with change, and it is imperative to remember that human beings as well as nature can foster massive environmental changes.

A Model of Ecological Systems

To unite the previous considerations, it is useful to construct a generalized model of an ecological system. Suppose, to begin with, that we may assume the system of interest is not changing significantly in order to concentrate upon a particular, fixed, ecological structure (we can do this by adopting the time unit over which changes are not evident; that is, our system is "frozen in time"). Provided that we could do it, an ecological system would look like Figure 1-9. The ecological system is defined as the contents of the box. The system's environment is represented by "controlling factors" and "heat loss and material loss to other systems." The various "controlling factors" at work, though external to the system, affect it by influencing both "energy-material inputs" and "environmental conditions." The "structural features of the ecological unit" that ultimately develop represent the results of complex interactions between "environmental conditions" and "energy-material inputs." An important feature of the system is the presence of feedback (see Chapter 3), which provides the intricate interactions that are regular features of complex ecological systems. For the sake of completeness, an arrow represents the "heat loss and material loss to other systems."

What is important in this model is not the exact identity of the major states specified, but the precise flows that determine the structural elements observed. To illustrate this, suppose that the system of interest is the human population of country X. This population has been growing so that the age distribution is disproportionately young. Inserting this supposition into the model, we see that this age distribution could result from (1) abundant supplies of energy and materials that promote reproduction or (2) favorable environmental conditions that minimize mortality. The population would respond by continuing growth. In the process, however, it would tend to limit its own growth because sooner or later it would exhaust the resources needed to sustain further reproduction (feedback); at that time a new equilibrium in population numbers would result.

While the dynamic nature of the interacting elements which shape the system are evident in this example, it clearly is defective when compared with the real world. Real conditions are changing constantly, and so living systems must change themselves. This element must be incorporated into the model. A more complex version (Figure 1-10) incorporates the elements of changing conditions.

The dashed line encloses the system defined previously; it freezes time out of the model by recognizing only one set of "controlling factors" and one system resulting from it. By relaxing this assumption, we can portray the passage of time as two large arrows, within which both environmental conditions (the upper arrow) and the system (the lower arrow)

FIGURE 1-10 *Model of an ecological system incorporating changing conditions. The passage of time is represented by two large arrows, within which environmental conditions (the upper arrow) and the system (the lower arrow) can change.*

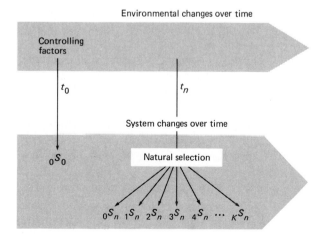

can change. If we postulate that conditions have changed over n time and represent the consequences of this change with an arrow labeled t_n, several outcomes are possible:

1. Either change is not marked, or the system has sufficient resistance. In this case there may be a temporary perturbation and a change in structure, but the system would rapidly restore the status quo. Hence $_0S_n$ is essentially the same as the original state $_0S_0$. (The subscript to the left of the S specifies a state, with 0 representing the original state. The subscript to the right of the S indicates the time, with 0 being the original time.)

2. The system could change in response to conditions. Depending upon the effects of natural selection, any number of new system structures is possible; in this model we assume K possible states.

Whatever state $(_0S_n \ldots _KS_n)$ is selected for, it then becomes the designated $_0S_0$, and a new round of selection proceeds. For visual clarity, feedback arrows and the heat-sink–material-loss pathways have been eliminated from this diagram.

This chain of cause-and-effect relationships is complex, but the model is still incomplete because the system S is only a single level of ecological organization. If we classify the ecological world into four categories, an appropriate diagram of their relationships at t_0 is that shown in Figure 1-11. Every lower level is contained within each higher level of organization, and each level interacts with every other one. To incorporate this into the previous model, we would need to replace each S with the above matrix of

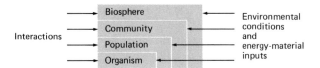

FIGURE 1-11 *Classification of the ecological world into four hierarchical categories at t_0. Every lower level is contained within each higher level, and each level interacts with every other one.*

hierarchical organization and add all the new flows needed to portray the differential effects of environment and natural selection upon each level. Such a model—difficult to draw and awesome in its complexity—would still not be the ultimate model. It could apply to a large area or to a small one, depending on environmental factors applicable to the situation; but if there were environmental gradients or place-to-place heterogeneity within these areas, the new model would have to account for variation from these sources, as well as for variation in time. The result would be confusing mazes of interactions, which is precisely the reality of ecology.

In Retrospect: Ecological Systems and Environmental Studies

This introductory chapter has examined the broad features of ecological systems and combined them into a series of models. The outstanding feature of these systems is their dynamic nature and its manifold consequences. Unfortunately, we are only beginning to gather the data and conduct the experiments needed to unravel even a few of the complex interactions in natural systems. This is why ecology remains a strongly empirical science that is still searching for a first set of broad but valid generalizations and useful principles.

Given these limitations, we might be tempted to conclude that for the present, at least, ecology can provide little help in understanding environmental systems. On the other hand, this relationship between ecological theory and an improved understanding of environmental systems can be seen in a more optimistic light: Though our present understanding of ecological systems is limited, it is sufficient to help us develop insight into solutions of environmental problems. Though there is much to be gained from this approach, we must remember that this use of theories and analogies is limited, a problem that receives more detailed treatment in Chapter 2. In particular, we need more detail about ecological systems before we can use this to answer some salient questions about the role of ecology in environmental problem-solving. Let us pose just three examples of the type of questions with which we are concerned:

1. Are there features of ecological systems that are directly applicable to human problems?
2. Can we gain conceptual advantages for environmental studies from ecology?
3. Will ecological studies prove to be a minor part of environmental decision-making compared with the exigencies of distributing political power and other nonecological considerations?

Questions like these dominate the next two chapters. Ultimately you will decide how useful an intimate knowledge of the dynamics of ecological systems is for environmental studies, but it is helpful to remember that until organizing elements are found, environmental studies will remain a diffuse subject. Whether or not we redefine ecology to include human beings and the socioeconomic, political, and technological factors that distinguish us from other organisms, it is difficult to envision any other subject that can confer all the organizational advantages as well as show us how to avoid becoming entangled in extraneous detail at the expense of understanding what causes a specific problem.

Questions

REVIEW QUESTIONS

1. What is the relationship of an energy-flow diagram (such as Figure 1-7) and a food web (Figure 1-4)? How may one be constructed from the other?

2. Reconcile the concept of biochemical (mineral) cycling and energy flow from two viewpoints:
    a. the first two laws of thermodynamics
    b. the energy content of minerals

3. Examine humanity's role in the global ecosystem using the four fundamental properties outlined in this chapter. Are we subject to these ideas, or have we managed to escape them?

4. How are the concepts of the "niche" and "natural selection" related? In particular, is the realized niche a product of natural selection?

ADVANCED QUESTIONS

1. Examples are known where the biomass (living weight) or the numbers of a producer may be *less* than those of the herbivore feeding upon them, but this is never true when the pyramid is constructed in energy units. Why? See E. P. Odum (1971), fig. 3-15.

2. How have human beings circumvented natural selection and what might be the consequences? See Dubos (1965) for one thesis applicable to this question, after you have thought about it awhile.

3. Jordan (*Bull. Ecol. Soc. Amer.* 56(2):3) has redefined ecology as "the study of *entire* ecosystems . . . the *total* flows of chemical elements, energy, and water . . . the interaction of these . . . populations . . . and *all* the effects of man's activities." Critique this definition in light of the more classic one (p. 1).

4. Throughout this book you should continue to apply ecological and systems concepts to environmental problems as you encounter them. This would be a good time to set down some initial written impressions to review as you progress.

Further Readings

Note: Additional suggestions for further reading will be found at the end of Chapter 2.

Dubos, R. *Man Adapting*. Yale, New Haven, Conn., 1965.
   A shorter and earlier version is *Mirage of Health*, Perennial Library, Harper & Row, New York, 1959.

Heatwole, H., and R. Levins. "Trophic structure, stability, and faunal change during recolonization." *Ecology*, 53 (1972), 531–534.

Odum, E. P. *Fundamentals of Ecology*, 3rd edition. Saunders, Philadelphia, 1971.

Odum, H. T. "Trophic Structure and Productivity of Silver Springs, Florida." *Ecological Monographs*, 27 (1957), 55–112.

Whittaker, R. H. *Communities and Ecosystems*. 2nd edition. Macmillan, New York, 1974.

2. Ecological Systems: Combinations and Recombinations

Now we shall consider the structure and function of ecological systems in more detail. This subject occupies entire textbooks and would seem a preposterous undertaking for a single chapter, especially if the ideas are to be applied later to environmental problem-solving. At the least, rigorous criteria are needed to keep this discussion from expanding into textbook length.

One way to decide which details are relevant to environmental analysis is to consider which aspects of natural systems are important to human interests. For instance, natural systems are sources of food and fiber that we wish to exploit, but we cannot afford to overexploit them. Even if we choose to replace nature with food-fiber systems of our own fabrication, the same ecological limitations apply, for we still must obtain all necessary materials. Let us look at our relationship to nature in another way: Perhaps our wellbeing and survival as a species are bound to the fates of other species. Or perhaps it is merely for aesthetic or psychological reasons that we want to preserve nature for our enjoyment and that of future generations. For whatever reasons, it is increasingly necessary that we learn to manage these systems to our ends or, at least, to prevent their wanton destruction.

Suppose all these goals are important to significant numbers of people. Can we use the model presented in Chapter 1 to help attain each goal? By stressing hierarchical organization and dynamic interactions, the model poses general principles into which we may fit more specialized knowledge. More important, the model erects an interesting hypothesis: that different levels of organization in the ecological hierarchy share many fundamental properties and hence are more similar to one another than we might suppose.

	Dynamics	Structure	Change
Individuals	Homeostasis	Shape, form, size	Adaptation
Populations	Growth and regulation	Age, sex, other characteristics	Evolution and coevolution
Communities	Regulation and stability, diversification, repair, and formation of interrelationships	Diversity, trophic structure	Succession

TABLE 2-1 *The framework of ecological organization used in this chapter.*

Despite their many different names, each of the three levels of organization listed in Table 2-1 has a characteristic structure, is subject to change, and reacts to change in roughly predictable ways. In the remainder of this chapter we shall examine these things.

Systems Structure

INDIVIDUALS

Individual organisms inherit various characteristics from their parents which largely determine their appearance. Thus it is obvious that giraffes reproduce young that are recognizable as giraffes. However, environmental influences also affect the size and form of organisms. A young giraffe that survives repeated attacks by lions may grow to be stouter and stronger than other giraffes. Similarly, human children raised in a poor nutritional environment may suffer stunting and inadequate intellectual development. Environmental influence is most dramatic in plants and various invertebrate animals that lack the ability to buffer environmental change; these groups can show remarkable differences in response to differing environments. One classic example is the physical form of the perennial herb *Achillea lanulosa* (Figure 2-1), which illustrates the extent to which environments can affect the physical appearance of individual organisms.

POPULATIONS

A population can be said to have a structure equivalent to that of an individual body, although such a structure is a property of many individuals rather than just one. For a population, the three most important structures are

1. The age distribution, which shows what proportion of the total population is found in each age category. This classification is important because the number of individuals in each age class can vary, and for most organisms reproduction is very strongly related to age. For example, human beings are most fertile between the approximate ages of 18 and 30 years; hence a population with a larger proportion of its individuals in this age category will be able to increase more rapidly than one with a smaller proportion.

2. The sex ratio is important for species that either are not strictly monogamous or do not mate for life,

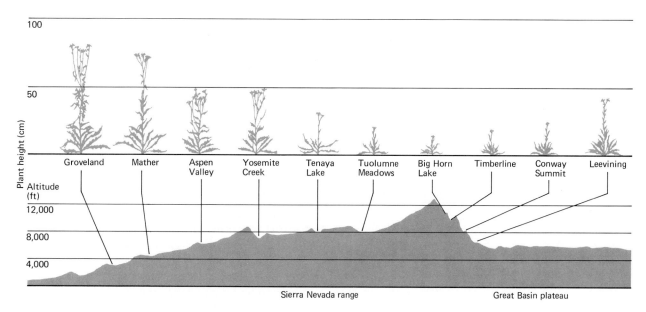

FIGURE 2-1 *Ecotypes in the yarrow plant* Achillea lanulosa. *Seeds were collected from the sites named on the trans-California transect. Mean plant size and the range of variation in size are both related to altitude. The higher and colder it is, the smaller and less variable the plants are.* (Redrawn from Clausen, Keck, and Hiesey, 1948, Carnegie Institution of Washington, Publication No. 581.)

because where mating between individuals is unrestricted, the greater the number of females, the greater the reproductive capacity of the population.

3. The proportion of genetically different types can be important if these types either have different reproductive potentials or have different reactions to natural selection. A population consisting of several genetic types rather than only one has a broader range of potential successful adaptations to changing environmental conditions.

The size and shape of individuals are influenced by environmental conditions, as is population structure. For example, a favorable environment may stimulate rapid population growth, which in turn leads to a youthful age structure, which itself can stimulate even more growth (Figure 2-2) since a greater proportion of individuals in the population are of reproductive age.

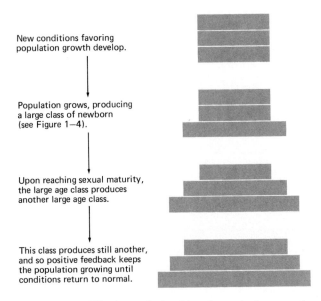

FIGURE 2-2 *The interrelationship of population growth and age structure. The horizontal bars on the right represent a hypothetical population consisting of three age classes undergoing the changes given to the left.*

COMMUNITIES

The community equivalent of genetic types in a population is called *species diversity*. This is a measure of the number of different species found in a community. Diversity may be measured by a simple count of the number of different species present, or it also may take the number of individuals of each species into account. It is intuitively reasonable that a community of 100 species having 2 individuals each is more diverse than a community of the same size with 99 species having 1 individual each and the last species having all 101 remaining individuals. In its simplest terms, however, species diversity is a measure of biological variety in a community.

Species diversity in itself is not a very meaningful concept; it has several contrasting interpretations. But since it is relatively easy to measure, it can be a useful indicator of other trends in the community. For example, suppose we required that a particular community (for example, a grassland) produce at least x pounds of cattle a year, or that the species composition of the grassland not fluctuate much from year to year. If we know exactly how species diversity is related to these two properties in the grassland system, we could base our management practices upon trends in diversity. Furthermore, there is some evidence that diversity is more than a passive indicator of community trends. Thus diversity is said to enhance community stability and, as a result, is a natural means to buffer environmental changes. More will be said about this later, but the problem with species diversity at present is that its relationships to other measures of community structure and function are too poorly known to permit many predictions or generalizations precise enough to use for a particular situation.

System Dynamics

The word "dynamics" as it is used here refers to the ways in which characteristic structural elements are first developed and subsequently maintained in the various levels of ecological organization.

INDIVIDUALS

The structure of the body develops in response to inherited and environmental factors. While the former are expressed early in development, the latter

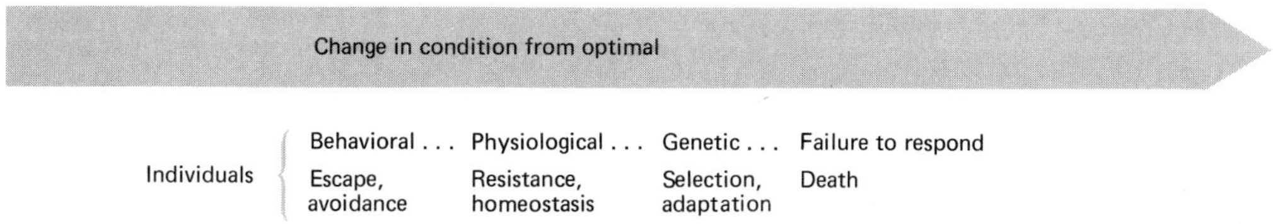

```
                    Change in condition from optimal →

              ⎧ Behavioral... Physiological... Genetic...  Failure to respond
  Individuals ⎨ Escape,       Resistance,     Selection,   Death
              ⎩ avoidance     homeostasis     adaptation
```

FIGURE 2-3 *Reactions of ecological systems to changing conditions. The simplest response is behavioral—the organism can run away, hide, hibernate, or enter dormancy. Physiological responses may involve increased resistance to adverse conditions, such as changes in metabolism to produce more or less body heat or changes in the amount of body insulation. As conditions grow more extreme, these responses no longer suffice; some individuals are eliminated and natural selection favors appropriate adaptations. Finally there is a point beyond which there are no suitable responses except death.*

are characterized by continuing and often drastic changes that challenge the continuing existence of the organism. If we had an infinitely flexible organism, we might expect it to adapt well to any conditions. Such an organism might be found, for example, not only in both cold and hot climates, but also in areas that fluctuate between the two. In the real world, however, such a creature does not exist. One finds instead that organisms can display one of four general reactions to environmental change: avoidance, resistance, adaptation, or destruction (Figure 2-3). Although they are really elements in a continuous series, only avoidance and resistance will be considered here; the latter two are discussed in the section on system evolution.

The simplest thing an organism can do to preserve itself in the face of hostile conditions is to flee or to enter some state of dormancy in which it becomes insensitive to those conditions. Particularly dramatic examples include the migration of terns from the northern polar regions to Argentina during the winter to avoid the bitter cold (Figure 2-4), the burrowing of insects and toads into the soil to avoid heat and

FIGURE 2-4 *Greybacked terns* (Sterna lunata) *on Mokumano Island, Hawaii. Unlike other species in the genus, this species does not undertake hemispheric migrations.* (Courtesy of Russ Kinne from Photo Researchers, Inc.)

dry conditions, and the hibernation of mammals during cold winters. All these specific examples share one common facet: The organism is forced neither to disappear permanently from the scene nor to adapt (although these reactions themselves probably resulted from adaptation in earlier times).

For organisms that can neither flee nor enter dormancy, some form of active resistance still remains. For mammals, this might be manifested as greater heat generation during cold periods or enhanced heat loss (sweating, panting) during hotter times. For lizards, this may instead involve behavioral adjustments that enable the organism to stay within the narrowest range of temperatures (that is, seeking sunlight in cold weather and shade in hot weather). For purposes of our classification, resistance here can mean physiological or behavioral changes, but not anatomical ones, although it is obvious that such a separation is not clearly valid. This question is taken up below.

POPULATIONS

Individual homeostasis finds its population counterpart in the numerical changes in a population size that result from environmental influences affecting the birth and death processes. An appropriate analogy is a pail of water which has continuous drainage and continuous inflow: The level of water remains constant only as long as inputs and outputs are balanced (Figure 2-5). For a population, this means that unfavorable conditions lead to population decrease, and favorable conditions lead to increase. Population growth can result from net immigration, increased births, decreased deaths, or any combination of these factors—just as increasing the rate of inflow over the

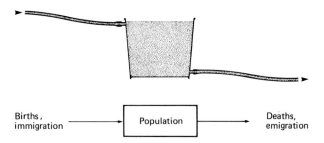

FIGURE 2-5 *The pail analogy for population growth. Imagine a pail with one hose for input of water and another for outflow. By varying the rate of input it is possible to attain an unchanging volume of water in the pail as a function of input rate. Conversely, controlling the rate of outflow with a hose clamp with a constant inflow will produce another set of steady state water levels, as would simultaneous operations on both hoses. This analogy is directly applicable to population dynamics, for which the water level is population size and inputs and outputs are births, deaths, and migration.*

rate of outflow in the pail of water will increase its volume. Insofar as environmental factors can influence the balance of the equation

Population changes = (immigration + births) - (emigration + deaths)

in dramatic or in subtle ways, it is clear that the difference between explosive growth and disastrous decline often can depend upon nothing more than small, subtle shifts in the environmental conditions controlling population inputs and outputs.

Despite its simplicity, the pail analogy is appealing because it illustrates the dynamics of the process of population change. To make it more realistic, we need to specify the biological parameters that trigger growth, delay decline, and determine the rate of potential or actual growth. For example, it would be useful to know whether certain kinds of populations (such as crop pests) are able to respond to favorable conditions faster than others, or whether they are more resistant than others to poor conditions that inhibit growth. However, to specify this level of detail seems pointless here; it seems sufficient to summarize the major findings about population growth that are important for environmental problem-solving.

1. Populations with a large growth potential usually are characterized by an equally high potential for decrease. These kinds of populations are adapted to take advantage of opportunities when they occur, but they may otherwise maintain only small numbers. Conversely, populations that are numerically stable tend to have only a small capacity for increase.

2. Even populations with relatively low birth rates, however, can grow fast when favorable conditions persist long enough. For instance, suppose that a pair of elephants produces only a single calf each year and that all survive to reproduce themselves. If young elephants mate and reproduce each year beginning at age three, then by the tenth year that single pair and their accumulated descendants would number 31 (Figure 2-6). This example underscores the importance of environmental checks on population increase to offset the intrinsic capacity for reproduction.

3. Most populations rarely grow or decline very much. Instead, they usually are found in a state of relative stability. Nevertheless, the potential for growth is there, waiting for an appropriate set of conditions to release it. For human beings, the most significant implication of this is that most populations are stable under some regime of natural control—a regime that we disrupt advertently and inadvertently throughout our planet.

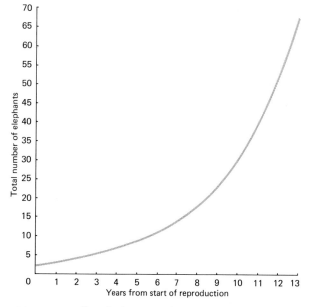

FIGURE 2-6 *Population growth for a pair of elephants. In year 1 the first pair reproduce. In year 4 the first young reproduce; if we graph their continued reproduction, we get the classic exponential growth pattern.*

The subject of natural control has been a central topic in population ecology for decades and is marked by controversy and disagreement. In recent years, however, a rough consensus has emerged about how population stability (restricted fluctuations in numbers) is attained.

Some populations exist only in environments that are harsh and unpredictable, while others are subject to irregular or periodic environmental fluctuations. If such populations are unable to overcome the destructive effects of the unfavorable conditions to which they are subjected, they may appear to be controlled by these random environmental conditions and not by any well-ordered set of biotic interactions. It has been argued that these kinds of populations actually escape control because they are constantly growing at their maximum rate just to compensate for continuing high mortality from environmental fluctuations. These kinds of populations will be characterized by large fluctuations in size from generation to generation, by high birth and death rates, and by poorly developed mechanisms for controlling their population sizes.

Other populations have developed means to escape the effects of environmental variation. These populations usually are free to grow until they are limited by other populations that feed upon them, parasitize them, or compete for food or space. Suppose, however, that natural selection produces a population that is a superior competitor and that also can escape its predators and parasites. It then will be released from control by these agents to reach the ultimate limitation: its own numbers. Even if no other agent proves effective, eventually a growing population will exhaust its resources, and further growth will cease.

The principal difference between a population controlled by harsh environments and one whose size reflects complex biotic interactions is the way in which the fluctuation of numbers is governed. A population responsive to harsh environments will display no trend in its changes: A decrease in numbers is just as likely to be followed by another decrease as by a compensatory increase. Consequently, such a population will be unstable, growing and crashing violently, until ultimately it goes to extinction. On the other hand, a population whose size is controlled by biotic interactions and availability of resources will tend to show a stabilized mean size because a spurt of population growth will be cancelled by increased mortality or depressed reproduction, while a decline either will release enhanced reproduction or will suppress mortality. For example, a population that grows too much may overuse its resources and will crash, while a population that is decimated by unfavorable conditions may experience reduced competition and thus a spurt of growth that quickly recoups the loss. The population will show a characteristic density governed by these forces.

Though these are the major types of population regulation, some ecologists have argued that certain birds and mammals with highly developed social hierarchies have a more advanced form of population control. Populations of this kind are said to be controlled neither by weather nor by natural enemies; they are restricted by resources in a unique way. The somewhat anthropomorphic argument is that such populations have adopted control mechanisms which minimize the chance of overexploitation of resources. Thus there is an advantage in that this mechanism confers on the population some means of birth control that prevents the unrestricted competition which would lead to insufficient resources and a subsequent population crash.

The proponents of one proposed mechanism argue that the social structure of the population determines the birth rates of individuals and that in these social groups the response to the population control mechanism is uniform among all individuals. Natural selection then operates upon the group as a whole, and groups that overexploit their resources are eliminated sooner or later. Other scientists oppose this theory and argue that ordinary selection operating upon specific individuals is sufficient to produce the same result. While the details of the mechanism are clearly controversial, there is general agreement that the social structure of a population is a critical element. For example, territorial behavior in many bird species

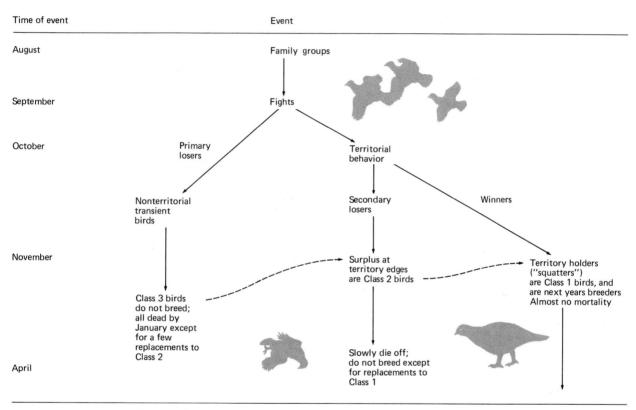

FIGURE 2-7 *The population regulation mechanism in red grouse. Birds may be divided into three classes, only one of which will breed in the next year. The others are normally eliminated. The essence of the system is that breeding success in any one year is inversely proportional to that of the previous year, with territorial behavior and territory size as the mediating variables.* (Redrawn from Watson and Moss in *Behavior and Environment*, A. H. Esser, ed., 1971, Plenum Publishing Corporation.)

can be viewed as a social mechanism to limit the number of breeding adults and to insure that each breeding pair has enough resources to raise a successful brood. (Territorial birds of a given species that do not have territories do not breed, and other pairs with insufficient territories rarely, if ever, nest successfully). The red grouse of Scotland's heather highlands provide one specific example (Figure 2-7). While it appears that the red grouse does maintain a reasonably stable breeding population, the *total* population is not very stable because the mechanism of breeding limitations is not perfect and all excess birds are ruthlessly eliminated. In the extremely "class-conscious" red grouse population, social mechanisms underlie population control, foreshadowing means of population control relevant for human populations. This topic is discussed in Chapter 5.

COMMUNITIES

Community dynamics is probably the poorest-known aspect of ecological theory; at this level of aggregation there is a very high cost for any information comprehensive enough to explain the causal mechanisms involved. We shall approach this problem by examining how populations form interrelationships that foster community development and diversification. There are three kinds of interrelationships: *predation*, in which one species eats another (Figure 2-8); *competition*, in which two or more species attempt to control a particular resource; and *cooperation*, in which two or more species act toward the benefit of either or both.

Natural selection is the mechanism that preserves lasting interrelationships of these kinds. For example, natural selection eliminates all relationships in which a predator overeats the prey species and in which the prey is exterminated; moreover, in cases where

FIGURE 2-8 *A pair of lions* (Panthera leo) *about to bring down a wildebeest* (Connochaetes taurinus) *on the East African veldt.* (Courtesy of Thomas D. W. Friedmann from Photo Researchers, Inc.)

overeating does not lead to the prey's extinction, natural selection strengthens the predator-prey relationship by improving the future resistance of the survivors since the predator eliminates the most susceptible prey first. Thus the predator-prey relationships that survive tend to be self-perpetuating and part of the basis for further community development. Similarly, a competitive interaction can be preserved by natural selection if neither competitor eliminates the other completely before natural selection has a chance to operate. No matter how imperfect they may be, relationships that persist can be improved, because natural selection will tend to reduce the intensity of competition to the benefit of both parties. Each competitor will predominate in its areas of superiority, but competition will *not* be eliminated; instead, there will always be the potential for competition. If one competitor attempts to displace the other from its favored areas, the potential competition will become very real. As long as this potential for competition is present, natural selection will favor the existence of many specialized species with a narrow range of competitive superiority. This is true because competition favors species that are superior for some relatively narrow range of environmental conditions for which they are best suited. If given the opportunity, these kinds of specialized species will eliminate less specialized ones. Thus diversification of the community also would be favored by competition.

Just as natural selection acts to perfect predator-prey and competitive relationships, it is also important in the development of successful cooperative relationships. Cooperative relationships require a mechanism that prevents opportunists from exploiting the other member or members and eventually destroying the relationship. Although the mechanism that allows these cooperative relationships to persist is not now totally agreed upon, the existence of some remarkable coevolved relationships is living proof that such cooperation can evolve in nature. Two examples are the "cleaning symbiosis" in coral reefs, where certain small fishes clean larger ones of parasites for the mutual benefit of each; and the relationship between ants and acacia trees in Central America, in which the ants obtain nutrients from the tree and in turn provide some protection to the tree from predators and competitors (for details, see Feder, 1966, and Janzen, 1967).

It seems clear, then, that natural selection favors lasting relationships and thus biotic diversity—if we may assume that environmental conditions remain suitable long enough to permit the selection process to operate. If we agree that tropical conditions provide the time needed for the operation of natural selection, high diversity in these areas is easily explained by competition and rich resources. While there is no unanimity about which factors control diversity, there is general agreement that diversification is favored by environmental stability and abundant, predictable supplies of energy. Conversely, simpler, less diverse communities should occur in areas of environmental instability and unpredictable resources.

The development of interrelationships leads to assemblages of species that we call communities. These interrelationships also confer properties on the community, one of the most important being *community stability*. This community concept is an especially important parameter for communities that supply resources to human beings; instability would introduce major economic disruptions, at the very least. Unfortunately, operational use of the concept of community stability is plagued by definitional problems. Community stability can be measured in at least three ways, none of which is perfect:

1. As an aggregate index of the numerical fluctuations of key populations in the community. In a forest, for example, community stability might be measured as a composite index of the population changes of dominant trees, shrubs, and animals.

2. As the ability of the community to resist disturbance, to preserve its characteristic structure, and to function despite loss of some populations. In the forest example, this index might be defined as the ability of the forest to withstand logging without destruction.

3. As the ability of the community to recover from disturbance. In this case, the characteristic structure and function of the community may be lost for a while, but in time its resilience enables it to regain its original structure and function. This differs from item 2, which contains no necessary provision for eventual recovery. Stability of this type might be measured as the time required for the forest to repair flood or hurricane damage and return to its "normal" pattern.

It is difficult to apply stability concepts to the management of ecological systems because of our ignorance, but there clearly is a need to understand how to encourage community stability. One idea that has appeared repeatedly is that high diversity will engender high stability. The reasoning is as follows: The higher the diversity of the system, the less dependent it will be upon the fate of any one population; for this reason, it is less likely that the system will change significantly when subjected to stress. For this assumption to be true, we require that stability not be measured as some form of collective population stability and that any species dependent upon a disturbed population be able to switch freely to another. At present there is little solid evidence that "diversity engenders stability" is a universal truth.

Systems Evolution

INDIVIDUALS AND POPULATIONS

The reactions of individuals to changes in their environments were discussed in Chapter 1. Individuals are inherently variable because their genetic constitutions differ from one to another. Natural selection then may act upon the existing range of variability to favor the individual best adapted to the environment. Insofar as environmental conditions may vary, the results of selection also may vary from place to place and from time to time.

The mechanisms for change in populations are much the same as for individuals. Populations also may be inherently variable, reflecting the individuals they comprise. Natural selection then can act upon the aggregate of similar individuals much as it acts upon individuals, with the same conditions about prevailing circumstances. As an illustration, let us examine Krebs' hypothesis for the cyclic alteration of types in a small mammal population (Figure 2-9). When the population is small, the "docile, highly fertile" male is favored because there is plenty of food and space and no particular reason to quarrel. When food and space are more limited, selection favors the "aggressive, all-fight, low-fertility" male who defends a territory vigorously. But as the population declines because of low reproduction, the cycle is completed, and selection once again favors timid males.

At these two levels of organization the mechanisms of change are relatively well known. In planning a project, especially a large-scale project with a potential for major ecological impact, what *could* happen is not nearly so important as a more situation-specific prediction of what is *likely* to happen. Hence the relevant questions are

1. Is genetic variability sufficient to absorb the deleterious ecological effects of the project?
2. If there is an initial setback, can the system adapt to or recover from the effects?
3. If another ecological system replaces the original one, will it have qualitatively different effects upon human interests or upon nearby, less disturbed systems?

SYSTEMS EVOLUTION 35

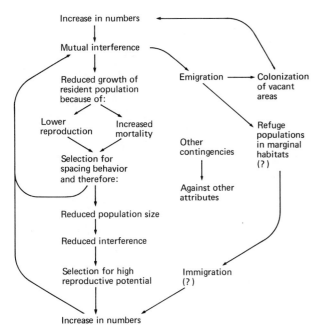

FIGURE 2-9 *Krebs' hypothesis for population regulation in small mammals. The system of regulation is similar to that of the red grouse. This proposal makes selection pressure a function of population density; at high densities aggressive behavior is selected for at the expense of viability and reproduction. Consequently population density falls until selection favors individuals that are not aggressive but who reproduce rapidly. The population continues to cycle in this manner.* (Redrawn from C. J. Krebs et al., *Science*, 179, pp. 35–41, January 5, 1973. Copyright © 1973 by the American Association for the Advancement of Science.)

Of course, examples exist for all three possible outcomes, and it is impossible to predict which option will hold without more specific details. This dependence upon specific biological detail makes ecology an empirical, exciting science.

COMMUNITIES

As we have seen, favorable environmental conditions permit natural systems to develop structural and functional complexity through relatively slow evolutionary processes. Major instabilities—ice ages, volcanic eruptions (Figure 2-10), significant human interference—produce the opposite effect and favor simpler communities over more complex ones. The condition of complete, unchangeable stability does not occur. The dynamics of ecological assemblages normally produce a sequence of changes which are directed by physical conditions, changing biological relationships, or even by human beings; this sequence is termed *ecological succession*.

Ecological succession consists of a series of changes in which one set of organisms (called a *sere* or *seral stage*) replaces another in a predictable sequence. If a glacier creates a new lake, we know that in as little as ten thousand years the lake will become a marsh and eventually an upland forest. Many other examples of succession have been described, including some in arctic environments where succession is cyclic rather than unidirectional.

One facet of ecological succession that requires attention is the extent to which we influence the nature and speed of these events. While ecological succession would proceed without human interference, we frequently have the power both to start and to arrest successional sequences as well as to regulate their rates. Farmers who clear their land start a new successional sequence (if they allow the ground to revegetate naturally) or frustrate it (by planting and tending crops). Foresters stop succession when they burn the seedling hardwoods on the floor of a pine forest. And septic tanks or sewage treatment plants that leak nutrients into a cold, biologically poor lake accelerate its progress toward the forest that ultimately will occupy its site.

FIGURE 2-10 *An illustration of the contrast between the heat of a volcano and a cold climate. The locale is the Westman Islands, Iceland, where winter snows have been coated with a fine ash of cinders from the volcanic eruption in the background.* (Courtesy of Frank Siteman from Stock, Boston.)

Another facet of this subject that deserves analysis is the question of what actually controls the succession of events. Ecological succession usually is understood to imply a situation in which an earlier sere prepares the environment for its successor. Recent evidence, however, suggests that this supposition is not always true. For instance, research in old field communities in North Carolina has indicated that, in this system at least, competition may be more important than the preceding seral stage in determining succession, and data from rocky intertidal areas indicate that in this system there may be no particular succession at all. The existence of evidence contrary to prevalent theory does not deny that the original definition is correct for some examples, but it again points out the lack of a well-substantiated general theory. This reflects the empirical nature of ecology and further underscores the need to fully understand the causal mechanisms involved in a particular situation. This need is crucial in cases where a goal can be accomplished only through active management of the system. In the case of ecological succession, this goal may well be attained only by deliberate interference with succession, especially when the goal is preservation of a species that normally would be eliminated during succession.

One Viewpoint on Ecological Systems

We end our brief look at ecology with four conclusions:

1. Ecological systems are not composed of discrete elements called "structure," "dynamics," and "change," although it was convenient to break them down in this way. More realistically, ecological systems are interacting assemblages in which any two elements affect the third. One element cannot be modified without changing the other two.

2. Living systems are in dynamic balance with their environments; that is, their structural and dynamic patterns are determined by an interaction of environmental forces that constantly change. Consequently, real systems constantly adapt in response to changing conditions. It is not a simple matter to draw the line between an ecological system in steady state and one that is ready to collapse because survival depends upon many dynamic alternatives, few of which are evident without study.

3. While general features seem to characterize most ecological systems, as a result of specific differences, there can be great differences in how systems react to the same disturbances. For example, it is possible for an environmental change such as fire to destroy one system while benefiting another. The same conclusion applies to the same system at different times if its structure varies over time.

4. The principles of ecology that have been developed are most relevant to nonhuman systems because biological limits to the behavioral patterns can be shown; that is, the organisms involved are limited in their capacity for social decision-making. Therefore, when human beings and their creations are a large part of the system being studied, the broader definition of ecology (Chapter 1) is more applicable. Conversely, caution is required when classic ecological theory is used for problems of human environments.

Questions

REVIEW QUESTIONS

1. Consider the following statement: The structure of any ecological system reflects basically the same processes whether the system of interest is an individual or a community.

How is this statement true, and how is it false?

2. What would be the result if the breeding success of one year for a red grouse population
    a. did not affect breeding success the next year
    b. led to even greater success the next year

3. In what different ways can community stability be defined? Which definition is consistent with "diversity engenders stability"?

4. Why is human ecology only partly based upon the ecological principles discussed in this chapter?

ADVANCED QUESTIONS

1. Fire is a useful tool for managing some ecosystems. Using the *Annual Proceedings of the Tall Timbers Fire Ecology Conference* series as source books, outline the ecological conditions that make fire management practical, and the ones that make it impractical.

2. Using the elephant example, work out the numerical consequences of the assumption that when there are 40 elephants, reproductive rates per pair
   a. drop to 0 at > 40 individuals
   b. decrease 20 percent for each 5 individuals > 40.

3. Read Keever, *Ecological Monographs*, 20 (1950), 229, and interpret her results on the effects of individual dynamics on community structure.

Further Readings

Note: This list is relevant to both Chapters 1 and 2, and readers are strongly urged to use it to delve more deeply into the detailed literature of ecology. This will further some of the objectives of this text. The recent proliferation of textbooks in ecology (quite apart from books on environment) has forced me to edit this list heavily. The emphasis is upon readily accessible textbooks.

Boughey, A. S. *Contemporary Readings in Ecology.* Dickenson, Belmont, Calif., 1969.

Brookhaven National Laboratory. "Diversity and Stability in Ecological Systems." *Brookhaven Symposia in Biology*, No. 22, BNL 50175 (C-56), 1969.

> A collection of viewpoints from various ecologists about the meaning of diversity and stability. This volume is uneven and is intended for advanced students.

Clausen, J., D. D. Keck, and W. M. Hiesey. "Experimental Studies on the Nature of Species. III. Environmental Response of Climatic Races of *Achillea*." Carnegie Institution of Washington Publication, No. 581, 1948.

Colinvaux, P. *Introduction to Ecology.* Wiley, New York, 1973.

> Unique organization and excellent first text.

Cox, G. W. *Readings in Conservation Ecology*, 2nd edition. Appleton Century Crofts, New York, 1974.

Emlen, J. M. *Ecology: An Evolutionary Approach.* Addison-Wesley, Reading, Mass., 1973.

> Mathematically oriented and best for graduate students and professionals.

Emmel, T. C. *An Introduction to Ecology and Population Biology.* Norton, New York, 1973.

> A simple, inexpensive introduction to ecology.

Feder, H. M. "Cleaning Symbioses in the Marine Environment." In *Symbiosis*, vol. 1 (S. M. Henry, ed.). Academic, New York, 1966.

Giesel, J. T. *The Biology and Adaptability of Natural Populations.* Mosby, St. Louis, 1974.

> Perhaps the best of the paperbound, concise treatments.

Grieg-Smith, P. *Quantitative Plant Ecology.* Butterworth, London, 1957.

> One of the best texts in plant ecology despite its age.

Janzen, D. H. "Interaction of the Bull's-horn Acacia (*Acacia cornigera* L.) with an Ant Inhabitant (*Pseudomyrmex ferruginea* F. Smith) in Eastern Mexico." University of Kansas Science Bulletin, 47 (1967), 315-558.

Jeffers, J. N. R. (ed.). *Mathematical Models in Ecology.* Blackwell, Oxford, 1972.

> See also Chapter 3 bibliography.

Krebs, C. J. "The Lemming Cycle at Baker Lake, Northwest Territories, during 1959-62." Arctic Institute of North America Technical Paper, No. 15, 1964.

Krebs, C. J. *Ecology.* Harper & Row, New York, 1972.

MacArthur, R. H. *Geographical Ecology.* Harper & Row, New York, 1972.

   The last synthesis of many of MacArthur's most important works.

McLaren, I. A. *Natural Regulation of Animal Populations.* Atherton, New York, 1971.

   Papers on the population regulation controversy.

McNaughton, S. J., and L. L. Wolf. *General Ecology.* Holt, New York, 1973.

   One of the best books for the intermediate to advanced student.

Odum, E. P. *Fundamentals of Ecology,* 3rd edition. Saunders, Philadelphia, 1971.

Oosting, H. J. *The Study of Plant Communities,* 2nd edition. Freeman, San Francisco, 1956.

   A classic view of the nature of communities.

Phillipson, J. *Ecological Energetics.* E. Arnold, London, 1966.

   The best primer available on energy flow in ecological systems.

Pianka, E. R. *Evolutionary Ecology.* Harper & Row, New York, 1974.

   One of three entries in evolutionary ecology, the others being Emlen and Ricklefs.

Ricklefs, R. E. *Ecology.* Chiron Press, Newton, Mass., 1973.

Slobodkin, L. B. *Growth and Regulation of Animal Population.* Holt, New York, 1961.

   Now old, but Slobodkin's thinking will always remain stimulating.

Smith, R. L. *Ecology and Field Biology,* 2nd edition. Harper & Row, New York, 1974.

   For beginning students.

Sutton, D. B., and N. P Harmon. *Ecology: Selected Concepts.* Wiley, New York, 1973.

   A self-study guide that works well.

van Dyne, G. M. (ed.). *The Ecosystem Concept in Natural Resource Management.* Academic, New York, 1969.

Vernberg, F. J., and W. B. Vernberg. *The Animal and the Environment.* Holt, New York, 1970.

   Specialized toward physiological ecology, that is, toward the attributes of individuals.

Watson, A., and R. Moss. "Spacing as Affected by Territorial Behavior, Habitat, and Nutrition in Red Grouse (*Lagopus l. scoticus*). In *Behavior and Environment* (A. H. Esser, ed.). Plenum, New York, 1971.

Watt, K. E. F. *Ecology and Resource Management.* McGraw-Hill, New York, 1968.

   Directed toward aspects of ecological management.

Wiens, J. A. (ed.). *Ecosystem Structure and Function.* Oregon State, Corvallis, 1972.

Wilson, E. O., and W. H. Bossert. *A Primer of Population Biology.* Sinauer Associates, Stamford, Conn., 1971.

   A short paperbound text uniting ecology and genetics in a workbook.

Wynne-Edwards, V. C. *Animal Dispersion in Relation to Social Behavior.* Hafner, New York, 1962.

Wynne-Edwards, V. C. "Self-regulating Systems in Population of Animals." *Science,* 147 (1965), 1543–1548.

3. Systems Concepts for Environmental Analysis

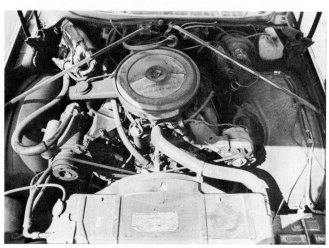

In the first two chapters the term "system" denoted an ecological unit that consists of a series of identifiable components connected in certain ways. Thus a population is an ecological system because it consists of recognizable components, in this case, different individuals of the same species that interact in specific ways. A population meets our definition of "system." A community is also a system because its components are the various populations of different species that it comprises, and these populations interact in specific ways. Although the community is at a higher level of aggregation, it is no less nor no more a system. At an even higher level of organization, the planet, all its creatures, and all its inanimate resources might be considered a system.

But "system" has a second meaning. Through mathematical tools, we can formalize the concept of a system by constructing an analogue. This analogue usually is called a *model* or a *mathematical model;* it is meant to be a mathematical representation of all essential elements of the system. A model that is an acceptable analogue of a system can be used to predict possible future behavior of that system and to experiment with its reactions to a variety of conditions that might be impossible to manipulate in reality.

Systems ecology is the branch of ecology that unites biological research and model-building in an attempt to develop a new level of synthesis. Systems ecology is a relatively new activity; as a result, systems ecologists have so far been more successful in pointing to limitations of ecological modeling than they have been in demonstrating its problem-solving capacities. In particular, attempts to build comprehensive models of ecological systems have repeatedly revealed how little is known about most systems. This lack of reliable information handicaps construction of high-quality models. Furthermore, results of a particular model-building process are usually highly specific to that system, and so the applicability of information gleaned by this method to other systems is likely to be limited. Finally, systems ecology has not yet progressed to a point where new principles of ecology have emerged from comparisons of different models; in fact, many believe that systems models will more contribute to the testing of tentative principles than act as a source of new ideas.

Though all these problems are important, another difficulty is encountered because of the enormous amount of information needed to construct an accurate model of a system. To illustrate this, suppose we wished to construct a model of a grassland community: We would need data about all the grass species present as well as information about their population dynamics in relation to weather, to soil, and to each other. To get a more comprehensive model that also included the animals of the grassland system, we then would need information about the dynamics of all animal species in relation to changes in the grassland as well as data about the effects these animals collectively had upon the grasses. Finally, to predict the effects of human use of the system, we would need the results of experiments in which the effects from the contemplated uses were produced; or if this were not possible, at least effects produced from some perturbation thought to be similar to human interference. New concepts that emerged from the modeling process would have to follow its completion, and this could not occur until all data had been collected and synthesized. This obviously would be extremely time consuming and quite expensive, in most cases prohibitively so.

If systems ecology can be confirmatory but not particularly inventive, one may wonder why we look at it. The reason is that though many difficulties are met, they partly result from the lack of systematically gathered, high-quality information. While data problems undoubtedly will continue to plague us, this does not detract from the conceptual advantages that systems methods offer, particularly for complex systems that demand methods which reduce confusion to simple, readily grasped concepts.

Systems Methods and Environmental Decision-making

Suppose we wished to use systems ecology for environmental decision-making. How might we do it? When possible, we try to use existing systems models, which can be superior policy-making tools—if we can assume conditions have not changed greatly since the model's formulation, and if the model realistically represents the system. However, such high-quality models are rare indeed, and so policy changes seldom will arise from a well-documented mathematical model. To use systems methods productively, given this limitation, we must emphasize the broader conceptual understanding of environmental problems that systems methods can give us. In particular, we want to anticipate the consequences of our activities upon environmental systems better than we have with more conventional methods.

Suppose we decide to use "the systems point of view" to improve environmental decision-making. This decision raises several questions, the most important being the goal, which in turn dictates the strategy. If the goal is to develop a comprehensive and accurate model, we face the same problems as when we use an existing model—only more so. On the other hand, if we want a general understanding of a problem without demanding a complete and detailed solution, we may have to put together only a *conceptual* model—or, as a next step, provide insight into the construction of a mathematical model.

The imposing requirements for formal model building and the youth of the field of ecology dictate the latter strategy as our only choice at present. Whatever the advantages of systems concepts, they do not come easily, for decisions are still required about what information we need. For example, can ideas derived solely from classical ecological systems permit us to understand environmental problems? The first two chapters show that ecological systems do not possess all behavioral patterns relevant to problem solving, especially for human-dominated systems. Whenever humans are involved, it is likely that a wide range of subjects will apply. The lessons are twofold: Great care must be taken in drawing analogies between one system and another, and the choice of appropriate fields of knowledge used to define the system must be made as broadly as possible. For most environmental problems, ecology is only one of many fields that may make important contributions; and compared with anthropology, sociology, chemistry or public health, its help may be minor.

So far we have assumed that our use of systems methods would be directed solely toward a specific problem. However, one area of research attempts to minimize the differences between specific systems by finding elements that characterize all of them. This is known as *general systems theory*, and its goal of ultimate generality stands in stark contrast to models designed for specific situations. It is premature to claim, however, that one approach is more useful than the other. Both approaches must be studied thoroughly. In the following sections, a pair of familiar examples will serve to illustrate the differences and similarities more systematically.

Defining a Systems Model

We may define a systems model by specifying what components are to be part of it. The first example is familiar: an internal combustion engine. The engine itself is the system; everything beyond its limits constitutes its environment:

Environment

Engine

All the other systems (chassis, transmission, brakes, etc.) upon which the engine depends to do useful work are part of its environment, just as is the road surface, the air, the weather, and the user. Upon further examination of the system, we find that the engine is itself made up of specific components:

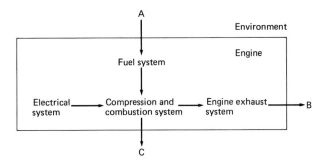

Each box within the engine is also a system. These boxes are called *subsystems* only because they are an integral part of the system (the engine) we are interested in; otherwise, they are systems just as the engine is. Because the subsystem definition is arbitrary, each of them can further be broken into component parts. Consider the fuel system:

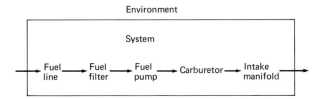

In these diagrams, the arrows that extend beyond the system merely indicate the specific relationships between the system and its environment. Looking at the diagram of the engine and its four subsystems, we discover that only three environmental relationships need be considered:

A: Input energy from the environment (fuel plus atmospheric oxygen) required for the engine
B: Combustion products from the engine that must be disposed of into the environment
C: Waste heat from the engine that is absorbed by the lubricating oil and whatever cooling fluid is used and subsequently disposed of into the environment

These diagrams define a specific system that can be further developed with precise equations. This is the dividing point between general systems approaches, which require comparisons between systems to discern similarities, and specific models, which logically are the sets of equations developed from a single conceptual framework. The process of system definition can be applied to any example, no matter which approach is used.

To illustrate this, let us contrast our physical system and a grassland community. For the hypothetical grassland, we may identify the relevant subsystems in energy terms just as we did for the internal combustion engine:

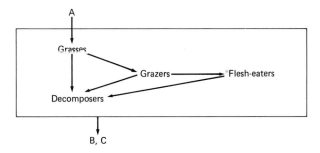

This diagram should be familiar, for it is the type of trophic system discussed in Chapter 1. If we consider the system as having a function (if nothing else, it has the function of maintaining itself), the trophic web is analogous to the internal combustion engine. Like the engine, the ecological system absorbs energy (in this case, flow is in the form of sunlight). Eventually, all the energy and waste materials are eliminated from the system, as are some useful materials that happen to be removed by factors beyond the control of the grassland (for example, wind, erosion, and flooding remove soil and nutrients). All these losses are combined to form a joint loss composed of flows B and C. The analogy between these systems is based upon the fact that both systems are energy users, and their structures are arranged so that they can obtain useful work from whatever energy they use. These two systems, then, are examples of a general system in which the capture and conversion of energy into useful work obeys physical thermodynamic laws applicable to all systems. Of course, there are many specific differences between the two systems, the one being purely mechanical, the other purely biological. But in the general systems viewpoint, all these differences are purely secondary characteristics.

Properties Shared by All Systems

We have seen that all systems using energy obey certain thermodynamic rules and that these common behaviors constitute one general systems property. Several other properties are useful in describing all systems, and we will now discuss them. The effects of these properties is implicit in all the systems used in this book, although specific features may sometimes obscure them. It is useful for this reason to review them occasionally.

The Nature of System Boundaries

At least in one sense, all systems are artifacts of the human brain and imagination. In other words, there is only one real system on the planet, and that is the entire planet; if we recognize any part of this as a subsystem, we do so by establishing arbitrary boundaries, no matter how natural the assemblage may seem. In some cases, a system we define may be fairly "natural," whereas other systems are not. Without being very specific about it, the difference may be measured as a function of the importance of the environment (that part excluded from the system). If the environment has many components that significantly affect the behavior of the system, the system is said to be *open*; but if the system is more "natural"—that is, its behavior is mostly determined by its own components—it is said to be *closed*.

Let us return to our previous examples. The engine is a relatively closed system because, given a source of fuel, it can produce output with little or no influence from external factors. Natural systems, such as communities, are a different matter. The nature of the community, and indeed of all natural systems at any level, is such that the vast number of interrelationships that exist keep the system relatively open. The grassland, for example, can be affected by many factors: climatic variability, frequency of fires, soil structure, and the nature of surrounding communities, including the use made of the grassland system by human agents (Figure 3-1). Obviously, these things make a grassland less than a discrete, easily identified system. As an example of problems one might encounter, consider the situation of a forest at the edge of the grassland. Since the grasses extend over the forest floor and the trees invade the meadow, where is the boundary between the two systems? The internal combustion engine, on the other hand, is a conveniently discrete piece of hardware with obvious boundaries.

Goal Orientation

The organization of all systems reflects some set of goals, whether explicitly stated or not. For ecological systems, the primary goal of survival is implicit, and the agency of natural selection favors species or systems that are best able to use or at least withstand their conditions of energy, climate, and biological surroundings. In this light we may interpret the position of grasslands in Figure 1-2 to be the set of environmental conditions of sufficient seasonal warmth, light, and precipitation where natural selection favors grassland ecosystems. Although natural systems have the primary goal of survival, secondary goals are also governed by natural selection. In the grassland example, just as in other ecosystems, there are such implicit goals as diversification, the maintenance of the largest possible standing crop, and dominance of the landscape.

The goal structure is most explicit in human-dominated systems in which natural selection can even be overridden. We can manipulate and even create grassland systems that are intended only for their yield of food energy. We do this by using various technological means to suppress alternatives that would be favored by natural selection, such as diversification of a system by organisms we consider pests as well as those producing disease in our food crops. The example of the internal combustion engine shows explicit goals even more clearly; it is a system built specifically for purposes of work output from fuel-energy conversion.

FIGURE 3-1 *A western cattle pasture, showing the effects of trampling and heavy grazing. Note the large areas of bare ground.* (Courtesy of Bruce Coleman, Inc.)

Hierarchical Organization

Since all systems comprise other systems, there must be a hierarchical organization involving all of them. The grassland is but one part of a region such as the midlands of the United States, and it is composed of many species of plants, as well as soil, nutrients, water, sunlight, and other components. Similarly, the internal combustion engine consists of subsystems and is in turn part of a larger system—the vehicle.

Basic Components of Systems

One of the most effective ways to analyze a real system and construct a model of it is to work with its subsystems ("functional blocks") rather than attempting to deal with the complex whole at one time. It has proved to be easier to analyze a system by breaking it down and relinking the functional blocks later. Some of the decisions on the location of boundaries are easy to make because the subsystems of interest are naturally defined; others are more difficult because the system is open and the boundaries are indefinite. For example, the engine-exhaust system is discrete; it can be unbolted from the rest of the engine and neatly put aside. On the other hand, it would be very difficult to uncouple a grazing animal from the grassland system, especially if many of the animals were omnivorous or otherwise not totally dependent upon just the grasses.

No matter how defined, the subsystems eventually designated are called *compartments*, *levels*, or *states*. Usually they are the tangible elements, the physical and biological entities that can be observed, measured, and used as indicators for the performance of the system. In fact, elements that fulfill these requirements usually end up as the subsystems of greatest interest. Once the compartments are specified, the rest of the system will be defined when the subsystems are linked together with the appropriate kinds of *flows* (also commonly called *rates*). The state of each compartment is controlled by the rates, which actually determine the magnitude of the inputs and outputs for that level; hence specification of all relevant rates determines the values characterizing the various levels. One useful illustration is the population-bucket analogy presented in Chapter 2. In the two examples in this chapter the arrows represent the rates of fuel flow from the fuel system into the combustion process or of dead grasses to the decomposers. In part, the rate of fuel input to combustion controls the work output, and the rate of death among the grasses is one determinant of decomposer activity. Of course, in any complex system the entire set of rates and levels determines overall behavior. The important difference between a relatively open system (the grassland) and a relatively closed one (the engine) is that the former will present additional problems of definition. In addition to these definitional problems, the open system has further complications in that the dynamics of the environment—a significant component of this type of system—are poorly understood. For all these reasons it is even more difficult to predict and understand overall behavior of an open system.

System Feedback

To this point we have treated systems as though they had stereotyped behavior, every action leading to a predictable reaction. While a few simple systems (such as a thermometer) behave like this, most systems feature interrelated rates with collective behaviors that are neither simple nor predictable. These interrelated systems are known as *feedback* or *control systems*, because they commonly involve compartments that react to changed input or output by acting themselves (literally, feeding back) either to counteract the initial effect or to enhance it. *Positive feedback* is the name applied if the initial effect is enhanced; the opposite case, where the initial effect is reduced, is called *negative feedback*.

To illustrate these important concepts, let us again consider the internal combustion engine. If we increase the rate of fuel input and the engine simply runs faster proportional to the increase in fuel, no feedback is exhibited. But suppose the additional fuel results in improved combustion and more work output. This is an example of positive feedback because the additional fuel has an effect proportionately greater than would be expected per unit of additional fuel. If we invent some numbers for a hypothetical experiment comparing two engines, we might find these results:

Engine A			Engine B		
Fuel supplied per hour (liters)	Work output per hr (brake horsepower)	bhp per liter	Fuel per hr	Work per hr	bhp per liter
5	15	3	5	15	3
10	30	3	10	40	4
15	45	3	15	75	5

Engine A clearly shows no feedback: the output/fuel unit is fixed at 3 no matter what the input. Engine B shows increases in output/fuel unit; however, each extra liter of fuel produces a larger output in work. Whatever the factors responsible, this is a situation of positive feedback because a positive change in fuel produces a positive change in output/fuel unit. To complete the story, suppose a third engine produces the following results:

Engine C		
Fuel/hr	Work/hr	bhp/liter
5	30	6
10	30	3
15	30	2

This is a case of negative feedback because increased fuel produces decreased output per fuel unit.

For the example of the grassland ecosystem, suppose our experiments looked at reproduction of three grass species in relation to fertilizer:

Species 1	
Pounds of fertilizer/month	Seeds/parent
1	60
2	60
5	60

Species 2	
Pounds of fertilizer/month	Seeds/parent
1	60
2	120
5	150

Species 3	
Pounds of fertilizer/month	Seeds/parent
1	60
2	40
5	20

The parallels with the engines are obvious, and the results are the same. A mathematical treatment is given in the appendix to this chapter, but otherwise the specification of the example is left to the student.

Predictability of System Behavior

One final property of complex systems requires our attention. Are complex systems predictable? To some extent, major behavioral patterns of complex systems are predictable—otherwise there would be no justification in trying to model them. However, practically all are inherently unpredictable in the sense that their state at any one time may be any of a large number of alternatives. This kind of unpredictability can arise in several ways. It may result because any one of a large number of alternative states have a low probability of occurrence, and so the probability of selecting the correct alternative is small. Another source of low predictability in a system is the sheer complexity of the rates which control a process: We may simply not know enough to understand how a change in this complex control process will be expressed. Some systems may just be highly variable and contain elements of randomness that can never be assessed more closely than with a probability value. The net effect of all these sources of low predictability is uncertain behavior in most complex systems.

The ecological systems clearly relevant to environmental decision-making are prime examples of low-predictability systems, for all the reasons given above, but also because specific data about their structure and behavior are generally scarce. Equally clear, the complex systems of human beings and nature plus the social and technological institutions that mediate most human activities are subject to the same considerations. Hence it should be no surprise that our ability to predict future events is severely limited and likely to remain so.

What, then, is the utility of systems methods for low-predictability systems? If we could be satisfied with a very general level of predictability that minimizes both the amount of data we would need to collect and the magnitude of potential error, we could construct models with the foremost goal of predicting future events. But it seems likely that systems model building will prove more productive for other purposes—improving our understanding of environmental systems, establishing tentative limits to the number of alternative behaviors, generating new hypotheses and new experiments. Since most systems of interest to us are both complex and unpredictable, this is an important limitation upon our current use of models.

Complex Models of Particular Systems: Global Dynamics

Let us now take an opposite approach: Instead of defining general systems properties, we will examine models of one system—the world models constructed by J. W. Forrester, D. L. Meadows, and their colleagues.

We begin with a simplified schematic diagram of the world model (Figure 3-2) employing the flow-chart techniques used by Forrester and Meadows. The principal compartments (enclosed in rectangular boxes) are pollution, population, natural resources, capital resources, and land use. These compartments are linked together by a maze of rates (here named in the circles); only the principal rates are shown. To understand how the model represents events on the planet, follow the effect of population growth through Figure 3-2. The model postulates that population growth will have direct effects upon such things as the use rate of nonrenewable resources, the rate of pollution generation, the amount of food available per capita, and the size of the labor force. If we consider just the pollution pathways, we see that the

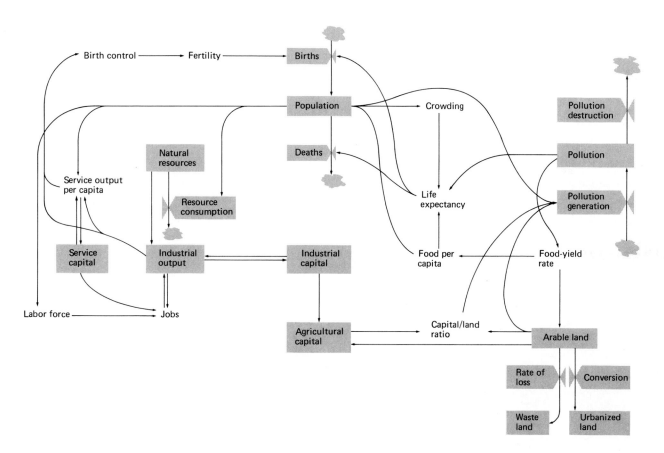

FIGURE 3-2 *A simplified version of the Meadows "World 3" model. The rectangles are major state variables; the fish-shaped boxes are major rate variables; and the cloud-shaped symbols represent unspecified sources and sinks in the systems environment.* (Adapted from *The Limits of Growth: A Report for the Club of Rome's Project on the Predicament of Mankind*, 1972, by D. H. Meadows, D. L. Meadows, J. Randers, and W. W. Behrens III, A Potomac Associates book published by Universe Books, N.Y.)

model links pollution level to reduced life expectancy, lowered birth rates, and increased death rates for a doubled feedback effect upon population level. This is just one example of the kind of detail built into the world models that the reader may discover by tracing through the flow charts. This exercise will show the validity of the complex systems properties discussed above (boundaries and feedback in particular); but the models also will reveal the basic beliefs of the authors about how the world works. This specificity is an advantage—there is no room for fuzzy concepts or ambiguity—but it is also a source of controversy. The principal strength of the global model is the strong development of intimate feedback loops in great detail. In order to develop this level of detail, the model must incorporate statements that may apply to no other system and certainly not to all others. Hence the trade-off for complexity is a loss of generality.

The main objective of the global models was to specify the range of possible futures for this planet given particular assumptions. For example, if the human population and the collective national economics continue to grow indefinitely, the models predict catastrophic events: overwhelming levels of pollution (Figure 3-3), exhausted natural resources, and population crashes; these events will recur in pronounced cycles. If we make enough assumptions such as these and then trace their consequences using the model, we can discern that some general patterns appear repeatedly. The Forrester-Meadows models were used to predict a period of lasting global instability unless there was very stringent control of population growth and resource use.

The publication of the Forrester-Meadows global models stimulated a storm of controversy. We shall not attempt to evaluate the models here, except to set forth the type of considerations involved in such an analysis:

1. Is the structure of the model an adequate picture of reality? To formulate an answer, we must be able

FIGURE 3-3 *A panorama of air and water pollution near an industrial plant in Brunswick, Georgia.* (Courtesy of Grant Heilman.)

to evaluate the accuracy of all the equations, a task that requires specialized knowledge of all salient features of the real system as well as an ability to check the equations themselves.

2. Even for a model of global dynamics, an unspecified environment still could play an important role in determining the behavior of a system. In this particular case, the system simulated by the Forrester-Meadows models is relatively open, and either predictions will apply only for the specific conditions examined (subject to the other potential problems), or the effects of all likely environmental changes will also have to be evaluated.

3. The Forrester-Meadows models have fixed answers for a particular set of circumstances and allow for no variation of response within the particular set of conditions. Real systems usually do not behave this way, although it is not necessarily true that a model which does not allow for random variation will be useless. The assumption does, however, require verification before its results are fully accepted.

Most controversy over the global models has centered upon the question of reality. The global models themselves were completed over a fairly short time, but if verification of all the specific beliefs built into the selection of levels and the interconnections with rate variables were to be attempted systematically, work would probably take decades. Is there, then, a middle ground between a specific, expensive, and controversial model and a set of systems properties that generally are applicable but not very useful as policy formulation tools? The last section examines this question.

The Systems Point of View

To apply systems ideas to environmental problems, we first must make a strategy decision. Do we need specific models or not? For those inclined toward systems analysis the only satisfying approach is a comprehensive network of overlapping models applicable to all problems. Unfortunately, since problems are already pressing us and sophisticated analyses are not immediately forthcoming, the utopian solution is not viable.

For individuals interested in using systems ideas in environmental studies but not prepared to devote weeks or months to specific examples, the most productive approach is to use the properties of systems outlined above to gain an appreciation of the scope of the problem under examination. The systems orientation, or "the systems point of view," consists essentially of the realizations that (1) systems are arbitrarily defined, (2) there is always a systems environment, and (3) unexpected interactions can always produce unexpected effects. Appreciation of these facts trains us to include variables beyond those usually considered, although there may be wide variation in what this statement means: It could range from something as simple as the post facto wisdom of hindsight after something unusual or unexpected has happened to attempts to build a quantitative model of the system of interest.

Recall that the process of systems model building proceeds in stages, starting with the delineation of the system, then proceeding to the conceptualization of the interrelationships found in the system, and finally resulting in the quantitative specification of these relationships in a model. We can perform the first two stages simply by organizing our thinking properly, but the final stage demands greater effort if the end result is to be a nontrivial quantitative model. The volume of work needed to attain the third stage is one reason why the development of high-quality models has been slow. As a result, models have developed unevenly in different areas; models relating air pollution and transportation are well developed, but models of social phenomena are not. For this reason, too, we cannot present comprehensive models here for all important environmental problems.

Nevertheless, the enormous scope of the word "environment" demands some sort of systems orientation, even if we can proceed only through the relatively qualitative aspects of systems analysis. Not only are the advantages of this approach abundantly clear, they also are obtainable at low cost; all that is required is an attempt to systematically broaden one's thinking. The implications of the systems point of view will be drawn in the chapters that follow, but it is up to you to decide how useful the systems orientation really is.

Questions

REVIEW QUESTIONS

1. Define the term "system." If an automobile is a system, is its driver a part of the system or a part of its environment?

2. Show how an internal combustion engine and a grassland are similar with respect to
   a. feedback properties
   b. hierarchical organization
   c. goals

Show how they differ with respect to these three points.

3. Construct a flow chart of levels and rates to show how recreation in a national park is related to population, travel distance, and accommodations and recreational opportunities.

4. Work through some of the examples in Churchman's book, or in Forrester's *Principles of Systems*, if problem 3 was difficult for you.

ADVANCED QUESTIONS

1. How might the systems point of view enter into analysis of environmental impact?

2. By now you are ready to progress beyond the scope of Chapter 3. This means detailed work with a specific system, which in turn means familiarity with specific data for which you may well have to increase your expertise in the area. Select one or more of the following examples and analyze the models that have been built for a particular problem:
   a. An aquatic ecosystem model
      R. B. Williams. "Computer Simulation of Energy Flow in Cedar Bog Lake, Minnesota, Based on the Classical Studies of Lindemann." In B. C. Patten, *Systems Analysis and Simulation in Ecology*, vol. 1. Academic Press, New York, 1971.
   b. A model of criminal behavior
      R. L. Kyllonen. "Crime Rate vs. Population Density in United States Cities: A Model." *General Systems*, 12, (1967) 137–145.
   c. A model of an urban area
      J. W. Forrester. *Urban Dynamics*. M.I.T., Cambridge, Mass., 1969.
   d. A model of forest production
      E. M. Gould and W. G. O'Regan. *Harvard Forest Papers: Simulation*. Petersham, Mass., 1965.

Further Readings

Ackoff, R. L. "General Systems Theory and Systems Research: Contrasting Conceptions of Systems Science." In *Views on General Systems Theory* (M. D. Mesarovic, ed.). Wiley, New York, 1964.

See comment under Mesarovic (1964), below.

Boulding, K. E. "Environment and Economics," In *Environment: Resources, Pollution and Society* (W. W. Murdock, ed.). Sinauer Associates, Stamford, Conn., 1971.

An updated and extended version of one of Boulding's favorite themes—difficulties with throughput economics. Also see his papers in *Future Environments of North America* (F. F. Darling and J. P. Milton, eds., Natural History Press, 1965); and in *Environmental Quality in a Growing Economy* (H. Jarrett, ed., Resources for the Future, Johns Hopkins, 1966).

Churchman, C. W. *The Systems Approach.* Delacorte, New York, 1968. (Paperback edition, Delta, Dell, New York, 1968.)

If you read just one additional book, this should be the one. Concise, balanced, and well written; contains an unusual bibliography and a set of problems.

Forrester, J. W. *Principles of Systems.* Wright-Allen, Cambridge, Mass., 1968.

Best volume for the method of system dynamics.

Forrester, J. W. *Urban Dynamics.* M.I.T., Cambridge, Mass., 1969.

Forrester, J. W. "Counterintuitive Behavior of Social Systems."
Technology Review, 73 (1971a), 52–68.

Short and highly readable preview of *The Limits to Growth and World Dynamics.*

Forrester, J. W. *World Dynamics.* Wright-Allen, Cambridge, Mass., 1971b.

Volumes (Industrial, Urban, and World Dynamics) applying Forrester's systems dynamics to different systems. Readers are advised to read the one whose topic is the most interesting, but are reminded that the results should be read carefully and interpreted cautiously.

Gordon, G. *System Simulation.* Prentice-Hall, Englewood Cliffs, N.J., 1969.

A broad survey of the use of simulation for various problems using a variety of computer simulation languages. Recommended for readers interested in languages other than those they might know and for beginning systems analysts.

Meadows, D. H., D. L. Meadows, J. Randers, and W. W. Behrens. *The Limits to Growth.* Universe, New York, 1972.

The now-famous Club of Rome report on experiments with simulation models as a tool for predicting future world events. There is also a technical version called *Towards Global Equilibrium* (Wright-Allen, Cambridge, Mass., 1973).

Meadows, D. L., and D. H. Meadows (eds.). *Toward Global Equilibrium:*
Collected Papers. Wright-Allen, Cambridge, Mass., 1973.

The technical volume to accompany *The Limits to Growth.*

Mesarovic, M. D. (ed.). *Views on General Systems Theory.* Wiley, New York, 1964.

Papers presented at the Second Systems Symposium

at Case Institute. A highly varied and interesting volume containing some very diverse viewpoints that should be of interest to engineers, biologists, and sociologists alike. See particularly the papers by Ackoff, Boulding, Gerard, and Churchman.

Odum, H. T. *Environment, Power, and Society.* Wiley Interscience, New York, 1971.

The most recent compendium of Odum's particular viewpoint. This book develops policy through the analogue system models Odum favors. Parts of his writings are highly speculative, but the whole book is extremely stimulating.

Patten, B. C. (ed.). *Systems Analysis and Simulation in Ecology,* 2 vols. Academic, New York, 1971, 1972.

Two advanced volumes containing the latest "state of the art" of simulation and systems in ecological research.

Ramo, *Cure for Chaos.* McKay, New York, 1969.

Written by a top executive of TRW-Systems, this is a highly readable essay by a committed systems analyst. Nontechnical and not directed toward any area but the solution of massive social problems, this book is reasonable but still is highly optimistic about the applicability of systems techniques. For a more cautious view, see Roszak (1969) in the section bibliography.

Appendix

This is a mathematical analysis of the effects of feedback on populations. It supplements elements of Chapter 2 as well as Chapter 3.

Consider a population whose number is represented by the symbol N. We shall assume that the potential for increase of the population depends upon the product rN, where r is the total number of offspring produced by all N in the population for a specific time period $\Delta t = t_2 - t_1$, where the actual unit between 1 and 2 is arbitrary. Then we may write an equation for the growth of the population ΔN in Δt:

$$\frac{\Delta N}{\Delta t} = rN$$

The important fact is that N is on both sides of the equation. Hence, if $\Delta N/\Delta t$ is positive, both r and N must be positive. (If both r and N are negative, there is no corresponding real situation.) But if r and N are positive, $\Delta N/\Delta t$ must also be positive, and, as long as there is no provision for deaths (loss from N), N must increase from t_1 to t_2. Even if r is constant, $\Delta N/\Delta t$ in the next time interval must be larger because the product rN must be larger. This circularity is positive feedback, and it is the basis of explosive potential population growth.

Now consider a population which has a relationship of $\Delta N/\Delta t$ and N of the general form:

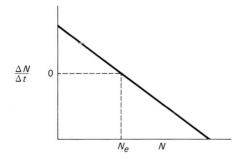

There will be a value of N (N_e) at which $\Delta N/\Delta t$ equals 0 and the population will not change. At values of N greater than N_e, $\Delta N/\Delta t$ becomes negative; conversely, with N less than N_e, $\Delta N/\Delta t$ will be positive. Thus if the population declines ($N<N_e$), it will tend to counteract this effect by growing; and if the population grows such that $N>N_e$, it will tend to decline. By the definition presented in this chapter, this is negative feedback.

For useful exercises to test your understanding, verify that (1) the graph for positive feedback is of the form

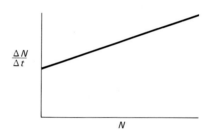

and (2) negative feedback tends to stabilize population size at N_e, while positive feedback does not.

Section B
To Cover the Earth?

SECTION B / SELF-STUDY ORGANIZER

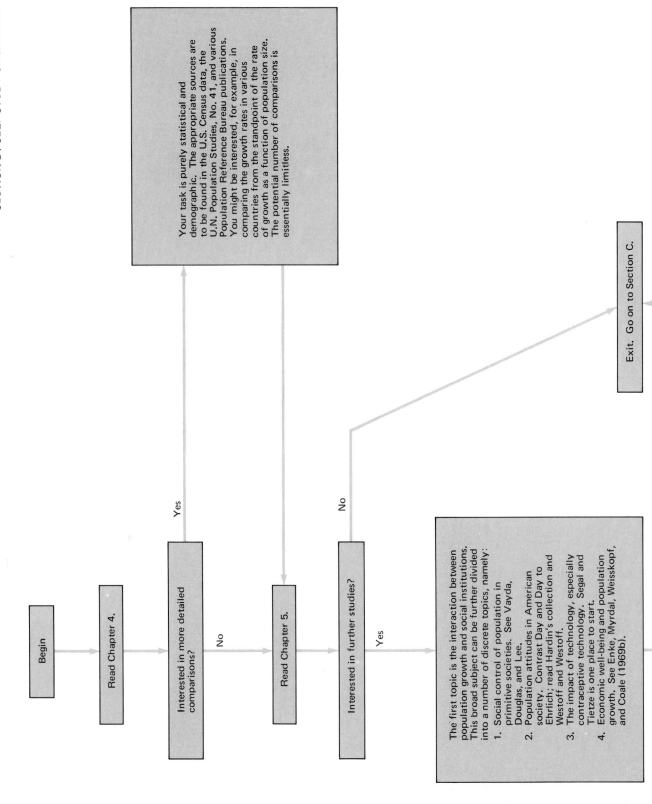

```
┌─────────────────┐     ┌──────────────────────────┐     ┌─────────────────┐     ┌─────────────────┐     ┌─────────────────┐
│ The next topic  │ →   │ Several topics are       │ →   │ The last two    │ →   │ There have been │ →   │ Finally, there  │
│ is population   │     │ discernible in this area │     │ topics are more │     │ arguments about │     │ is some         │
│ stabilization.  │     │ as well. Choose those    │     │ specialized.    │     │ the relative    │     │ interesting     │
│ If you are      │     │ that interest you.       │     │ Follow them at  │     │ importance of   │     │ work on         │
│ interested, go  │     │  1. Biological           │     │ your option.    │     │ population and  │     │ population      │
│ on; otherwise   │     │     precedents for       │     │                 │     │ technology on   │     │ optima. See     │
│ exit.           │     │     population control.  │     │                 │     │ the             │     │ Odum, Singer,   │
│                 │     │     Read Wynne-Edwards   │     │                 │     │ environment.    │     │ and Taylor.     │
│                 │     │     and compare this to  │     │                 │     │ To follow this  │     │                 │
│                 │     │     the papers by        │     │                 │     │ debate, see     │     │                 │
│                 │     │     Douglas and Lee.     │     │                 │     │ Holden,         │     │                 │
│                 │     │  2. Demographic          │     │                 │     │ Commoner, and   │     │                 │
│                 │     │     transition. See      │     │                 │     │ Ehrlich and     │     │                 │
│                 │     │     Coale, in Behrman    │     │                 │     │ Holdren.        │     │                 │
│                 │     │     et al., or Benedict, │     │                 │     │                 │     │                 │
│                 │     │     in Harrison and      │     │                 │     │                 │     │                 │
│                 │     │     Boyce.               │     │                 │     │                 │     │                 │
│                 │     │  3. Family planning and  │     │                 │     │                 │     │                 │
│                 │     │     determinants of      │     │                 │     │                 │     │                 │
│                 │     │     family size.         │     │                 │     │                 │     │                 │
│                 │     │     Appropriate          │     │                 │     │                 │     │                 │
│                 │     │     references are       │     │                 │     │                 │     │                 │
│                 │     │     Rainwater, Whelpton  │     │                 │     │                 │     │                 │
│                 │     │     et al., Berelson,    │     │                 │     │                 │     │                 │
│                 │     │     and Davs.            │     │                 │     │                 │     │                 │
│                 │     │  4. Involuntary methods. │     │                 │     │                 │     │                 │
│                 │     │     See Ehrlich and      │     │                 │     │                 │     │                 │
│                 │     │     Ehrlich.             │     │                 │     │                 │     │                 │
└─────────────────┘     └──────────────────────────┘     └─────────────────┘     └─────────────────┘     └─────────────────┘
```

The only species on this planet that has shown continuous population growth over many uninterrupted decades is *Homo sapiens*—mankind. As the population swells onward toward 4 billion people, the controversy about how and when population growth is to cease sharpens. The more vociferous proponents of immediate population control advocate such measures as changing tax laws to create both new incentives for birth limitations and new penalties for failing to limit births, making abortion freely available, providing incentives for sterilization, and even withholding aid from countries judged to be overpopulated. Arrayed against these suggestions are the proposals of the equally vociferous proponents of a continuation of population growth. They regard population growth as necessary for a healthy economy and as an important force for social harmony. Some even regard population control by anything other than voluntary means as a racist attempt to suppress minorities.

The heated controversy generated over the population problem has produced bewilderment and frustration. On one hand, personal experience with crowding, traffic congestion, and social impersonalization are powerful signs of population effects; but on the other hand, personal considerations, family pressures, and intrinsic biological forces oppose reproductive abstinence. Nevertheless, the past few years in the United States have brought a decline in mean family size and in the population growth rate, although the decline is not fast enough to suit some. In the rest of the world the situation is mixed: A few countries (mostly in Europe) enjoy population stability, but most continue to show yearly population increases with little sign of change.

The Population Change System for Human Beings

Figure B-1 is a diagram expressing the relationship of population change to society and environment. Two of its major elements require further explanation.

1. The most important idea contained in Figure B-1 is the pervasive interaction between population growth and the entire social system. Population change can be positive or negative, depending upon the influences of the three factors given. The social institutions and mores include pressures from family and friends to bear children (commonly called *pronatalist*), and these are reinforced by such factors as tax laws favoring children and disfavoring unmarried persons, laws against sex education and abortion, and social attitudes indicating that having children is "good" and childlessness is "bad." On the other hand, antinatalist social pressures (while slow to develop) can be an equally powerful counterpressure. One sign of this is the now widely held position that two children is the perfect family size.

The impact of technology upon population change can be direct, such as the development of effective and widely available contraceptives, or it can be indirect, through its effect upon economics. The advocates of demographic transition see *reduced* mortality, aided by technological advance and improving economic conditions in the society, as the key to a cessation of population growth. *Demographic transition* is the name applied to the observation that declining death rates have been followed by declining birth rates and subsequent population stability. (See Chapter 4. The paradox that reduced death rates lead to population stability is explained by demographic transition.)

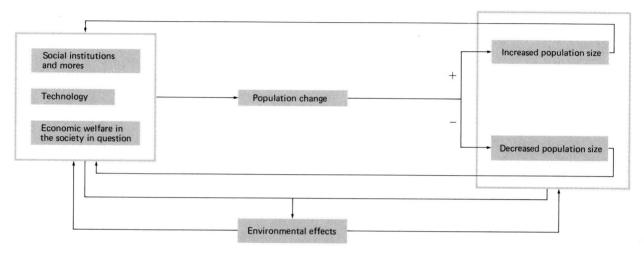

FIGURE B-1 *Interaction between population change and social factors.*

Together these factors operate through a series of positive and negative feedback loops, and their effect is expressed as some degree of population change—which in turn feeds back to affect the social system. That there should be contradictory, widespread, frequently unpredictable effects is an expected consequence of the nature of complex systems.

2. The same ideas apply to environmental effects of population change. These incorporate another whole set of feedback loops running in both directions. Population changes interact with technology, production machinery, and sociopolitical institutions that support them to determine such things as the rate of uptake of land, the effects of resource exploration and utilization, and the impact of such acts as dumping wastes in the environment. In turn, these things affect the social institutions, place constraints on technology, change the rate of economic growth, and in extreme cases directly affect population change. These topics are covered in the later parts of this book, and no further attention will be given to them here.

Questions before Us

The existence of a complex system centered around human population dynamics and the social system, replete with feedback loops that make the behavior of the system difficult to predict, raises a practically infinite series of questions, two of which we shall concentrate on in this section:

1. Since human population growth must inevitably cease, are there any clues in natural systems to indicate how population growth may be ended? In other

words, can we apply a major interest of animal ecology—namely, population stabilization—to human situations? Or are we so unique that our control mechanisms must also be unique?

2. We know there is an ultimate solution to the "population problem." As the population outgrows its resources, it would crash as the death rate rose; then the population would undergo a series of violent fluctuations. Assuming that a conscious choice to avoid this kind of fate is preferable, what mechanism is likely to end population growth in the United States and in the world?

Selected General Sources on Human Populations

Berelson, B. "Beyond Family Planning." *Science*, 163 (1969a), 533-543.
Berelson, B. *Family Planning Programs: an International Survey.* Basic Books, New York, 1969b.
Coale, A. J. "Decline of Fertility in Europe." In *Fertility and Family Planning.* (S. J. Behrman, L. Corsa, and R. Freedman, eds.). University of Michigan, Ann Arbor, 1969a.
Coale, A. J. "Population and Economic Development." In *The Population Dilemma*, 2nd edition (P. M. Hauser, ed.). Prentice-Hall, Englewood Cliffs, N. J., 1969b.
Coale, A. J. "Man and His Environment." *Science*, 170 (1970), 132-136.
Commoner, B. "The Closing Circle." *Environment*, 14, No. 3 (1972), 25, 40-52.
Daedalus, 102, No. 4 (1973). "The No-growth Society." See in particular Davis and Ryder.
Davis, K. *Population Policy: Will Current Policies Succeed? Science*, 158 (1967), 730-739.
Day, L. H., and A. T. Day. *Too Many Americans.* Delta Books, Dell, New York, 1964.
Douglas, M. "Population Control in Primitive Groups." *British Journal of Sociology*, 17 (1966), 263-273.
Ehrlich, P. R. *The Population Bomb.* Ballentine, New York, 1968.
Ehrlich, P. R., and A. H. Ehrlich. *Population, Resources, Environment,* 2nd edition. Freeman, San Francisco, 1972.
Ehrlich, P. R., and J. P. Holdren. "One Dimensional Ecology." *Bulletin of the Atomic Scientists*, 28, No. 5 (1972), 16, 18-27.
Enke, S. "Zero Population Growth: When, How, and Why." *Tempo*, General Electric Company, Santa Barbara, Calif., 1970.
Frejka, T. *The Future of Population Growth.* Wiley, New York, 1973.
Hardin, G. (ed.). *Population, Evolution, and Birth Control,* 2nd edition. Freeman, San Francisco, 1969.
Harrison, G. A., and A. J. Boyce (eds.). *The Structure of Human Populations.* Clarendon Press, Oxford, 1972. See in particular, the papers by deJong, Benedict, and Boydon.
Holden, C. *Science*, 177 (1972), 245-247. Summary of the Commoner-Ehrlich debate.
Lee, R. B. (1968). "What Hunters Do for a Living, or How to Make Out on Scarce Resources." In *Man the Hunter* (R. B. Lee and I. deVore, eds.). Aldine, Chicago, 1968.
Myrdal, G. *An Asian Drama*, vol. II. Pantheon, New York, 1968.
National Academy of Sciences. *Rapid Population Growth.* Johns Hopkins, Baltimore, 1971.
Odum, E. P. "Optimum Population and Environment: A Georgia Microcosm."*Current History*, 58 (1970), 355-359, 365.
Rainwater, L. *Family Design.* Aldine, Chicago, 1965.
Segal, S. J., and C. Tietze (eds.). "Contraceptive Technology: Current and Prospective Methods." Reports on

Population/Family Planning, No. 1. The Population Council, New York, 1971.

Singer, S. F. (ed.). *Is There an Optimum Level of Population?* McGraw-Hill, New York, 1971.

Taylor, L. R. (ed.). *The Optimum Population for Britain.* Academic, New York, 1970.

U.N. Department of Economic and Social Affairs. "World Population Prospects as Assessed in 1963." *Population Studies,* No. 41, 1966.

U.S. Department of Commerce, Bureau of the Census. *County and City Data Book* series. Washington. A single-volume summary of the decennial census. Latest issue, 1972.

Vayda, A. P. (ed.). *Environment and Cultural Behavior.* Doubleday, Natural History Press, Garden City, N.Y., 1968.

Weisskopf, T. E. "Capitalism, Underdevelopment, and the Future of the Poor Countries." In *Economics and World Order* (J. N. Bhagwati, ed.). Crowell Collier & MacMillan, New York, 1971.

Westoff, L. A., and C. F. Westoff. *From Now to Zero.* Little, Brown, Boston, 1971.

Whelpton, P. K., A. A. Campbell, and J. E. Patterson. *Fertility and Family Planning in the United States.* Princeton, Princeton, N.J., 1966.

Wynne-Edwards, V. C. "Self-regulating Systems in Populations of Animals." *Science,* 147 (1965), 1543–1548.

Selected Periodicals

Demography
Population Bulletin, Population Reference Bureau
Science
Studies in Family Planning and Studies in Population/Family Planning, The Population Council

Many original studies appear as books or as edited conference proceedings rather than in periodicals.

4. Populations Marching Onward to an Uncertain Future

For human beings, the ultimate source of most environmental problems is our growing population: the displacement of other species, the pollution of air and water, the disintegration of urban centers. To these processes we can readily link population, both to numbers and to distribution. This does not mean that the human population, and population growth in particular, is evil. Nor does it imply that population growth is the sole cause of environmental or other problems. Both statements are grossly oversimplified. Nevertheless, today the world faces a burgeoning population and an increasing realization of its limits. The magnitude of the crisis has been argued violently in public forums and in scientific circles. This chapter examines the history of human population growth, our present state, some guesses about what the future holds, and some methods formerly used to curtail population expansion. Some features in the history of human population growth may supply clues about how to slow population growth in the future, or at least establish some conditions in which this can occur. Surely no help is needed to foster more babies.

The Early History and Development of Human Populations

There is no doubt that the evolution of early people, from early anthropoid ancestor to hominid, did not modify their livelihood as hunters and gatherers. They killed animals for meat, dug roots and edible tubers, picked leaves, and gathered fruit. From this kind of living we have inferred, using contemporary but primitive cultures for comparison, that these early populations must have been small and highly dependent on resources. At this stage in their evolution human beings had not yet developed their brain capacity fully, and their numbers were controlled by processes similar to those applicable to almost any animal population. For example, Birdsell (1953) has shown that the aborigines of Queensland, Australia, are highly dependent on game, which are themselves dependent on adequate water. Hence, the numbers and distribution of aborigine groups are well correlated with average rainfall. Douglas (1966) has reported that the Western Shoshoni Indians of the Southern Great Basin (northeastern California) distributed themselves in relation to their potential food supplies. Those in wetter areas could get enough food locally and could afford to live relatively densely (2 per square mile) in permanent villages. Those in drier areas had a less dependable food supply, the villages were only semipermanent, and the population fluctuated year by year. Their density averaged 0.5 per square mile. The Shoshoni of the most arid areas could not maintain permanent winter villages at all. They wandered sporadically in small groups and averaged only 0.02 per square mile—60 times sparser than their most fortunate kin. However, another interpretation is equally possible. The Shoshoni may not have settled permanently in the more arid areas at all, but may have passed through only to hunt and obtain materials not available in their usual foothill areas (the piñon-juniper belt). If this interpretation is correct, then this type of estimation of densities in poorer environments is misleading. But even if this particular method provides misleading information, it should not obscure the basic fact that population adjusts to its environment.

Many primitive cultures in relative harmony with their environment practiced slash-and-burn (swidden) agriculture; patches of tropical forest were cut and burned periodically, and then the land was used for a few years for small gardens before being ceded back to the encroaching forest. However, the real opportunity for human population growth probably came not with the mere practice of agriculture, but with the advent of fixed permanent agriculture. The rise of the Sumerian civilization was due in part to irrigation of agricultural lands in the fertile soils of the Tigris-Euphrates Valley. The important difference was that permanent, intensive cultivation could produce more than was necessary to feed the people working the land, freeing them to develop permanent urban places based on something other than agriculture. The loosened link between human beings and food resources caused an explosive reproductive response and began a broad expansion of human presence over the earth.

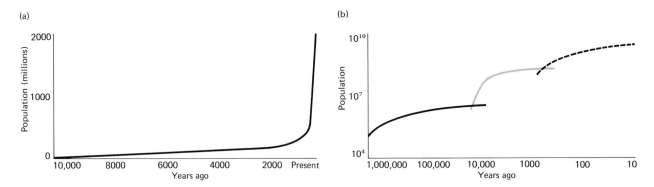

FIGURE 4-1 *Growth history of the human population. (a) Growth plotted on an arithmetic scale, resulting in the usual exponential growth curve. The time axis is plotted in years before present (BP). This curve implies a fairly stable population for most of human history with a sudden spurt in the recent past. The beginning of the rapid increase is generally held to be around* A.D. *1650. (b) Growth plotted with population on a logarithmic scale. This facilitates direct comparison of growth rates at different times. In this figure, population is shown to have three periods of rapid growth with longer, intervening periods of relative stability. The dark solid line represents the advent of toolmaking and culture; the light solid line the beginning of fixed agriculture; and the dashed line the scientific–industrial revolution. (Redrawn from "The Human Population," by Edward S. Deevey, Jr.,* Scientific American, *September 1960. Copyright © 1960 by Scientific American, Inc. All rights reserved.)*

The second major influence was the equally explosive opportunity afforded by the development of technology and increasing communication between various parts of the world. Both of these acted to further decrease the influence of limited local resources on human population growth.

A third change, of a different character, has occurred more recently: the spread of an advanced and sophisticated technology to places formerly passed by. Improved public health and the introduction of modern agricultural techniques dramatically lowered the death rates in many countries. In addition, they frequently have disrupted the cultural mechanisms responsible for effective population stabilization; as a result, historically high birth rates are no longer matched by the high death rates so characteristic of the population before these introductions. Unfortunately a dramatic increase in population growth has been the inevitable result.

The history of human population growth can be, and usually is, depicted as a smooth exponential curve[1] with a relatively recent explosion of numbers (Figure 4-1a). Deevey's (1960) logarithmic plot (Figure 4-1b)

---

[1]Strictly speaking, Figure 4-1a only approximates an exponential curve (see Equation 4), the plateau of relative stability is too long.

emphasizes much more clearly the discontinuities of growth rates. Animal populations do not live in constant environments; continual changes are reflected in shifting birth and death rates, which are subsequently mirrored in irregular population changes. Deevey's representation shows human population growth as a discontinuous phenomenon, with long periods of relative stability being interrupted by short bursts of intense growth. The release points have been the development of new tools and new cultures, such as permanent agriculture, or the destruction of established mores. Several anthropologists have maintained that the latter has been more important because social conventions are what really regulate family size and reproductive rates in human societies.

For perhaps a million years primitive people scattered sparsely across the landscape much as undisturbed hunter-gatherer societies still do today. In the few thousand years since the advent of permanent agriculture, many of the isolated civilizations—Mayan, Roman, Greek, Chinese, Egyptian, Sumerian—have fallen and been replaced, or grown, clashed, and coalesced with others. The exact causes of past events will never be known but will remain fruitful areas of speculation and argument for historians, social scientists, and human ecologists. We can be sure, however, that in the course of history human societies have grown, mingled with their neighbors, and disappeared in a continuous process of change and adaptation. Through all these changes in various human societies, however, the size of the human population has inexorably continued to grow (Table 4-1). In the United States at least, perhaps we are entering a new cycle, another period of relative population stabilization. But even if this is so, we have no guarantee that growth has stabilized; as terrifying as it may seem, another period of rapid growth may lie in the future.

TABLE 4-1 *Estimated world population throughout human history, according to various authorities.*

Year (Period)	World population estimate (Millions)	Authority
Paleolithic (300,000 B.C.)	1	Deevey
Mesolithic (8000 B.C.)	5	Ehrlich & Ehrlich
	5.32	Deevey
Neolithic-Europe (8000–7000 B.C.)	10	Hauser
(6000 B.C.)	87.5	Deevey
Birth of Christ	250	Dorn
	133	Deevey
A.D. 1000	190	von Foerster et al., from Mills
1650	465	Carr-Saunders
	500	Dorn, Hauser
1750	660	Carr-Saunders
1850	1098	Carr-Saunders
1900	1610	Deevey
1950	2497	Hauser
1960	2991	United Nations
1970[a]	3592	United Nations[b]

[a]Estimate.
[b]Medium variant projection.

Demographic Parameters: The Measure of Population Dynamics

Human populations are no exception to the general rules of population dynamics, and so the bucket analogy presented in Chapter 2 is appropriate here as well. We shall now extend the analogy in a particular mathematical equation for population growth. Let B = the number of births in time period t, D = the number of deaths in the same period, I = the number of immigrants, and E = the number of emigrants. Then the simplest assumption to make is that the population N at time t will be given by

$$N_t = [(B + I) - (D + E)]N_{t-1} \qquad (1)$$

This equation states that the population size will be a function of the relative magnitude of the inputs $(B + I)$ and outputs $(D + E)$ and therefore will reflect the influence of the system environment, specifically those factors which affect the inputs and outputs. For example, consider the resources which influence all these parameters. With plentiful resources, the birth rate might be high, and the population would grow; with scarce resources, the death and emigration rates might increase, and the population would decline.

Now let us assume that we want a model which portrays the maximum rate of growth. If we could control environmental conditions such that $(B + I)$ are maximized and $(D + E)$ are minimized, clearly the balance will be shifted so that $[(B + I) - (D + E)]$ will be the largest possible positive number. If we assume that this number is constant, we have

$$\Delta N/\Delta t = [(B + I) - (D + E)]N_t \qquad (2)$$

By setting $r = [(B + I) - (D + E)]$, we have the same equation analyzed in the appendix of Chapter 3, and the same interpretation applies here. Using the calculus, we may obtain an equivalent equation to that of the appendix:

$$dN/dt = [(B + I) - (D + E)]N_t \qquad (3)$$

where the d notation means that the change in N with respect to t is over an instantaneous time interval. Equation (3) permits us to obtain an overall equation which expresses population growth under the assumptions made:

$$N_t = N_0 e^{[(B + I) - (D + E)]t} \qquad (4)$$

where N_t = population size at time t,
N_0 = population size at the beginning,
e = exponential constant (2.1718).

Equation (4) is the well-known equation for *exponential growth*. Exponential equations have a characteristic form (Figure 4-8b) that is extremely important for human beings because it means that our populations can grow explosively. Looking at equation (3), note that N appears on both sides of the equation and is therefore a positive feedback factor, as pointed out in Chapter 3. This equation shows why population growth is such an important problem: With the passing of time, it tends to get out of control at an accelerating rate.

We have already noted the major effect of the system environment upon the rates which affect population change. The second major factor is the fact that different segments of the population may have different rates. A population that is mostly female can increase much faster than a mostly male one, if strict monogamy is not observed. Social conditions may dictate different rates of reproduction or death rates for different racial groups. More important, however, is the age distribution of the population. The greater

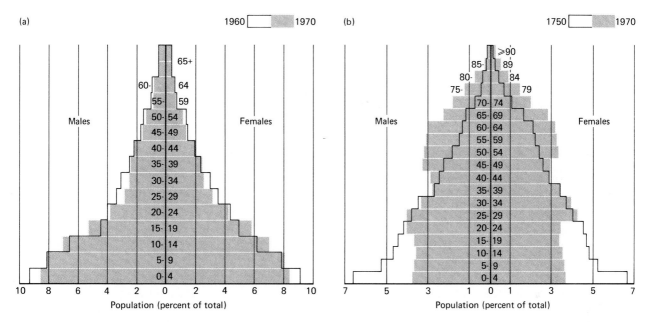

FIGURE 4-2 *Examples of regional population-age structures.* (a) The Dominican Republic in 1960 and 1970, showing an age structure very heavily weighted with young age classes. (b) Sweden in 1750 and 1970, showing the effects of demographic transition. (Redrawn with permission of the Population Council (a) from "Dominican Republic" by H. P. Montas, *Country Profiles,* 1973, p. 3, (b) from "Sweden" by G. Svala, *Country Profiles,* 1973, p. 5.)

the proportion of the population that is of reproductive age, the greater the population's growth potential. For instance, in the human population, the reproductive performance of a woman between the ages of 18 and 30 will likely be very much higher than that of girls less than 18 or women older than 30. Thus, a population consisting predominately of women in prime productive ages or of women who have yet to reach prime reproductive age is likely to have a large potential for further growth for the same reason as a population composed mostly of women.

It is important to remember that the population is not a static assembly of parts. A stable population must produce exactly the number of young needed to replace the number of deaths (Figure 4-2a), whereas the growing population will have more births than that needed for replacement so that the base of the age pyramid tends to expand (Figure 4-2b). But since an expanding base means a greater proportion of people who have yet to experience their greatest reproduction, the growing population is building up the potential for even more future growth. Here we have still another way to view positive feedback.

To integrate these ideas into a single example, let us examine the impact of population growth upon public funding for education. We start by assuming that a population has characteristic rates of demand for public education (between ages 5 and 18) and for the ability to pay tax monies to support it (highest in the working ages). These rates are specified in Table 4-2. If we allow hypothetical populations to grow exponentially at certain fixed rates, we find that the ratio of the ability of the population to pay education taxes to the rate of demand for education money declines exponentially as a function of population growth rate (Figure 4-3). The reason for this is familiar: The faster the population grows, the greater the proportion of youths requiring education dollars and the less the proportion of people to pay for this service. This is one example of the implications of processes that are linked to population structural char-

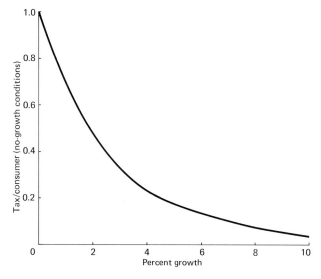

FIGURE 4-3 Sample output from education taxation-population growth model, to reveal one cost of population growth. This curve relates tax-producing ability standardized to a zero population growth rate to higher rates of population increase. Note that the ability to generate taxes for public education at fixed rates per producer declines exponentially with increasing rates of population growth. Hence, high rates of natural increase mean increased local taxes or declining educational quality. (Redrawn from *Is There an Optimum Level of Population?*, S. F. Singer (ed.). Copyright © 1971 by the McGraw-Hill Book Company. Used with permission of the McGraw-Hill Book Company.)

TABLE 4-2 *Parameters used in the education tax-population model. Rates are expressed as proportions of a theoretical maximum figure.*

Age	Tax-producing rate	Tax-consumption rate
6–16	0.00	1.00
17	0.03	0.54
18	0.03	0.23
20	0.04	0.15
26	0.06	0.10
30	0.15	0.07
36	0.30	0.02
40	0.50	0.00
50	0.91	0.00
60	1.00	0.00
65	0.99	0.00
75	0.00	0.00

acteristics. For instance, the approaching population stabilization in Japan has decreased the number of young people available as workers. A population that is declining may have to be concerned about the generation of innovative ideas and the cost of health care when a large proportion of the population is in failing health.

TABLE 4-3 *Crude estimates of the distribution of mankind on the major inhabited continents (excludes Antarctica).*

Continent	Area (Square kilometers)	Population (Midyear 1971 est.)	Crude density
Africa	30,320,109	354,000,000	11.68
North America (including Caribbean and Central America)	24,246,904	327,000,000	13.49
South America	17,833,331	195,000,000	10.93
Asia (except USSR)	27,532,130	2,104,000,000	76.42
Europe	4,936,247	466,000,000	94.40
Oceania	8,510,946	19,800,000	2.33
USSR	22,402,200	245,000,000	10.95

SOURCE: *The World Almanac*, 1974; published by Newspaper Enterprise Association, Inc.

Demographic Statistics: The Present

In 1974 the estimated population of the United States was approximately 212 million. In the same year the world population was estimated by the United Nations at 4 billion people. However, these statistics probably mean very little to most readers, for whom a doubling or even a tripling would seem to be inconsequential compared with the distribution of the population. People are more aware of local crowding and its effects than they are of an overpopulation crisis somewhere else.

The distribution of the world's population is given in Table 4-3. Only Asia, the continent with the greatest overall density, approaches having its population spread uniformly over the land. In countries with a primitive technology (Indonesia, the Philippines, and Ceylon, for example), subsistence agriculture may keep many people on the land; but in Asia as elsewhere the tendency is toward concentration of people into urban areas (Figure 4-4). With the advent of the Green Revolution (Chapter 6), India has experienced an accelerated shift from the rural areas into the already overcrowded cities, due to drastic and rapid changes in agricultural practices. Hong Kong,

FIGURE 4-4 *The intense crowding of Hong Kong is illustrated in this photograph. The high-rise housing of Hong Kong yields to shanties on the waterfront, which in turn yield to simple houseboats floating in the harbor itself.* (Courtesy of George Bellerose from Stock, Boston.)

a tiny enclave of South China, teems with people (9744 per square mile). The islands of Japan, especially Honshu, have an enormous density (704 per square mile),[1] but the country still possesses large areas of relative wilderness where solitude is possible, even near Tokyo, the world's largest city. In Australia, 60 percent of the population is concentrated in the country's five largest cities. The rest of the country is practically unpopulated, a matter of concern to the government in World War II, when Japanese invasion into these vast empty spaces seemed likely.

On other continents people settle where others already are, just as is the rule in Asia. Though Brazilians talk eagerly of the possible settlement of the great Amazon basin, they remain concentrated along the coast. Most Egyptians are to be found only along the Nile River Valley, Europeans in their cities, and Americans in theirs. In the United States, 95 percent of the population currently is urban, compared to 66 percent in other developed countries (Table 4-4). Some south Florida counties have sustained staggering population increases for a decade, while California continues to pack new residents into coastal cities with a moderate climate.

The present distribution of human beings illustrates that the expected increases in population probably will be absorbed at or near present urban areas, intensifying whatever crowding effects may already be present. The dream of conquering and settling new territories will probably always be with us, but it is likely to remain unrealized unless we are forced to disperse uniformly across the countryside. Otherwise we need to find some new factor of attractiveness or to remove an existing resource limitation, for such areas would have been settled previously had there

TABLE 4-4 *Trends and patterns in worldwide urbanization.*

Year	Developed countries (Percent urban)	Developing countries (Percent urban)
1950	51	16
1960	60	20
1970	66	25
1980	71	30
1990	76	36
2000	81	43

SOURCE: U.N. *World Population Situation in 1970*, 1971.

not been something to prevent it. In tropical forests the soils are poor and unable to stand long-term cultivation. In the fertile crescent of the Tigris and Euphrates, irrigation brought excess salinization and an end to Sumerian civilization. The USSR tried to build a new agricultural state in the arid regions of Kazakhstan, but limitations in water supply and distribution proved to be its downfall. Americans, however, are trying to do the same in the dry Southwestern deserts.

The worldwide tendency of people to crowd together reveals dramatically how meaningless simple counts of population are and how insensitive the populace is to orders of magnitude differences. This is not equivalent to saying that population size is not an important influence on many processes; we have discussed one example above. But it does point out that the perception of the "population problem" is dependent on local sensors. People sense the density of people in their neighborhoods, not the total population density of their cities. Only if future increases

---

[1] Compare these figures with the gross planetary density of 60 per square mile (1968).

in population must be accommodated in rising densities may the simple arithmetic of population increase have a significant impact.

Population growth in different parts of the planet is rather variable, ranging from a low rate of 0.8 percent in Europe to 2.9 percent in South America. Thus, all major regions of the planet are still experiencing human population growth. In some regions—notably in Japan, North America, and Europe—populations are approaching the steady state. For the entire world, however, von Foerster and his colleagues (1960) have shown that the population has grown even faster than predicted by the simple exponential. Given the rapid growth of the world population, an important question at the present time is whether or not the number of children in a completed family is being reduced at a rate fast enough to achieve a stable world population in time to prevent some sort of overpopulation catastrophe.

We previously have seen that for many countries, especially underdeveloped ones, the positive feedback inherent in population growth has been augmented by increases in r, more through the lowering of death rates than through rising birth rates. Several demographers have pointed out, however, that this situation is unstable and that r will automatically be readjusted toward zero. For example, when death rates in Europe fell and populations grew between 1850 and 1900, there was a corresponding decline in the birth rate. In Sweden, death rates fell from 20 per 1000 in 1860 to 10 per 1000 in 1970; but in the same period the birth rate fell more dramatically, from 35 per 1000 to 16 per 1000. This situation, where a later decline in birth rates compensates for falling death rates, is called *demographic transition*. Those that propound this theory believe that psychological and societal mechanisms are automatically triggered when appropriate prior conditions are attained, for example, when the standard of living has risen enough so that future additional children represent a liability rather than a potential benefit, at least to prospective parents. A strong belief in demographic transition as a mode of population stabilization also requires faith in a relative laissez-faire attitude toward family size and timing—that the population will stop growing when it is large enough. Revelle (1967), for one, clearly thinks this is so. He has suggested that one critical variable controlling human fertility is the parents' perception of the probability that one of their sons will survive to propagate the family name; if they believe the first will survive, there will be less reason to have others. In an apparent paradox, Revelle suggests that death rates must decline further (thereby at least temporarily increasing the rate of population increase) to give that extra bit of assurance to parents which might set off the process of demographic transition.

Demographic Projection: The Future

The best estimates available about future human numbers are those provided by the United Nations (1966). The estimates contain four projections for each area based on various extrapolations of fertility and mortality rates for the population of the area of interest (see Table 4-5). Long-term migratory movements also were considered in three of the four projections.

"Auxiliary" estimate This is the baseline projection that assumes continuation of trends in declining mortality, but without net migration or declining natality.

TABLE 4-5 *Estimates of populations for various regions in the year 2000: four projections.*

Area	Projection (Billions)				Census (1960)	Projected percent increase (Median projection over 1960)
	Auxiliary	Low	Medium	High		
World	7.52	5.45	6.13	6.99	3.00	2.04
Developed areas	1.58	1.29	1.44	1.57	0.98	1.47
Developing areas	5.94	4.16	4.69	5.42	2.02	2.32
East Asia	1.81	1.12	1.29	1.62	0.79	1.63
South Asia	2.70	1.98	2.17	2.44	0.87	2.49
Europe	0.57	0.49	0.53	0.56	0.42	1.26
Soviet Union	0.40	0.32	0.35	0.40	0.21	1.67
Africa	0.86	0.68	0.77	0.86	0.27	2.85
North America	0.39	0.29	0.35	0.38	0.20	1.75
Latin America	0.76	0.53	0.64	0.69	0.21	3.05
Oceania	0.033	0.028	0.032	0.035	0.016	2.00

SOURCE: U.N. *World Population Prospects*, 1966.

In particular, the U.N. report used estimates of observed fertility (age-specific birth rate) combined with projected changes in life expectancy to estimate overall fertility. For mortality trends, it was assumed that people up to age 55 in all areas would gain 2.5 years of life expectancy in each 5-year period, with the gain in life expectancy slowing above age 55 until there were no further increases beyond age 73.9. Clearly, this population estimate is a "no demographic transition" projection.

"Medium" estimate This projection is intended as "the most plausible in view of what is known of past experience and present circumstances in each region" (United Nations, 1966). This variant assumes that mortality rates will continue present trends in each area (which means falling further), but that birth rates will also drop to some extent depending on the region. Long-term observed trends in migration were extrapolated.

"Low" and "high" estimates These two variants were calculated to provide informal estimates of the probable degree of error surrounding the medium projection. The difference between them reflects the stability (predictability) of the region. The United Nations did not intend these estimates as absolute limits, however; nor is it necessarily true that the medium variant is half the difference between high and low. In cases where the United Nations felt an overestimate or underestimate more likely, the medium variant is closer to one or the other. All three estimates are based on the implicit belief that family planning programs will lead to reductions in fertility.

All four projections share the assumption that global conditions will be stable, that there will be no major wars or worldwide natural catastrophes. Additionally, it was assumed that technology, social change, and economic growth would be sufficient to allow the planet to absorb the increased population.

The U.N. projections made in 1963 for the major regions of the earth are listed in Table 4-5. Additional detail including projections for smaller regions as well as the actual birth and death rates may be consulted in the original publication. Certainly the most striking element in these projections is the contrast between "developed" and "developing" regions. The latter have much higher expected population growth, and the gap between birth and death rates is larger. The significance of this is revealed in the fact that under the auxiliary projection the developing areas together would contain 79 percent of the world population in the year 2000, compared to only 67 percent in 1960. Although these projections are useful indicators, they cannot be regarded as necessarily true. Previous U.N. projections (as well as those of most other demographers) have consistently underestimated the population, which demonstrates how difficult it is to predict the performance of a human system.

Despite this cautionary note, it is valuable to compare expected trends in different regions. South America is expected to more than triple its population, adding to already-crowded urban slums, perpetuating economic hardship, and increasing official susceptibility to such glamorous panaceas as fixed farming and booming industrial cities in the Amazon basin. Africa is projected to be close behind, with a population in the year 2000 that is 2.85 times as large as that in 1960. On the other end of the scale, Europe is projected to increase only 1.26 times. Possibly the sole surprise is the region of East Asia, projected to increase only 1.47 times its 1960 population. Perhaps this is not so surprising: In this case, the positive-feedback nature of population growth may be balanced by the negative feedback imposed by local perception of an environment unsuitable for more people. For example, all four countries included in East Asia are heavily and officially involved in family planning and population control. Birth rates have been reduced in Japan, Taiwan, and Korea in the last decade (and probably in China as well, though no one knows with certainty). The free availability of abortion in Japan has reduced the rate of natural increase to the point that the government has begun to worry about a decline in the labor supply. Furthermore, all but China have the relatively low mortality rates characteristic of developed economies (and China's has probably decreased considerably). In Taiwan, for instance, the mortality rate for 1965–1970 calculated by the United Nations was only 5.4 per 1000, the lowest for any country with more than 5 million people. It is impossible to say conclusively that this is the result of classic negative feedback on a group of populations approaching equilibrium, but in crowded South China and incredibly dense Japan, it seems likely that officially condoned policies for the limitation of family size are but the means enabling people to implement their desires to stop reproducing in an undesirable environment.

The von Foerster Growth Model and the U.N. Projections

One possible explanation for why demographers have consistently underprojected actual increases in the world population was provided by von Foerster and his colleagues (1960). To explain their work requires

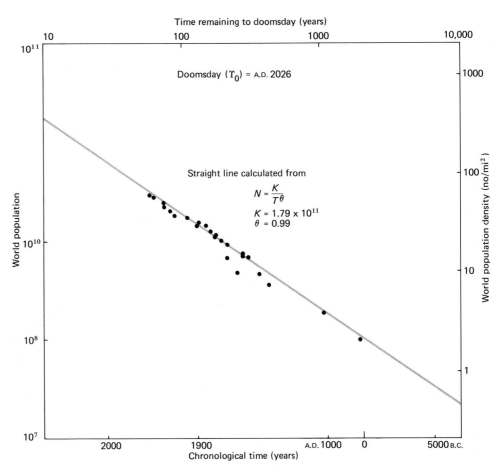

FIGURE 4-5 *The von Foerster et al. growth equation. The parameters K and θ were obtained by least squares analysis from data points linearized on a log-log plot; the data themselves were obtained from various authors listed in the original paper. The doomstime (τ) is an inverse measure: it measures the time remaining to the time when the population N goes to infinity and squeezes itself to death. (Adapted from H. von Foerster, "Doomsday, Friday, 13 November, A.D. 2026," in* Science, *132, pp. 1291–1295, 4 November, 1960. Copyright © 1960 by the American Association for the Advancement of Science.)*

a return to mathematics and the exponential growth model. Recall that in an unlimited environment, one can hypothesize that a population will grow at some fixed maximum rate times its current numbers. This hypothesis leads to the exponential growth model discussed previously. What von Foerster observed is that instead of r (the rate of natural increase) having a fixed value as it would in the simple exponential growth curve, environmental influences on the human value of r would not be fixed in space or time; as a result, r would be dynamic and evolve toward higher and higher values. For human populations, "steady social buildup during historical time" was cited as one example of this type of evolution. Hence von Foerster et al. proposed that r was a variable whose development was of the form

$$r = r_0 N^{1/\theta} \qquad (5)$$

where r_0 = initial rate of potential increase
θ = constant to be determined from data

Using data derived from various sources, von Foerster and his coworkers fitted the modified growth equation (Figure 4-5). Among other interesting results, they found that if the process continued without check, the population would grow until every square inch of the earth was occupied. The time when this would occur was found to be A.D. 2026.9. Thus von Foerster concluded that only 26 years after the turn of the century, human beings would annihilate themselves by crowding. A more important observation is the fact that this model fits the data for past human population growth. This fact alone does not deny the chance that a simple exponential might not be equally suitable; but if von Foerster's modified exponential were correct, it could indicate one possible reason why past demographic projections have usually been low. In Table 4-6, the projections derived from the von Foerster model are compared with those of the United Nations. The estimates from von Foerster were calculated from the equation

$$N_T = K/\tau^\theta \qquad (6)$$

where N_T and θ are previously defined, $K = (\frac{\theta}{r_0})^\theta$, and τ = doomstime (or the time remaining before the population grows to infinity and fills the world).

The von Foerster projections were made on precise, albeit tentative premises based on mathematical structure. The U.N. projections were made on the basis of statistical extrapolation of current trends with attempts to assess the effects of present and future programs to curb births as well as deaths. The von Foerster growth curve has built into it a steadily increasing rate of growth, while the U.N. projections do not. Hence the results in Table 4-6 are all the more surprising, because von Foerster's estimates are uniformly lower than present U.N. high projections and do not surpass the medium ones until 1995. The

TABLE 4-6 *Projections of world population according to (a) U.N. medium projections, (b) U.N. high projections, (c) von Foerster et al. growth model; for the years 1965–2000. Estimates in billions.*

Year	(a)	(b)	(c)
1965	3.28	3.31	3.02
1970[a]	3.59	3.66	3.28
1975	3.94	4.07	3.59
1980	4.33	4.55	3.97
1985	4.75	5.10	4.44
1990	5.19	5.69	5.03
1995	5.65	6.63	5.81
2000	6.13	6.99	6.88

[a]World population, 1970, U.N. estimate = 3.63 billion.

1970 U.N. population estimate was already intermediate between the U.N. 1963 projections and well in excess of von Foerster's prediction, even though the latter was made only 10 years earlier.

Thus we see that von Foerster's growth model, which implies an ineffective family planning program and accelerating rates of growth, itself shows signs of underpredicting the size of the population. The U.N. projections, which have made similar errors in the past, are also showing signs of repeating these mistakes. In 1970 the United Nations released interim estimates while revising its projections. Scheduled in these revisions is a moderate *upward* revision for parts of Africa, Asia, Europe, and Oceania, versus a slight *downward* revision for North America, the USSR, and parts of Latin America.

What does this mean? The von Foerster model assumes that there will be no effective feedback to slow the growth of the human population until it crowds itself to extinction in 2026, an assumption which we can easily show to be unrealistic. Nevertheless, its basic form indicates a growth rate higher than the standard exponential form usually assumed to apply. The fact that the global population has already surpassed von Foerster's estimates for 1970 indicates either that the current growth phase is far from over, or, even worse, that the acceleration in growth is even greater than von Foerster and his associates could detect. Neither case, when underscored by the latest U.N. revisions, is cause for celebrating a trend toward population stabilization. Although several developed and a few less developed nations are making progress in population stabilization, and in some cases (the United States, for example) even are making rapid progress, much of the world's population continues to grow explosively toward some future limit and unknown consequences.

Extensions and Implications

The growth process and the future of the human population reflect the same basic laws that govern all populations. In this section we shall discuss some implications of these laws for the human population.

Popular views about the size of the "optimum" population range widely. At one extreme are those who view any further increases in population as unnecessary, evil, and immoral. There is a valid point to this reasoning, in that some social problems ultimately are rooted in population size. One such problem is related to the crime rate. For example, if a population P can be characterized as having p_i age classes and c_i age-specific propensities to commit a serious crime, then the overall rate can be calculated as $\Sigma p_i c_i$. If the population is growing, the age distribution will shift toward younger ages, and so, if the more youthful are more prone to commit a serious crime (which, for some crimes is the case), $\Sigma p_i c_i$ will be higher both because there are now more people to commit crimes and because there is a differential concentration of numbers in younger age classes. This, of course, assumes that the age-specific characteristics (or propensities) c_i are true constants: If a youthful age structure serves actually to increase the propensity to commit a crime (as seems quite likely), the problem would be exacerbated. In this specific case population growth bears a high social cost which can be avoided only if the age-specific rates c_i are reduced at a rate proportional to the rate of population growth, all other factors being constant.

At the other extreme are those who believe that a growing population bears multiple benefits for human progress or, at worst, that cessation of growth imposes a heavy cost in lost opportunities, youthful ideas, and human dignity. Several authors have commented that any supposed state of overpopulation will bear chiefly nonhuman (and therefore, by implication, rather trivial) costs. Others have pointed out the economic and social costs of any rapid transition from growth to stability—costs such as a stagnating economy, a high dependency load,[2] or an insufficient labor supply.

The difference between these extreme views is the relative emphasis their proponents place on the seriousness of the population problem. No one seriously doubts that growth, *any* growth, must stop sooner or later; but many believe that when the optimum level is reached, population growth will cease through entirely natural mechanisms. Others support this belief but also support more active measures to discourage growth because "natural mechanisms" may be triggered only after density produces an intolerable, low-quality life. This is interesting, and the difference is clearly significant. But what does 7 billion or 5½ billion people mean? Perhaps the best way to translate these projections into one's own perception is to partition the increase under certain assumptions into local areas. The key variable is distribution, a nonrandom, nonuniform process which magnifies the growth process. The U.N. projections do not reflect migration effects because they are minor both between continents and even between most countries. (There are exceptions such as Australia, where migratory influx has been heavy.) The influence of migration is most easily seen at regional and smaller scales. For example, most population growth in California is attributable to in-migration (Figure 4-6). Growth

---

[2] Defined as the ratio of the young plus the elderly over the population in wage-earning ages. This is a simple way to express a population that is unusually young or old.

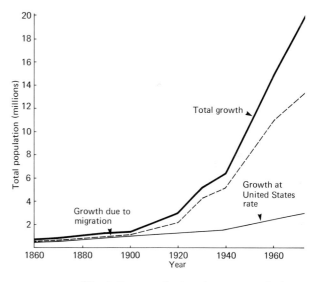

FIGURE 4-6 *The influence of migration on population growth in California from 1860 to 1960 is seen clearly in this graph. The total growth curve for California is accounted for mostly by migration. Note that the curve "Growth due to migration" not only accounts for most of the total, but also displays the same trends. The "Growth at U.S. rate" curve depicts the expected total growth curve for California if rates were equal throughout the United States.* (Adapted from *The Magnitude and Character of California's Population Growth*, Davis and Styles (eds.), Institute of International Studies, University of California, Berkeley.)

rates in Arizona are now even higher than for California (in central Arizona, the rate is 500 percent per decade), most of which is attributable to migration. In the United States the trend has been the movement from the colder, more densely settled Northeastern–North Central areas to the warmer or milder areas in the South and Southwest.

TABLE 4-7 *Proportion of the population classified as urban[a] in 41 countries.*

Percent urban	Developed countries	Developing countries	Percent urban	Developed countries	Developing countries
80–84	Australia Netherlands		40–44	Hungary	
75–79	United Kingdom Germany		30–34	Portugal Romania	Iran UAR Philippines
70–74	East Germany United States Argentina		25–29	Yugoslavia	Korea Turkey
65–69	Canada Belgium		15–19		India Thailand South Vietnam
60–64	Japan France				Congo China Burma
55–59	Spain		10–14		Indonesia
50–54		Mexico			Pakistan
45–49	USSR Czechoslovakia Italy Poland	South Africa Colombia Brazil	5–9		Nigeria North Vietnam Afghanistan Ethiopia

SOURCE: U.N. report, "Growth of the World's Urban and Rural Population, 1920–2000." *Population Studies*, No. 44, 1969, by permission.

[a] Defined in context of each nation, but generally treated as larger groupings of people.

A worldwide phenomenon of greater significance has been the movement of people off the land into the cities. In Australia, the great waves of in-migration have been almost totally absorbed into the urban areas; as a result, Australia has the highest proportions of urban population (Table 4-7). While this phenomenon is accentuated in more developed nations, it is also occurring to a more limited extent in developing ones. Though these trends have not been occurring for the extended timespan seen in the developed region of the world, they have been established for several decades in the developing countries and show no sign of abatement or change. The United Nations predicts that the process will continue and that urbanization will intensify with the appearance of several megalopolises in the near future. This prediction is not really very controversial, either, as numerous others have discussed urbanization as a means for accommodating people. Paolo Soleri has been experimenting with vast vertical cities—whole, self-contained urban systems built for millions of people. Constantine Doxiadis, the famous Greek planner, foresees huge urban areas surpassing even the megalopolis (he calls his urban area the *ecumenopolis*). His visions are the end point, the ultimate in aggregation, the total plan for the world. Under his plan, the world's population would be distributed into ecumenopolises covering only 5 percent of the land surface of the earth. The rest would be carefully organized for life-support systems accompanied by vast transportation and distribution networks.

These kinds of trends—increasing concentration, with probable population pressure to continue increasing these residential densities as the world gets more and more crowded—are the reasons for concern over the 1.5 billion difference between a world population of 5½ billion and 7 billion people. To envision the future as a continuation of current lifestyles is fallacious. Since the world in general and the developed countries in particular are urbanizing at increasing rates, further population increases are being highly localized, increasing the impact of each increment on the population. This is especially true for the areas of highest growth rates, like the Phoenix metropolitan area (Figure 4-7). One consequence of rapid population growth is the equally rapid growth in environmental problems, in the increased demand on urban services, and in the difficulty of maintaining the quality of life. And these problems are intensified, not relieved, by developing horizontally over the land rather than being concentrated vertically. Consider, for example, the contention of Seaborg and Corliss (1971) that the problems of concentrating so many people

FIGURE 4-7 *A striking aerial view of Sun City, a radially concentric suburb in the Phoenix, Arizona metropolitan area.* (Courtesy of Georg Gerster from Rapho/Photo Researchers, Inc.)

in so little space—such as in Calcutta, Tokyo, or New York City—are insurmountable. Hence they maintain that the answer is not a superconurbation (such as that proposed by Doxiadis) but really maximum dispersal; they want no cities at all. However, their suggested policy itself bears costs. First, there is the loss of qualities for which a city exists: economic activity,

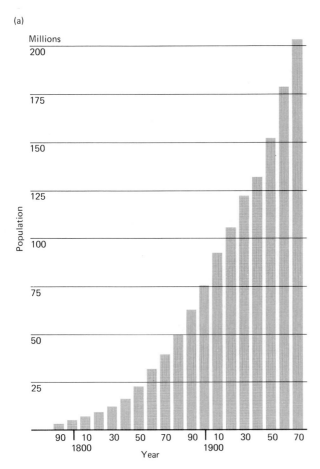
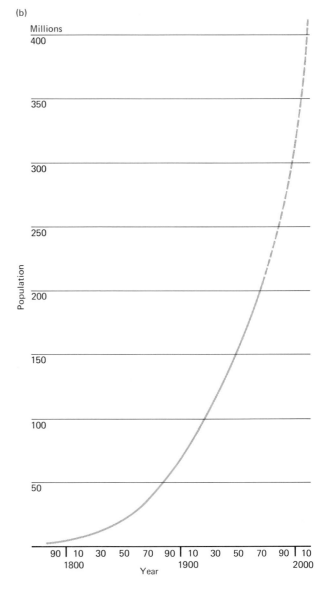

FIGURE 4-8 *United States population growth 1790 to 1970: (a) population curve, and (b) the population increase curve with a projection for the future.* (Redrawn from U.S. Bureau of Census, 1970 *Census of Population*, Vol. 1, U.S. Summary, Figure 14.)

cultural opportunity, human diversity, and social upgrading. Second, dispersing people over the land converts more and more of nature into human ecosystems, an act with incompletely known but probably serious consequences. Such things as land use, air pollution, and transportation become increasingly severe problems, possibilities for solution are more and more constrained. The answer then, would seem to be to move toward urbanization. In land use, this would mean building more intensely, restricting suburbanization; in air pollution, moving power plants to isolated locations (becoming more and more the practice in the Southwestern U.S.) and restricting the use of automobiles; in transportation, increasing the reliance on subsidized public transit and establishing vehicle-free business districts. What is behind all these problems? It is population growth that is limiting our choices and leaving us with solely unpalatable alternatives.

A Case Study: Patterns in the United States

The demography of the United States is typical of developed countries in general and illustrates the trends discussed above. From the first census in 1790 to 1970 the population grew in exponential fashion (Figure 4-8a), from 3.9 million to 207 million. Projections made by the U.S. Bureau of the Census indicated continuation of the long trend (dashed lines, Figure 4-8b). In 1967, the Bureau of the Census made four projections based on the completed family size expected by Americans (Table 4-8). Series A used 3.35 children per completed family per woman; it is the highest projection. Series D is the lowest, with 2.45

TABLE 4-8 *U.S. Bureau of the Census projection of U.S. population, reflecting the dramatic decreases in American fertility. Estimated population, millions, for the year 2000.*

Series	1967 projection	1972 projection	Fertility estimate
A	361.4	abandoned	3.35
B	336.0	abandoned	3.10
C	307.8	300.4	2.78
D	282.6	286.0	2.45
E	not made	264.4	2.10
F	not made	250.7	1.80

children. It projected a population only 78 percent of the A series. The intermediate B and C projections used 3.1 and 2.78 children respectively.

But in recent years, the situation in the United States has been changing rapidly. Evidence is now in hand that Americans have dramatically decreased their reproductive performance. The rate of population increase has been dropping approximately 0.1 percent per year for the last few years; this observation is supported by a series of surveys made by the Bureau of the Census. In 1955, the average family size was 3.2; by 1967, 2.9, a small drop. But in 1972, the mean family size was only 2.0, less than that required to maintain a stable population; by 1974, the mean family size had dropped to 1.9.[3] As a result, the Bureau has revised its projections of U.S. population downward toward a population no greater than 300 million by the year 2000 (Table 4-8).

Despite the declining rate of population increase, growth in some regions remains high because of migration. Between 1959 and 1960 only one region showed

---

[3] A completed family size of 2.0 should lead to zero growth in theory, but 2.1 accounts for early mortality.

an increasing share of the American population: the West, up from 12.5 percent to 14.8 percent. Of this, nearly all was accounted for by California and southern Arizona. The South lost 1 percent, but one of its subregions—Florida and the Gulf Coast—was up to 6.6 percent in 1960 from 5.3 percent in 1950. The trend of movement from colder regions to warmer ones is continuing, although California immigration declined greatly in the early 1970s, and people seemed less inclined to move about.

Urbanization and suburbanization are also characteristic behaviors of the American population. Little will be said of these trends here (they are treated in more detail in Section F), except that the movement is rural to urban and, in urban areas, from older, central areas (the urban core) to the suburbs. This is not the case in more densely populated Europe, but parallels exist in Australia, Canada, and other developed countries with relatively large land areas and low densities.

Questions

REVIEW QUESTIONS

1. Calculate doubling times for a human population (*doubling time* is the time required for a 100 percent increase in the population) for a series of values of r in equation (4). What value of r is required to have a doubling time of exactly 50 years?

Hint: Solution and sample answers are to be found in Ehrlich and Ehrlich (1972), but try to avoid consulting this reference without a thorough effort at an a priori solution first.

2. Search the U.N. statistics and list in order of decreasing value the countries showing the highest growth rates for 1950–1960. Can you find any patterns in this list, specifically geographic, economic, or political?

Comment: The answers are mostly known, but original discovery is always useful. This is a case in point.

3. In Chapter 2, we explored some distributions that organisms have in their habitats and some of the reasons underlying these patterns. Extend this knowledge to explain why people are moving predominately to the South and Southwest in terms of basic physiological phenomena.

4. Devise a critical experiment or design a survey to reveal whether or not American reproductive behavior has really changed. (In other words, how can you tell whether or not current declines are not just short-term fluctuations?) What might be the effect of the current age structure of the U.S. population?

5. How realistic is the von Foerster equation? List at least four points that could be questioned as improbable, unrealistic assumptions, or inadequate data.

ADVANCED QUESTIONS

1. Comment on the meaning of "demographic transition" in the dynamics of national populations. Is this term meaningful, or does it simply describe a more fundamental phenomenon?

2. Assume that human reproduction begins at age 16 with 0.2 births per year per female, increases 0.1 for each year to age 24, then declines in constant increments to 0 by age 45. Calculate, for one year, the total number of births for a county having the age structure of Great Britain, the United States, and Mauritius. (*Sources:* Mauritius and Great Britain, *Population Bulletin* 18. No. 5, 1962; U.S. Bureau of the Census, 1960 Census, 1962.) See Figures 4-2a, b.

3. If the relationship between the costs for medical services per individual increases as a function of the number of potential patients and the service area in accordance with the function

$$\text{cost/individual} = 150.00 + 0.05\ (P_T - P_{T-1}/\text{year})$$

How many years would it take for a population growing at 1.0 percent per year to raise the cost of medical services to 1000 dollars per individual?

4. Given the 1970 age distribution in the United States from census records, how much would the growth rate have to be reduced to attain zero population growth immediately? In 10 years?

Further Readings

Note: Human demography is a field by itself. As might be expected, it has generated a large literature, only a small part of which is represented here. In this case it is especially true that these entries are only keys into the literature. Be sure to consult the bibliography of Chapter 5, preferably at the same time you enter this one.

Birdsell, J. B. "Some Environmental and Cultural Factors Influencing the Structure of Australian Aboriginal Populations." *American Naturalist*, 87 (1953), 171-207.

    A complete, if not unique, study of the intimate relationship between primitive society and environment.

Bogue, D. J. *The Population of the United States*. Free Press, New York, 1959.

    A standard text in the demography of the United States.

Bogue, D. J. *Principles of Demography*. Wiley, New York, 1969.

    A recent text offering useful entries into the literature. See also Keyfitz (1968, below) for a more mathematical and rigorous treatment of population dynamics.

Carr-Saunders, A. M. *World Population*. Oxford University Press, Fairlawn, N. J., 1936.

    Now old and badly out of date in a field that evolves rapidly, Carr-Saunders is still widely quoted. It is a refreshing historical perspective for its now-quaint ideas.

Davis, J. B. "Reproductive Motivation and Population Policy." *Bioscience*, 21 (1971), 215-230.

    This recent paper is typical of the viewpoint of the Davises of Berkeley. It could have been cited as appropriately in the next chapter.

Day, L. H., and A. T. Day. *Too Many Americans*. Houghton-Mifflin, Boston, 1964. Also Delta Books, Dell, New York, 1965 (paper).

    Cogent, lucid summary of the demography of the United States. Especially recommended for intermediate readers who want a good capsule summary of several of the larger texts represented here. Contains full annotation and full references to guide readers to other works.

Deevey, E. S. "The Human Population." *Scientific American*, 203, No. 3 (1960), 194-204.

Douglas, M. "Population Control in Primitive Groups." *British Journal of Sociology*, 17 (1966), 263-273.

    Comparison of the dynamics of four primitive societies in qualitatively different habitats.

Ehrlich, P. R. *The Population Bomb.* Sierra Club, Ballentine, New York, 1968.

> This small, quickly read book started the public debate on population dynamics. Ehrlich has a biased but persuasive argument. All students of demography should be aware of his reasoning, and this is a good place to start out.

Ehrlich, P. R., and A. H. Ehrlich. *Population, Resources, Environment.* Freeman, San Francisco, 1970. (2nd edition, 1972).

> An expanded and better version of the preceding citation. The basic argument is unchanged—this is still a political tract.

Hardin, G. *Population, Evolution, and Birth Control,* 2nd edition. Freeman, San Francisco, 1969.

> This is a reader, a collection of abstracted articles by numerous authors. Its main value is to expose readers systematically to some population controversies, favoring the pessimistic slant. The historical perspective and the section on birth control are especially good, but the book suffers from its lack of references to further reading.

Hauser, P. *Population Perspectives.* State University Press, Rutgers, N.J., 1960.

> An earlier but highly readable series of essays on the demography of the world and the United States, including valuable ideas on what can be done. Fully referenced.

Hauser, P. (ed.). *The Population Dilemma,* 2nd edition. Prentice-Hall, Englewood Cliffs, N.J., 1969.

> This collection represents the views of the Chicago and Princeton schools of demography and the economists of Resources for the Future. Integrates demography and resource needs in one volume. Compare this volume especially to Davis (1971) and to Ehrlich and Ehrlich (1970).

Hinrichs, N. (ed.). *Population, Environment, and People.* McGraw-Hill, New York, 1971.

> This volume fits as easily into Chapter 5 as here, but it is included because it mixes more different viewpoints than any other, adds a foreword by the distinguished humanist Rene Dubos, and binds it together with good editorial comment. Its only weakness is that it does not also provide references to additional readings.

Keyfitz, N. *Introduction to the Mathematics of Population.* Addison-Wesley, Reading, Mass., 1968.

> A rigorous book delving deeply into the mathematical foundation of demography. Requires considerable background in calculus and matrix algebra to be useful. See also Bogue (1969, above).

Keyfitz, N., and W. Flieger. *Population.* Freeman, San Francisco, 1971.

> This volume is for mathematically inclined demographers and would-be demographers.

National Academy of Sciences. *Rapid Population Growth.* Johns Hopkins, Baltimore, 1971.

Revelle, R. "Population." *Science Journal,* 3, No. 10 (1967), 113–119. Reprinted in *The Survival Equation,* (Revelle et al.). Houghton Mifflin, Boston, 1971, pp. 4–15.

> Considerably more optimistic viewpoint than that expressed by the Ehrlichs (1970); a rather more decentralist viewpoint than the other way around.

Seaborg, G. T., and W. R. Corliss. *Man and Atom.* Dutton, New York, 1971.

Taeuber, C., and I. B. Taeuber. *The Changing Population of the United States.* Wiley, New York, 1958.

Thompson, W. S., and D. T. Lewis. *Population Problems,* 5th edition. McGraw-Hill, New York, 1965.

Two more standard texts in the field.

U.N. Department of Economic and Social Affairs. *World Population Prospects as Assessed in 1963.* Population Studies, No. 41, 1966.

U.N. Department of Economic and Social Affairs. *A Concise Summary of the World Population Situation in 1970.* Population Studies, No. 48, 1970.

The sources of specific U.N. projections and population analyses.

von Foerster, H., P. M. Mora, and L. W. Amiot. "Doomsday, Friday, 13 November, A.D. 2026," *Science,* 132 (1960), 1291-1295.

5. Shutting Off the Population Valve, or Who Does What to Whom?

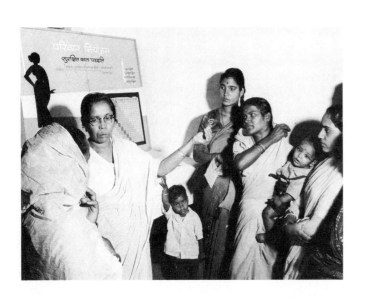

This chapter considers two major questions raised in the introduction to this section. In particular, we shall want to know what aspects of human population stabilization are uniquely different from other animals'. Second, we shall want to explore mechanisms that are likely to be substituted in the human population for the classic environmental and biotic controls explored briefly in Chapter 2.

The Biological Background

From even the slight detail provided in Chapter 2, it is obvious that most known factors in population stabilization theory do not apply very well to human beings. We have an enormous technology with which to buffer environmental extremes, combat enemies, and suppress possible competitors. Consequently, none of these classic stabilizing factors seems to have much relevance to human population dynamics.

That leaves us with some form of resource limitation as the sole remaining mechanism for human population stabilization, from which we may eliminate unregulated competition for resources as an important human mechanism—at least to date. Thus we are left only with some regulated form of resource limitation, some socially mediated mechanism. This topic, known generally as *social population regulation*, was introduced in Chapter 2; it forms the departure point for this chapter. You may find it useful to review the earlier material before going further.

Social Regulation in Animals and Human Beings

In 1962, V. C. Wynne-Edwards proposed that group selection was the mechanism responsible for social regulation. He argued that natural selection could operate on entire social groups as well as on single individuals. He reasoned that selection, if it operated solely on individuals, could never favor decreased reproduction, when the successful organism was defined as the one that fostered the most descendants. Therefore, there had to be some sort of group control, some means to put the welfare of the entire group before the welfare of individuals; selection between groups met this requirement where individual selection could not.

As noted in Chapter 2, group selection turned out to be a highly controversial idea, even if social regulation did not. Other scientists have contended that individual selection did not always favor maximum reproduction and that invoking group selection as a mechanism for reduced reproduction was unnecessary. Another major difficulty with group selection is that there seem to be few effective ways to force adoption of a particular behavioral pattern throughout the group.

Thus territorial behavior, which Wynne-Edwards interpreted as a group behavior, is also a property of individual males. In fact, it is easier to interpret the population-limiting effects of territoriality as the result of individual defense of territory and individual inability to reproduce more than it is to imagine how the group might regulate territory size to suppress excess reproduction.

For such reasons, Wynne-Edwards's ideas of group selection as a general mechanism for population control have been rejected by most ecologists; a highly developed social structure would seem to be a prerequisite to allow group selection to operate. Human beings, however, do have a highly evolved social structure, and so it is reasonable to postulate that group selection, if it applies to any species, would apply to human populations. In defense of this idea, one argument is that societies with laws, cohesive social organization, and a political structure can restrict individual freedom to move about and even to breed. Second, human beings are social animals whose actions are determined by many external forces as well as internal ones. Extensive communication networks keep people aware of technological developments in birth control and, more importantly, expose individuals to ideas. These things are educational at least, and in many instances they set up intensive peer pressure to conform, to change laws, and to change attitudes. Group selection appears in the behavior of nations, ethnic groups, and regions that have a patterned reproductive behavior set by the society and enforced by its agencies. Hence even if one does not embrace group selection as a mechanism for population regulation in other species, it is eminently reasonable to hypothesize that human societies have progressed toward such an advanced form of social regulation aimed at halting population growth. As we shall see, even primitive human societies had developed ritualized and highly effective forms of social population control, although many of these methods have fallen into disuse.

FIGURE 5-1 *A class for Indian men in family planning. This is part of a program organized by the Asian Trade College, with assistance from the International Labor Organization (ILO) and the United Nations Fund for Population Activities (UNFPA) for landless rural laboring-class couples.* (Courtesy of the United Nations and the International Labor Organization.)

Population Stabilization in Primitive Human Societies

Even a cursory examination of anthropological studies will reveal the striking influence of human institutions on population regulation, including the institutionalized execution of unwanted individuals. Infanticide has been used widely for a long time, especially in Australia, in Africa, and by American Indians. In some of these societies, perhaps half the newborn, and nearly all female babies, were killed. Even in Europe infanticide persisted into the Middle Ages. Prevention of births is equally common. Abortion is one method that has been common in the past and is being rediscovered and legitimated even in "civilized" countries today. Its incidence has increased with legalization, notably in Japan and the United States; in fact, in the contemporary world it has become so widespread as to be typical of no specific society. But abortion is not the only mechanism for the prevention of births. To this end, many Indian castes and Near Eastern societies have rigidly enforced various preconditions before marriage. This includes such prohibitions as the strict segregation of sexes before marriage, frequently combined with delayed marriage. The prohibitions might be carried even further—to probable celibacy for all daughters and sometimes for all but the eldest son. In addition to these long-standing methods of birth limitations, we also have the more technical methods such as mechanical and chemical contraception, but they are a more recent development (Figure 5-1). In addition to the prevention of births and the execution of individuals as population controls, a third parameter of population change, migration, has been described, particularly for Polynesian societies. When overpopulation threatened, a proportion of the population would elect or be forced to migrate. In ancient

Society or area	Mechanism(s)	Authority
Polynesia	infanticide	Danielsson, *Love in the South Seas* (1956)
Tikopia	infanticide, migration, expulsion	Firth, *We, the Tikopia* (1957)
Australian aborigines	subincision to weed out weak, territorial behavior	Birdsell, *American Naturalist*, 87 (1953), 171
Ao Naga (Assam)	female celibacy through work	Smith, *The Ao Naga Tribe of Assam* (1925)
Caribou Eskimos Pelly Bay Eskimos	female infanticide	Birket-Smith, *Report of the Fifth Thule Expedition, 1921–1924* (1929) Balikci, *Man*, 2 (1967), 615
Rendille (Kenya)	infanticide, delayed marriage, expulsion	Spencer, *The Samburu* (1965)
Hunter-gather societies	entire range, plus war, cannibalism, ritual sacrifice	Dunn, in Lee and DeVore, *Man the Hunter* (1968); see also discussion, part V

TABLE 5-1 *Some examples of social regulation in human populations.*

Polynesia, hard times frequently caused groups of fertile young people either to launch outriggers in a nearly hopeless pursuit of new land or to leave their home villages and relocate in other, less depressed villages. In either case the resources of the home village would no longer have to sustain them. Examples of these modes of population regulation are given in Table 5-1.

Economic influences on population size appear even in the most primitive societies. Douglas's (1966) thesis that edible luxuries, say oysters and champagne, are a more important resource limitation than more basic items of food may initially appear to be foolish. However, this type of limitation may actually reflect a remarkably sensitive attunement to the quality of the whole resource base. Moreover, these kinds of economic considerations can strongly influence social norms for reproduction because economics is frequently the means whereby population size and available resources are harmonized. For example, consider the Nambudiri Brahmins of Southern India. These people have enormous wealth and influence because they are a land-holding caste. To maintain their status, the Nambudiri Brahmins practice rigid control of reproduction and inheritance within their caste. Because their power and wealth are vested in large tracts of family land, their strategy dictates that family holdings shall not be divided into smaller holdings. Hence even though his brothers will benefit from the family's wealth, only the eldest son actually inherits the land, and he ultimately bequeaths it to only his eldest son. Not only do the brothers inherit no land, they are not allowed to marry within their caste. Of the sisters, only the eldest may marry—the rest are kept in celibate seclusion for life. This example very clearly demonstrates social population regulation and the population homeostasis that results because the land would support more Nambudiri Brahmin families only at the cost of lost wealth, power, and prestige for all. The political, social, and economic benefits of keeping estates large are valued so highly by these families that ruthless reproductive bans have been instituted to ensure their continuation.

The most primitive societies have a more direct linkage to their environment and especially to the resources available. Birdsell has shown rather elegantly, for example, that Australian aborigines living in territories of a given size in the northern part of Queensland will adjust their population very neatly to the size that can be sustained by these territories at any given time.

The limitation of population growth has been practiced throughout human history, but the methods have changed. We have seen that birth control was practiced early, and even that most of its specific forms have continued into modern times. As we shall see, technology has supplied more alternatives in birth control, but earlier societies conpensated for their lack of technology by using methods that involved individual deaths. Therefore, it should be remembered that even now if resource limitations could not be overcome because there was no available technology and attempts at birth control were unsuccessful, then other methods of population limitation such as forced migration, abortion, infanticide, and ritualized warfare would eventually become unavoidable alternatives.

What picture emerges from all this? There are some clear parallels between population control in human and nonhuman species. Chief among these is the strongly empirical nature of the mechanism and the demonstrable link of population control to resource supply. Thus in human societies, whether the perceived limitation is a basic commodity such as land or a more sophisticated item such as oysters, the population that is too large will use whatever means it can to stop growth. If the population is unable to make birth control work or to overcome the resource shortages, there is always a way to increase the death rate. Chief among the contrasts between human and nonhuman populations is the unprecedented human ability to increase our resources, which thus far has permitted the world's population almost totally to escape population stabilization. On the other hand, any disruption of the structure of a given society may produce unanticipated and frequently serious ramifications. In this light, it is easy to see why interference of societies that practice execution as a population stabilization device by other societies that shun execution but do not replace it by other immediately practical methods historically has led to population disruptions that end in a local population explosion, just as surely as would a real increase in resources or resource technology. This should not be surprising, given that human beings, like all other creatures, also have a reproductive capacity far beyond what is needed to replace individuals. The present population spurt in developing countries, where medical technology has again unbalanced the social mechanisms of population control by greatly lowering the natural death rate, is a modern example of what can happen when our population stabilization mechanism has been disrupted.

The Tikopia of the Pacific: A Classic Case

The Tikopia people, who live on a small island of the same name, separated by more than 75 miles of open ocean from the nearest island and located 200 miles from Espiritu Santo in the New Hebrides to the west, have been studied extensively by the renowned anthropologist Raymond Firth. On Tikopia the people were agriculturalists, living on root crops cultivated in their gardens as well as on practically anything growing on the island; they supplemented their diet with fish protein from the reefs (Figure 5-2). So far as is known, the Tikopia were acutely aware of their limited food supply, especially the high-status coconut cream they used to make their food more palatable and thus, to them, more suitable for important ceremonies. To maintain at least a rough balance between food supply and population, the Tikopia practiced contraception, abortion, hopeless migration, celibacy, infanticide, and on at least two occasions massive, if selective, slaughter.

Before significant contact with Western societies, the population of Tikopia was small and growing rather slowly. In 1828, the population was estimated at 400 to 500. By 1858 permanent missionaries had moved onto the island, and by 1929 the population had reached 1300. As in other places, the missionaries also brought Western diseases, but the effects were minor compared with the effects of the iron tools and religious practices they brought. The population changes wrought by these introductions had dramatic and far-ranging implications for the entire society.

FIGURE 5-2 *A Tikopian man summoned from his farming to be photographed.* (From Raymond Firth, *We, the Tikopia*, American Book Company, 1936, plate XXV, by permission.)

With a more dependable food supply made possible with the use of iron tools and without the ability to control population (infanticide, warfare, and abortion had been strongly discouraged or banned outright by the missionaries), the population increased another 35 percent from 1929 to 1952, resulting in a population of 1750 persons. Then a famine resulted from successive typhoons; deaths from starvation were minimized only by emergency relief supplies. The population–food-resource system was clearly upset.

Firth was able to demonstrate that the Tikopia had not suffered high mortality from disease before the missionaries came, and so medical death control was not the cause of the population explosion. As a result of his studies, Firth was able to discover the roles played by abortion, infanticide, war, migration, contraception, and male celibacy in this society's population stability. Firth blamed the intervention of the Christian church for completely upsetting the system. In his own words:

As the result of European contact these checks are no longer operative to the same extent as formerly. Fear of the government forbids the overt expulsion of any considerable section of the people . . . the government may forbid the emigration of the young men in canoes, as has been done in other parts of Polynesia. The other checks are also affected. Owing to the attitude of the mission . . . celibacy is being virtually discouraged. The Tikopia young man, unused to the foreign ideal of prenuptial chastity, "sins" and is forced by indignant mission teachers to marry the girl. As a celibate . . . he is careful not to cause his mistress to conceive; as a married man he does not exercise the same restraint and produces children. Abortion and infanticide are frowned upon likewise by the mission. The result is that the former equilibrium is being upset, and there is a threat of congestion of population on the lands of many families. Moreover, there has been a tendency to plant more crops in the woods, with the result that the reservoir of supplies which these afforded in time of drought has been diminished.

The really regrettable feature of the situation is that but for the moral preconceptions of the interpreters of the Christian religion the old checks would act in a perfectly satisfactory [and discrete] manner.[1]

In reconsidering his earlier refusal to intercede with the government in response to a plea from Tikopia leader, Firth then concluded:

But I felt then, as I do now, the injustice of enforcing our European moral attitudes on a people who before our arrival had worked out a satisfactory adjustment to the population problem—particularly when we can offer them no adequate solution to the maladjustment which we thus create.[2]

It is clear that the Tikopia were a society upset by outside interference. In other cases the society either was destroyed or is at present in the process of being destroyed through these same processes of acculturation. (For examples, see the cases of the Miskito Indians [Nietschmann, 1974] or our discussion of Geertz's studies of colonial Java in the next section of this chapter). The Tikopia also illustrate the operation of social regulation through a primitive economic system. There is growing awareness among ecologically oriented anthropologists of the subtle interpretation of resources and resource limitation. With the Tikopia, coconut cream was "oysters and champagne"; long before shortages were sufficiently severe

---

[1] From R. Firth, *We, the Tikopia*, 2nd ed., George Allen and Unwin, London, 1957, pp. 415–417, by permission.
[2] Op. cit., p. 417.

to result in starvation, this type of shortage—which threatened the quality of the Tikopian lifestyle—was critical. Like the Nambudiri Brahmins, the Tikopia had emphasized quality of life over quantity. One of these societies, then, is quite willing to sacrifice reproductive freedom for more economic power, and the other sacrifices it for assured supplies of quality food. In both, social regulation was well developed; and while it still is in one, the Tikopia system has been destroyed by interference from another human society.

Cross-cultural Comparisons among Contemporary Societies

In modern societies the influence of social organization and economic development upon human population dynamics can be seen easily. Clifford Geertz's studies of the impact of Dutch colonialism upon Indonesia provide one excellent example. Geertz demonstrated that Dutch colonialism caused Indonesia to split into two easily recognized segments, an "inner Indonesia," consisting of Java and South Bali, and "outer Indonesia" (the remainder of the archipelago). The latter part is characterized by shifting agriculture, nongrowing populations, and an intact social structure that can absorb slow economic growth through native leadership. In contrast, Java and South Bali were subject to intensive development in commerce, resources, and agriculture directed by the Dutch colonists. Geertz also showed that while per capita incomes for the Javanese increased slightly as a result of this colonial development, the long-term costs far outweighed these short-term income benefits; the profits from commerce and resource developmet flowed into Dutch pockets, and the new style of intensive plantation agriculture actually reduced the Javanese to little more than peasant labor for which every additional helper was a bonus. The disruption of the existing social structure left the people with no chance to evolve their own leaders and stimulated massive population growth that is unsurpassed in history (see Table 5-2). This dualism left the Javanese with massive population problems but without a growing economic base to compensate for it.

The Javanese example introduces a new element in the relationship between population and economics: the idea that economic development must take

TABLE 5-2 *The history of population growth on the island of Java, 1830–1900.*

Year	Population	Percent increase from 1830	Percent increase over previous decade
1830	7.0×10^6		
1840	8.7×10^6	24	24
1850	9.6×10^6	37	10
1860	12.7×10^6	81	32
1870	16.2×10^6	131	28
1880	19.5×10^6	179	20
1890	23.6×10^6	237	21
1900	28.4×10^6	306	20

Java population, 1960 = 41.7×10^6

SOURCE: Geertz (1963).

certain forms before it can significantly slow population growth. The contrast between "outer" and "inner" Indonesia also reaffirms the importance of social structure, because economic development that disturbs the social norms will upset the highly evolved, socially mediated mechanisms of population control. Polgar (1972) has generalized Geertz's argument for all modern societies. Polgar argued that strong social organization suppresses high birth rates and that colonial economic dominance of underdeveloped countries was central in disrupting societies, thus setting off the population explosion that affects most of the world today. As a specific example of this phenomenon Geertz cited as the population trigger in colonial Java an "externally imposed rise in labor output, accompanied by blocked economic mobility," which he says decreased "the net costs of large families without increasing the utility of small ones." Thus it seems likely, given present evidence, that major social disruptions that raise death rates are less important for population change than those that release social checks on family size. Sporadic wars and epidemics are ineffective checks because these are neither widespread nor regular enough to compensate for the reproductive capacity of the population; reproduction goes on even in a disrupted society. In contrast, once the social mechanism for population control has been disrupted and the population begins to increase, the mechanisms of social control may be so thoroughly disrupted that the population cannot check its growth until a new mechanism is instituted.

Polgar's belief in the importance of social mechanisms for population control is underscored by an American example. The American population has been experiencing a dramatic drop in the growth rate because of the steady decline in the desired number of children per family. The Hutterite sect of South Dakota, Montana, and Canada and the Amish of Ohio and Pennsylvania are conspicuous exceptions. Their high fertility rates result from unique cultural institutions that promote large families: an agrarian lifestyle in which large numbers of children are used as inexpensive farm labor; strict religious beliefs including a literal interpretation of the biblical instruction to "go forth and multiply"; early marriages; and refusal to practice any form of birth control. At the

FIGURE 5-3 *Population growth rates of the Hutterite and Amish communities compared to that of the United States. The calculations include migration. The Hutterite rate has remained at a historically high level, while the Amish have shown a long-term downward trend, approaching the U.S. rate by 1966.* (Redrawn from *Population Bulletin* 24(2), p. 33, with permission of Population Reference Bureau, Inc.)

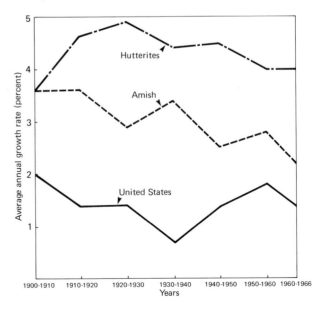

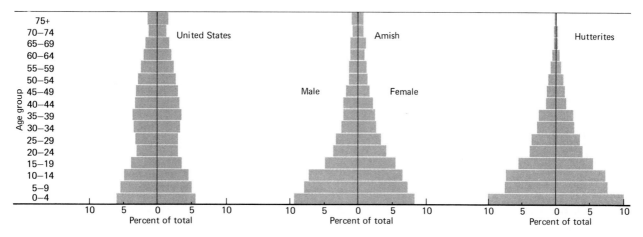

FIGURE 5-4 *The high growth rates for the Amish and especially the Hutterites are reflected in their age distributions. The age pyramids are very heavily skewed toward the young compared to that of the United States.* (Redrawn from *Population Bulletin* 24(2), p. 47, with permission of Population Reference Bureau, Inc.)

same time, both sects have not hesitated to take full advantage of the high level of health care in the United States, and both populations have skyrocketed. The Hutterites have averaged over 4 percent annual growth from the turn of the century, while the Amish rate has declined slowly until at present it is just over 2 percent (Figure 5-3). The growth rates between the two sects differ for several factors. The Hutterite women, like the Amish, marry young and begin reproducing at an early age, but they extend their reproductive years right up to the onset of menopause, thus producing more children per woman. Also, the entire Hutterite community shares the responsibility of children, and so the two parents are not solely burdened by these large families they produce. The Amish, on the other hand, place the burden of raising children directly on the parents, and this results in a few less children per woman. Finally, though the Hutterite sect remains strong, the Amish are being assimilated slowly into the American population, as many young adults are defecting from Amish traditions and beliefs and leaving the settlements. Still the Amish growth rate is twice as large as that for the United States, which now has declined to less than 1 percent. The long period of sustained growth has produced typically skewed age distributions for the two sects (Figure 5-4).

We have already discussed some economic influences on population control in the Tikopia and the Javanese, and it is not necessary to consider specific

TABLE 5-3 *Population statistics of major regions of the world.*

Region	Population (Millions, mid-1971)	Annual birth rate (per 1000)	Annual death rate (per 1000)	Percent annual growth rate	Percent population <15 years old	Doubling time[a]
World	3706	34	14	2.0	37	35
Africa	354	47	20	2.7	44	26
Asia	2104	38	15	2.3	40	31
North America	229	18	9	1.2	30	58
Latin America	291	38	9	2.9	42	24
Europe	466	18	10	0.8	25	88
USSR	245	17	8.1	1.0	28	70
Oceania	20	25	10	2.0	32	35

SOURCE: Population Reference Bureau, Inc., Washington, D.C., *World Population Data Sheet*, 1971. For a country-by-country breakdown see the original publication.

[a] The number of years required to double the size of the population at current growth rates. This may be calculated

$$\frac{0.6931}{\text{annual growth rate}} = \text{doubling time}$$

case studies to discover the strong link between economic influence and the growth of contemporary populations. The discrepancy in growth rates between developed and developing countries (see Table 5-3) can be traced directly to the differences in birth and death rates; these in turn reflect economic well-being (measured as per capita Gross National Product). Even casual inspection of Figure 5-5 shows a reasonably clear relationship. While the reasons are not fully clear, there is no doubt that economic development and population stability are related. What may be more surprising is that each additional increment of economic development may further enhance progress toward stability because the increasing welfare of the population may sharpen its perceived dependencies upon "oysters and champagne." The greater the amount of material goods each individual has, the more investment he needs to protect it and the less willing he may be to share it with offspring; he is likely to feel a developing shortage more quickly. This is an extension of Douglas's argument for primitive societies. It has been raised in the form of an argument that at least a minimum level of well-being is required to sustain any population control, and thus this argument is the economic basis for a mechanism by which demographic transition would be realized.

The Looming Crunch

At last we are able to provide a partial answer to the two questions raised in the introduction to this section. Barring some future calamity that reduces us to an aboriginal state, our methods of population control are unique among animals. Our complex society permits subtle pressures, overt laws, and a pervasive economic system to determine our reproductive performance. However, if we should fail to arrive at a sustainable population, then sooner or later natural laws will be imposed, and we shall rediscover that we can suffer and die like any other animal. If we cannot check our population growth ourselves, it inevitably will be done for us.

Because we cannot escape this choice, some difficult decisions lay ahead. When the Tikopia ran short of resources, they did not hesitate to sacrifice individuals to preserve their social fabric. Only when well-meaning missionaries prevented normal sacrifices from occurring did Tikopian society suffer; their most effective method of population control was no longer available. For the first time contemporary societies face food shortages that are global and not purely local; water is scarce, good farm lands are disappearing under houses, and fuel energy is both expensive and increasingly difficult to get. Like the Tikopia, every nation must choose a method of population stabilization knowing that the wrong choice could be disastrous.

Why is the choice so difficult when the goal is both so clear and so important? Because, as the Tikopia knew, group welfare and individual welfare are not always compatible. Population stability is a group attribute, and if it is to be sustained, individual choices must be limited. Individuals cannot reproduce freely unless they also die freely. But on the other hand, human beings are not mere numbers to be categorized and programmed for stereotyped responses. Therefore, if human dignity is to remain a basic tenet of free societies, then the private and fundamental freedom to bear and raise children can be restricted only cautiously.

These considerations raise questions which will be explored in the remainder of this chapter:

1. Will some form of involuntary fertility control be necessary, or are voluntary programs sufficient?
2. If involuntary control is needed, how will we know when to institute it?

FIGURE 5-5 *The relationship between per capita GNP (gross national product, a measure of well-being) and annual population growth rate for some developed and developing nations.* (Data from Population Reference Bureau, Inc., Washington, D.C. *World Population Data Sheet* (1971).)

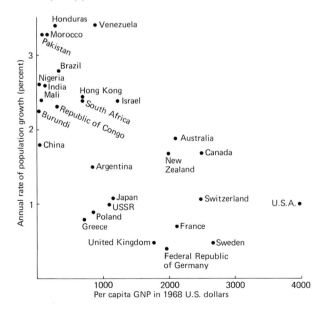

3. Who is to make this decision, who pays the cost, and who receives the benefits?

The answers to these questions will indicate the probable policies that ultimately will be formulated to control world population.

The Controversy over Family Planning

Family planning is the prevention of unwanted births by contraceptive methods. It is sanctioned officially by the United Nations and by many nations of the world. At present it is practically the only method of population planning that is practiced. While family planning organizations have enjoyed moderately good success (notably in the Republic of Korea, Taiwan, Singapore, and Hong Kong), they have been severely criticized. Now we shall examine why.

The goals of family planning have been summarized by the International Planned Parenthood Federation (IPPF) and the U.S. Agency for International Development (AID). These goals are strongly conditioned to voluntary participation. For those who participate, family planning provides the latest information about timing births, controlling fertility, using contraceptives, and developments in birth control technology. A successful family planning program will reduce the birth rate and slow the population growth rate, but it is a different matter to claim that family planning can stabilize a population. For this to occur, we must assume all births in excess of replacement to be unwanted; otherwise the population will continue to grow. Proponents of family planning argue, however, that eventually all births in excess of replacement will be unwanted, once a population feels it should grow no larger. This is the major advantage of family planning: It respects the ethical and moral beliefs of citizens. By taking a passive educational and technological role that supports private decision-making regarding family size, family planning avoids prescriptive measures and thus is compatible to most countries except those where religious belief against contraception is official policy. Thus family planning has high political feasibility because it offers large potential benefits at negligible costs.

Critics of family planning view its strengths as weaknesses. If one starts from the premise that the world is overpopulated and that every additional increment multiplies our problems, then family planning is not very helpful; it offers no guarantees that it will halt population growth. It follows that the only acceptable solution is one that promises the fastest possible halt to further growth. Hence we require some means to control fertility, such as fertility licenses, involuntary abortion for all pregnancies above a state-approved number, or economic sanctions. Such involuntary controls represent the antithesis of family planning, for with them the ethical and moral issues are less important than immediate and decisive action is. For societies that value human dignity and individual freedom it is necessary to demonstrate the truth of the initial premise in order to justify the sacrifices of individual freedom needed to implement these authoritarian policies. Even when enough people agree that a new policy is needed, there remain the problems of selecting the best policies and implementing them. Is it then surprising that involuntary control of fertility is unlikely to become official policy soon?

At present noninterference with reproductive freedom seems the most important property that any proposed program can have.

Social Factors Affecting Reproductive Attitudes

The evidence suggests that population growth is primarily under social control, and so we can expect present and future social conditions to determine reproductive performance. Let us examine a few trends that may affect world population growth.

Perception of resource adequacy History offers evidence that birth rates decline when economic conditions worsen. Perhaps the best example is the dramatic decline during the Great Depression (1929–1940), when American birth rates plunged to a low of 75.8 births per 1000 women. At present American fertility is falling again, and it is likely that economic and resource considerations are involved, although it is tempting to believe that environmental motives are at work, too.

National priorities for economic development Many developing nations seek industrial and economic development above all. While some nations encourage population growth in the belief that it will encourage economic development, the argument that a growing population may prevent attainment of a better life is beginning to be heard more often. Perhaps more and more nations will be actively engaged in various birth control programs in the coming years.

Alternative roles for women The challenge that the women's liberation movement has put to traditional roles of unadorned wife and mother promises population benefits. If this movement helps create more diverse opportunities (for jobs in particular, but in more general terms for the use of time), women may want fewer children. Each child would represent lost opportunities for productive employment or for other activities requiring freedom of action.

Awareness of population impacts on the global system A major accomplishment of the ecology movement has been to increase public awareness of the widespread effects of population growth on the planetary system. While present evidence for environmental considerations such as global resource shortages, environmental quality control, and overcrowding entering into family-size decision-making is weak, the facts that demographers are including it in surveys and that part of the populace is expressing environmental concerns are encouraging signs.

Changing social norms and moral-value systems Reproductive performance is partly a function of norms established by prevailing social attitudes. The United States is experiencing a remarkable decline in birth rates, where the norm is two children per family, and to remain childless is perfectly acceptable. Less than a decade ago, three to four children were the norm, and to remain childless was interpreted widely as a sign of psychological or physical abnormality. However, this decline in population growth may be counterbalanced by another modern trend. The moral attitudes among younger generations have changed so that illegitimate children no longer are considered a stigma; consequently the number of illegitimate births has risen sharply in the past few years as various

FIGURE 5-6 *Communal living is a popular alternative to the traditional family structure. These people are part of a 600-member commune called "The Farm" in Summertown, Tennessee.* (Courtesy of Rick Smolan.)

alternatives to classic family structure have been explored (Figure 5-6).

Changes in moral and social norms have been made possible in part by advances in medicine and contraceptive technology. Child survival in the United States is so high that there is little need to bear children to ensure having a surviving son to carry on the family name. Government programs that provide adequate care to the elderly also make it less important for a couple to have children who can care for them in old age. These conditions do not exist in most developing nations—one reason why population growth in such areas has been rapid.

Perhaps the most remarkable developments are related to contraceptive methods and routine abortion. Many chemical and mechanical contraceptive alternatives are now available. In addition to methods that have been the women's responsibility, today men can share the burden of contraception since vasectomies are routine and completely effective. Even when a failure results in conception, abortion is available as an inexpensive, painless, and a routine procedure. An interesting return to abortion as a birth control technique seems to be in progress.

Governmental legitimation of policy The reproductive responses of populations to changing conditions means that family planning may lead eventually to population stability. The movement toward stability can be influenced strongly by governmental policies. Various economic measures, such as lower taxes for single persons and married couples without children and reduced deductions for large numbers of children, are fairly easy to implement; systems of birthrights (wherein each woman is given rights to bear two children but must buy someone else's to bear more) or an outright ban on more than a fixed number of children are more difficult. The individual freedom to reproduce and governmental rights to infringe on this freedom are such controversial subjects that little policy-making beyond family planning seems likely in most, if not all, democracies. For example, former President Nixon commissioned the first panel under John D. Rockefeller III to officially study population policy. Most of their recommendations, contained in the report of the Commission on Population Growth and the American Future, are conventional. The Commission recommended that a new office of population be set up as a regular federal agency and that family planning efforts be strengthened; and it suggested that we must continue economic development to achieve population stabilization. (This is a reference to the observation that a rising GNP seems to be accompanied by a falling population growth rate.) The one controversial recommendation—that inexpensive abortion be made available upon demand—was rejected immediately by Nixon, presumably because many of his supporters disagreed. It is ironic and illustrative of the political problems involved in most population stabilization practices, that inexpensive abortion upon demand is already available in the United States. Even if the government refuses to legitimate that fact, its reality will not disappear. However, if something so widespread as abortion cannot become formal policy in this country, what chance do other, more controversial and prohibitive policies have in the political arena?

Zero Population Growth

Population stability (zero population growth) can be attained only when several demographic conditions have been met. Now we shall consider the demographic conditions required to reach ZPG and how long that ultimate goal will take.

As a first step we shall assume that immigration equals emigration, which is a realistic assumption for most countries. Hence to attain ZPG, the number of births must equal the number of deaths at any one time. Thus—given generally declining death rates —some specified, time-dependent, decline in the birth rate must be maintained (see demographic transition, Chapter 4). Since the number of children being added to the population is the important consideration, we can influence either the number of surviving children or the number of children born. To achieve either aim, many alternatives are available (delayed marriage, delayed childbearing, legalized abortion, and so on), although the exact means is not important to this discussion. All we need do is hold the average completed family down to 2.1 children. Though the means of this reduction in the average completed family may seem straightforward enough, how to induce the population to cooperate in the endeavor is neither simple nor straightforward. To the various problems in reducing the birth rate to ZPG we must add the age distribution of the population.

A population that has been growing in the past will have a large potential for future growth because there will be a large number of women whose reproductive potential lies ahead. Thus to prevent future growth further reductions in fertility must compensate for the large numbers of potentially fertile women entering adulthood. In human terms, this means that if the population suddenly decides to stop growing immediately, it must cut its fertility to less than 2.1 children per family.

Frejka (1968) analyzed how long it would take to reach ZPG if the U.S. population decided to seek it in 1965. First, he assumed that mortality would continue to decline at a rate such that the mean expected life span would increase to 77.5 years over a 30-year period and then increase no further. Second, he assumed no net migration. Third, he assumed that women would continue to show patterns of fertility (births between ages 15 and 44) proportional to those of 1965. From these assumptions Frejka calculated what fertility rate would attain ZPG immediately and hold it indefinitely. The relevant equation is

$$GFR_x = \frac{B_x \cdot P_{15\text{-}44}}{\delta} \quad (1)$$

where GFR_x = general fertility rate,
B_x = total female births in x,
δ = 1/female sex ratio at birth
$P_{15\text{-}44}$ = female population in reproductive years 15–44.

This equation yields the number of individuals to be born in each year. After solving this equation for GFR_x and dividing by the number of women in the reproductive years 15 to 44, Frejka calculated the average family size permissible. He found that the average completed family size would have to be 1.2 children for 20 years to compensate for the current age distribution; it then could increase slowly for about 70 years to 2.8 to compensate for the smaller number of women of reproductive age that would result from the smaller family size in the initial 20-year period. In this particular case, the cost of ZPG is an unrealistic reduction in family size. Because further delay in attempting to reach ZPG increases the imbalance in the age distribution, the reduction in family size that is needed to achieve immediate ZPG gets larger as the population grows.

Frejka rejected immediate ZPG as unrealistic. He constructed a more realistic model by assuming that some motivational change leads to a gradual decline in fertility, eventually leveling off at the ZPG level. (This, in fact, is happening over most of the developed world.) This mathematical model requires B_x to be a variable instead of a constant, since the decline in fertility would be gradual by assumption. Frejka defined

$$B_x = k_x \sum_x l_x(a) \cdot m_{1965}(a) \quad (2)$$

where k_x, = fertility coefficient set by investigator as rate of decline in fertility from 1965 levels,
Σ = summation,
$l_x(a)$ = average proportion of age group (a) in x,
$m_{1965}(a)$ = age-specific rate of births for a in 1965.

This means we assume the number of births will decline at k_x from 1965 levels. Equation (2) is substituted into equation (1), and the resulting equation is solved for GFR_x. By using a computer to solve this equation for as many years as it takes to obtain a stable population, we get Table 5-4. Inspection of Table 5-4 reveals that the effects of age distribution inevitably lead to continued expansion of the population to some higher level. If the decline to ZPG levels had started in 1965 and it required 50 years to achieve ZPG, the population would number 398 million in 2015, assuming an overall 1:1 sex ratio. Even if the decline to ZPG took only 5 years, the stabilized population would be 284 million, 44 percent higher than the 1965 population.

This example demonstrates that ZPG probably will be reached only after considerable further growth of the population. Since many young women will yet reproduce, how fast we reach ZPG depends very much on attitudinal change and social institutions that favor reduced fertility. In this regard, events in the United States are heartening: The age at marriage and, more important, the mother's age at the birth of the first child (which affects generation time and therefore the rate of population growth) are increasing, and desired family size is declining. The net result is decreased fertility, which even now is at a level that will result in ZPG eventually.

But the world situation is less optimistic. At the recent world population conference in Bucharest, Romania revealed the schism in world opinion. Nations that have achieved or are in the process of achieving a steady population also happen to be fairly well-off economically. And nations that have not achieved a steady population happen to be worse-off economically. Some nations regard suggestions that population growth be discouraged as unwarranted interference or, worse, as colonial imperialism. Others hold that rapid economic development must precede population stability. The resulting inaction promotes continuing growth.

TABLE 5-4 *The predictions of Frejka's model under the assumption of declining fertility at various rates sufficient to reach ZPG in n years (column 1).*

Fertility assumed to reach ZPG level[a] by:	Size of final equilibrium, females (Millions)	Total population (Both sexes, 1:1 ratio)	Percent of known 1965 population
1965	135.9	271.8	137.6
1970	141.9	283.8	143.7
1975	147.9	295.8	149.0
1980	152.7	305.4	154.6
1985	158.8	317.6	160.8
1990	165.0	330.0	167.1
1995	171.1	342.2	173.3
2000	177.9	355.8	180.2
2005	184.4	368.8	186.7
2010	192.3	384.6	194.7
2015	198.9	397.8	201.4

SOURCE: Frejka (1968).
[a]NRR = 1.000.

Gazing into a Clouded Crystal Ball

With the controversy and uncertainty that surround the future course of the world population, it is risky to predict what will happen. Things are changing so quickly that anything is likely to happen—a statement both trite and elusive. Thus this chapter concludes with a series of predictions—guesses to some!—that may at least stimulate critical thinking and perhaps even may lead to the generation of alternative schemes.

The evidence for social control of human reproductive behavior justifies the tentative conclusion that behavioral systems will halt population growth. It seems logical that when resources are so limited and conditions are so bad that it is not worth raising a child, people will refuse to do so. This philosophy is the basis of the rational model—population stability will arrive when people want it to, assuming that family planning is ready to supply the necessary advice and aid. Of course, there are no guarantees that people are rational, or that they can take effective action, or that they will take action in time. Hence there is no guarantee that the world we or our descendants leave behind when world population stops growing will be a pleasant place. Life might become automated and dreary; worse, it might be precarious to live in a world that has been so thoroughly disrupted or that so depends on a precise sequence of events that the quality of existence depends on slender threads.

The basis for believing that human beings ultimately will control their numbers automatically is found in the concept of negative feedback which is so well rooted in ecology. A growing population is aware of its environment, even if only dimly, and as it senses that the environment is deteriorating, its awareness is translated into strong social pressures to decrease fertility. Sooner or later as awareness builds, small families become the majority, and a stable population is realized. When this attitude becomes the norm, the situation is reminiscent of Wynne-Edwards's ideas about group selection. Ecological research has provided theory; it is now up to social scientists—anthropologists, economists, and demographers especially—to provide proof that human societies will behave like this. For example, it is important to know whether modern societies will prevent resource overuse as the Nambudiri Brahmins and Tikopia did. Are those societies at all comparable to our own?

Clearly we do not know what is likely to happen, and we are unlikely to get better information soon. As the controversy rages on, we shall have to choose between two alternatives:

1. We can pursue the philosophy of family planning. In return for the minimal governmental interference associated with this strategy, we must accept the risk that if the approach fails, the planet we end up with probably will not be worth having.

2. We can accept the philosophy of direct population control. Here the trade off is reversed: We might be able to save the quality of life as we know it, but we no longer can decide for ourselves how many children to have.

There is reason for optimism, at least for the developed parts of the world. People are becoming more aware of the systemwide impacts of human population growth on such diverse factors as other living systems, resources, personal freedom, personal finances, and environmental quality. I have faith that people thus educated will be rational and will act in their own interests. But the problem of the populations in undeveloped parts of the planet remains, populations that continue to grow at a rate which would double the world population by the year 2000. These people may be unable to control their fate. What then? The probable costs of the global population explosion are staggering, but since the population involved is largely unaware of the consequences, the dismal prospects facing them can have little impact in bringing about action to reduce the growth rate. In addition, the costs in human dignity of having one's reproductive behavior controlled are perceived as large and highly undesirable. All these factors contribute to official inaction. If governments do not act in concert—and that seems improbable—and if feedback does not occur in time, there will be no alternative to the population crashes that will result inevitably as bloated populations are brought into equilibrium with their environments. Certainly *Homo sapiens* will not become extinct. But if resource limitation is ultimately the mechanism that brings a population into equilibrium, the quality of life and human dignity for survivors could hardly be lower. Unless we assert our willingness to act beforehand, this fate is inescapable.

Questions

REVIEW QUESTIONS

1. The Commission on Population Growth and the American Future has claimed that illegal immigration should be resisted vigorously because it is part of the population problem. How could you determine whether it is or not? What is its major impact likely to be?

2. In animal populations reproduction is not the right of every individual. Does this set an ecological precedent for vesting reproductive decisions in the state? Why or why not?

3. What are the prerequisites to make family planning equal population stabilization?

4. Are there "oysters and champagne" equivalents for contemporary American society? What might they be, and how could they be used to encourage smaller family size?

5. Rainwater (1965, p. 150) claimed that his sample showed: "One shouldn't have more children than one can support, but one should have as many children as one can afford." With the growth rate and desired family sizes both decreasing, do people want fewer children, or are they less able to afford them?

ADVANCED QUESTION

1. Interpret Tikopian society and population trends in terms of Wynne-Edwards's concept of group selection and social regulation. Pay particular attention to the economic aspects of the Tikopian system. Are there parallels in animal social systems, for example, the red grouse of Watson and Moss (1971; citation in Chapter 4)?

Further Readings

Note: The bibliographies for Chapter 4 and Chapter 5 should be consulted together. Many references cited in Chapter 4 are pertinent here, and vice versa. References below, however, concentrate on family planning and population stabilization. References cited in Chapter 4 are omitted here.

Andrewartha, H. G., and L. C. Birch. *The Distribution and Abundance of Animals.* University of Chicago, Chicago, 1954.

Berelson, B. (ed.). *Family Planning and Population Programs.* University of Chicago, Chicago, 1966.

Berelson, B. "Beyond Family Planning." *Science,* 163 (1969), 533-543.

Bernard Berelson is president of Planned Parenthood, and there is no better spokesman for the explanation and defense of family planning institutions in population control. Contrast his views with those of Davis (1967).

Bogue, D. J. (ed.). "Progress and Problems of Fertility Control around the World." *Demography* 5, No. 2 (1968), viii, 539-1002 (the entire issue).

A review of family planning programs over the world—one of many available. This one contains the papers of Deverell (1968), Jaramillo-Gomez (1968), and Ravenholt (1965), cited below, and a particularly useful bibliography is appended.

Callahan, D. "Ethics and Population Limitation." *Science,* 175 (1972), 487-494.

This paper attempts to develop a mechanism for implementing population limitation in a holistic fashion that would minimize ethical and moral infractions on individuals. It sheds light on some of the issues Berelson (1969) raises.

Crowe, B. L. "The Tragedy of the Commons Revisited." *Science,* 166 (1969), 1103-1107.

See comment under Hardin (1968).

Davis, K. "Population Policy: Will Current Programs Succeed?" *Science,* 158 (1967), 730-739.

An explicit denial that family planning is population control, or will ever be. See also Judith Blake Davis (1971, cited in Chapter 4) for a similar view using slightly different arguments.

Deverell, C. "The International Planned Parenthood Federation: Its Role in Developing Countries." *Demography*, 5 (1968), 574-577.

    The strategy and tactics of IPPF in various areas.

Firth, R. *We, the Tikopia*, 2nd edition. G. Allen, London, 1957.

    The results of the first study, when the social institutions of the original society could still be observed.

Firth, R. *Social Change in Tikopia*. G. Allen, London, 1959.

    The second study, detailing the drastic changes in a society exposed to Western culture in a single generation.

Fogarty, M. *Sex, Career, and Family*. G. Allen, London, 1971.

    The most recent study on the changing role of women and its effect on human fertility.

Freedman, R., and J. Y. Takeshita. *Family Planning in Taiwan*. Princeton, Princeton, N.J., 1969.

    The complete details of the Taichung experiment; the complete Taiwan experience is reported in Potter et al. (1968).

Frejka, T. "Reflections on the Demographic Conditions Needed to Establish a U.S. Stationary Population Growth." *Population Studies*, 22 (1968), 379-397.

    One of many papers concerned with ZPG, this one is mathematically precise, and the arguments are developed systematically and neatly.

Frejka, T. *The Future of Population Growth*. Wiley, New York, 1973.

    An update and extension of the 1968 paper.

Geertz, C. B. *Agricultural Involution*. Association of Asian Studies, Monograph 11. University of California, Berkeley, 1963.

    This chapter relies heavily on anthropological case studies. This is the one that details the process of change in Indonesia following Dutch entrance.

Harcourt, D. G., and E. J. Leroux. "Population Regulation in Insects and Man." *American Scientist*, 55 (1967), 400-415.

    The extended views of two entomologists on the relevance of insects to human population stabilization. Stimulating and thought-provoking.

Hardin, G. "The Tragedy of the Commons." *Science*, 162 (1968), 1243-1248.

    Reprinted widely in environmental readers, this paper bemoans the fact that the consequences of free reproduction are common property and thus accountable to no one in particular. Hardin calls for an "extended morality" to cover the problem of common property; Crowe (above) denies that such will happen.

Harrison, G. A., and A. J. Boyce (eds.). *The Structure of Human Population*. Clarendon Press, Oxford, 1972.

    An excellent collection of papers focusing particularly well on the role of ecology.

Jaramillo-Gomez, M. "Medellin: a Case of Strong Resistance to Birth Control." *Demography*, 5 (1968), 811-826.

    The history of family planning in Medellin, Colombia: political resistance, the role of the church, and subversion of the movement.

Lee, R. B. "!Kung Bushmen Subsistence: An Input-Output Analysis." In *Ecological Essays: Proceedings of the Conference on Cultural Ecology* (D. Damas, ed.). National Museum of Canada, Bulletin No. 230, 1966. (Reprinted in Vayda, 1969.)

Lee, R. B. and I. DeVore (eds.). *Man the Hunter.* Aldine, Chicago, 1968.

Malthus, T. R. *An Essay on the Principle of Population as It Affects the Future Improvement of Society.* Printed for J. Johnson, in St. Paul's Church-yard, 1798. Reprint by A. M. Kelley, New York, 1965.

> The classic, the center of former controversy, and of great historical interest.

Nietschmann, B. "When the Turtle Collapses, the World Ends." *Natural History,* 83, No. 6 (1974), 34-43.

> A film by Brian Weiss and James Ward called *The Turtle People* (B&C Films, Los Angeles) covers the same subject.

Polgar, S., "Population Theory, Anthropology, and Family Planning." *Current Anthropology* 13 (1972), 203-211.

Population Reference Bureau. "Pockets of High Fertility in the United States." *Population Bulletin,* 24, No. 2, (1968), 25-55.

> The story of the Hutterites and the Amish.

Potter, R. G., R. Freedman, and L. P. Chow. "Taiwan's Family Planning Program." *Science,* 160 (1968), 848-853.

Rainwater, L. *Family Design.* Aldine, Chicago, 1965.

Ravenholt, R. T. "The A.I.D. Population and Family Planning Program: Goals, Scope, and Progress." *Demography,* 5 (1965), 561-573.

Singer, S. F. (ed.). *Is There an Optimum Level of Population?* McGraw-Hill, New York, 1971.

Taylor, L. R. (ed.). *The Optimum Population for Britain.* Academic, New York, 1970.

> These two volumes represent the latest on the question of optimum population size.

Vayda, A. P. (ed.). *Environment and Cultural Behavior.* American Museum Sourcebooks in Anthropology. Doubleday, Natural History Press, Garden City, N.Y., 1968.

> The Vayda and Lee-DeVore (1968) volumes constitute the basic, widely available, library in ecological anthropology. Many of the papers show this interdisciplinary nature very clearly. In the Vayda volume, see particularly the papers by Lee, Stott, and Geertz.

Westoff, L. A., and C. F. Westoff. *From Now to Zero.* Little, Brown, Boston, 1968.

> A neat summary of demography and family planning in the United States. Concise and reasonably balanced but not highly rich in information.

Wiens, J. A. "On Group-selection and Wynne-Edwards's Hypothesis." *American Scientist.* 54 (1966), 273-287.

Wynne-Edwards, V. C. *Animal Dispersion in Relation to Social Behavior.* Hafner, New York, 1962.

Wynne-Edwards, V. C. "Intergroup Selection in the Evolution of Social Systems." *Nature,* 200 (1963), 623-626.

Wynne-Edwards, V. C. "Self-regulating Systems in Populations of Animals." *Science,* 147 (1965), 1543-1548.

> The idea of self-regulation is not new, but the suggestion of voluntary birth control was at the time. Readers are urged to read all three of these papers if they wish to do some original thinking on social regulation in human populations. In any case, Wynne-Edwards has started a healthy controversy. Probably the most thoughtful reply is that of Wiens, cited above.

Section C
Matching
Present Resources
and Future Needs

SECTION C / SELF-STUDY ORGANIZER

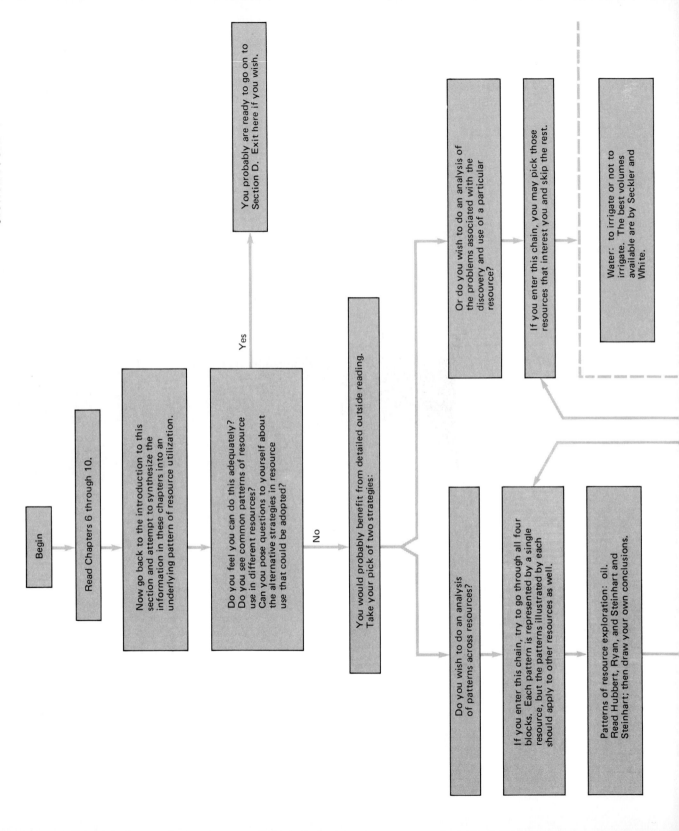

Food: the potential of the land and of the sea. A study in contrasts between the optimists and pessimists is particularly valuable. For the land, contrast Brown to Paddock and Paddock. For the sea, contrast Ryther and Idyll (1973).

Land: The control of its use. The best documentation is for the Brandywine basin and for the special case of coastal marshlands. For the history of the former, see Thompson. For the latter, see Wenk, Chapter 4, and references therein.

Minerals: coal resources in the United States are not in short supply, but there are problems associated with high-sulfur coals and with getting the coal out of the ground. Read Stacks and Hubbert.

Renewable resources: timber and the clear-cutting controversy. Read Wood and Ford-Robertson.

After finishing your sampling program, you are urged to proceed through the other line of inquiry. If you do not wish to do that, go on.

You should now be confident that you understand something about the flow of resources in human society. Go on to Section D.

Patterns of exploitation: fisheries. Read Paulik and Idyll on the Peruvian anchovetta, and follow the recent proceedings of the International Whaling Commission.

Economic factors in resource use. There is a very large literature here. Almost any of the publications by Resources for the Future will serve as an introduction. For an analysis of the changing nature of conservation, contrast Macinko on one hand and Krutilla on the other. For a classic work on the resource economist's viewpoint, see Landsberg.

The impact of technology and resource substitutability: nuclear power generation. The extent to which an expert believes in the growth of technology and the ability to substitute an abundant resource for a scarcer one has created some interesting controversies, especially in nuclear power generation. See Weinberg and contrast to Alfven.

It would be useful to tackle a specific kind of resource after completing this set of readings. If you are inclined to do so, go back. Otherwise continue.

The term "resources," like "environment," is not very precise. *Resources* are simply the material requirements of an organism; the definition is all-inclusive and implies no judgment about the need itself. Minimal requirements of all organisms include space for living and breeding and some form of usable food energy. Certain plants require specific insects in order to reproduce, and certain animals must have particular kinds of other animals or plants for food. The study of the evolution of biological interrelationships is fascinating in itself, but in the context of resource requirements it serves to emphasize how evolution and natural selection can favor the development of a more diversified strategy of resource utilization. We are the ultimate example, of course, because not only do we have the same biological limitations as other animals, but we have a complex society interwoven with the growth of technology and the production of goods. These generate additional new resource demands, for both material and energy; satisfaction of these demands creates additional demands—a classic example of an economic growth pattern that is self-generating. The result is dependence upon a diversity and quantity of resources unknown for other organisms.

In 1798 Thomas Malthus perceived that population growth could inherently exceed the human capacity to produce enough food. Although this observation is indisputably true, its ultimate truth is obscured by human success in raising food production enough to stay abreast of population growth. Euphoria and a denial of Malthus are not warranted, however, as long as the world population continues to grow. Even if we are successful in meeting resource needs over the short run, we still must bear the weight of the dependence we have deliberately developed upon a steady and uninterrupted flow of critical energy and material sources. Economic and population growth are easy to encourage and sustain as long as the euphoria of limitless resources is not denied. Perhaps only a cynic or a pessimist would view our burgeoning growth as all the more likely to collapse when resources fail, but it remains an unassailable truth that each increment of growth creates additional structure that must be maintained by a greater reliance upon resources.

Like a drug addict's struggle to feed a growing habit, our complex society's vulnerability to resource shortages creates strong pressures to ensure future supplies. Even if economic collapse and social disruption did not result from resource shortage, the size of the population, the strength of the economy, and above all the well-developed interdependencies within the economy would ensure that the consequences would be felt throughout the country. A single experience like this would grant the search for resources highest priority. The question is whether so much is at stake, so much structure is dependent upon resources, that any sacrifice becomes worthwhile. This possibility is important for the improvement of environmental quality, but it is not restricted to it.

In any system with extensive positive feedback (in this case continuing population and economic growth), the way to escape escalating dependency on resources is to stop growth by instituting mechanisms that lead to ZPG (Chapter 5) and a stationary state economy (Chapter 19). However, this obvious and dramatic solution may be neither easy to attain nor free of its own problems; feasible alternatives may be considerably more mundane and less satisfactory. The observation that growth must at some point cease in order to relieve the need for more resources does not tell when this will occur. Effectively, then, growth and increasing demand for resources are likely to continue until it becomes clear that this is no longer possible. At that time, reduction of demand becomes inescapable.

The future behavior of our population, economy, and resource system is a matter of conjecture. A Malthusian crisis is the ultimate consequence if we ignore resource limitations. At the other extreme, the system could slowly and systematically approach steady state with minimum perturbations. Of course, it is to our interest to follow the second model as much as possible (unless you are a technological optimist who believes in no limits at all). In this section, we shall examine two aspects of resources that have implications for the future: the recent history of resource utilization, which should tell us something about patterns that could continue into the future; and an assessment of the present status of the major categories of resources and of their future adequacy. We shall consider food, water, land, mineral resources, and biological populations.

Selected General Sources for Resource Sufficiency

Brown, L. R. *Seeds of Change*. Praeger, New York, 1970.
Brown, L. R. *In the Human Interest*. Norton, New York, 1974.
Ford-Robertson, F. C. (ed.). *Terminology of Forest Science, Technology, Practices and Products*. Society of American Foresters, Washington, 1971.
Hubbert, M. K. "Energy Resources." In *Resources and Man* (P. E. Cloud, ed.). Freeman, San Francisco, 1969.
Idyll, C. P. *The Sea against Hunger*. Crowell, New York, 1970.
Idyll, C. P. "The Anchovy Crisis." *Scientific American*, 228, No. 6 (1973) 24-29.
Krutilla, J. V. "Conservation Reconsidered."*American Economic Review*, 57 (1967), 777-786.
Landsberg, H. H. *Natural Resources for U.S. Growth*. Johns Hopkins, Baltimore, 1964.
Macinko, G. "Conservation Trends and the Future American Environment." *The Biologist*, 50 (1968), 1-19.
Paddock, W., and P. Paddock. *Famine—1975!* Little, Brown, Boston, 1967.
Paulik, G. J. "Anchovies, Birds, and Fishermen in the Peru Current." In *Environment* (W. W. Murdoch, ed.). Sinauer Associates, Stamford, Conn., 1971.
Ryan, J. M. "Limitations of Statistical Methods for Predicting Petroleum and Natural Gas Reserves and Availability." *Journal of Petroleum Technology*, 18 (1966), 281-287.
Ryther, J. H. "Photosynthesis and Fish Production in the Sea." *Science*, 166 (1969), 72-76.
Seckler, D. (ed.). *California Water*. University of California, Berkeley, 1971.
Stacks, J. F. *Stripping*. Sierra Club, San Francisco, 1972.
Steinhart, J. S., and C. Steinhart. *Blowout!* Wadsworth, Belmont, Calif., 1972.
Thompson, P. "Brandywine Basin: Defeat of an Almost Perfect Plan." *Science*, 163 (1969), 1180-1181.
Weinberg, A. M. "Social Institutions and Nuclear Energy." *Science*, 177 (1972), 27-34. See also the exchange of letters, *Science*, 178 (1972), 933; and Alfven, *Bulletin of the Atomic Scientists*, 28, No. 5 (1972), 5.
Wenk, E. *The Politics of the Ocean*. University of Washington, Seattle, 1972.
White, G. F. *Strategies of American Water Management*. Ann Arbor Paperbacks, Ann Arbor, Mich., 1971.
Wood, N. C. *Clearcut*. Sierra Club, San Francisco, 1971.

Selected Periodicals

"Natural resources" is a poorly defined subject having few journals of its own. Here are a few examples:

Fisheries Bulletin, U.N. Food and Agriculture Organization
Land Economics
Natural Resources Journal
Oil and Gas Journal
U.S. Geological Survey Bulletin
Water Resources Research

Be sure to consult the state, federal and U.N. lists of publications, because various agencies sponsor many committees, reports, analyses, and forecasts of resources under their jurisdiction. Resources for the Future is one active foundation that publishes largely through its own books. One of its most recent publications is by Freeman, Haveman, and Kneese: *The Economics of Environmental Policy*. Wiley, New York, 1973.

6. Difficult Choices: Strategies for Food Production

Basic biological requirements make an adequate diet essential to the support of any population. In the wealthy and well-fed United States, where only 3 percent of the population currently supports the remainder, this fact may be remote and hard to visualize. The powerful, highly technological, and energy-supplemented style of agriculture in the United States blurs Malthus's maxim—at times, increase in food production may seem to be as fully exponential as population growth. Nevertheless, the spectacular successes of agricultural technology do not invalidate the basic truth of Malthus's proposition, because while population growth is inherently exponential, technological progress is not. The problem of keeping up with population growth through land-based agricultural production is complicated by competition for land, because a growing population may demand the best agricultural land for urban uses.

The more developed countries generally are well fed, but statistics show this to be otherwise for many less developed ones. The problem is not so much caloric as it is a matter of adequate nutritional quality, especially in the protein needed for full physical and mental development. In 1963, the United Nations

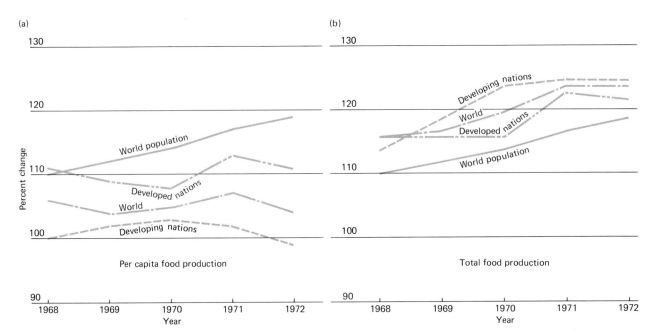

FIGURE 6-1 *The population-food race, as traced through FAOUN data: (a) per capita food production, and (b) total food production. In these figures the numbers are index numbers that represent weighted percent changes from a baseline figure of 100, the mean of the 1961–1965 period. The per capita food production indices for 1968–1971 show that we held our own, but 1972 was a bad year, especially in developing countries. FAOUN region definitions are as follows: Developed areas: Western Europe, North America, Oceania, Japan, Israel, and South Africa. Developing areas: Latin America, Near East (except Israel), Africa (except South Africa), and the Far East (excepting Japan, China, and the socialist states). (Data from Food and Agriculture Organization of the United Nations, 1972.)*

estimated that 20 percent of the residents of developing countries were undernourished (received insufficient calories), but 40 percent more were malnourished (insufficient quality). Food shortage is a real problem for contemporary human society; it is one of the great gaps that separates rich and poor nations.

In 1973 and 1974, for the first time global food shortages threatened human welfare. At the end of 1973, A. H. Boerma, director of the U.N. Food and Agriculture Organization (FAO), reported that world cereal stocks were depleted, leaving nations vulnerable to local crop failures. And 1974 was marked by a gloomy picture in which the Green Revolution (discussed later in the chapter) no longer offered such bright hope, the great drought of the Sahel (sub-Sahara) continued to lead an entire region of Africa toward disaster, and even the United States faced drastically reduced yields of corn and soybeans.

Historically the United States has been farmer for the world. In past decades, we gave food to nations in need, particularly to India and Pakistan, and we paid farmers not to produce as much food as they could. Thus it is easy to believe that we need only to grow more food to avoid the global Malthusian dilemma. But can we do it? At this time it seems impossible for a number of reasons. Population growth threatens to outstrip increases in food production (Figure 6-1); our surpluses are depleted, and we can sell our stocks only by bearing higher prices as in the Russian wheat sales; we cannot control the vagaries of weather; and for a variety of reasons—economic, technological, and ecological—other nations are unable to raise production enough to meet their needs.

This chapter concentrates on a single resource: food. The next few decades are likely to bring chronic food shortages and complications caused by their effects, and it is time to develop a global strategy to meet food requirements by some means that avoids drastically reducing the world population through starvation. If we do not act, a reduction eventually will be imposed upon us whether we like it or not. Here we shall examine the difficult, sometimes impossible choices that we must make under the pressure of a burgeoning population. We start by assessing the current distribution of food and food production capacities in relation to population around the world.

Population and Food across the Planet

The most extensive technological support and the most advanced, mechanically intensive agriculture are practiced in the developed countries that can afford them; these countries are found exclusively in temperate climates. Conversely, the great concentrations of people are found in tropical or subtropical areas, where arable land is more limited by soil quality and climate, where economic development is slight, and where technology has not penetrated. In many densely populated countries in Asia, Africa, South America, and Europe, it is either impossible or not economically feasible to become calorically self-sustaining. This situation is tolerable only if enough areas continue to produce food for the world and if food-deficient countries have the capital to purchase food from producers.

The countries for which such a highly integrated international scheme would be intolerable are the heavily populated underdeveloped countries with slow economic growth at best, for example, India, Pakistan, Bangladesh, Indonesia, and Egypt. These countries have large populations to feed but no capital with which to feed them because each increment of additional population makes it more difficult to attain economic growth. Some of these countries, sadly, may adjust population through war, emigration, starvation, or some other means of death control. Paddock and Paddock (1967) have forecast this future for several developing nations. They believe that the situation is so hopeless that there is no use wasting resources on countries they consider to be overpopulated beyond hope. They proposed that aid be concentrated solely in countries where eventual self-sufficiency is possible, a "triage" system named for the military method of treating only the wounded who are thought to have a chance of recovery and ignoring those who will die anyway. This solution is cold-blooded but may well turn out to be ecologically realistic. It raises an interesting question: Do our normal ethical, moral, and human standards apply if some nations are in fact hopelessly overpopulated? Do we save individuals now only to postpone the crisis?

There seems little question that economically advanced countries and many developing countries (Taiwan, Korea, Venezuela) will be able to secure adequate food to prevent widespread undernourishment. What then for countries that are widely undernourished, where hope is dim? If we examine the land pool per se, we see that about one-quarter of the 34 billion acres of the continents is suitable for cultivation, with another quarter being limited to grazing, and most of the remainder being useless for agriculture except for a small part under clearable forests. About half of the arable land is now being cultivated, about one acre per person on average. Table 6-1 shows the distribution of arable lands and their utilization for the continents. The greatest remaining acreages are in Australia, the U.S.S.R., and North America, not areas of great present need. Of the generally less developed continents, Latin America has the largest pool of arable acreage, and Asia the least. With over half the arable lands remaining to be farmed, the gross view supports the contention that land is not the critical factor limiting food production today.

Even if the agricultural land supply is not a critical limiting factor, other influences condition this conclusion. The basic factor is the distribution of land; Table 6-1 shows that 73 percent of such lands are found in the more developed countries. Hence acreage could well be limiting in certain countries, and current evidence indicates this is true. Such inequities must be compensated for by food export and import, but it is questionable that the developed countries want to be the breadbasket for the world. In India, most agricultural land is held in small parcels (the average being less than 1 acre per family), from which a family is expected to produce enough food. Since this is unlikely under any circumstances, either supplements from the outside or a highly cooperative distribution system within the political unit will be needed. For Japan, this means sharing a small tractor among several families; for India, this means sharing between regions—or, more frequently, sharing wheat from the United States or elsewhere.

Also influencing the quantity of arable land to be put into production are ecological limitations upon the intensive agricultural use of this land. The term "arable" refers to lands that have soil and climate conditions which can support crops. However, many arable acres require a dependable source of inexpensive irrigation to support intensive agriculture, or they lack long-term yield potential; some have both limitations. To solve the first problem large capital invest-

TABLE 6-1 *The distribution of arable land and its use across the world. All land units in 10^9 acres.*

Continent	Total land area	Presently arable land	Percent arable land cultivated	Remaining potential arable land
Europe	1.24	0.43	88	0.41
USSR	5.63	0.89	64	1.53
Asia	6.97	1.34	83	0.74
Africa	7.56	2.63	22	0.81
North America	4.93	1.11	51	1.49
Latin America	5.19	2.49	11	1.40
Oceania	2.13	0.46	2	4.68
Totals	33.65	9.35		11.07

SOURCE: Calculated from Borgstrom (1969) and Hendricks (in Cloud, 1969).

ments are needed for irrigation, and subsidies for agriculture may be required. In this case, the economic costs involved may make it unfeasible to put the land in production. Regarding yield potential, particularly in the tropical forest soils of South America, Asia, and Africa, there are several limitations in the soil's capacity to hold minerals and retain a structure suitable for agriculture. Thus, much potentially arable acreage is not really arable without improved economic and technological support, and even then the land may not be able to continue producing crops.

Thus where immediately arable land is most plentiful, population pressures are minimal. This is to be expected, since population and agriculture compete directly for land when a growing population demands more food. In some small countries (notably Japan) the explosion of urban growth has decreased the supply of arable land. Before World War II, Latin America, Asia, and Africa actually managed to export grains; afterward, they became net importers. The distribution of untilled arable lands away from areas of greatest need increases the demand for more intensive production in areas presently under cultivation, with dietary supplements from the seas and other nations. Attempts to compensate for an arable land pool that is growing less adequate with each population increment focuses attention upon other factors of production: technological progress for agricultural applications, economic impacts on crop management and decision-making, and ecological impacts of farming.

The Population-Food System

Before we examine the strategies available, let us look at the population-food system from which such choices must be derived. If we exclude other resources and all matters of environmental quality from consideration, we have the model shown in Figure 6-2.

This model exposes the sensitive points which would be expected to affect food supply per capita (a measure of food resource adequacy). These are the competition for arable land between the human population and agriculture, the impacts of food production on biological energy production and vice versa, possible future supplies of synthetic foods, and the pressures generated by net population growth. Each point constitutes an area of decision—choices that determine our food production strategy. Each is characterized by the manifold interactions and unexpected consequences that are properties of complex systems. For example, suppose someone suggests that total arable land could be increased by massive desalinization of sea water for irrigation. These modifications could start an escalating chain of events: more area farmed, greater food production, larger food supply. But other forces may oppose the execution of this chain at any step. It is likely to be costly to supply the irrigation water, and there are no guarantees that even with irrigation the land will remain fertile for long (both excessive build-up of salts and degradation of soil structure are known to occur). In addition, increased crop production may be canceled by increased interference from "pest" species, crops grown in that area may be more sensitive to variable climate, or increased food supplies may stimulate a new wave of population growth and increased competition for that newly opened arable land. The large number of opposing forces makes the outcome uncertain; increasing arable land may or may not increase food supply.

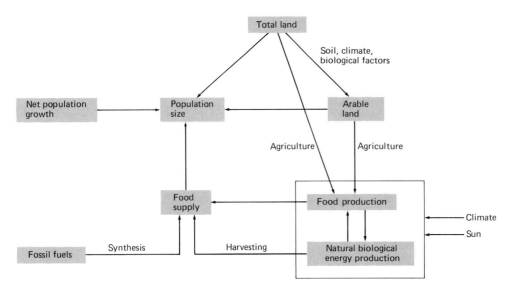

FIGURE 6-2 *Population-food system model, excluding other resources and all matters of environmental quality.*

Suppose we were able to reach an instantaneous end to population growth. Then the pressure for increased food production would be removed, and competition for land would be eased. In a bold stroke the Malthusian dilemma is avoided, and the food resource problem evaporates. However, the practical difficulties in reaching population stability usually dictate some compromise between a slowly declining population growth rate and the fastest possible increase in food supply. Against this background, we may now examine alternative strategies of food production.

Current Strategies of Food Production

Two main strategies of food production are currently in use. One is to exploit natural food resources; the other is high-intensity agriculture.

STRATEGY 1: EXPLOITATION AND MANAGEMENT OF NATURAL FOOD RESOURCES

One of the cheapest ways to increase the food supply is to harvest from natural ecosystems because the potential gain may be obtained at minimum cost compared with intensive agriculture. Probably the best examples of harvesting come from international marine fisheries, which historically have involved little more than pure exploitation of the resource (Figure 6-3). In fact, these fisheries perfectly illustrate the problem of common property resources: Since they are owned by no one in particular, they usually are overused by everyone because no one has a vested interest in protecting them. For innumerable cases in marine fisheries—baleen whales, fur seals, North Sea herring fisheries, Peruvian anchovetas, and California sardines—population crashes and ruin of the fishery have occurred because there was no economic

FIGURE 6-3 *Soviet flagship towing sperm whales, soon to be hauled aboard and processed. This is a prime, and increasingly controversial, example of resource exploitation.* (Courtesy of United Press International.)

reason, short of altruism, to protect the populations from overexploitation. Figure 6-4 illustrates some typical cases. The world's baleen whale populations show the effect of systematically overfishing one species only to switch to the next-smallest whale species when the larger one has been depleted. It is an unrestrained example of exploitation for a rapidly dwindling resource, and it continues today despite an international commission to control whaling.

On the other hand, when national and limited international interests have taken long-term considerations into account, exploitation of the fisheries have incorporated more elements of real management. In an attempt to protect marine resources upon which it is dependent, Ecuador has established a 200-mile coastal waters limit that effectively removes its tuna fishery from international exploitation. Attempts by American tuna fishermen to abrogate this arbitrary limit have not succeeded, and at a recent conference on marine resources in Venezuela there were signs that other countries also plan to extend territorial limits so that coastal fisheries may become national property worldwide. The fur seals of the Pribilof Islands have been protected and a sustained yield assured by a bilateral pact between the United States and the USSR, which have nearly exclusive interest in this fishery.

Nevertheless, such examples of enlightened, scientific management to promote sustained yield indefinitely are scarce for marine and natural terrestrial ecosystems, especially when many national interests are involved. Each year, as pressure upon marine resources increases, more and more fisheries are overexploited. Not only is there little hope for feeding more

124 DIFFICULT CHOICES: STRATEGIES FOR FOOD PRODUCTION

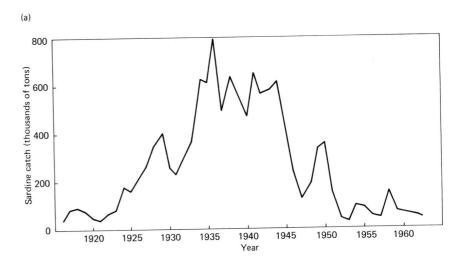

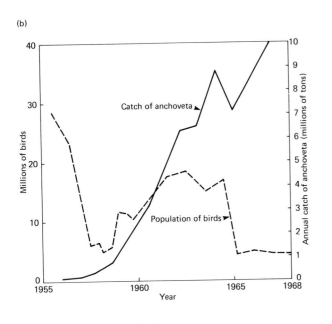

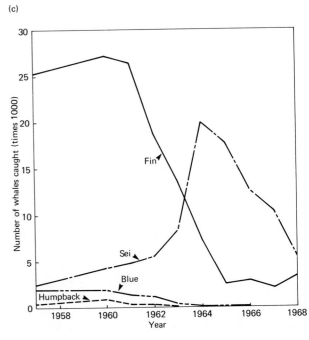

FIGURE 6-4 *Three examples of overexploited or potentially overexploited fisheries. (a) Sardine catches off the Pacific Coast, 1916-1962. This illustrates the collapse of a fishery. Consequently, anchovy populations have grown to replace those of the sardine. (b) A fishery nearing trouble: anchovetas off Peru. The competition between human beings and guano birds for the anchovetas is probably responsible for the decline in the bird populations. (c) The recent history of Antarctic baleen whale stocks, 1956-1968.* ((a) Redrawn from Murphy, 1966, *Proceedings of the California Academy of Sciences:* 34(1), p. 2. (b) Redrawn from G. J. Paulik, Fig. 6 in *Environment*, W. W. Murdoch (ed.), Sinauer Associates, Inc., 1971. (c) Redrawn with permission of the Trustees of the Australian Museum from "Man and a vulnerable earth: the need for ecological sense," T. C. Foin, *Australian Natural History*, 1970.)

people this way, there is a distinct danger that ultimately the number that can be fed will be drastically reduced.

How much yield can the harvest of marine resources yield? A number of estimates have been made, including one ecologically based review provided by Ryther (1969), who estimates that the world's oceans ultimately could yield 110 million tons per year. This is relatively conservative compared with those who believe that the expanse of the world's oceans (70 percent of the surface of the earth) and the relatively high protein content of the yield justify more optimistic projections and even the hope that the seas may solve the food problem. For example, Idyll (1970) believes that a minimum of 200 to 250 million tons of useful food eventually will be taken from the seas but that with better management and improved technology 400 million tons is attainable and 1 billion is an outside possibility, compared with the present catch of 66 to 72 million tons annually.

Ryther arrived at his estimate by applying the concepts of thermodynamics and trophic pyramids (Chapters 1, 2). He broke the seas into three major regions differing greatly in their rates of energy fixation by plants (primary productivity). These areas are (1) the deep open oceans; (2) the coastal areas, including shallow banks and estuaries; and (3) the limited areas of upwelling where nutrient-rich deep waters rise to the surface and are combined with abundant sunlight for particularly enhanced productivity. Using estimates of the extent of each of these areas, Ryther then calculated productivity values based on the literature, as shown in Table 6-2. This gave him the energy contained in the first trophic level for all the oceans; to derive the rest, he had to assume that the world marine system is characterized by certain efficiencies of energy transfer. With these values, it was possible to calculate the production of fish biomass supported by annual primary productivity. Table 6-2 shows that nearly all fish production occurs in the coastal and upwelling regions, about half in each, emphasizing the importance of the latter despite their limited occurrence. The figure of 240 million metric tons arrived at by this method is not, however, the final estimate. Clearly, if all this biomass were harvested, the fisheries would collapse—they would be eradicated. And the interrelationships of complex marine ecosystems mean that some part of the production will be absorbed within the system. Subtracting these considerations, Ryther arrives at a harvest estimate of 100 million metric tons or 110 million tons.

Idyll claims that the increased production he feels can be obtained from the seas should greatly aid in feeding the world's population, or at least in helping

TABLE 6-2 *Estimated fish production from the seas based on trophic dynamics. Note that despite the large extent of the open oceans, total fish production is two orders of magnitude below the other areas.*

Province	Percent of oceans	Mean primary[a] productivity (gms C/m²/yr)	Total primary productivity (10⁹ tons organic C)	Number of trophic levels	Efficiency of energy transfer	Fish production (10⁵ tons)
Open ocean	90	50	16.3	5	10	16
Coastal zone[b]	9.9	100	3.6	3	15	12×10^2
Upwelling areas	0.1	300	0.1	1.5	20	12×10^2
Total						24×10^7

Source: J. R. Ryther, "Photosynthesis and Fish Production in the Sea," *Science,* 166 (1969), 72–76. Copyright 1969 by the American Association for the Advancement of Science.

[a] Primary productivity = fixation of organic carbon = photosynthetic activity.
[b] Includes shallow offshore areas (such as the Grand Banks).

end chronic protein deficiencies, but he explicitly denies that the sea can feed an increased population by itself. Ryther's estimate, less than double the world's current production and only half of Idyll's most conservative projection, does not support even these modest expectations. If the population increases according to the U.N. medium projection, this means about 0.9 ounce of marine catch per person per year if the yield of 100 million metric tons is achieved even as early as 1980. Unless we are willing to continue gross overexploitation, or to forgo fish in favor of direct use of algae, or to convert all shallow oceans into vast farms, it is unlikely that the seas will yield much more food. Unless we adopt sound management practices to insure continuing harvests, we may actually end up with less yield than we have now.

STRATEGY 2: HIGH-INTENSITY AGRICULTURE

Aside from soil erosion, salinization from irrigation, and similar problems, the strategy of terrestrial agriculture is less exploitative than strategy 1 and more directed to the intensification of yields from any given acreage. Hence we have a whole complex of supporting technologies, best exemplified by Western agricultural practice, the major intent of which is to increase yields. We employ genetics and plant breeding to attain more productive strains; a whole battery of deadly pesticides to reduce and, if possible, eliminate competition from other organisms for the yield we desire; irrigation and fertilization; highly specialized machinery; and economic incentives such as property taxes and various agricultural subsidies to increase the utility of high yields.

Perhaps the most recent significant development has come from joint efforts in plant breeding by the Rockefeller and Ford foundations. Their successes in developing new strains of wheat and rice for use in underdeveloped areas have been called the *Green Revolution.* In the late 1960s, these successes were spectacular, and the new grains spread rapidly in countries where they had been planted. Much has been written about the Green Revolution, and one of its pioneers, Dr. Norman E. Borlaug, has even received the Nobel Prize for his accomplishments. Basically, the Green Revolution has substituted new varieties of wheat and especially rice (dwarfed and very strong stemmed) for traditional strains. Provided that sufficient fertilizer and water are available, these varieties can double yields quickly because their shorter, stiffer stalks can bear the heavier yield that fertilizers stimulate; traditional strains actually broke and reduced yields when provided with these inputs. For example, the Philippines, home of the International Rice Research Institute and site of development of the phenomenal new rice strains known as IR-8 and IR-22 (Figure 6-5), was able to export rice for the first time after 50 years of foreign reserve-draining imports. The per capita cereal production of Mexico, where the wheat research center is located, rose 37 percent (see Table 6-3). Total wheat yields were more than doubled. The increase even stimulated greater meat production because for the first time surplus grains could be spared for stock herds. The most convincing success of the Green Revolution is seen in the acceptance of the new grains by local farmers. In Pakistan, the acreage planted to IR-8 rose an order of magnitude to 1 million acres in a single year (1967–1968). Throughout Asia, 34 million acres were planted to the new strains by 1969—10 percent of all available cropland. At least initially, tremendous increases in yield were so highly visible that the new grains sold themselves.

FIGURE 6-5 *The rice plants on the left (H-4) are the traditional variety of tropical rice plants. By crossing short varieties from Taiwan with disease resistant tropical varieties, the International Rice Research Institute was able to develop a dwarf plant (IR-22) which can stand up with two and three times the grains of traditional rice.* (Courtesy of United Press International.)

The Green Revolution resulted in such dramatic increases in yield that food production temporarily outstripped the rate of population growth in many developing nations. The increased yields inspired some wildly optimistic predictions that the food problem had been solved. But in the last few years, the potential problems of the Green Revolution have emerged with a vengeance. The Green Revolution absolutely demands Western technology: The yield can be obtained only with copious applications of fertilizer and water, and then it must be harvested

Year	Country/crop (Per capita production, pounds)				Total production (Thousand tons)		
	India/wheat	Pakistan/wheat	Ceylon/rice	Mexico/all cereals	Mexico/wheat	Philippines/rice	India/rice
1945					347		
1950					587		
1955					850	3273	41336
1960	53	87	201	495	1190	3705	51861
1962	59	87	213	525	1432	3967	47871
1964	46	83	213	611	1527	3993	58098
1966	46	71	188	649	1609	4094	45660
1968	76	116	247	680	1864	4445	59642
1969	80	121			1968	5233	60645

TABLE 6-3 *Yield increases due to the Green Revolution, measured in per capita production (Brown and Finsterbusch, 1971), and for total production (Athwal, 1971). Recall that the effect of the Green Revolution has been recent (1968 and later), except in Mexico, where the program began by 1962.*

quickly before it is destroyed by floods or other natural catastrophes. Furthermore, it must be protected from pests, and this requires liberal applications of various pesticides. The market and distribution systems must be able to absorb the surge in productivity, and so must the social institutions built upon the agricultural sector. As Ladejinsky (1970) has shown for India, this is not a simple matter, and the benefits of the new grains are neither easily nor automatically spread throughout society. One major problem has been that the benefits of increased production are absorbed most readily by large landowners who have the capital necessary to finance the technological support. As a result, the poor get relatively poorer and have still less security, as the new grains favor large land holdings and the hiring of cheap temporary labor rather than the maintenance of sharecropping tenants. Thus the Green Revolution has generated pressure to drive tenants from the land into already overcrowded cities, and social injustice is increased. Nevertheless, it is important to point out that these trends are derived from the social, political, and economic structure; the Green Revolution has only aggravated existing ills. For these particular problems, better mechanisms for assimilating the new productivity are needed.

Another problem with the success of the Green Revolution is that it turns attention away from the basic population problem. Since it is much easier ethically and morally to increase productivity than to interfere with free reproduction, the Green Revolution is an example of a spectacular technological advance that is favored by society even though its social effects are marked. With yields declining and gains less spectacular, even the most enthusiastic supporters are admitting once again that the Green Revolution merely buys time in which to solve the population problem; it is not a permanent solution. The central question is then whether or not "buying time" is damaging in the long run; when the world population is finally stabilized, will we be worse off because we have to work the land harder to support the greater number of people? Admittedly, this is a difficult question, but it is also being asked about water and freeway improvement and other large-scale projects that irrevocably shape society and the nature of the planet.

At present much attention is devoted to intensive production in aquatic systems, and spectacular productivity has been recorded in some highly controlled situations. The Japanese have cultured small eels and pearl oysters for many years in controlled ponds, and U.S. catfish aquaculture has been spreading rapidly in California and the South. In Mississippi catfish ponds may become an important part of agriculture, since catfish grow fast, have an expanding market, and can be cultured both profitably and easily.

Two other outstanding examples of freshwater aquaculture are the production of fish in rice paddies and canals in Southeast Asia and the cultivation potential of *Tilapia* (mouthbreeders—commonly found and sold as tropical fish) in small African ponds. In the former case the growing of carp and other fish feeding low in the trophic chain is a natural method of further intensifying production from rice paddies. Though not absolutely necessary, the fish do benefit from fertilizer application, and in the Philippines yields of about 1 ton per acre have been reported.

Progress to date has been spotty from country to country as different economic, political, dietary, and technological considerations have become important, but a number of governments are experimenting actively with cultivation of various species of fish in different environments.

Species of the genus *Tilapia* are widely cultivated in Africa (Figure 6-6), and their cultivation is being tried extensively in Asia and the Near East as well.

FIGURE 6 6 *The male Tilapia incubates the eggs in its mouth and retains the fry until they are ready to take care of themselves. Here the male has just spit out the fry to their new lives. Tilapia are rapid and careful breeders, making their cultivation a potential source of inexpensive protein.* (Courtesy of Wallace Kirkland, Time-Life Picture Agency © Time, Inc.)

These fishes are much smaller than Asian carps and American catfish, but turnover rates are much higher as a result. Common practice is to put a stock of young fish in small, very shallow, ponds, allow them to grow (usually without fertilization) for 8 to 10 months, and then drain the ponds and harvest the entire population. Yields of this species in Africa generally have been lower than Asian cultures of carp and their allies: In the Malagasy Republic, yields in the tropical lowlands have been up to 500 pounds per acre; Uganda and Zambia have attained yields up to 900 pounds per acre. Authorities believe, however, that yield could be increased through more extensive management; it is known, for instance, that without thinning, the small ponds tend to produce stunted fish much as unmanaged ponds in the United States do.

In contrast to freshwater aquaculture, mariculture (marine aquaculture) is still relatively primitive. Marine environments are less easily controlled than small freshwater ponds, and so the results of culture are less predictable. Because of its undeveloped state, mariculture may mean different things under different circumstances. For instance, it may involve trying to harvest a species at a sustainable rate or enhancing the growth rate of a desirable species. Then again, however, it might entail learning how to culture a species in an artificial marine system. Examples of mariculture include harvesting the giant kelp *Macrocystis* as a source of fuel, using waste heat from nuclear power plants to increase growth rates of fish and shellfish in coastal waters, and culturing lobsters, oysters, and abalones in salt water ponds. One variant of the nuclear heat proposal calls for addition of sewage as a source of nutrients to increase biological productivity even further.

While progress toward large-scale mariculture continues, the productivity of marine systems is threatened by activities having little or nothing to do with food production. The estuaries and shallow coastal waters that are so highly productive are threatened by development in the coastal zone and by the effects that accompany it. The marine creatures that grow so rapidly in this zone's combination of high nutrients and plentiful light are sensitive to the pollution, landfilling, and impoundment practices which affect the rivers leading into the ocean and the shallow coastal basins. This again is an example of a complex system: One part of society raises the funds and works hard to increase marine food production, while another sector destroys the natural productivity that occurs in the shallow coastal waters adjoining the shores.

Economic Influences in Agriculture

Economics dominates agricultural strategy. Here are some examples of how economics affects decision-making in agriculture.

COMMODITY PRICES AND PLANTING STRATEGY

Since agriculture is run by people, there must be adequate profit in the enterprise. Hence profit will dominate production strategy. The profit incentive will determine what crop is planted, how much of it is planted, and what management policy is followed. One example in California illustrates the importance of prices at the highest level of decision-making: At present the quantity of agricultural land in California is relatively stable. Dean et al. (1970) discussed the agricultural implications of the California State Water Project (Chapter 7), a project designed to transport water from the relatively wet northern part of the state to support agriculture in the more arid southern part. One prime objective of the plan is to supply inexpensive water for agriculture in the presently dry western half of the San Joaquin Valley. Dean and his colleagues were able to show that supplying cheap water to the western San Joaquin Valley would be counterproductive, in that the new acreages opened would only increase competition in certain specialty crops, thereby depressing their prices to low enough levels to threaten the existence of many smaller enterprises unable to survive intense competition. Hence the state water project would not be justified by its agricultural benefits and could even harm the industry.

Nationwide we are in an unstable period for agriculture, and this clearly has upset the production strategy of the nation's farmers, most dramatically, of the meat ranchers. When Peruvian anchoveta practically disappeared in 1973, their role as cattle feed was taken over by soybeans; but that drove soybean prices so high that beef became too expensive for many Americans. The resulting drop in beef demand caught producers between falling profits and large cattle stocks that had been generated during times of higher prices (Figure 6-7). This maladjustment between supply and demand continues, and since the producers are so poorly insulated from fluctuations in profits, they have demanded government support to buffer them from wild fluctuations that have introduced instability into planned production. Clearly these are economic matters relatively distant from any national or international problem of food shortage.

FIGURE 6-7 *Caught between low profits and high stocks, members of the National Farmers Organization killed calves with revolver shots and dumped their carcasses in trenches to dramatize their plight in an unstable market.* (Courtesy of United Press International.)

SUBSIDIES AND SHUFFLES

Although the same economic dependencies apply, many crops are not planted strictly in accord with supply and demand but are insulated from it by government intervention. For example, California rice finds little demand in this country, and most of it has to be exported. To prevent intense competition, low prices, and the bankruptcy of rice growers, direct subsidies for export are given, strict acreage allotments are made, farmers get large volumes of subsidized water, and they are allowed to burn their stubble in the fall—a form of air-quality subsidy. All these in combination act to keep the price of rice up and the farmers' production costs down; without them, most growers would be out of business.

One law that has had a substantial international impact is P.L. 480 (The Food for Peace Program). Faced with famine during years of crop failure, India did not pay the price of starvation for an overgrown population because P.L. 480 provided the food to survive the crisis. But while humanitarian on the surface, P.L. 480 only treated symptoms, and it aggravated larger social problems. Free wheat sustained the overgrown urban population and undercut the markets for tenant farmers, who because they were unable to sell their wheat were driven off the land or left it voluntarily to get free wheat themselves and to seek other work in cities, swelling an unskilled labor pool that was already too large. The program also freed the Indian government to redirect scarce economic resources away from agricultural reform into industrialization. Thus the social effects of the free wheat were a weakened agricultural sector and a nation even less capable of self-sustenance.

A third example shows how subsidies can lead to a shuffling of resources that can strengthen agriculture. The University of Arizona–University of Sonora laboratories at Tucson, Arizona, and Puerto Penasco, Mexico, have been working on self-contained desert agriculture, a fine example of the coupling of systems (Figure 6-8). Basically, the system takes advantage of the intense sunlight available in tropical deserts year round. Desirable crops such as lettuce, peppers, tomatoes, melons, and strawberries are grown in plastic greenhouses; diesel generators supply electrical power, and their waste heat is used to desalinate sea water. The water is used to make nutrient solutions that are recaptured and used again and again in the greenhouses. The first application of this method is a 5-acre facility located in Abu Dhabi[1] that is estimated to be producing 2 million pounds of produce per year in plastic greenhouses. This tremendous yield is not cheap, but the oil that runs the generators is so plentiful and generates the capital so freely that high production costs still are less than the cost of importing produce. This is a special case, however, and the option is not available to many poorer countries.

TECHNOLOGICAL SUPPORT: PROTECTING AN INVESTMENT

Still another economic influence is found in cases where an expensive investment in equipment must be made before returns can be obtained. These influences tend to reinforce a strategy decision once it has been made. In our previous examples, Sheikh Zaid of Abu Dhabi had to invest 1.5 million dollars for the initial physical plant of his produce-growing facility. Presumably he is not eager to dismantle it

---

[1] A small, oil-rich sheikdom of the Trucial States located on the Persian Gulf.

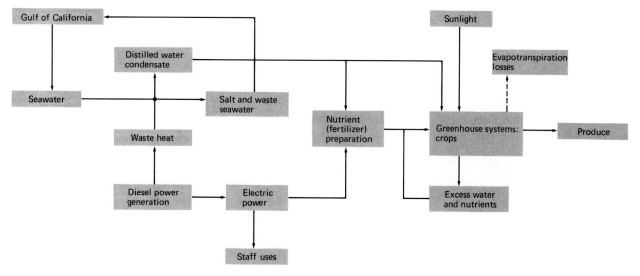

FIGURE 6-8 *The Puerto Penasco facility as an example of a closed system in agriculture. The closed nature of the system conserves fresh water originally obtained by utilizing waste heat. Note that the evapotranspiration component that would occur in open fields is lacking (dashed lines). (Constructed from* Power/Water/Food Experiments at Puerto Penasco, *courtesy of the Environmental Research Laboratory.)*

and lose that enormous investment. Similarly, rice farmers in California need expensive equipment because the timing of planting and harvesting is critical. With expensive machinery to purchase and maintain, rice growers could ill-afford a variable market. In fisheries, the development of more complex, efficient, and expensive gear is self-serving: The more invested in the latest and best gear, the better the competitive advantage; but one must work harder to recover the initial costs, and the fishery becomes harder pressed as the cycle continues.

We can see the impact of investments for gear quite easily in the whale fishery. Today a whaler who wants to stay in business does not sail off in a three-masted schooner with an ample supply of rum, dories, and hand harpoons. Whaling gear now includes fast catcher boats equipped with explosive and electric harpoons and massive floating factories that can swallow and process whales in bunches (Figure 6-9). A primary consequence of escalating technology has been the systematic decimation of each whale species in turn, because the investment in gear has been made, and the two whaling nations feel they cannot afford to let the equipment sit idle. For whales in

FIGURE 6-9 *Whales are literally swallowed up through the massive chute of this floating processing plant.* (Courtesy of Rolt Randall, Black Star.)

particular, and common-property fisheries in general, not fishing means no return at all for a fixed and depreciating investment; hence it makes economic, if not ecological, sense to continue until the fishery collapses and the returns in catch can never meet the costs of daily operation. Unfortunately for baleen whales, this point is probably well below the minimum population necessary for continued survival.

DEALING WITH NATURE: PEST CONTROL

The intrusion of carefully tended, energetically rich fields into natural landscapes simultaneously destroys natural complexity and inevitably creates golden opportunities for populations that can utilize the human desire for crops. These populations are known as pests. This is another example of an economic influence on agriculture both because the measure of the importance of a pest is the economic damage it does and because growers cannot afford the losses they might incur, whether or not their agricultural practices are responsible for the pest in the first place.

In modern agriculture the major method of minimizing pest damage is application of chemical agents. There is general agreement that this measure is effective for the short term. Chemicals are easily applied, are highly toxic, can persist for various lengths of time, and often act immediately. The net effect is an instantaneous decrease in the pest, much as an episode of bad weather affects any population. However, as evidence continues to appear that some chemicals have unacceptable hazards or side effects, that some pests are resistant to or even aided by chemical treatment, and that in the long term chemicals do not work, more attention is being given to alternative methods of pest control.

Biological control is the introduction of species that can control a pest. Success requires either a lot of research to find species that might be effective or a lot of luck in trial introductions of possible control agents. Successful biological control is also highly sensitive to economic factors. Not only must control agents be cheap enough to use, the pest must be controlled easily at a level below economic damage. Turnbull and Chant (1961) evaluated the success or failure of biological control attempts in Canada. They found that pests fell into two categories: direct pests, which attack the harvestable product (for example, apples, wheat grains, corn, spinach), and indirect

FIGURE 6-10 (a) *Example of the damage to apples caused by the codling moth. (b) A close-up of the codling moth* (Carpocapsa pomonella). (Photos courtesy of Grant Heilman.)

pests, which generally weaken plants so that fruit is reduced in size or number. Using a rating system they developed, Turnbull and Chant found no success for "good" biological control of direct pests but over 54 percent success for indirect ones. The lack of success with direct pests results from the fact that natural selection will seldom if ever favor a population which eradicates its prey, as is often required for direct pests. The best example given by Turnbull and Chant is the case of the Ontario apple grower and the apple codling moth (Figure 6-10). To quote the authors:

In the apple-growing industry for example, present market standards demand virtually perfect fruits. The apple codling moth is almost a universal pest of apples, and, when uncontrolled, often completely destroys the crop. Biological agents could conceivably reduce the damage to fractional levels and ensure the production of a crop. But a single moth larva attacking an apple will render that apple valueless, and a loss of from 4% to 5% of apples in a crop often eliminates the grower's margin of profit. To achieve satisfactory control, therefore, codling moth numbers must be reduced to fewer than one larva for each 20 apples on the tree. It is doubtful whether the most efficient of biological agents could maintain the moth population at this level.[2]

---

[2] From A. L. Turnbull and D. A. Chant, "The Practice and Theory of Biological Control of Insects in Canada." Reproduced by permission of the National Research Council of Canada from the *Canadian Journal of Zoology*, Vol. 39, No. 5, pp. 697–753, 1961.

Other techniques offer an alternative to biological and chemical control. The logical successor is *integrated control*, a combination of biological control agents and spot chemical control. But integrated control remains largely unproved, and its potential will emerge only in the future (see Conway, 1971, for a review of this and other methods). Finally, one outstanding example is eradication of the screwworm using the "sterile male" technique. This method uses x rays to sterilize males, which then are released to mate with fertile females. Because their eggs will be sterile, the screwworm population collapses. The method—applied successfully in Caracas, Venezuela, Florida, and Texas—depends upon the facts that fertile males can be overwhelmed by large releases of sterile ones and that females mate only once. (Further details can be found in Bushland, 1960.) Can such technology be transferred verbatim, or are its supporting institutions possible only in highly developed societies? What local adjustments, if any, are necessary, and will they prove to be adequate?

These are important issues for developing international agriculture. The successes and failures of the Green Revolution illustrate that a set of social institutions, market mechanisms, and technological support are necessary to absorb such gains. One tactic, therefore, is to build a society and its economy before these innovations are introduced. For example, Stout (1968) has calculated that the benefits of the Green Revolution in India could be realized fully only if adequate power generation to pump water and manufacture fertilizer is made available. The size of the areas to be served, plus the low availability of fossil fuels, indicates that nuclear power plants are the cheapest source of large power. In the past, the higher priority of industry over agriculture in India often has meant that agricultural development received inadequate power. To apply nuclear energy now could produce a second crop in one year and an increase in yield up to 6000 pounds per acre. Stout's analysis shows that India desperately needs energy to pump water and to make fertilizer; otherwise complex machinery and crops with high demands for management make no sense.

Agricultural engineers are beginning to recognize that technological innovation must be accommodated to the economic and social institutions of various countries. American farm machinery was designed for an agricultural sector that is becoming concentrated into fewer hands. For example, chemical solutions can be mixed automatically and applied at proper rates by airplanes, covering thousands of acres in a short time. Most important, in this country, new discoveries are linked directly to agriculture by experiment-station and extension personnel. In contrast, in most other countries farm plots are small, capital is scarce, and information and problem-solving support are weak or absent. Moreover, local preferences in food crops may be stereotyped and resistant to change in some developing countries, which even may adhere unshakably to traditional methods. Hence technology in these regions should emphasize smaller, cheaper machinery that can adapt to local political and economic realities and on other modifications such as the use of local wastes for fertilizers instead of refined but more costly chemicals. But even with such modifications, some of the most powerful aspects of Western agriculture will be of little use in developing nations; if Western machines are to contribute much to the world's food supply, they will have to do it at home.

MONOCULTURE, POLYCULTURE, AND PRODUCTIVITY FROM THE LAND

Monoculture, the cultivation of a single crop, is the predominant mode of operation in high-yield agriculture. Besides economic incentives for maximum yield, there are biological reasons for monoculture. Yields of wheat are maximal when planting is done in thick stands right after weeds are plowed under, for wheat then can outcompete and eliminate other species. Corn yields are superior in monoculture, something which is usually true for wind-pollinated species. In addition, monoculture is suited to mechanical agriculture. On the other hand, dense monoculture clearly favors the multiplication and spread of pests and diseases. In spite of any disadvantages, however, monoculture is a dominant tactic of production in most situations.

Although monoculture remains the rule, polycultural techniques sometimes are used when problems associated with monoculture appear. Strip cropping is gaining in popularity, particularly when a less valuable plant attracts insects that might infest a more valuable plant. Some legumes may be interplanted profitably with other crops because they can fix and supply nitrogen, thus cutting fertilizer requirements. On a smaller scale, organic gardeners who eschew all inorganic fertilizers and pesticides have turned to polyculture as a way to minimize pest damage.

The Environmental Costs of Intensive Agriculture

To increase the total supply of food, more land must be put into production, or productivity per acre must be raised. With the first alternative, less land will be available to support natural systems; they are replaced by simplified systems that are inherently unstable. On the other hand, increasing yields per acre bears a different kind of environmental cost: the need to use more pesticides and higher rates of fertilizer application, and, as a result, runoff of water loaded with salts and pesticides that can then pollute other environments. However, because environmental costs either can be ignored or passed on to someone else, it is not surprising that these kinds of major costs have not been included in economic analyses of decision-making regarding agriculture and population growth. As long as a city is growing, supply and demand for land to build upon generates rising values that promote conversion of agricultural land to urban uses; farmers then can take part of their profits and move their operations to cheaper land farther from the city. Higher transportation costs can be passed on to consumers, while other costs would be expected to be the same. The real loser is the environment, if only because another piece of marshland, prairie-chicken habitat, or native wild grasses is irretrievably lost, or because large quantities of potent pesticides are used to the detriment of organisms in surrounding habitats.

Ecological Alternatives to High-production Agriculture

There are numerous ways to farm and harvest, but to rival the successes of Western, high-production agriculture, the method must produce sufficient yield to meet the demands of the population while reducing the environmental impact. High-production agriculture in ecological terms is accomplished inevitably at the cost of species diversity, because by definition

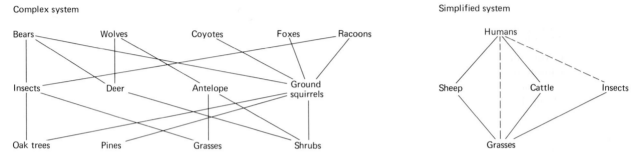

FIGURE 6-11 *Diagrammatic representations of complex and simplified trophic webs, to clarify a major dilemma in preserving nature in the face of growing food needs. Dashed lines are potential food pathways.*

it diverts more of a fixed solar-energy pool into a few selected pathways (Figure 6-11). All agricultural policies show a tacit understanding of this principle and a willingness to accept the environmental costs that may accrue. Are there better alternatives? Let us examine some practices and proposals in this light.

SWIDDEN AGRICULTURE

Swidden agriculture is the practice of shifting agriculture, usually in tropical forest environments. It relies on natural processes to restore soil productivity after a short period of cultivation. While the old patch recovers, a new patch is cut, burned, and planted. The farmers return to the regrown forest sooner or later, when the burned trees have provided nutrients for crops. After two or three years, soil productivity falls, and the forest is allowed to rebuild the site.

Swidden agriculture is not very productive because it depends on the addition of nutrients to the soil by burning vegetation. Since the forest requires some minimum time (about 8 to 10 years) to rebuild the nutrient supply in the area and because the process of regrowth and nutrient regeneration overwhelms the swidden gardens, the populations are limited by their food supply without much hope of expanding production to support a larger population. Attempts to shorten the swidden cycle are dangerous. Zinke and coworkers (unpublished) have examined the swidden cycle of the Lua people in northern Thailand, from which we can infer what might happen when population pressures force shortening cycle times. Swidden agriculture depends on practices that promote high soil fertility. In northern Thailand, the cycle takes 10 years for plots of about 100 hectares (1 hectare = 2.47 acres). The trees are cut and allowed to dry for several months before burning. When they are burned, they release nutrients into the soil. (In tropical forests, most nutrients are bound in vegetation, not soil). In the first year, rice is planted in the ashes and soil to coincide with the monsoons. An important consideration allowing the Lua to clear such large areas is that the cut stumps are seldom killed, and they begin to sprout during the first year. By the second year, cotton and tobacco succeed rice, and the trees slowly become dominant. By the third year, the field is abandoned to the forest, not to be reclaimed for another eight years.

Suppose that the Lua people increase their population or encounter competition from other swidden agriculturalists. An easy way to accommodate a larger population is to cut the length of the swidden cycle. Zinke's data show that the consequences of this would be drastic. In the tropics, soil nutrients—carbon, nitrogen, phosphorus—are minimal anyway, and the risk of severe soil erosion is increased as the organic matter and litter that bind the soil have less chance to accumulate. At some point, the cycle would be shortened so much that the soil would break down and ultimately destroy the productive capacity of the land.

The advantage of swidden agriculture is that its small scale and dependence on natural regeneration minimizes environmental impact. Other than weeding by hand, little attempt is made to exclude pests. No foreign chemicals are introduced, and the disturbance of the natural ecosystem is hardly more severe than might result from a natural disturbance. The price lies in production, for a given plot will produce crops only 2 out of 10 years at normal cycle times, and even then the yield is limited to not more than a subsistence level. Such a system can hardly meet the food needs of a burgeoning world population, but it handsomely illustrates the trade-offs available to us between productivity and environmental disturbance.

ORGANIC FARMING

The term "organic gardening" was coined to refer to agriculture that excludes the use of unnatural pesticides and inorganic fertilizers. This is actually a rediscovery of traditional subsistence agriculture that exists over great parts of the world today. Both depend on organic wastes, plant or animal; one has few pesticides available, while the other shuns them. To date, there is little evidence that these methods are much superior to swidden agriculture for large-scale agriculture. Productivity is higher but at the cost of permanent conversion of the land to agriculture. Hence organic gardening is more valuable psychologically and for use on small garden plots than as a major hope for feeding the world.

POLYCULTURE AND DISPERSED AGRICULTURE

Monoculture is highly favored by farmers because it permits large-scale application of specialized machinery, thus substituting energy and technology for labor. Fewer workers are needed to produce vast quantities of food, and labor released from food production is then available to build an industrial sector in the national economy.

Successful monoculture has been a prime achievement of American agriculture, but it has not been achieved without problems. Single, highly productive varieties may be susceptible to variations in weather and to a range of pests and diseases, and hence they require regular chemical sprays to insure a dependable, regular yield from year to year. Since these sprays generally cannot be confined solely to the system for which they were intended (that is, to the crop itself), we find effects radiating out into other parts of the system (Chapter 14).

Polyculture is a popular name for a form of dispersed agriculture in which different crops are planted together without necessarily causing a decrease in yield. If polyculture can reduce pest problems, disease, and fluctuations in yields without the need for chemical intervention, then it promises to have a new and more ecologically sound meaning. However, if polycultural techniques are to be worked out on a large scale and then adopted widely, certain operational problems must be solved.

The first concern is whether or not pest and disease problems can be reduced sufficiently by polycultural practices. We require a polycultural pattern that, in effect, spaces potential host plants (the crop) enough to prevent explosive population growth of a pest or disease. The dispersal abilities of a pest would be critical; a pest capable of wide dispersal probably would defeat any polycultural pattern that is economically feasible, but a poorly dispersed pest could be controlled. A second problem is that each crop cultivated must have a considerable pool of genetic diversity; that is, we would need many genetically different varieties of each crop, at least in regard to their resistance to disease and pests. In this case the objective is to have a selection of crop varieties each different in relation to disease and pest resistance. This would prevent natural selection within the disease and pest organisms from destroying the entire crop, and it would ensure at least a moderately consistent yield of any given crop from year to year. A fascinating example is the possibility of manipulating genetic varieties of wheat systematically in order to keep specific wheat rusts from becoming established and destroying the crop. It should be possible to devise a rotation system so that a rust which is just becoming a problem at the end of a growing season is presented later with a variety of wheat it cannot attack. By the time another variety of rust, well-adapted to the wheat being grown, has a chance to build up, the crop is rotated again—we create a kind of "superinstability" for the rust community, which continually evolves the wrong way. A final problem is to devise a polyculture system that can be operated successfully within the framework of the economically complex systems in developed countries today. Crops with similar environmental and resource requirements must be matched, and even then society must face the prospect of greater labor requirements followed by rising prices for commodities grown in this manner.

We have a microcosm of environmental decision-making in this example. Polyculture offers the potential for pest control and sufficient agricultural production without the environmental costs of persistent pesticides. On the other hand, monoculture has a proven track record and is an integral part of a larger economic system that helps perpetuate it. Unless polyculture turns out to be more productive in the long run in conjunction with decreased environmental costs, it is difficult to see how it can obtain much attention given all the economic incentives to continue monoculture and the economic disincentives to institute polyculture with potential economic loss. To assume that polyculture can be favored by legislation assumes that economic matters are secondary—and that is open to question.

HARVESTING NATURAL ECOLOGICAL SYSTEMS

As Watt (1968) points out, one strategy not usually considered is harvesting within a natural system—the replacement of the natural system with a carefully constructed agricultural one has always been assumed. Experiments have been performed, however, on the protein potential of American bison and African plains herbivores. Dasmann (1964) has documented the case of the Henderson ranch in Rhodesia. His studies and many others indicate that the mixed species of the plains have an overall productive advantage over cattle or other domestic animals because they are adapted to prevalent conditions. They do not suffer excessively from disease, they are able to use the vegetation better, and as a result their condition and meat yield is appreciably higher than for cattle in the same area (Table 6-4).

The clear ecological superiority that natural selection has conferred upon an assemblage of wild species over an artificial combination of herd animals would seem to demand a switch to large-scale game ranching. The implementation of game ranching is blocked, however, by a combination of economic and social variables despite its ecological rationality. There are questions of ownership of wild game, questions of management technique (fencing, harvesting, domestication, predator control, range management), and questions of marketing (refrigeration, transportation, selling, and public demand), none of which is trivial. While Dasmann minimizes the importance of operation and management problems, he believes social traditions to be so conservative that the long-term outlook for game ranching is bleak. Certainly the combination of circumstances reveals why it is so much easier, no matter what the ecological conditions, to replace an unfamiliar system with a more familiar one.

Other ecologically based schemes suffer from the same defects. While it is eminently reasonable to harvest algae from the sea (thereby saving at least one energy-transfer step over the harvest of fish), this idea has not been applied commercially because there is no market for the product. Odum (1969) has argued that many ecosystems produce large quantities of detritus (dead organic material) which could be collected and processed for food, but as for algal breads and wildebeest steaks, both social and technological barriers mitigate against such a proposal.

TABLE 6-4 *Yields in meat (pounds) of various wild herbivores on the Henderson ranch, southern Rhodesia, compared with yields for cattle on 50 square miles of ranch.*

Species	Estimated number in area	Total meat yield
Impala	2,100	34,125
Zebra	730	37,230
Steenbuck	200	480
Warthog	170	5,950
Kudu	160	10,800
Wildebeest	160	8,320
Giraffe	90	15,000
Duiker	80	560
Waterbuck	35	1,400
Buffalo	30	2,850
Eland	10	1,200
Klipspringer	10	42
Bushpig	10	350
Total	3,795	118,307
Cattle	1,155	94,500

SOURCE: Dasmann (1964).

High Productivity, Ecological Diversity, and the World Food Problem

The evidence overwhelmingly favors the idea that human institutions place a greater priority on high productivity than on high biological diversity. Population growth and social tradition have combined to force the strategy and tactics of food production away from goals associated with the maintenance of diversity. First, gathering edible plants was displaced by swidden agriculture, which in turn yielded to fixed cultivation. Agriculture in the last few centuries has slowly gathered the technological tools for protecting and nurturing its increasingly unnatural productivity.

Potential tragedy lies in the fact that world population growth gives us no alternative. If human welfare is to dominate all other considerations, agriculture cannot escape or significantly ameliorate its environmental costs, nor can less ecologically costly schemes produce the volume of food needed. Thus whether or not we can produce enough food is not really a relevant question; there are no real alternatives. We must try to do so, being as innovative as possible. If we fail, the population side of the equation will be adjusted. As we have seen, anthropological studies support the idea that populations have always adjusted to food supply, and there can be no question that these adjustments will have to be made again, probably on a global scale.

The prospects of an increasingly desperate race to avoid starvation is truly frightening because the consequences of failure are so great. It is especially so when we note that modern standards of high productivity are based upon large subsidies of energy from fossil fuels that also are limited. Thus we are now entered—like it or not—in an inescapable race to simplify and put order on this planet merely to maintain ourselves. Until the nations of the world face the population problem squarely, there will be no escape from this race and no help for the environment.

Questions

REVIEW QUESTIONS

1. Compare Idyll's (1970) account with Ricker's (1969) or Paulik's (1971). Do you see why Idyll is relatively optimistic? Does the history of marine fisheries support his opinion?

2. Some might claim that the achievements of the Green Revolution are an exception to the general rule that raising productivity bears environmental costs. Is this likely to be true or not? On what grounds?

3. Do you think polyculture could work for apples? How, or why not? Assume the apple codling moth to be a principal problem.

4. Read Commoner's (1971) account of the saga of Decatur. Do you agree that advanced fertilizer technology is at the base of the problem?

ADVANCED QUESTIONS

1. Ryther's (1969) paper has been criticized for its assumptions about trophic effects. Estimate total fish production if
    a. trophic efficiency in the open ocean was 20 percent
    b. all three areas had identical efficiencies and trophic chains
    c. 25 percent of the world's estuaries were destroyed

What primary productivity value for the open oceans would have to exist (all other factors being equal) to obtain double Ryther's estimate of fish output?

2. What are some of the difficulties of major diversifications in diet? Based on your knowledge of ecological systems, what would some of the compensatory reactions be?

3. Given that there are arable lands in the temperate areas in developed nations and that food could be produced there and then shipped abroad, why is this not a long-term solution to the world food problem? Compare the possible advantages and disadvantages.

Further Readings

Athwal, D. S. "Semidwarf Rice and Wheat in Global Food Needs." *Quarterly Review of Biology*, 46 (1971), 1-34.

> A technical but highly readable report on the development and use of the grains of the Green Revolution. Recommended to readers interested in the development of genetic strains for higher productivity.

Borgstrom, G. *The Hungry Planet*; and *Too Many*. Both Macmillan, New York, 1965, 1969.

> Two widely quoted volumes by a noted food technologist. Each treats the population-food problem from one side; the viewpoint is predominately pessimistic. Extensive bibliography applicable to this chapter.

Borlaug, N. E. "The Green Revolution, Peace and Humanity." *Population Reference Bureau Selection*, No. 35, 1970.

> The acceptance speech for the Nobel Prize. Succinct summary for a lay audience of the achievements of the Green Revolution.

Brown, L. R. *Seeds of Change*. Praeger, New York, 1970.

> Lester Brown is currently with the Overseas Development Council. His qualifications to write about the Green Revolution are unsurpassed. The viewpoint expressed in this book is for popular audiences and is not significantly different from that of others involved in the Green Revolution.

Brown, L. R., and G. Finsterbusch. "Man, Food, and Environment." In *Environment* (W. W. Murdoch, ed.). Sinauer Associates, Stamford, Conn., 1971.

> More than a short version of the preceding, this paper has a broader environment viewpoint more closely akin to that in this chapter.

Bushland, R. C. "Screw-worm Research and Eradication." *Advances in Veterinary Science*, 6 (1960), 1-18.

Chang, K. L. "The Agricultural Potential of the Humid Tropics." *Geographical Review*, 58 (1968), 333-361.

Commoner, B. *The Closing Circle*. Knopf, New York, 1971.

> This book is a statement of the primacy of technology as a source of environmental problems, and it offers some solutions based on this assertion. His argument can only be evaluated within the context of other viewpoints, for example, Ehrlich and Ehrlich (1972). Provides the documentation for the fertilizer problems of Decatur, Ill.

Conway, G. R. "Better Methods of Pest Control." Chapter 14 in *Environment* (W. W. Murdoch, ed.). Sinauer Associates, Stamford, Conn., 1971.

Dean, G. W., G. A. King, H. O. Carter, and C. R. Shumway. "Projections of California Agriculture to 1980 and 2000." California Agricultural Experiment Station Bulletin 847, 1970. (Available from University of California Agricultural Extension Service, Berkeley.)

Dasmann, R. F. *African Game Ranching.* Macmillan, New York, 1964.

> The Henderson ranch experiment and other data are presented within the context of the history of large mammals in Africa—and probable futures if ranching is not instituted.

Ehrlich, P. R., and A. H. Ehrlich. *Population, Resources, Environment,* 2nd edition. Freeman, San Francisco, 1972.

> The chapters on food production and sufficiency present a very different viewpoint than that of Brown (1970) or Borlaug (1970).

Food and Agriculture Organization of the United Nations (FAO or FAOUN). The State of Food and Agriculture. FAO, Rome; available through U.N. documents, New York.

> Issued annually, these FAO reports are the best sources of current world food statistics and short summaries of current trends available.

Food and Agriculture Organization of the United Nations (FAO or FAOUN). Proceedings of the World Symposium on Warm-water Pond Fish Culture. 5 vols. FAO, Rome, 1968.

> The discussion on aquaculture was largely abstracted from these volumes, especially vol. 1-3.

Harris, M. "The Human Strategy: How Green the Revolution?" *Natural History,* 81, No. 6, (1972), 28-30.

Harris, M. "The Human Strategy: The Withering Green Revolution." *Natural History,* 82, No. 3, (1973), 20-22.

Hendricks, S. B. "Food from the Land." In *Resources and Man* (P. E. Cloud, ed.). Freeman, San Francisco, 1969.

Hodge, C. O. "The Blooming Desert." *Bulletin of the Atomic Scientists,* 25, No. 9, (1969), 32-33.

> A description of the Abu Dhabi hydroponic facility. See also *Science,* 171 (1969), 989-990, or contact the Environmental Research Laboratory, University of Arizona, Tucson.

Idyll, C. P. *The Sea against Hunger.* Crowell, New York, 1970.

> Popular and highly optimistic assessment of the potential of the sea.

Ladejinsky, W. "Ironies of India's Green Revolution." *Foreign Affairs,* 48 (1970), 758-768. (Reprinted in Revelle et al., 1971.)

Paddock, W., and P. Paddock. *Famine—1975!* Little, Brown, Boston, 1967.

Paulik, G. J. "Anchovies, Birds, and Fishermen in the Peru Current." In *Environment* (W. W. Murdoch, ed.). Sinauer Associates, Stamford, Conn., 1971.

> Comprehensive and detailed analysis of the development and potential decline of a new fishery. While the tragedy of an anchoveta is not the same as that of a whale (see Small, (1971)), the circumstances are closely parallel.

President's Science Advisory Committee Panel on the World Food Supply. *The World Food Problem,* 3 vols., 1967.

> Information collected in the course of assessing U.S. policy alternatives on supplying food overseas.

Revelle, R., A. Khosla, and M. Vinovskis. *The Survival Equation.* Houghton Mifflin, Boston, 1971.

> One of the finest edited volumes available for the population-food problem. The editors have succeeded in balancing a number of diverse viewpoints.

Ricker, W. E. "Food from the Sea." In *Resources and Man* (P. E. Cloud, ed.). Freeman, San Francisco, 1969.

> A review paper; more scientific than Idyll (1970), broader than Ryther (1969) or Paulik (1971).

Ryther, J. R. "Photosynthesis and Fish Production in the Sea." *Science*, 166 (1969), 72–76.

Small, G. L. *The Blue Whale*. Columbia, New York, 1971.

> One might expect such a saddening account to be an emotional appeal, but this one is carefully documented and effectively written. There is no better account of the basic weakness of the International Whaling Commission relative to various national interests.

Stout, P. R. "Potential Agricultural Production from Nuclear-powered Agra-industrial Complexes Designed for the Upper Indo Gangetic Plain." Oak Ridge National Laboratory. Document *ORNL-4292*, 1968. See also a shorter, more popular version in *Bulletin of the Atomic Scientists*, (November 1968), pp. 26–28.

Turnbull, A. L., and D. A. Chant. "The Practice and Theory of Biological Control of Insects in Canada." *Canadian Journal of Zoology*, 39 (1961), 697–753.

Watt, K. E. F. *Ecology and Resource Management*. McGraw-Hill, New York, 1968.

> This book contains interpretations of the African game management operations and whale fisheries.

Zinke, P. J., S. Sabhasri, and P. Kunstadter. "Soil Fertility Aspects of the Lua Forest Fallow System of Shifting Cultivation." (Unpublished.)

7. The Dilemmas of Water Development

Water is probably our second most important resource after food. In fact, one could argue that water is the most important natural resource; it is required to sustain life, serves as the most common solvent, is a major means of transportation, generates electric power, is essential in industrial production, and is vital to agriculture.

Despite the fact that water is so vital, it is not distributed uniformly over the earth. Some regions are blessed with large quantities, and some are even inundated periodically. But other regions are chronically short of water. Many countries with a large arid area may have retarded economic development if they cannot alleviate their water limitations.

Since water development is generally popular, it is no surprise that large-scale programs have been developed to ameliorate distributional inequities. Solving the problem, however, requires large-scale applications of technology and capital. After all, we are talking about a resource that is thoroughly integrated into the global system; hence we should expect complex interactions to result from any attempt to distribute water. (Examples of these interactions and their environment effects are considered in the next section.) We also must consider the realities of human institutions when we talk about water development, because large organizations with a mission to supply water will not be neutral. Not only must they supply the water, they must control its flows, allow for its storage, cooperate with other agencies that are tied to its distribution and use, and manage the water after its primary use.

In effect, the problem of water development represents the behavior of large agencies, and so this chapter in a broad sense is about the development and management of large-scale resources, not only about water. Large agencies are created at large cost, usually in the public sector, and after a while they tend to become self-perpetuating. For a long time, water development and allied activities (flood control, recreation impoundments, coastline protection, and harbor development) have occupied a hallowed place in our political system. Since water development has had no serious opposition, there is enormous political profit for a congressional representative who can claim credit for a new dam, canal, irrigation system, or harbor breakwater. Thus in Congress itself we can learn why large water agencies are perpetuated indefinitely and why waterworks proliferate.

With the growth of the environmental movement, environmental interest groups gained enough strength to challenge the whole process of planning and developing water resources. Aided in particular by the requirements of the 1969 National Environmental Policy Act (NEPA, see Chapter 19) and its state-level equivalents, these groups have made legal delays a part of development. There are strong arguments for both sides of water development issues, though the strength of either side may vary with a particular use. With these newly awakened environmental concerns, our society seems less willing to give its blanket approval for each and every water-related project that is suggested, and conflicts over present and future projects are likely to continue.

The Natural Hydrological Cycle

The natural hydrological cycle is an example of a closed system of exchange involving a large pool of sea water and a much smaller one of fresh water (Figure 7-1 and Table 7-1). The driving mechanisms for these exchanges are evaporation and transpiration, which together act as water redistribution agents. Water vapor collects as clouds, which are blown by prevalent patterns of winds. When clouds cool by rising or by collision with cooler air, water condenses and falls as rain. The part that does not fall back into the seas may accumulate temporarily in ponds, lakes, bogs, streams, marshes, the ground, and living organisms; but sooner or later the water is reevaporated or transpired, only to fall again later as rain. Thus water circulates continuously over the planet.

The significance of the hydrological cycle lies in the dependence of all biological systems, including human beings, on a predictable water supply. This results in competition for water, and we usually have the edge over other creatures. More important, the hydrological cycle is a system, and changes in one part of it have important consequences for all other parts.

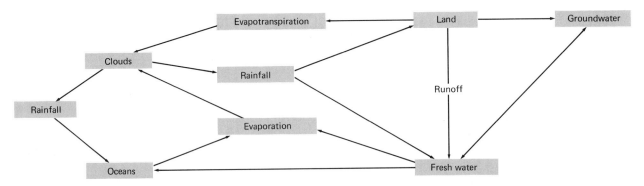

FIGURE 7-1 *Schematic diagram of the hydrological cycle expressed in block diagram form.*

Hence the hydrological cycle is a closed system conforming to the general principles set forth in Chapter 3. From this we can deduce, understand, and even predict some of the problems that arise from human modification of the cycle.

For example, consider a group of farmers who need additional water to bring their crops to harvest. They decide that the simplest solution is to dig wells on their lands, that is, to use groundwater supplies. The wells could be an acceptable solution if there is long-term balance between the withdrawal and the recharge of this water supply; but if the groundwater supply was stored during an era when the region had more plentiful water supplies, the wells may succeed only in mining a finite groundwater supply, like digging for limited resources of coal or drilling for oil. If we were to trace this example through Figure 7-1, reducing the supply of groundwater would drive the water table (the depth below the surface at which the soil is saturated by water) down and consequently reduce the volume of surface fresh water, since many of these bodies are supplied from these groundwater reservoirs. That water which is pumped to the surface through wells effectively reverses the normal flow between "groundwater" and "land," and at this point there are three ways to disperse that water withdrawn from below the earth's crust: run-off, return to the groundwater component, and evapotranspiration. The greater the demand, the greater the net reverse flow, since only part of this water would be returned to its source. That is why the water table would fall, as it has over much of the Southwest. The speculative consequences of this small-scale illustration foreshadow the possible effects of such large hydrological projects as nuclear power plants and the Aswan dam, and it indicates how valuable the hydrological cycle is as an educational tool to help us understand the systems consequences of its alteration.

A Base Line for Policy Considerations: Present Supplies and Demand

There is substantial agreement about the size of the world's water supply. Two assessments (Table 7-1) vary only slightly from one another, something that

TABLE 7-1 Estimates of (a) world water supply of various components and (b) the rates of major transfers between components, using Figure 7-1 as a reference point.

Component		Water (10^{15} gallons)		Percent of total	
		VH[a]	HS[b]	VH	HS
(a)	Total surface fresh water	7698.08	7733.3	2.15	2.16
	lakes	33.04	33.00		
	rivers and streams	0.34	0.3		
	glaciers	52.86	7700		
	ice caps	7611.84			
	Total groundwater	2131.58	2217.60	0.60	0.62
	soil moisture	17.18	17.60		
	shallow groundwater ($\leq$ 1000 m)	1057.20	1100		
	deep groundwater	1057.20	1100		
	Saline lakes	26.43	27.5	0.00	0.00
	Total on or in land	9856.09	9978.4	2.75	2.78
	Atmosphere	3.44	34.1	0.00	0.00
	Oceans and seas	348876.00	348700	97.25	97.22
	Total of all water	358735.53	358712.5	100.00	100.00
(b)	Annual evapotranspiration	105.72	111.7		
	from the land	14.80	16.4		
	from the oceans	90.92	95.3		
	Annual precipitation	105.72	111.6		
	onto the land	25.37	26.1		
	onto the oceans	80.35	85.5		
	Runoff from land to seas	10.57	9.8		
	Total	222.01	233.1		

[a] van Hylckama (1971) estimates.
[b] Harte and Socolow (1971) estimates, after Skinner.

might not be expected for measurements and estimations of such scale. The van Hylckama and Harte-Socolow estimates differ widely only for the amount of water held in the atmosphere; but even measured in terms of percentages, the difference in estimates is insignificant. Most of the earth's water (97.22 to 97.25 percent) is found in its oceans and seas; of the remainder, almost all water held on the continents is frozen, since most water that is not frozen either flows back into the seas or is evaporated rapidly. Since the

FIGURE 7-2 *Huge icebergs like the one in the background could be towed to population centers and melted down for water.* (Courtesy of Ira Kirschenbaum.)

bulk of the world's fresh water (77 percent according to Harte and Socolow, and 78 percent according to van Hylckama) is held in glaciers and ice caps, there have been several serious suggestions that they be used as sources of fresh water (Figure 7-2). To this end, the glaciers would be mined and melted, and icebergs broken off from the ice caps would be towed to population centers. The largest potential supply of fresh water in an aqueous form is groundwater, which constitutes about 22 percent of the total freshwater supply. There are two problems, however, with this assessment of the total groundwater supply. First, fully half of the groundwater in both estimates is very deep in the earth (more than 1000 meters) and practically inaccessible by any economical means. Second, the mere expression of the total amount of surface groundwater says nothing about its distribution over the earth or its recharge potential (the resupply process by which groundwater is restored).

Areas of abundant rainfall have sufficient recharge of groundwater and plentiful surface waters. Drawing on either pool has little effect unless demands are extraordinarily heavy. On the other hand, dry areas with neither water supply suffer from rapid depletion of water since there is often no source to recharge the depleted groundwater. The water table may be surprisingly high so long as demand is very light, but even a few wells can drive it far down into the underlying soils. Without a recharge capacity, the groundwater supplies in dry areas cannot be drawn upon very much, despite a superficial indication of abundant supply.

Although prominent, the surface fresh water of the world constitutes only 2.1 percent of the total water on the planet. Nevertheless, this still amounts to 7700 quadrillion gallons. If we assume (1) that technology can solve problems of inequitable distribution, (2) that about 10 percent of the water can be used by society, and (3) that the present world population is about 4 billion people, we still would have about 192 million gallons per capita. Even with world population doubled to 8 billion, there would be 96 million gallons per capita. Thus even with the worst population projection, we seem to have more than adequate water for future needs—given our assumptions.

Figures for current world use of water are unavailable but can be estimated roughly from published patterns in the United States. Water requirements for human beings and livestock are most visible, but the data (Table 7-2) show that these uses are only a small fraction of the water used by irrigated agriculture and industry. Less than 9 percent flows into cities for public use, and the total is raised to only a fraction more than 10 percent if we include all public use plus livestock. In comparison, the volume used by irrigated agriculture amounts to 96.8×10^9 (96.8 billion) gallons per day, or 42 percent of all water used in human ecosystems. Industrial uses, 75 percent of which is for cooling power-generating plants, account for 108.1×10^9 (108.1 billion) gallons per day, or 47 percent of all water used. This large figure conceals the efficiency with which industry can use and reuse water, since quality standards are often much lower than for drinking water, and a sizeable quantity is reused by industry. For power plants, even salt water can be and is used for cooling purposes.

A second trend in the United States is the predominant use of surface waters for all but one category in Table 7-2. This reflects the ease and economy of tapping rivers and lakes compared with sinking deep wells, but it also indicates the distribution of water users close to surface waters and the predominant policies that move surface water to areas of need. Though the United States has always used surface waters, there are regional differences. As might be expected, the states drawing on eastern watersheds have high annual average rainfalls and consequently more lakes and perennial rivers. Hence water use in these areas is predominately from the surface. In the more arid states west of the Mississippi River, the situation is reversed, with ground and surface water being used in a ratio of 15:1 in some areas.

Total annual water consumption amounted to 0.1×10^{15} gallons per year for the United States in 1965. Even allowing for a doubling of this figure every decade to the year 2000 (surely an unrealistic assumption), the magnitude of total water use is insignificant compared with an estimated U.S. supply of inland water and groundwater that totals 185×10^{15} gallons. Simply in terms of the diversion of water into human ecosystems (and therefore excluding such problems as those concerned with water quality—the return of these waters to environment in a polluted state), this figure is insignificant when compared with the diversion of land resources into food production.

TABLE 7-2 *Water usage in the United States divided among three major categories, in 10^9 liters/day. Data are for the year 1965.*

Category	Surface waters	Ground-water	Total	Percent of total fresh water use
Urban public use	60.2	30.9	91.9	8.8
Rural use	283.2	172.1	455.3	43.8
livestock and people	3.2	12.1	15.3	1.5
irrigated agriculture	280.0	160.0	440.0	42.3
Industry	461.4	29.9	491.3	47.3
electric utilities	348.0	4.2	352.2	33.9
all other industry	113.4	25.7	139.1	13.4

SOURCE: van Hylckama (1971).

Alternatives to Insure Continuing Water Sufficiency

Unfortunately the analysis of supply and demand in water resources is not very helpful because information at this level does not reflect the distributional inequities between regions that is the crux of the water problem. These distributional inequities for

152   THE DILEMMAS OF WATER DEVELOPMENT

FIGURE 7-3   *The use of water stored for firefighting purposes for ad hoc summer recreation is a unique form of multiple use.* (Courtesy of Jack Prelutsky from Stock, Boston.)

water, combined with a continuously increasing demand for it, insure that future policy will be directed toward ameliorating differences between supply and demand. Five general alternative ways to do this are discussed below.

REDUCING PER CAPITA USAGE

Urban areas are inefficient users of water. Moreover, water of high quality is used for all purposes from water for drinking and washing to water for sewage, although in places there have been attempts to use water meeting lower-quality standards for some urban needs (Figure 7-3). Because of this inefficiency in water use, it should come as no surprise to learn that under certain circumstances urban water consumption can be reduced considerably. From time to time there have been droughts, and major cities have had to ration water. During a major drought in 1949, for instance, the use rate in New York City fell 40 percent, from 220 gallons per capita to only 132 gallons. This shows how flexible the public can be about the volume of water it uses. However, because public use of water is only a small part of total water use, reducing per capita usage would have a significant impact only for certain local situations where demand is mostly urban and where it equals or exceeds the available supply. Whereas small per capita savings in public use might have a major impact in New York City, they would be unnoticed in heavily industrial Chicago, where the household use of water is small compared with industrial use. They might not even be as important in New York if the water of the Hudson River were not so heavily polluted.

It is more important to evaluate the potential for savings in water usage for areas of major agricultural or industrial need—as we have seen, these are the major users. In agriculture, most water used for irrigation is recycled into the atmosphere by evapotranspiration and is lost to human control; without covering the world with plastic and thereby drastically modifying the hydrological cycle, cutting use rates here seems rather unlikely. Irrigated agriculture would have to be cut back as world demand for food escalates. For industry, Table 7-2 tells the same story. Seventy-five percent of industrial needs are for power plants which use the water for cooling and which recycle it back into the environment, available for other uses with little except thermal change. For other industrial processes—such as the manufacture of paper and pulp, industrial chemicals, and others—recycling of water is practiced widely already. The problem here is that water which finally is returned to the environment usually is polluted badly with a whole array of chemicals; it therefore is no longer useful for most other purposes. It thus seems highly unlikely that any significant savings in industrial water usage can be realized unless we are willing to curtail industrial output. In conclusion, then, it seems improbable that we can significantly reduce our agricultural and industrial water consumption; and while reducing per capita usage may be justified for certain local situations and for psychological reasons, in most cases it would have little effect and in fact pales in comparison with other actions that can be taken. We will discuss these actions in the remainder of this section.

REDISTRIBUTION OF WATER SUPPLIES

To reduce water usage requires social readjustments—changing economic patterns, values, and habits that have been based on supplies of abundant and inexpensive water to patterns appropriate for conserving and recycling it. On the other hand, to move water from areas of surplus to areas of need requires only public approval and suitable technology; hence to this point redistribution has commanded higher political feasibility than reduction in water usage has.

When we examine what water we can move and what is worth moving, we rapidly discover that only surface waters possess an appropriate combination of characteristics. Without desalinization, sea water is not worth shipping for great distances, and moving groundwater is impractical; this leaves only surface waters and ice as potential candidates for redistribution. Speculation and controversy still reign over the feasibility of mining glaciers or towing icebergs to coastal cities. Towing large icebergs to warmer coastal cities is possible, but the losses incurred due to melting in transit when combined with the energy costs of such operations probably make it economically unfeasible unless the future price for water is much higher. Mining glaciers would be easier, but they are so rare that they would be exhausted rapidly. In addition, consideration must be given to the environmental effects of such activities—removing large parts of glaciers might seriously affect the ecology of the high mountains. Calculations indicating effects of this sort have been carried out for polar ice; the removal of a significant proportion of Arctic pack ice, say 10 to 20 percent, would open up large areas of dark water. Being dark, the water has a lower albedo (reflectivity) than ice, and so it soaks up more heat from the sun, which raises its temperature. The higher temperatures then act to prevent new ice formation and accelerate the melting of still more ice—still another example of a reinforcing, positive feedback loop. Hence the removal and shipment of very much ice could melt the remaining polar ice and have a number of environmental consequences, the most drastic of which is the probable effect on world climate. Clearly, ice-mining policies designed to supply large quantities of fresh water to users are potentially dangerous and must be considered carefully because climatic stability of the region and the ice itself depends upon a finely adjusted heat budget. In conclusion, these schemes seem useless unless there is real need. Even then, the plans might be useless or counterproductive.

Only one alternative remains: the large-scale movement of surface waters. Typically, the first step toward this end involves the construction of one or more reservoirs to accumulate and store the water entering a given drainage basin (Figure 7-4). The purpose of these reservoirs is to ameliorate the influence of wet and dry cycles in precipitation, thus guaranteeing a reasonably dependable water supply. The main advantage of reservoir construction is simplicity—all three phases (planning, construction, and operation) are straightforward engineering problems. Reservoirs are usually constructed by impounding streams, a method at least as old as the first beavers. Thousands of small farm ponds are built by putting a check dam across a small creek. Such ponds can be used for watering livestock, for irrigation, and for recreation. On a larger scale many rivers that are dammed primarily

FIGURE 7-4 *The Quabbin Reservoir in central Massachusetts was created in 1937 by damming the Swift River. The reservoir covers 39.4 square miles of previously inhabited valley and serves as water supply for metropolitan Boston. (Courtesy of Jeff Albertson from Stock, Boston.)*

for flood control or hydroelectric power generation also may provide ample water supplies for industries or cities located near them. In some river basins, the rivers have been impounded at so many points that they resemble a chain of lakes more than a river.

Though the simple construction of reservoirs may be all that is necessary under many circumstances, frequently further elaboration may be required. If, for instance, the demand for water in an area exceeds the supply available from the reservoir serving that drainage basin, it will be necessary to import water from another drainage basin where supply exceeds demand. This usually is accomplished through the construction of a series of aqueducts equipped with pumps to move the water. One example is the Hetch-Hetchy dam in Yosemite National Park. This dam creates a reservoir that accumulates water from the drainage basin and ultimately supplies the city of San Francisco some 150 miles away. New York City depended solely on the nearby Croton-on-Hudson reservoir for a long time, but as its population grew it had to develop additional sources from drainage basins farther away.

From simple systems like these, it is mainly a problem of scale in graduating up to the really large systems such as the California State Water Plan–Central Valley Project or the Texas Water Plan. These projects are complex engineering feats, totally integrated to impound, pump, and control water flows, generate electrical power, and even supply recreational resources. Despite their complexity, however, their basic purpose remains the same.

The most grandiose schemes include enormous public works projects such as the North American Water and Power Alliance (NAWAPA), which would link the water resources of the United States, Canada, and Mexico. In an engineering sense at least, nothing is too large to be impossible. In recent years, however, environmentalists and even some economists have questioned the need for large projects such as these on the grounds of their probable environmental costs and their distribution of benefits—in particular agriculture. The future scale of redistribution projects, and the fate of several now being built or planned for, may hinge upon the strength of this environmental concern. If it proves to be strong enough, economic analyses, which always have shown redistribution to be the least costly alternative, may have to be revised to include environmental impacts previously excluded. This raises the possibility that other methods for supplying water may become feasible, not because a higher level of technology lowers their costs, but because considerations other than economic ones will render redistribution more costly than it has been.

REDISTRIBUTION OF NEED

If there are serious environmental costs associated with moving water supplies, one logical solution might be to encourage social development at the sites of adequate water supply; that is, we might encourage the redistribution of demand rather than of supply. For example, one might discourage population growth and irrigated agriculture in arid regions, a condition for the use of these areas being acceptance of prevailing environmental conditions with an official policy that no additional water supplies will be imported into the area. While schemes like this are superficially attractive, we will consider a number of serious issues related to them that must be resolved before we can make final judgment.

Concentrating development at or near sites of water adequate for projected needs for some number of years into the future does not necessarily solve water problems as much as it shifts them to other sites. The history of American cities, and indeed of human settlements in general, shows that abundant water typically is wasted first and then polluted beyond use, New York City and the Hudson River being a case in point. Moreover, extensive human competition for water in areas where the ecosystems are adapted to abundant supply is likely to have more striking and significant effects than the same process in more arid areas where ecosystems are already adapted to little water. Hence we cannot be sure that manipulating demand rather than supply will have significantly lower environmental costs in the end.

A good case in point is development in the state of Florida relative to development in Los Angeles. Whereas Los Angeles is located in an arid area and would be expected to have limited freshwater supplies, Florida is a humid state that receives over 50 inches of rainfall yearly; it certainly would seem reasonable that considerable development could be supported on this particular resource base. But Florida is peculiar in that it is a peninsula surrounded by salt water and in past ages most of its area was inundated repeatedly by the sea. The major part of Florida exists in its present state only because the groundwater pool is maintained by rainfall at a rate sufficient to prevent salt water from seeping into the porous rock underlying the state. Thus, the freshwater pool underlying the state of Florida requires a considerable volume of recharge to prevent saline intrusion from the surrounding ocean. Development in Florida, and particularly competition for water supplies between agriculture and public use, is the root of many of Florida's problems. Competition for water is certainly a problem in the Everglades, but in a larger sense the entire state's environment and economy are involved. Each additional population increment with its accompanying development puts additional demand on the very water supply that keeps salt water from contaminating the entire groundwater supply. At the same time, buildings, roads, concrete, and asphalt progressively cover more of the land, preventing the seepage of rainwater back into the ground to replace water that is removed at an ever-increasing rate. Clearly, if this cycle goes too far, the Florida peninsula will have serious salinization problems. If this occurs, there will be a critical shortage of fresh water, and one would be entitled to wonder (in retrospect) if it would not have been better to locate the people elsewhere, even if it meant shipping water supplies to them.

Another problem is the political feasibility of moving people rather than water. Freedom of movement is strongly rooted in American society, and it is presently not possible to bar a person from moving to Los Angeles or Miami. It is sheer folly to supply additional water to an area simply because projections say it eventually will be needed, for frequently this becomes a self-fulfilling prophecy. A chronic water shortage might cause potential immigrants to think twice before moving to an area; but as long as water is abundant, they will continue to move in. We need a flexible way to discourage excess development that requires massive water imports without foreclosing options on redistribution of water. This is not a simple task; the engineering and planning of water development is complicated because it is so difficult to predict how much water is the "right" amount. These are, in fact, the problems facing planning and zoning boards concerned with development of urban land (Chapter 8).

DESALINIZATION

The prospect of tapping the enormous saltwater component of the hydrological cycle is highly attractive. Salt water itself can be useful under certain circumstances. For instance, nuclear power plants require large volumes of water for cooling, and salt water can be used if these plants are located along coastlines. Under these circumstances, the plants will have a supply of cooling water that is plentiful and inexpensive; and the salt water can be returned to the sea with minimal environmental damage. However, the greatest potential of sea water is for public and agricultural use, but only if large volumes can be rendered salt free.

FIGURE 7-5 *Point Loma, California sea-water desalinization plant distills a million gallons of water a day. The purified water is pumped into the San Diego water system.* (Courtesy of Monkmeyer Press.)

The processes used to separate water and its dissolved salts are collectively called *desalinization* (Figure 7-5). One way to remove salts is through distillation. It would be nice to do this with sunlight, but in practice it is difficult to capture sufficient heat to yield fresh water economically. Many small installations, however, use coal, petroleum, or some other fuel to distill sea water. A notable example is the vegetable and fruit facility at Abu Dhabi, where small diesels produce enough heat to provide all the desalinated water needed. Water also can be desalinated by osmotic and electrolytic methods, both of which depend upon the differences in chemical properties between salt ions and water. *Electrolysis* requires a positive terminal at one end of a tank of salt water and a negative terminal at the other; the electric gradient between terminals causes all the salt ions to move to one pole or the other, depending on their charge, leaving the water in the center relatively salt free. By using a membrane through which water can pass but not salt ions, it is possible to desalinate water by a technique called *reverse osmosis*. This requires the sea water to be pressurized to force water through the membrane, leaving a highly concentrated salt solution. Another method that is conceptually simple is to freeze sea water; only the water freezes, and the salts are excluded in crystalline pockets. Subsequently the ice can be removed and remelted to obtain fresh water.

Experts seem to agree that the technology for large-scale desalinization plants has already been developed or will be available in the near future. As of 1973, there were 800 desalinization plants in operation or under construction in the world, mostly in arid regions along the seacoast where the need for fresh water is great and the sources of salt water are plentiful. These facilities are small, with individual capacities all under 5 million gallons per day, but there are plans to build plants having capacities up to 1000 million gallons per day, enough to supply a city of 1.5 million persons or 60,000 acres of agricultural land. The possibility of coupling nuclear power plants and desalinization is a particularly strong hope for large-scale plants in the future, even far from coastlines. Under such a scheme, the waste heat from the power plant would be used to distill sea water. This not only would provide a cheap source of heat to distill sea water but also would provide a means to deal with the waste heat that can be a problem with nuclear power plants.

If desalinization has a problem, it is concerned with the question of economic competitiveness with other sources of water. Some plants have proved competitive for public supplies, especially where natural sources of fresh water are scarce and expensive anyway. But so far desalinated water cannot be produced cheaply enough for use in irrigation on a large scale. Whether one is to be optimistic or pessimistic about the future of desalinization as a source of water depends on what technological developments lie ahead. The optimists have faith that new technology can lower the cost of desalinization, while the pessimists argue that the techniques to do this are still unknown, and so there can be no guarantees that large-scale, economically competitive desalinization will be realized.

There seems to be little doubt, however, that desalinization plants will continue to increase in number and in output, especially in arid coastal regions where economic limitations are not a problem. On the other hand, it is strictly a matter of faith to believe that desalinization will replace water diversion over large parts of the world or that it ever will be a plentiful source of irrigation water.

ATMOSPHERIC MANIPULATION

A final alternative is to manipulate the hydrologic cycle by influencing precipitation patterns. The objective is to obtain precipitation when and where it is needed by seeding clouds. While this initially would seem to be attractive, experiments to verify its practicality have not been uniformly successful; and even if successful, large increases in surface water or groundwater would not result, since this would succeed only in removing moisture from the atmosphere and could in no way increase that moisture. A significant problem associated with weather modification is that weather patterns are poorly known; for this reason it is often difficult to put rain where it is wanted, even if the seeding process can be made more successful. Moreover, some areas of greatest need are so dry that there are few, if any, clouds to seed. In seeding to date, rainfall has been increased no more than about 20 percent, and in some cases rainfall actually may have decreased as a result of seeding. Attempts such as this demonstrate yet again that technological innovation alone cannot necessarily solve our problems. But perhaps this particular limitation is not so unfortunate as it seems. The large-scale manipulation of weather and climate has such an awesome potential impact on ecological systems that it might be far better if this power is not realized.

Three Case Studies of Water Redistribution

When we examine the alternatives to water diversion we can understand why it has been so important. Desalinization and atmospheric manipulation are beset by technological or economic problems, while attempts to manipulate demand in accordance with local supply limitations runs headlong into entrenched habits and social norms that resist change. Since water diversion is feasible economically and technologically and does not conflict with human behavioral patterns, it is not surprising that it will continue. The enormous projects that have been proposed (diversion of water from Canada over the United States into Mexico, the damming of the Amazon, reversal of the flow of two major Russian rivers) are essentially the same practice on an even larger scale.

Past decades saw no substantial opposition to water development. People who needed water got it, and new lands were opened to use and provided with water from other areas where it was regarded as surplus and otherwise wasted. In addition, water impoundment provided cheap electrical power, new fishery resources, and recreational potential around the reservoirs. But a gradual change is occurring, and at present some people regard additional impoundments of our rivers and streams as socially useless at best and environmentally harmful at worst. Also, as American society enters the first phases of transition to population stability, there is increasing awareness that if this trend continues, demand for water development will decrease, since the population no longer will be expanding at its previous rate. And fewer funds will be available for massive projects, since another consequence of population stability is a slowly growing economy. Because of these factors, plus growing environmental awareness in society, the systemwide consequences of water development have begun to enter more and more into future planning. These new constraints on water diversion emerge in the three cases we shall consider now.

CASE 1: THE ASWAN HIGH DAM

The Aswan High Dam project was designed to open new irrigated lands for food production and to stimulate new industrial development with cheap hydroelectric power. Hence the Aswan dam (Figure 7-6) was intended to be a single bold stroke in economic development for Egypt. Construction of the billion-dollar project was started in 1960, and was completed after a decade of hard work and massive financial aid from the USSR.

The possibility of moving Egypt from a state of unending poverty and international dependency to a state of economic independence, industrial health, and political leadership in the Middle East must have been an irresistible argument for the Aswan High Dam. Now that the project is completed, electric power is indeed being produced, and irrigation water is being pumped into the fringe of the desert, bringing new lands into production for food and cotton. But the other side of the balance sheet also is being revealed, since the Aswan dam has modified the total ecological system of the Nile River Valley and the eastern Mediterranean Ocean. If the promise of a bright Egyptian future closed official eyes to the other effects of the High Dam, they no longer can be ignored. Because more water is being diverted for irrigation purposes, less water can flow into the Mediterranean, which has resulted in higher salinity in its eastern portion. In addition, since the outflow at the mouth of the Nile delta is reduced, there has been an increase in salinity intrusion from the Mediterranean into the agriculturally rich delta. Silt-laden water flows much slower because of friction among the various silt particles. Since most silt accumulated by the Nile now is left behind in Lake Nasser, the water that does flow from the dam flows much faster

FIGURE 7-6 *The Aswan High Dam as photographed in the 1960s.* (Courtesy of Black Star.)

than it did before the dam was constructed, and this causes increased bank erosion downstream. None of these physical effects should surprise anyone knowledgeable in the field; they are predictable from the known physical and chemical properties of lakes, rivers, and estuaries.

The biological effects of the Aswan dam, however, could not be predicted so easily. The complexities of the systems involved are such that the best we can do is guess what the various biological consequences might be and then wait to see what materializes. Moreover, biological changes frequently occur more slowly, and even after several years all of them may not be evident. However, we can discuss some anticipated effects, the greatest of which result from salinity changes. The higher salinity in the eastern Mediterranean, for instance, could result in the replacement of existing ecosystems by others better able to tolerate higher salinities. The intrusion of salinity into the Nile delta already has made some coastal lakes brackish and now is threatening agriculture in the region. Moreover, the threat that salinity may increase sufficiently to stop growth of crops completely is developing farther inland.

There are other biological consequences from the tendency of silt to be accumulated behind the dam in Lake Nasser. This nutrient-rich silt has been the source of the fabled agricultural fertility of the Nile delta, and without this influx of nutrients from the deposit of silt throughout the delta during periodic floods (which are eliminated by the dam), the fertility of the area is likely to decrease even if salinization were not a problem.

The creation of a huge lake, myriads of irrigation canals, and miles of marshlike habitat along banks also opens the Nile Valley to the threat of disease epidemics, because the insect and snail vectors and the diseases they carry are more at home in sluggish lake and canal environments rather than in actively flowing rivers. Bilharzia is a blood disease caused by a parasitic fluke that requires a lake-dwelling snail to complete its life cycle. Experts from the World Health Organization (WHO) contend that the dam system will yield at least 1 million new cases, although Egyptian health officials deny it. Even more serious, malaria could become established permanently in the Nile Valley. Before the High Dam, malaria was resident in the Sudan to the south because the environment was favorable for the vector mosquito; only in favorable years could they spread north to the Nile River Valley. Lake Nasser provides an ideal breeding habitat for the mosquito, and the next northward invasion could make malaria endemic to Egypt. The same possibility exists for other vector-borne diseases that prefer the lake and marsh habitats that Lake Nasser provides.

The 200-mile, developing Lake Nasser has inundated a vast store of terrestrial nutrients, leading to the classic reservoir pattern of rapid growth of the fish population. The fishermen of Aswan have benefited from this growth and reportedly are taking a catch of between 10,000 and 20,000 tons of fish per year. However, poor transportation makes it difficult to get much of the catch to markets, and it is possible that Aswan fishermen are reaping the harvest that delta fishermen used to take in sardines. The nutrients in the silt that settles in the lake no longer reach the Mediterranean, and they are missed by local sardines, whose populations have dwindled—much to the dismay of the fishermen who once depended on them.

These are a few environmental and ecological problems that are emerging as some of the systems consequences of the Aswan High Dam. Most are obvious possibilities in hindsight, although we still are uncertain about the timing of events. For example, siltation behind the High Dam is inevitable, but how long it will take for this phenomenon to reduce power-generating capacity or to affect the storage capacity of Lake Nasser is unknown. Some experts believe it will take as little as 50 years, while others insist that it will be no less than several centuries. For this specific problem as well as for the future incidence of bilharzia, the size of the sardine catch, the industrial development of Egypt, the fate of eastern Mediterranean ecosystems, and even peace in the Middle East, we simply cannot predict the outcome.

The Aswan High Dam was built for water, to prevent floods, and to generate power. It does all these things. But it also has effects that were not intentional but which were inevitable just the same. And it could have consequences we do not recognize yet. There is no better example of the complex interactions and ramifying consequences that a single decision—to dam the Nile River—can have on a system. It is important to remember that such ramifications cannot be avoided in any complex system and that the wisest alternative by far is to avoid decisions that might affect the system. However, the proposed modifications frequently are deemed so important that this is not considered a viable alternative; under these circumstances policy-makers must be prepared to deal with the ramifications of their works as well as with the technological modifications themselves. Furthermore, even inaction constitutes a policy that frequently will be deemed worse than any positive action, no matter what its consequences.

CASE 2: THE CALIFORNIA WATER PLAN

There are striking parallels between the water situation in California and the Aswan dam. For instance, in California as in Egypt, power generation and water supply have been important considerations in the diversion of the state's major rivers that would otherwise flow out into the Pacific. Though even a brief perusal of the California system reveals a situation of greater technological complexity than the Aswan project, in both situations the same basic hydrological effects emerge.

The movement of water from northern California to the south has been under way only since mid century (Figure 7-7). Earlier development followed the patterns exhibited elsewhere around the country: Cities developed their supplies with simple dams and aqueducts (Figure 7-7b). As California grew and urban areas came to occupy whole regions, and as irrigated agriculture became the major industry in the state, simple reservoirs and the already diminishing groundwater supplies were no longer adequate. The development of the Colorado River Aqueduct and All American Canal (Figure 7-7c) were early efforts to tap a major river, and these canals served the metropolises of southern California and supported the growth of irrigated agriculture in the Imperial and Coachella valleys. Not long after, the federal government went into the water redistribution business when it became involved in irrigating California's Central Valley and generating electric power with the Central Valley Project (CVP). This system was based on the Shasta dam, which was constructed on the Sacramento River, and also involved the construction of a series of canals to carry the waters of Shasta Lake and the Sacramento River southward. Subsequent development of the branches of the Sacramento and San Joaquin rivers is well advanced (Figure 7-7d).

Even the CVP is dwarfed, however, by the scale of the California State Water Plan. Under this plan (Figure 7-7e) nearly all rivers of the north coast would be poured into the Sacramento River, from whence they would be channeled south via the California Aqueduct to the San Joaquin Valley and southern California. Not all portions of the State Water Plan have been authorized, but major sections of the authorized part of the plan (the State Water Project, or SWP) have been completed. The major difference between the plan and the project is that the project excludes all the plan's proposed diversions of the north-coast rivers into the Sacramento River.

Supporters of the project have cited several benefits: The proposed water programs would be instrumental in flood control, would provide inexpensive hydroelectric power, and would be popular, heavily used recreational resources. More important, the water sent south would be vital for irrigation and future urban development. Furthermore, supporters

FIGURE 7-7 Chronological development of California water resources from natural state to proposed state, as envisioned by the California Department of Water Resources and the Central Valley Project. (a) The natural pattern of surface flow in California—most of the smaller rivers and the drainage basins of Southern California are omitted—and three examples of the development of single-purpose water supplies for specific areas. See key. (b) The addition of the major water diversions from the Colorado River. Most of the water is used for irrigation in the Imperial and Coachella valleys, but some finds its way to the South Coastal cities. (c) Finished, proposed, and in-progress projects of the Central Valley Project. Major developments are concentrated on the upper Sacramento River, in the Delta region, and on both sides of the San Joaquin Valley. Most of the water is used for irrigation. (d) The projects of the State Water Plan, including the development of the rivers of the North Coast. This uses the Sacramento River as the major funnel (as the CVP does). The SWP serves the San Francisco Bay area, the South Coast, and irrigation in the San Joaquin Valley. The proposed Peripheral Canal is a joint federal-state project. Many projects, proposed and constructed, have been omitted from these diagrams. For a complete diagram see the source. (Adapted from *Annual Report of the California State Water Project, 1971*.)

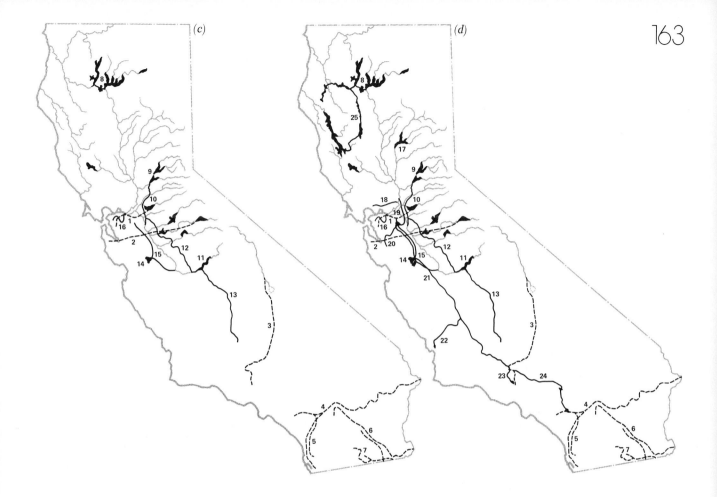

1 Contra Costa Canal and Camanche Reservoir
2 Hetch Hetchy Aqueduct and Hetch Hetchy Lake
3 Los Angeles Aqueduct
4 Colorado Aqueduct
5 San Diego Aqueduct
6 Coachella – East Highline Canals
7 All American – West Canals
8 Shasta – Engle – Whiskeytown Reservoirs
9 Auburn – Folsom Reservoirs
10 Folsom South Canal
11 Millerton Lake – Friant Dam
12 Eastside Canal
13 Friant – Kern Canal
14 San Luis Reservoir
15 Delta – Mendota Canal
16 Contra Costa Canal
17 Oroville Dam and Lake Oroville
18 North Bay Aqueduct
19 Peripheral Canal with Clifton Court Forebay
20 South Bay Aqueduct
21 California Aqueduct with Los Banos Reservoir
22 Coastal Branch
23 West Branch
24 East Branch
25 Reservoirs and aqueducts of the North Coast river system (empties into Sacramento River via Engle Lake)

warned that delays would be intolerable, since inflation would increase costs and long time periods were required for planning and construction. Opponents, on the other hand, believe that further diversion of water is both unwarranted and harmful. They cite known effects of present diversion as indicators of what would happen when diversion is even more substantial. For example, diversion is so great that the flow of the San Joaquin River has been reversed, with the result that salmon runs up the river have ceased. The salmon were confused and would not migrate up a river that was flowing the wrong way. Since delta outflow is projected to decline to less than half its present volume, opponents predict dire effects, especially if agricultural waste water from the south is returned and dumped into San Francisco Bay.

The expected effects on the physical and chemical environment of the bay-delta region are similar to the Aswan experience. Water diversion would reduce the volume of flow to the point where increased siltation in delta channels is expected. Furthermore, reduced flow would mean potential salinity intrusion farther up the delta with a consequent reduction in the volume of brackish water. (This brackish water gradient is vital for migrating salmon and other organisms, and a reduced volume could upset the entire biological system.) Finally, reduced outflow could mean a reduction in outflow from San Francisco Bay, an increase in the concentration of wastes, and a reduction in water quality. These things are known qualitatively; but to find the critical minimum flows, maximum pollutant inflows, and other factors that affect the bay-delta system requires detailed research.

The biological consequences that could result from each of these possible physical and chemical changes is not difficult to envision. Like the changes themselves, they are merely shifts in balances. But they differ in that the point at which the whole system is endangered will often remain unknown until it is too late. For example, decreased water flow into San Francisco Bay combined with a steady or increasing pollution load there could cause the crab population of the bay to decline even further than it already has, perhaps even to the point of local extinction. Exact biological effects resulting from physiochemical changes are difficult to predict, however, and the information needed to evaluate the prospects is difficult if not impossible to get. For example, salmon migrating upstream might find their situation little changed even if the bay did become more polluted, because they would be leaving the affected areas behind; organisms of the delta estuaries also might be able to accommodate to the possible salinity changes in their habitat resulting from the decreased flow of fresh water into the bay simply by migrating upstream until they reached their "preferred" salinity. Though we can make educated guesses about the outcome of situations such as these, we usually can be sure of the exact effects only after they occur. On the other hand, we can make some predictions with assurance. For example, nearly all would agree that most or all of the populations of organisms that live in the bay itself could be expected to show rather drastic changes if the waters of the bay became increasingly polluted.

In general, then, some environmental costs of diverting northern California water southward will result from hydrological changes in the diverted rivers and especially in the bay-delta system. Other problems have been raised. There is the possibility that irrigation water will increase the concentration of surface salts by capillary action. (The evaporation of irrigation water draws salts upward to interfere with the crop production for which irrigation was sought in the first place.) If enough water is used to leach out these salts, it must be drained off somewhere, and the extra water must come from the north. These factors initiate a whole set of economic, social, and political problems (see Boyles et al., 1971, and Marine, 1970).

Public concern over the potential environmental effects of the State Water Project center upon the last but vital link in the project: the means to transport water from the Sacramento River into the California Aqueduct, by-passing the delta, where most of it would merely flow into the sea rather than into the aqueduct. The Department of Water Resources (DWR) prefers to accomplish the task by routing water from the Sacramento River through the so-called Peripheral Canal (Figure 7-8) which would run around the delta and then into the California Aqueduct. Strong opposition to the Peripheral Canal has caused the federal government to delay its share of funding pending satisfactory water-release plans that will protect water quality in the bay-delta system. The construction of the Peripheral Canal undoubtedly will be delayed pending a court decision on the acceptability of the water-release plans submitted to meet these needs.

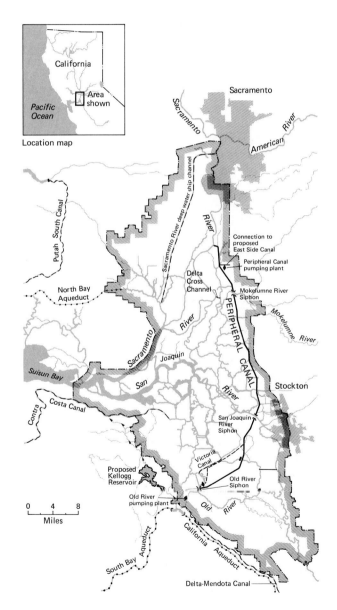

FIGURE 7-8 *The planned Peripheral Canal to transport Sacramento River water east around the Delta to the California Aqueduct, the Delta-Mendota Canal, and the South Bay Aqueduct. To maintain minimum standards of water quality, releases of water to the Delta would be made at many points along the canal. "Siphons" refer to tunnels where the Peripheral Canal passes under natural stream courses.* (Adapted from *Peripheral Canal Unit, Central Valley Project,* U.S. Bureau of Reclamation, 1968.)

The delays imposed upon construction of the Peripheral Canal are particularly galling to water-development agencies involved in the CVP-SWP construction as well as to other agencies concerned with wildlife and fisheries. The improvement in bay-delta water quality that is likely to result from the regular releases of sufficient water volume as guaranteed by the Peripheral Canal, if coupled to more stringent waste-release standards for the bay, is a strong argument for the construction of this canal. The State Department of Fish and Game and the U.S. Fish and Wildlife Service both support the Peripheral Canal because it could improve the marshland habitat for fish and water-fowl in particular. A case in point is that the Peripheral Canal would end reversed flow in the San Joaquin River and help restore the southern salmon runs. Opponents, on the other hand, see the Peripheral Canal as the last major element needed not only to complete the State Water Project but to put the entire State Water Plan within reach. They reason that the canal would provide the final essential link to get Sacramento Valley water around the delta and into the California Aqueduct. Hence north-coast river diversion, which would pour additional water into the Sacramento River, would come several steps closer in technological feasibility with the completion of the Peripheral Canal.

The similarities between Aswan and California are inescapable. Large exports of water from northern California mean changes for the hydrology of the area, just as exports in the Nile Valley modified its hydrology. Some of these changes may prove deleterious to existing biological systems, systems that are adapted to present conditions, and most of these effects cannot be predicted easily. While potential problems should be monitored and alleviated in advance if at all possible, there simply is no alternative to water redistribution for California at present. While future diversions may be halted in California and elsewhere, just as it is impossible to destroy the Aswan dam and pretend it never existed, it also is unrealistic not to conclude developments already partially completed. The two biggest reasons are (1) that the state and federal governments lack the authority to deny resource development for a public need since they are merely servants of the public and (2) that irrevocable decisions have been made already, and it would be counterproductive to reverse decisions at this point. The State Water Project is 96 percent built or started, often on funds paid in anticipation of future water deliveries. Decisions to terminate the project should have been made long before construction began or before water-delivery contracts were signed.

The DWR estimates that 90 percent of the water exported south will be used for irrigation. However, many people feel that to build the State Water Project for these agricultural demands—which in effect amounts to providing very cheap subsidized water for opening new agricultural lands in the western San Joaquin Valley—is quite another matter. If the economic analysis of Dean and his colleagues (Chapter 6) about the need for water in the western San Joaquin is correct, the opening of these new lands would actually be agriculturally counterproductive for a number of years. Thus a major support for the project—to open new land for agriculture—would turn out to be not very important after all. In any case, the State Water Project is delivering water to Southern California and eventually will be completed in some form; further impact by its protagonists may affect the final details of the project and the degree of environmental protection that will be provided, but it will not modify the basic fact that water will be delivered southward.

CASE 3: THE TEXAS WATER PLAN

The Texas Water Plan (TWP) has objectives similar to those of the California Water Project. The estimated diversion for the California project by the year 1990 is 21.5 maf (million acre-feet) per year, while the Texas plan would move 17.3 maf per year, of which 12 to 13 maf would be imported from the lower Mississippi River in Louisiana by 2020. Of that portion imported from the Mississippi, 1.5 maf would be pumped through Texas to supply future needs in New Mexico.

In Texas, 75 percent of the rain falls on the eastern quarter of the state, while much of the state's expansion of cities and agriculture has been in the more arid western portion. Neither surface waters nor shallow groundwaters have proved adequate for the western part of the state; monitoring of major aquifers has indicated that the water table is experiencing a drastic decline. Projections show agriculture in west Texas declining from lack of water, for the falling water table, unlike that in California, indicates that agriculture already is overextended relative to the long-term water supply. In 1964, the Texas government funded planning studies which led to the release of the TWP in 1968. Like the California plan, the TWP was based on a forecast of population and agricultural needs; but unlike California's situation, water in Texas itself was found to be inadequate for future supply.

The plan (Figure 7-9) calls for intrabasin transfers in eastern Texas river systems with storage in a large number of reservoirs. A major addition to the water accumulated in this manner (up to 13 maf by 2020) would be imported from the Mississippi River. From the major wet areas in the northeastern section, long concrete-lined canals would carry the water west as far as El Paso and New Mexico. Hence this particular portion of the canal system (the Trans-Texas) would serve in part as a pipeline from Louisiana to New Mexico. The second major branch of the system would be the Coastal Division, which would supply water for the coastal cities and south Texas in general. A major function of this branch would be to release water into estuaries and the Rio Grande River to counteract effects from local diversions of these rivers and thus prevent saline intrusion. A third branch, the

FIGURE 7-9 *Diagrammatic representation of the Texas Water Plan.* (Adapted from *Texas Water Plan Summary*, 1968.)

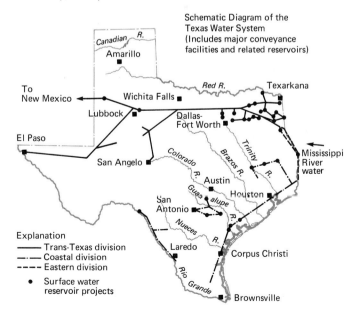

Eastern Division, would serve mainly to link the other two divisions and to function as the input point for water from the Mississippi River.

The environmental impacts of the TWP should be like those for California's State Water Project, because both projects are similar in terms of the hydrological cycle. Water releases into the Rio Grande and the coastal estuaries might alleviate effects such as saline intrusion there, especially relative to the Aswan experience. However, other problems remain: Large volumes of waste water from agriculture still must be disposed of or recycled, soil salinization remains a potential problem, and biological readjustments in the systems are bound to be involved. The Texas Water Plan actually may have its major effects out of state, however, in a manner strongly reminiscent of the Aswan dam and the Nile River. The Mississippi River delta and the outflow of the river are very important features to shrimp fisheries in the Gulf of Mexico. The TWP threatens the important spawning and growth areas for the shrimp in Texas estuaries, for unless the water poured into these areas from the coastal canal succeeded in maintaining the proper salinity, these estuaries might no longer function as a major nursery area for shrimp. Moreover, the TWP could seriously affect Louisiana nurseries as well if it modified the outflow of fresh water from the Mississippi.

As of 1974, the Texas Water Plan has not been authorized. The major bond issue to build the project was defeated in 1969, and since then Congress has refused to authorize the majority share of the financing that the federal government must supply. We should question whether or not irrigated agriculture should have been developed in west Texas in the first place or—now that it has been developed—whether it should be supported in the future. Certainly without the TWP, west Texas agriculture is in serious trouble, for the groundwater supplies continue to decline. The projections of the Texas Water Development Board indicate that water will have to reach the high plains of northwest Texas and the Trans-Pecos (western Texas) regions by 1985 and 1990 respectively to sustain irrigated agriculture there. From a 1964 level of 7.7×10^6 (7.7 million) acres of irrigated agriculture in the southern high plains, there would be a drop to 2.2×10^6 (2.2 million) acres by the year 1990 if the TWP is not built. However, these projections do not address the question of whether or not irrigated agriculture should be practiced in parts of Texas that could not sustain it without the influx of water from other regions or without an artificial input of water.

The Texas Water Plan was planned and submitted for approval just in time to suffer from the California experience and to meet an awakening environmental consciousness. Current federal disinterest in massive multipurpose water-diversion projects is an effective block to construction of the TWP. Proponents of the plan may be comforted in some small measure, however, by the fact that the Peripheral Canal in California is stalled for the same reasons.

Water Development, Present and Future

The importance of water to human beings is so obvious that it has permitted a large "establishment" to develop, charged with the mission of supplying water wherever it was needed. Legislators have responded only too well to the public's demand for continuing water development because, after all, they receive credit and thus votes for such projects. Moreover, this gives technological and engineering staffs plenty of work, and the public shares the benefits. Water development offers something for everyone, with no one harmed, and so it is no surprise that it would gather increasing momentum and finally culminate in projects stretching across continents.

The momentum accrued from public support and successful completion of large projects is evident in the history of American water development. Gilbert White (1971) has identified six stages in this development sequence:

1. Single-purpose construction of one to a few rather simple and similar structures by private managers. This includes projects such as the damming of streams to create farm ponds for watering livestock, irrigation, or fishing.

2. Single-purpose construction of perhaps greater proportions but of similar complexity by public agencies for navigation, flood control, or power generation.

3. Multipurpose construction by public authorities similar to stage 2 but perhaps of still greater complexity. This includes projects such as the California and Texas water plans which claim to provide for such diverse needs as irrigation, power generation, flood control, and recreation.

4. Construction that remains multipurpose but with the emphasis shifted so that there is no longer the strong need to make the project serve as many purposes as possible. Instead, the emphasis is toward using a combination of means to best accomplish the goal deemed most important. In a system like this, instead of employing a single means to accomplish a goal—say the construction of an extensive system of levees for flood control—we might have a more comprehensive system including levees, diversion canals, check dams, the use of natural floodplains, and even the prohibition of development in areas where occasional flooding is likely to occur. If this project serves purposes other than flood control, it is all the better; but this is a bonus rather than a requirement for the construction of the system.

5. Like stage 4, with a directed research program to improve the effectiveness of the program.

6. The merging of multiple purpose and multiple construction so that the methodology as described in stage 4 is retained but a greater effort is made to make the project multipurpose. Though a greater effort is exerted in this regard, the emphasis is still on the best means to accomplish the most important goal, a strategy which White states is not yet perfected, "but toward which there is groping on several fronts, particularly in metropolitan areas." This strategy is clearly the systems approach discussed in Chapter 3.

White felt that the newly developing strategy reflects a basic change in human attitudes, not so much toward meeting water needs as toward the whole of nature, followed by attempts to adjust policies and programs to these changing conditions.

At base, this emerging strategy reflects a shift in man's attitude toward nature and his concomitant role in society. The view of man the transformer or man the conqueror... is replaced by another. In the view of man as the cooperator, man the harmonizer, construction is only one means of coming to terms with an environment he never fully explores and that is constantly changing under his hand. With the adoption of this view, the means and instruments of handling water become increasingly complex, the concern with tracing environmental impacts more acute, the adjustments to human preferences increasingly sensitive, and the demand for citizen participation heavier.

The emphasis shifts from constructive to scientific probing, and from long-term commitment to short-term flexibility.[1]

It is easy to agree with White's contention that the main reason for these changes is a growing awareness of the pervasive environmental costs associated with large-scale manipulation of the hydrological cycle, and indeed, considerable evidence supports this contention. There seems to be significant and growing public doubt that all proposed projects are automatically worthwhile, and this shift in public opinion supports the efforts of well-established environmental interest groups and provides a favorable environment for the founding of still others. Well-organized, well-funded, and persistent opposition to projects of questionable merit in turn affects political support for water development, imposes costly delays on construction of projects, and finally focuses increased attention upon alternative means to supply water needs. This chain reaction was started by the growing realization that water development projects have significant environmental costs; and whereas in the past these same costs were considered unimportant relative to the demonstrable benefits, the public now is having second thoughts. The National Environmental Policy Act has formalized this shift in public opinion. It requires for the first time that environmental modification and damage expected from a project be listed in a formal report as a prior condition to approval of a project, and this requirement has gone far toward preventing the elimination of environmental considerations as irrelevant or unimportant even before the matter was brought before the public for discussion, an occurrence that had been all too frequent in the past. There has also been increasing skepticism toward the benefit-cost ratio, an engineering technique in which all the costs and benefits are estimated in terms of dollars. Typically the benefits must exceed the costs for a project to go forward. The benefit-cost ratio has been under fire because costs and benefits which are difficult to quantify, for example, the value of a solitary visit to open country or the continuing survival of a wild species, often have been ignored. A second problem is that the benefit-cost ratio can be manipulated easily to show net cost or net benefit, depending upon how the variables are quantified.

Long-range planning is beginning to enter water development. The Sahel (sub-Saharan Africa) has been suffering a long-term drought that threatens both the lives and livelihoods of the people in this vast region. The world response to famine in the Sahel has been twofold: to ship in emergency food and to supply the assistance to dig wells for groundwater. However, the latter has proved counterproductive, since every new well has resulted in resumption of the old life—temporarily—with livestock herds increased and the already drought-stricken sod overgrazed even more, accelerating soil loss and in the long run worsening the lot of the people. In recognition of this fact, the U.S. Agency for International Development (AID) has refused to help drill any more wells because the agency is convinced that they only worsen an untenable situation. In the long run this may be the more humanitarian policy, even if in the short term it seems not to be.[2]

Will we see more policies that place an emphasis on long-term interests? No one knows. If water cannot be transferred as freely as in the past, its rising price in water-deficient areas may stimulate more research in weather modification or make large-scale desalinization more competitive. Ultimately, we may even learn

---

[1] From G. F. White. *Strategies of American Water Management.* Ann Arbor. Paperbacks, Ann Arbor, Mich., 1971, p. 123.

[2] The choice has been postponed by the rains that relieved the drought in 1974 and 1975.

to accommodate to natural limitations and shape our land-use practices to conform more closely with water supplies in each area. However, in the background are the same concerns for human dignity, social welfare, health, happiness, and safety that started water diversion in the first place. If new environmental policy requirements lead to severe economic dislocation and drastic declines in health and welfare for some segments of society, or if it comes to a perceived choice between human beings and nature, the situation could revert to large projects with small concern for their impacts. However, at least in California the DWR is exercising the new caution as it searches for more adjustable policies; in short, it demonstrates the emerging strategy White identifies. We can be confident that all the complexities being built into the water development process as a result of this change in strategy are likely to benefit environmental quality, if for no other reason than that we shall be trying to anticipate environmental impacts rather than depending entirely upon hindsight and corrective action.

Questions

REVIEW QUESTIONS

1. To check your understanding of the hydrological cycle, assign the values given in Table 7-1 to Figure 7-1 as proportions adding to total water = 1.0 (for example, oceans and seas = 0.972). How much water would have to be taken from the seas to reduce the figure to 0.960? How might this water be distributed?

2. Compare the effects of the Texas Water Plan and the California Water Project on
   a. estuarine biological productivity
   b. freshwater fisheries
listing the costs and benefits of water diversion insofar as they affect these two aspects.

3. Why is surface-water diversion likely to remain the dominant form of water redistribution?

4. Show why the known and proposed environmental impacts of the Aswan High Dam could be predicted from an analysis of the hydrological cycle.

ADVANCED QUESTIONS

1. What environmental impacts might you expect if the Mississippi River were to be dammed at New Orleans? Hint: Read Graves's paper in *The Water Hustlers*.

2. Harte and Socolow (comment 1, pp. 263–264) present an exercise like question 1 in the review section above but somewhat more complex. Answer their question.

3. Construct a scenario, with the best documentation you can supply, based on the assumption that massive water-diversion projects will *not* be funded in the future. Concentrate upon the issues of
   a. human population dynamics
   b. social equity
   c. economic development
   d. impacts on natural ecosystems
   e. interactions among these four categories

Further Readings

Note: The bulk of volumes and literature available in this area is mostly technical in nature, but the environmental movement has stimulated a growing number of texts critical of the water-development process. Most of these are concerned with the politics of water development, an issue not treated in this chapter.

Borgstrom, G. *Too Many*. Macmillan, New York, 1969.

> Contains a number of chapters on water, the most valuable of which is the chapter on international conflicts over water diversion.

Boyles, R. H., J. Graves, and T. H. Watkins. *The Water Hustlers*. Sierra Club, San Francisco, 1971.

> Sharp attacks on water development in California, Texas, and New York. It is especially valuable for contrast to the viewpoints of water development agencies. The chapters on California and Texas contain additional bibliographies to supplement the case studies presented in this chapter.

California Department of Water Resources. *Annual Reports on the California State Water Project*. Available from D.W.R., Resources Agency, 1416 9th Street, Sacramento, Calif. 95814.

> Succinct summaries of the aims, progress, financing, and expenditures in the California State Water Project.

California Water Resources Control Board. Final report, San Francisco Bay-Delta Water Quality Control Program. Issued in June 1969 and available from the board at the D.W.R. address above.

Clawson, M., and H. H. Landsberg. *Desalting Seawater*. Gordon and Breach, New York, 1972.

> The authors are very pessimistic about desalinating agricultural water.

Goldman, C. R., J. McEvoy, and P. J. Richerson (eds.). *Environmental Quality and Water Development*. Freeman, San Francisco, 1973.

> A collection of papers forming a report to the National Water Commission. See the paper by Hagan and Roberts in particular.

Harte, J., and R. H. Socolow. *Patient Earth*. Holt, New York, 1971, chap. 16.

Marine, G. "The California Water Plan: The Most Expensive Faucet in the World." *Ramparts*, 8, No. 11 (1970), 34-41.

> A scathing attack on the California State Water Project, concentrating on political and population issues.

National Academy of Sciences/National Academy of Engineering. *Environmental Problems in South Florida*. Vol. II of the report of the Environmental Study Group to the Environmental Studies Board, 1970.

> Treats the problem of the Miami-Dade jetport and the Everglades National Park, in which water is intimately involved. See also the treatment by Harte and Socolow (1971, chap. 12).

Seckler, D. (ed.). *California Water: A Study in Resource Management*. University of California, Berkeley, 1971.

> The best volume available on the progress and problems of California water development.

Texas Water Development Board. *The Texas Water Plan: Summary*. Austin, Tex., 1968.

> See also comments by the executive director of the board in *Water for Texas*, vol. 2, nos. 1 and 2, 1972.

U.S. Office of Saline Water. *The A-B-Seas of Desalting*. Washington, 1968. (Updated and reissued periodically.)

van Hylckama, T. E. A. "Water Resources." In *Environment* (W. W. Murdoch, ed.). Sinauer Associates, Stamford, Conn., 1971.

> An excellent summary and perspective highly recommended to readers.

8. At the Heart of the Matter

Land is probably our most important natural resource because more than anything else the use of the landscape determines the pattern of society, patterns of economic activity, and local environmental quality. The nature of the land itself favors particular uses over others. For example, one certainly would not seek very mountainous areas for agriculture as long as sufficient fertile flat land existed. Moreover, human beings are strictly terrestrial; we breathe air and do not adjust easily to living on ocean floors (at least not yet and definitely not without assistance) or to floating freely in the atmosphere. Consequently, all our activities are anchored to the landscape, like it or not.

Yet, while we are securely tied to the land, we have not demonstrated an understanding of all the effects we have on this land or could have. Our present use of it normally reflects socioeconomic forces rather than a really comprehensive recognition of the interactions with all other resources that various land uses create. Nevertheless, these interactions are unavoidable, and their consequences will continue to appear whether we are aware of them or not. In Chapter 6, for example, we learned that land-based agriculture would have to supply most of the world's future food supply. As the population continues to grow, then, this would require vast tracts of land to be converted mainly to food production, which would reduce the land available for conservation of natural systems and, moreover, probably would require water development in the more arid regions of the world. But in Chapter 7 we learned that water development such as this would permit land uses which have an uncertain economic future and which actually might be environmentally harmful. Yet we are urbanizing rapidly and demanding still more land for housing, commerce, and industry. The fact that we do not yet know how best to urbanize land makes no difference, nor does the fact that urbanization itself creates the need for additional water and also competes for land with the agriculture needed to sustain urbanization.

Thus, a complex set of dynamic interactions between land use and human society have environmental consequences. Limitations of natural landscape determine policy, and human activities shape patterns of land use. For example, Environmental Protection Agency policy attempts to minimize vehicular traffic in mountainous valleys and heavily urbanized areas as one measure to help protect air quality (Chapter 12). Urban concentrations are one form of land use that threatens air quality, especially when topographical conditions favor pollution buildup. These interactions characterize all levels of land use from the neighborhood to the global scale, but in this chapter we shall concentrate upon land-use control and land-use planning in urbanizing regions, where the control of land use is a particularly important problem.

De Facto Land Planning, American Style

There is a great deal of land-use planning going on in the United States. Developers, landowners, federal, state, and regional agencies, and local governments are planning future land uses all the time, even if there is little coordination among any of them. Moreover, the land-use planning that does exist involves a chain of decision-making no less forceful than that in charge of water planning, and it is equally pro-development.

At the base of this chain are landowners, who normally regard their land as an economic commodity. Their land is just so valuable, and when someone offers the right price, they will sell it. Many farmers operate under this philosophy: Farm the land until rising prices make it profitable to sell, buy another piece with the handsome profit, and perhaps repeat the whole process. Land speculators operate in the same way. Buy the land, hold it until the price is right, and then unload it at a profit.

Buyers make up the second link in the chain by being willing to pay a price in excess of what the seller paid. The classic example of this process is development at the urban fringe, where farmers and speculators sell to developers, and subdivisions spring up to be bought by people eager to escape the central city (Figure 8-1). This shifts the urban margin farther into the countryside, where the process is repeated as long as there is unsatisfied demand to drive it onward.

To this point, we have a simple case of supply and demand: The higher values generated by urban and commercial land uses generally will insure the conversion of rural land to urban use. But other forces reinforce the trend and make it almost inevitable. The supply-demand marketplace itself encourages rising land values as long as inflation and demand keep prices moving upward. Just as important, large landowners and developers who are well known and powerful may form informal alliances with local governmental authorities, who, acting with the force of law, can insure rapid urbanization. Even without granting favors differentially, local governments long have favored development because it leads to a larger population and more taxable land, both of which could result in more political power and greater revenues for the city. Even the most conscientious and civic-minded administrator finds it difficult to resist this opportunity, and thus local governments generally favor development.

Other public actors involved in land-use planning include county assessors and the various resource and development agencies. The assessors are required legally to value land by its highest and best use, which means whatever use will generate the highest income from the land. Invariably this involves urban and industrial uses, and so the landowner who farms near the city may be forced to sell because taxes rise so steeply. (A similar practice directed toward harvestable timber is discussed in Chapter 9.) Willing or not, assessors contribute to the escalation of land values that favor development. Resource agencies in charge of water development and transportation in

FIGURE 8-1 *When one or more farmers sell their land to speculators and others do not, we often have scenes like this, where the urban margin slowly encroaches on farmland.* (Courtesy of Grant Heilman.)

particular promote development because they are facilitators—they provide part of the basic services and access needed to open an area to such changes as urban development, intensive range use, and irrigated agriculture. This sort of de facto, uncoordinated land planning has gone on for a long time, largely unwittingly; but at last the situation is changing, particularly at state and federal levels.

The U.S. Department of Interior is doing a great deal of future-oriented land planning as it endeavors to find suitable rivers to preserve under the Wild and

Scenic Rivers Act, to search for new land for new national parks, and, in common with other agencies, to seek additional lands suitable to designate as wilderness. This is particularly true in Alaska, where 125 million acres are being surveyed to select the best sites for future national forests, wilderness areas, national parks, and recreation areas. The California Department of Transportation is drawing up regional transportation plans specifically with the realization that the design of transportation facilities is a key determinant of land use. The most controversial agency in land use at present is the U.S. Environmental Protection Agency, which publicly has declared it will plan land use whenever local authorities are unable to draw up plans that insure air-quality standards will be met. Despite severe criticism, the EPA has pursued the role of local land planning quite aggressively in the belief that this is necessary to meet air-quality standards.

The only thing missing is interagency coordination. The EPA plans land use in urban air basins. The U.S. Bureau of Land Management, U.S. Forest Service, and U.S. National Park Service all have jurisdictions over parts of the federal land holdings, and each is free to manage its share in whatever ways it sees fit. Nevertheless, the time is approaching when even this problem will be dealt with. For several years Senator Henry Jackson of Washington has been sponsoring a National Land-use Policy bill, one of the main provisions of which would grant large funds to states that formulate statewide land-use plans. There are also signs that in many areas local forces which heretofore have favored development are beginning to reverse themselves because of growing public sentiment to slow or even halt urban growth. The idea that cities may limit their populations in the near future is no longer considered infeasible. As a matter of fact, some cities have already imposed such limitations.

Four Rationales for Land-use Control

Despite increasing signs of comprehensive land-use planning, some people feel more is needed. They dislike the fact that developers, often aided by local government, hold the initiative and are usually the real land-use planners. They dislike the extent of free-market control on land use and the pressures it creates for urban development. Most of all, they dislike the ugliness and discord of the resulting urban sprawl. Proponents of more comprehensive land-use planning base their arguments on four rationales:

1. The most frequently cited reason is couched in terms of efficiency and economy in urban development. Proponents maintain that urban land planning is necessary, especially at the urban fringe, in order to minimize public costs and to prevent chaotic, discordant development which wastes money, energy, and most of all land. It requires substantial capital expenditures to supply new streets and public utilities to new subdivisions, and when developments "leapfrog" (that is, appear beyond the urban fringe with undeveloped areas between them and the city), the costs for extending power, sewage, and new roads long distances increase even more rapidly until it becomes uneconomical to develop at all. The planning firm of Livingston and Blaney has found that it would be cheaper for the city of Palo Alto to leave large parts of its foothill areas in open space because the costs for utilities and sewage service would be greater

than the gain in municipal income from the increased property-tax base. This defense of the value of open space as being in the economic interest of the community is unusual, since the prevailing attitude of local authorities is that population growth benefits the finances of the community.

2. Another argument emphasizes the need to preserve sufficient open space for recreation (see, for example, Whyte, 1968). The assertions implicit within this argument are (a) that open space is a biological necessity for human health and (b) that planners can design open spaces which will fulfill these biological needs. Though both assumptions are unproven, we have some indications of their validity. For example, residents of large cities can and do appreciate the contrast of green parks with tall buildings, and most people find pleasure in visiting lakes, picnic areas, and mountain ranges. In fact, the lack of public agreement of what constitutes "open space" has made it easier to convince the public of the need for it—everyone has been free to define it individually. Fortunately, in addition to the unproven assumptions concerning our need for open space, some well-established reasons justify open-space land planning. In the first place, there are no acceptable substitutes for open space for some recreational activities (Figure 8-2), especially various active sports. Second, in cities, open spaces containing trees can exert a significant influence on urban climate. Cities form what is referred to as *heat islands;* that is, they are significantly warmer than the surrounding countryside. The reasons for this are not always clear, although the roadways and buildings are thought to absorb more solar energy and to reradiate the resultant heat more slowly than the soil surface of the countryside. Using a mathematical model, Myrup (1969) has shown that if the surface area of a city has as little as 20 percent tree cover, urban temperatures will be reduced by as much as 5°F during the day compared with a city having less than 20 percent tree cover; the difference is as much as 21°F for a city park. These differences result because trees transpire water into the atmosphere, which removes heat from the environment. (Heat invariably is lost when evapotranspiration occurs.)

Other reasons for the preservation of open space could be included. They are, however, not nearly as

FIGURE 8-2 *From this view, Central Park is truly an "open space" in the middle of New York City. For some New Yorkers, the park is the only place available for recreational activities.* (Courtesy of Carl Scholfield from Rapho/Photo Researchers, Inc.)

important as the fact that forces of supply and demand on land do not guarantee open space for recreation, cooling, and other purposes. Only comprehensive land planning will do so.

3. A third set of arguments support the preservation of natural landscapes and ecological systems. This needs to be separated from the open-space issue because the goals are quite different. In Chapter 6 we saw that natural systems have suffered from agricultural competition for land, especially in cases where extensive unbroken tracts are necessary to maintain the integrity of those natural systems. The agencies responsible for establishing national parks and federal wildlife refuges often recognize this fact as a primary objective, but not until Hawaii developed statewide zoning to accomplish the objective (conservation zones) was this recognized at the state level. (This subject is treated in more detail later in the chapter.) One outstanding example of the maintenance of open space solely for the preservation of a natural system can be seen in the recent history of a proposed development in the Sacramento area (see Peterson, in Goldman et al., 1973). The development has been stalled quite effectively because it would have brought the destruction (through flood-control canalization) of one of the few remaining marshes that serves the Pacific bird migratory flyway. The opposition, a group of well-organized conservationists supported by the cooperation of involved agencies and backed by a responsive local government, so far has succeeded in blocking the development despite the considerable economic power of the developer—something that never would have occurred only a few years ago.

4. The rationale most closely identified with the urban-land conversion problem is the protection of prime agricultural lands. Historically, cities also have been located at strategic places—along rivers and major highways and at the center of large agricultural regions. Cities that outgrow their original roles as agricultural service centers nevertheless are located amid the best agricultural land, and as they grow, they compete for lands most suitable for intensive agriculture. As we have seen, economic forces promote the conversion, and so land-use planning seems necessary to prevent waste of quality agricultural land.

Mechanisms for Land-use Planning and Control in Urban and Urbanizing Areas

If it is clear that land-use patterns are important, (Figure 8-3) it is equally clear that they are not well-planned with respect either to ultimate goals or to the ramifying effects on other parts of the systems involved. We shall assume (and thereby avoid the philosophical controversy over the loss of individual freedom) that more comprehensive land-use planning is required and even desired by the populace and that, therefore, it will be implemented. This does not, however, give us the means to accomplish the task. In order to simplify the discussion of a topic so complex, we will impose some limitations. First, we will consider only democracies, where the people affected can influence decisions about ways and means; otherwise land-use control is a simple matter of prescription and police enforcement. A second limitation is that we are talking about developed nations and certain developing nations where the society and law are advanced enough to effect social control over land use (usually by a governmental agency). In some countries where parts of the country are inhabited by more or less autonomous population groups, systemwide land-use planning is impossible.

When we consider political feasibility, equal treatment for all affected parties, enforceability, and economic cost, we see that the prospective land planner has remarkably few tools. In this section we shall examine the major plans that have been used and are being considered.

FEE-SIMPLE PURCHASE

Suppose the planning process indicates that certain parcels should be preserved as open space or that certain historically valuable buildings should be made a public heritage. One way to insure that desirable lands are preserved is *fee-simple purchase*, the outright purchase of land at its going market value by the state or by benevolent individuals and organizations. Governmental agencies sometimes have used this method, but in the United States more common practices involve purchase at reduced prices (in effect, a partial donation by the landowner or by eminent-domain condemnation), exchanges of property, or donation of the land to the public by private organizations. For example, parts were added to the new Redwoods National Park by exchange with private landowners of other forest lands in Nevada owned by the federal government. The exchange was reasonably equitable for both parties. The Nature Conservancy habitually buys land of special conservation value and subsequently holds it or deeds it to the public.

In conjunction with other mechanisms, fee-simple purchase of lands at the urban fringe has been used widely in Sweden and in the Netherlands. In these cases lands are bought at low prices so that, unlike in the United States, funds are not rapidly exhausted by inflated speculative values. And, of course, that is the main problem with fee-simple purchases in the free-enterprise context of the United States. As soon as the government wants to purchase land, its value is inflated. For this reason, such purchases have been restricted mostly to small parcels of highly desirable open space. Some states have attempted to circumvent the problem of inflated prices by exercising the right of eminent domain to expropriate lands for public use, usually with compensation; but the assignment of fair compensation has always been difficult. In addition to all these problems, local governments may resist extensive governmental purchase of lands, since these lands are not included in the tax rolls and therefore stop producing local revenues. For these and other reasons, fee-simple purchase is not likely to be a primary mechanism of land-use control in the United States.

DEVELOPMENT EASEMENTS

Open-space and conservation easements are less expensive allies of fee-simple purchase. An *easement* is the transfer of development or use privileges, usually for compensation, from landowner to some public agency. For example, sewage and utility easements give cities and utility services the right to install facilities on private property and, more important, allow continued access for maintenance. Conservation and open-space easements bar development even while the land remains in private hands. While they have the same effects on open space as fee-simple purchase, they are considerably cheaper. Nevertheless, even a modest scale of easement purchase would rapidly exhaust all public monetary resources, especially at the

FIGURE 8-3 *Four examples of land use: (a) tract housing in Daly City, California, (b) strip mining of coal in Ohio, (c) Pennsylvania farmland, (d) houses encroaching on marshland in Wisconsin.* ((a) Courtesy of Elizabeth Hamlin from Stock, Boston, (b) and (c) courtesy of Grant Heilman, (d) courtesy of Daniel S. Brody from Stock, Boston.)

MECHANISMS FOR LAND-USE PLANNING AND CONTROL IN URBAN AND URBANIZING AREAS 181

(c)

(d)

local level. The need to pay for speculative inflation is still present, though reduced, while the unfocused, poorly defined concept of "desirable open space" does not provide the operational criteria necessary to restrict the choice to a sufficiently limited number of parcels for the relatively scarce monetary resources available. These weaknesses negate the high political acceptability of easement as a conservation tool except in cases where public and private intentions are in harmony.

TAXATION AND ASSESSMENT PRACTICES ON REAL PROPERTY

State constitutions, backed by the equal protection clauses of the Fourteenth Amendment to the U.S. Constitution, specify that all real property shall be assessed and taxed at its highest and best use. For housing, the measure of highest and best use is market value based on comparable sales. In effect, this means that property taxes will be scaled to the most valuable *potential* use—which is invariably urbanization—whether the land is used for this purpose or not. Hence assessment practices tend to drive an area into development where there is subdivision and higher land prices in parts of the area (quite independently of whether or not anything is built on these sites). This is especially true when alternative uses cannot yield enough income to pay the property tax bills that result from elevated assessments. Owners of small parcels may find that because a neighbor sold land to a developer for a high price, the higher assessments placed on their properties insure that their taxes will exceed their incomes unless they sell or develop as well. Whichever they do, they succeed in transferring the tax pressure to other neighbors, and so the cycle continues. A large landowner may sell a small, scenic parcel to someone who wants to build a mountain cabin, only to find that the resulting increased tax bill on the remainder of the property forces the sale of more land—which increases the tax bill on the remainder even more.

To relieve these pressures, a growing number of states (19, according to Wagenseil, 1970) now offer taxation practices based on *current* use of the land rather than on its highest *potential* use. For example, Connecticut, Alaska, New Jersey, and Pennsylvania all have statutes providing for current-use taxation or for taxation based on agricultural yield. Hawaii has a tax-exemption plan for urban open spaces, recreation, and landscaping areas, provided that the use of the land for one of these reasons does not change for at least 10 years.

Perhaps the pioneering example of change in property-taxation practices can be seen in the California experience with the Land Conservation Act. After several abortive earlier attempts, in 1966 the electorate passed a constitutional amendment that permitted assessment on some basis other than "highest and best use." This cleared the legality of the 1965 California Land Conservation Act (also known as the Williamson Act, after its author). The Williamson Act specifies that landowners signing a contract restricting their development rights would be granted assessment on the basis of agricultural productivity rather than potential urban value, provided that the size of the agricultural preserve was 100 acres (or less by permission of the county supervisors) and that the county supervisors were willing to offer contracts in the county. Emphasis was placed on prime agricultural land and on the preservation of open space at the urban fringe. The 1969 constitutional amendments to the act broadened the requirements to include comprehensive local planning involving scenic highway corridors, important wildlife habitat areas, salt-evaporation ponds, managed wetlands, and submerged land.

Participation by landowners in the Williamson Act has been uneven in different areas. Wagenseil and numerous other authors have discussed some of the reasons for this, but the behavior of landowners is complex, and their reasons remain obscure. Nevertheless, it is difficult to see how any change in tax-assessment procedures of this type, requiring voluntary participation, could ever guarantee effective and comprehensive land-use planning. One could predict a priori that participation would be scattered and without a pattern necessarily corresponding to social requirements for open space and agricultural protection. Moreover, as is the case for easements and for fee-simple purchase, differential assessment cannot be expected to compensate invariably for lost speculative value. Where *potential* financial gain is thought to be large compared with *realized* tax savings, the option to sell is probably still more attractive to most landowners. Ultimately, therefore, the Williamson Act will appeal mostly to landowners who genuinely want to stay on their land and who need some protection from rising taxes in order to do so.

The value of differential assessment lies in its high political feasibility and low cost relative to various forms of easement and fee-simple purchases. However, successful preservation of open space through the use of preferential assessment requires that landowners appreciate the need for open space or other desirable social goals which might be attained through the use of differential assessment and that they be willing to forgo potential speculative profits. Even then, if they are unable to coordinate decisions concerning the future uses of their land with their fellow landowners' in the surrounding area, their decision, even if "desirable," could have little or no impact. Nevertheless, in spite of limitations, differential assessment practices do break the self-accelerating spiral by removing the pressure of escalating taxes from those who do not wish to sell land for urban development.

Many planners feel that use-value assessment is a useful device and that public participation will be good enough so that only amendments are needed to perfect the mechanism. How useful this kind of mechanism will be depends on the future of voluntary action. In this regard, Whyte has argued that Americans do not appreciate open-space values, and so he has concluded that only when open space is a scarce commodity can real institutional and attitudinal changes take place. He may be unduly pessimistic, because even now there are signs of change. Voters in California approved comprehensive coastal land planning and restriction of property rights despite the opposition of a powerful consortium of business and labor. When the time arrives that we see the population sacrifice profits from land sales for some social purpose, we shall be observing the arrival of use-value assessment as an important supporting mechanism to sustain open-space goals.

ZONING

Zoning is the prevalent American land-use control device. It dates back to the beginning of the century in the United States, but it was invented in Germany in the 1880s and is nothing more than the use of governmental police power to divide the landscape into pieces with variously restricted development rights. Zoning has been used mainly to restrict location and type of housing and industrial development, especially to establish uniformity in types of housing and to act as a suburban protection device by excluding nonconforming uses. This use of zoning underscores the fact that practices often have been arbitrary and directed toward purposes other than comprehensive land-use planning. In addition to this failing, at least one other major problem is associated with zoning. If an original zoning plan is arbitrary, there is conflict between what constitutes a valid but unforeseen use of the land (which requires an exemption called a *variance*) and what is actually an abrogation of the intent of the zoning plan. This is a difficult question with far-reaching consequences, for when variances are granted widely, zoning loses its planning significance. In practice the need for flexibility to deal with special cases has been abused so often that zoning has become a hollow mockery of what was intended. Even so, after an extensive survey of land use in the United States, the noted English planner John Delafons (1962) concluded that there was a need for even greater flexibility in zoning; but he felt that this must be accomplished without granting further discretionary power to local planning authorities. Unfortunately, he did not propose a means to accomplish this.

Since local governments generally control the zoning process, it is they that have made the greatest mockery of zoning. One of the best examples of this may be found in the history of urban development and the concomitant decline of specialized agriculture in Santa Clara County, California. In the 1930s, the county was thoroughly agricultural, concentrating mostly on wine grapes and fruit orchards. Servicemen on their way to the Pacific theater during World War II were attracted to the area, and when the war ended they settled in the county in large numbers, starting a developmental spiral in the area. As population growth continued into the 1950s, the cities of Santa Clara mushroomed across most of the county (Figure 8-4). The prevailing attitude of local governments was that their well-being would be improved by growing as large and as fast as possible, and the intensified competition for land among the cities alarmed farmers remaining in the area. These farmers banded together to have a country ordinance enacted for agricultural-exclusive zoning, but ultimately the ordinance proved counterproductive because the cities considered it a direct attempt to halt their growth. As a result, the cities raced with one another to annex as much land as possible before the ordinance became effective and accelerated the process of sprawling development. The farmers then obtained a state law for agricultural-exclusive zoning, but this only set off still another round of desperate annexation of land by the cities between the time the bill was passed and its scheduled implementation. Eventually the valley became so broken up by subdivisions that agriculture became uneconomical and virtually disappeared.

Despite the historical weaknesses of zoning, it remains popular as a land-use control device. While the problem of flexibility versus abrogating variance remains unresolved, the trend appears to be toward carefully drawn zoning plans and open public modifications following fixed guidelines to relieve any arbitrary and unwarranted decision. Thus agricultural-exclusive zones could be made and adhered to, and cities could delineate where development would be permitted and where agriculture and conservation would be supported.

A special form of zoning that has been useful is *floodplain zoning*, in which natural floodplains are zoned to exclude permanent buildings. Perhaps because risks of property loss and public danger are clear, floodplain zoning is used increasingly as an alternative to levee and dam construction. One related benefit is the conservation of marshland and riparian habitats in these areas.

RECAPTURE TAXES

If it is desirable to retain private land ownership but a state government wants to discourage unnecessary development through speculation, it can be accomplished by taxing away all profits made on land sales on the theory that speculative value is generated by public projects and therefore is unearned by a landowner. In Great Britain, the 1947 Town and Country Planning Act set up the mechanisms for taxing away 100 percent of this unearned profit. By 1959 the mechanism had been abandoned, however, partly because the government in power did not like the act and partly because there were a number of equity problems. English officials generally found that for tax purposes it was difficult to separate increases which resulted from inflation, increases earned by the landowner, and true unearned value. Another consequence of this policy was the destruction of incentive for landowners to sell land, and so there actually was a shortage of land for urban development. The British experience was, however, only an initial experiment that revealed the faults of recapture taxes. If recapture-tax systems could be set up to deal adequately with these problems, they could be a powerful tool in encouraging land use that would conform to plan. Recapture taxes could end the speculative profits that are so much a part of de facto land planning today.

DEVELOPMENT-RIGHTS TRANSFERS

If recapture taxes prove to be impractical, another alternative addresses the problem of the rights of private land ownership; it is called *development-rights transfers*. Land today is considered to confer to its owners the privilege of development without restrictions other than those imposed by external authorities; in the absence of restrictions owners may develop pretty much as they please. However, suppose any area of land has natural and social features that modify its inherent quality for development. If we discern all these factors, we could assign a certain number of development rights to each parcel of land. Thus a flat meadow with excellent soil and drainage may have a large number of rights that could be used for massive development such as the construction of a skyscraper, while a steep hillside could have so few that even a house is not permissible.

A development-rights transfer scheme requires a zoning plan (with faults of its own) to establish the number of rights inherent in each piece of land. Once this task is accomplished, rights could be transferred back and forth. Since there is a limited number of rights in each area, this scheme would automatically limit development. The mechanism whereby these rights are actually converted into development of the land would involve one of three things. The simplest is a free market in development rights, with rights being bought by owners who want to develop more than the inherent rights of their land would allow. They would buy these rights directly from owners of other parcels, and since the total number of rights cannot change, owners who sell some of their development rights are limiting automatically the potential development of their land unless they buy other

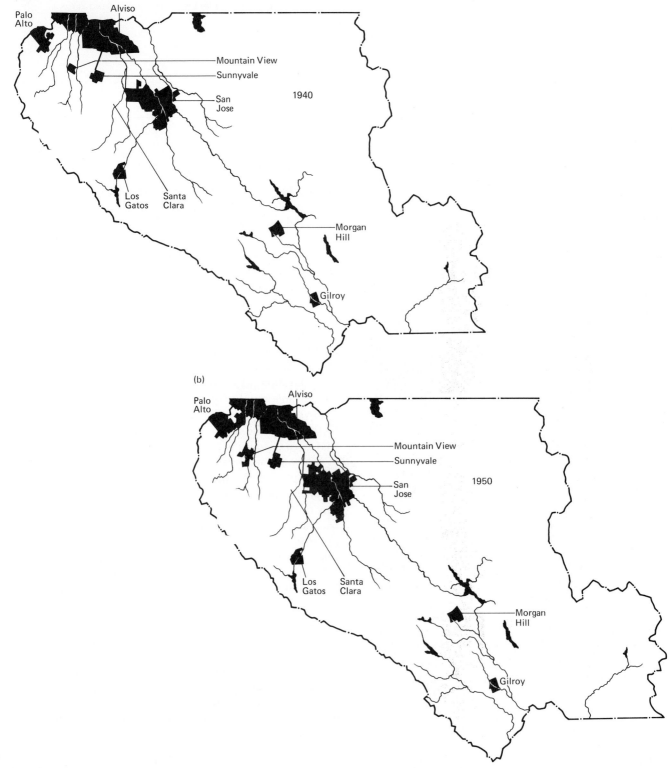

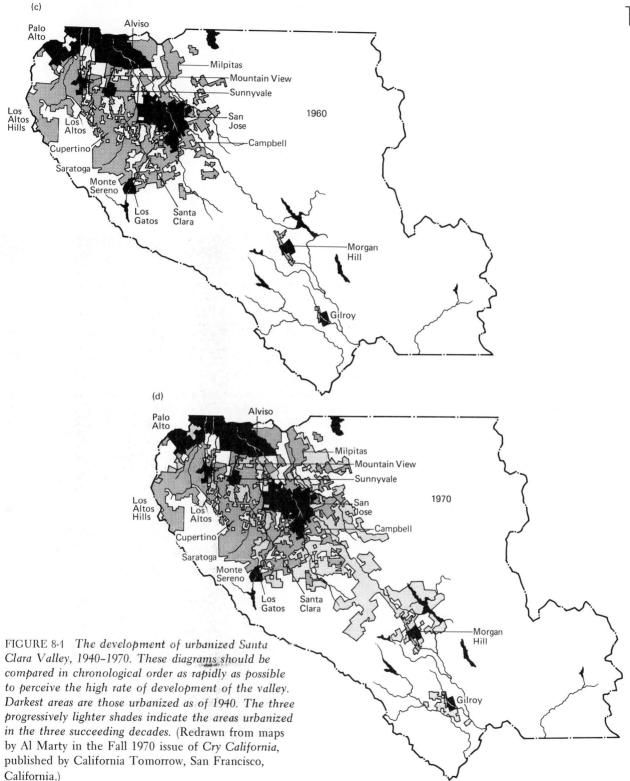

FIGURE 8-1 *The development of urbanized Santa Clara Valley, 1940–1970. These diagrams should be compared in chronological order as rapidly as possible to perceive the high rate of development of the valley. Darkest areas are those urbanized as of 1940. The three progressively lighter shades indicate the areas urbanized in the three succeeding decades.* (Redrawn from maps by Al Marty in the Fall 1970 issue of *Cry California*, published by California Tomorrow, San Francisco, California.)

rights at some future time. In this market, free enterprise would prevail, and supply and demand would set the value of a development right. A second possibility is the establishment of a governmental bank to control transfers and the price of rights; otherwise it would not interfere with development. A third way to limit development would involve the establishment of a governmental agency not only to control the transfer of rights and set prices, but also to make sure that some predetermined development plan is met.

Development-rights transfer schemes are attractive because they could be implemented easily and would require, in their simplest form, little attention. Questions remain, however, particularly about pricing of rights and establishment of inherent rights to each parcel. If either is improperly handled, development rights become extremely inequitable. Yet we have no easy answers to these questions. The third plan for management avoids these problems, but only by making the development rights a subsidiary means for implementing a preconceived plan. Hence this alternative is more closely allied with unadorned zoning.

CHANGES IN LAND TENURE: GOVERNMENT OWNERSHIP AND CONTROL

If all else fails, private ownership could be confiscated on certain parcels of land. Numerous tools are directed toward this end: eminent domain, easements, exclusionary zoning, and fee-simple purchase are a few. In the United States, the federal government already owns nearly half the land—in the West the proportion is over 75 percent. In Alaska and Nevada, the federal shares are 95 and 86 percent respectively, with most being under the management of the Department of Interior, Bureau of Land Management, and the Forest Service. In parts of Europe, such as Sweden and the United Kingdom, government ownership is extended even to urban areas. In Sweden, the government buys land in advance, determines when it will be developed, and then specifies what may be built upon it. In addition, the land is seldom sold to developers but simply is leased for the long term. It is Swedish national policy to encourage towns to buy land with at least a 10-year horizon; this policy is backed by the availability of funds for loans to municipalities for that purpose. All these provisions occur in the context of regional plans, often obligatory ones.

In the United Kingdom, municipal authorities and the central government buy and hold land for urban purposes, much as in Sweden. Land can be dedicated to open space, such as the famous Greenbelt of London, or it can be used for public housing. In no case is land sold back to private owners, although leases to users are made commonly. The management of land around the London Greenbelt is an example of this method of land use. The land has been used for completely planned communities, called "New Towns," that are built on municipal lands which were bought by the government years ago.

In other countries in Europe—France, the Federal Republic of Germany, Denmark, and Norway—similar policies hold. In Israel, a country populated by people predominantly of European origin, state ownership had reached over 92 percent by 1965. Thus in Europe land-use control programs are firmly under governmental control. In the United States, tradition is largely against this degree of governmental intervention, at least for single-family residences. Consequently, in the United States total regional planning like that seen in Europe (discussed further in Chapter 17) is not well developed except for a few special cases where people have ceded rights to the government. Floodplain zoning, discussed earlier, is one example. But perhaps the best example is related to public concern for beach access. On both the Atlantic and Pacific coasts, the population wants perpetual beach access even if it means that the shoreline must be mostly in government hands. Perhaps this exception to the pattern of private-land tenure may foreshadow changing public attitudes.

THE STATE OF HAWAII LAND-USE COMMISSION AND STATEWIDE ZONING

Hawaii was the first state to adopt statewide zoning under the aegis of a state master plan. In this plan, the state recognized four categories: urban, rural, conservational, and agricultural. These categories encompass all land in the state (Figure 8-5). The Hawaii State Land-use Plan is not a radical departure from land-use patterns established earlier by the usual forces. For example, the zoning for the southern shore of Oahu merely legitimates previously established urban patterns of land use. Nevertheless, the plan is a step forward because it recognizes that a small isolated state like Hawaii had an especially acute problem with rampant urbanization of severely limited agricultural land. The plan attempted to do something about potential problems while there was still time.

The statewide plan is subject to five-year reviews to adjust boundaries of zones and to more immediate appeals through petition and public hearings. All changes, except for some in agricultural and rural districts, must be approved by the commission and, in the case of special use permits, also by the county. After the regulations were passed in 1964, one review was made. The changes made reflect the continuing decline of agriculture and the increasing urbanization of Hawaii. In this sense, the zoning does react to the evolving patterns in Hawaiian society in a manner reminiscent of past examples of zoning. The revisions transferred agricultural lands that have gone out of production since 1964 to urban and conservational districts. Certainly the land-use plan is not inflexible.

While the commission was concerned primarily with agricultural land, the establishment of conservational districts has special importance. Conservational lands are generally those of steep slope and high scenic value, surviving forests, and major watershed areas. Development is not permitted in these zones, although a cynic might point out that many such lands are unsuitable for development. The declaration of conservation zones should insure that Hawaii will retain, at least for the present, some part of its natural ecosystems even while planning for expanding human activities. Hawaii does have explicit goals in wildlife and habitat management; the state has now developed a habitat-classification scheme to promote wildlife management for all the islands. Among other goals, the concept of the conservation district was intended to protect watersheds and water supply; preserve scenic areas; provide parkland, wilderness, and beach reserves; conserve endemic plants, fish, and wildlife; and prevent floods and soil erosion. Citizen advisory committees supported these goals by recommending that all conserved land be classified further as most suitable (1) for absolute conservation of native ecosystems (no overt management); (2) for modified conservation (managed wildlife habitat); and (3) for multiple-use practices that are conservation oriented.

190 AT THE HEART OF THE MATTER

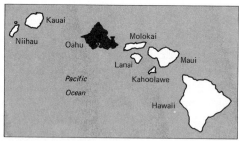

Topographic Map of the Island of Oahu

FIGURE 8-5 *The application of state-wide zoning to the island of Oahu, city and county of Honolulu, Hawaii. (a) The physiography of Oahu (darker shading indicates areas of high relief) (b) The land-use pattern in 1966. (c) The state zones for Oahu, as modified in the first review, 1970. ((a) Redrawn from USGS. (b) Redrawn from Part 5,* State of Hawaii General Plan Revision Program, *1967. (c) Redrawn from* State of Hawaii Land Use Districts and Regulations Review, *1970.)*

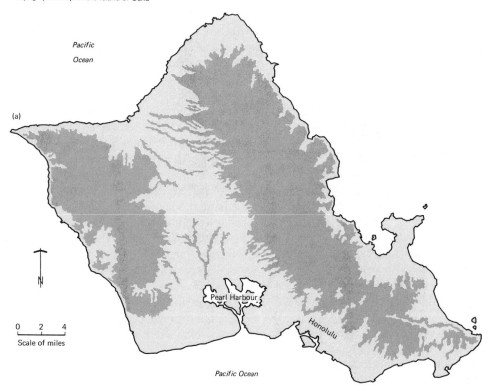

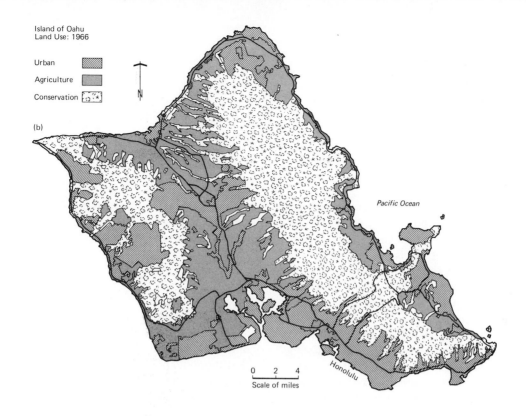

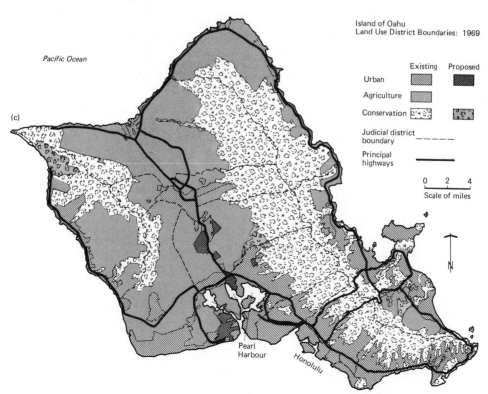

It is too early to judge how successful the zoning of conservation districts will be. Conservational zones are not always consolidated into large parcels of land, which merely reflects that the land's uses were determined long before it was zoned in this manner rather than measuring any lack of resistance of conservation districts to development. Many unique features of the Hawaiian landscape, fauna, and flora have been preserved; but as noted, much of the conservation area is practically unsuitable for development. Ecological theory does not suggest, however, that native species can be maintained in some areas but replaced by imported game species in others. Yet Hawaii is actively promoting programs for sports fisheries and hunting that bring in familiar species from the mainland (for example, blacktailed deer). Small islands are always susceptible to high rates of extinction (MacArthur and Wilson, 1967), even when there is no additional competition from other species, and the introduction of foreign species certainly has increased the danger of extinction for native forms. For Hawaiian land birds, at least, the picture is ominous. In the *Field Guide to Western Birds*, Roger Tory Peterson states that the Hawaiian Islands have lost more endemic birds than any other area their size except for the Mascarene Islands of the Indian Ocean. The state lists 15 species that are reasonably common, 7 that are nearly extinct, and 14 that probably no longer exist. Over 100 species have been introduced, 28 of which have become established, including the house sparrow, the pheasant, and various species of quail. The introductions continue to the probable detriment of remaining native species. It is reasonable to hypothesize from what we have seen up to now in Hawaii that land planning for the conservation of natural systems will repeat the European colonists' pattern, in which what really is preserved are native species and imported ones that have value to hunters and fishermen.

ECOLOGICAL PLANNING

The Hawaii State Land-use Plan is based on an ecological inventory, but it is a relatively crude one. The proponents of ecological planning all share the belief that more detailed information is needed to develop a really effective plan. In brief, the ecological planning process consists of five basic steps:

1. *An environmental resource inventory* During this opening phase of the planning process all relevant information on the environmental parameters which could (or should) limit development must be assembled.

2. *Value inputs and interpretations* At this critical phase the relative importance of various parameters must be determined—both in a scientific sense and in the value assigned by the public. Importance in the scientific sense refers to demonstrable evidence that certain types of development should be limited because of possible detrimental effects to the environment. Aesthetic matters may be important, but they are strictly social phenomena and in the long run should not carry nearly as much weight in the assignment of values as scientific evidence of possible environmental effects.

3. *Production of maps* A series of overlay maps is the logical way to summarize the results of information gathering and synthesis. At this point data are finally integrated and the spatial array is put together to yield a graphic portrayal of the landscape to be planned.

4. Identification of development constraints and the regional plan At this point the maps are interpreted, and the zones of interest to planners, developers, and conservationists are identified. Based on these interpretations, the land-use plan is drawn up.

5. Implementation In many respects this is the most difficult part of regional planning because the Fourteenth Amendment to the U.S. Constitution demands equal treatment of all parcels of land. In practice this means that any person affected by the plan has the right to ask a court to overturn the restrictions placed on the land. Like zoning, the regional plan must be defensible and subject to modification.

This basic framework underlies the various specific methods propounded by many, although different authors might treat certain aspects lightly or skip them altogether. Some principal examples follow.

Hills's method G. Angus Hills (1961) was one of the first planners to incorporate ecological parameters into the regional planning process. Hills defined areas as ecosystems and then analyzed the soil, climate, and geological characteristics of those ecosystems in order to produce maps showing capacities for development or agricultural production. For example, the type, depth, and texture of soil, the rainfall, and the type of underlying rock formations all are used to characterize an area, which then may be assigned a compatible use. The results of the entire assessment procedure are then put into a series of maps accompanied by a legend of recommended primary use as well as other permissible uses for each distinctive unit.

Hills's system neglects urban land uses and is particularly directed at open space and agricultural limitations. Furthermore, step 2 in the ecological planning schema above is implicit in his system, and steps 4 and 5 are ignored. For these reasons, Hills's method should be considered only a tentative step toward ecological planning, and it certainly could not be used in areas dominated by urban development. Lewis (1969) has developed a system much like Hills's, except that it is designed specifically to plan recreational and wildlands preservation in areas where human intrusion is inevitable. Otherwise it shares the basic flaws of Hills's method.

Ecological determinism Ian McHarg published his version of ecological planning in 1969 and subsequently became the most renowned of the ecological planners. His method, which he called *ecological determinism*, emphasizes the unity of natural systems and stresses the fact that natural processes should enter the planning process. In this spirit, McHarg proposed that all environmental parameters be studied in order of evolution—that is, first geological formations, then hydrology and climate, followed by soil and vegetation, and finally animals and human beings. Following this inventory of environmental resources the method calls for an analysis of variable interrelationships in order to establish prospective suitable uses; parameters especially sensitive to change and disturbance are given the most attention. The completed analysis yields a matrix of all permissible uses (Figure 8-6) that shows how compatible these uses would be as well as a series of maps incorporating the matrix into a single plan for primary uses of each area.

194 AT THE HEART OF THE MATTER

		Intercompatibility of land uses																							Natural determinants												Consequences				
		Urban	Suburban residential	Industrial	Institutional	Mining shaft-mined coal	active opencast coal	abandoned coal spoil	Quarrying stone and limestone	sand and gravel	Vacation settlement	Agriculture row crops	arable	livestock	Forestry even-stand softwood	uneven-stand softwood	hardwood	Recreation saltwater oriented	freshwater oriented	wilderness	general recreation	cultural recreation	driving for pleasure	Water management reservoir	watershed management	Slope 0-5%	15-25%	over 25%	Soils gravels	sands	loams	silts	Vehicular accessibility	Aquifer recharge areas	Water supply dependability	Climate fog susceptibility	temperature extremes	Air pollution	Water pollution	Stream sedimentation	Floor and drought control	Soil erosion
Urban		○																								○	△	□	▼	○				▼	○	□		▼	▼	▼	▼	▼	▼
Suburban residential		△	○																							○	○	△	□	△				□	○	○	□		△	□	▼	▼	▼
Industrial		○	▼	○																						○	□	▼	▼	○				▼	○	△	□		▼	▼	□	▼	□
Institutional		□	○	▼	○																					○	○	△	□	△				△	△	□	□		○	△	△	△	△
Mining	shaft-mined coal	▼	▼	○	▼	○																				□	△	○	△	○				▼	○	○	○		○	□	□	○	□
	active opencast coal	▼	▼	○	▼		○																			□	○	○	□	○				▼	○	○	○		○	▼	▼	□	○
	abandoned coal spoil	▼	□	○	△			○																		○	○	△	▼	▼				▼	▼	○	○		○	△	△	△	□
Quarrying	stone and limestone	▼	▼	○	▼				○																	□	○	○	▼	○				○	▼	○	○		○	○	△	△	△
	sand and gravel	▼	▼	○	□					○				▼												○	○	□	▼	○	○			○	△	○	○		▼	○	△	△	△
Vacation settlement		▼	▼	▼	△	▼	▼	▼	○	▼	○															○	○	△	□					△	○	△	△	△	□	□	□	□	□
Agriculture	row crops	▼	△	△	△	△	▼	▼	▼	▼	▼	○														○	△	▼	○	▼	○	○	○	△	△	○	▼	○	▼	○	▼	▼	△
	arable	▼	△	□	△	△	▼	□	▼	○	▼	▼	○													○	△	□	○	▼	□	○	△	△	△	○	△	□	△	○	△	□	□
	livestock	▼	△	□	△	○	△	▼	△	▼	△	○	○	○												○	○	△	△	□	△	○	○	□	△	○	△	○	△	△	△	△	△
Forestry	even-stand softwood	▼	▼	▼	□	△	△	○	▼	▼	□	▼	□	□	○											○	○	△	□	△	○	○	○	▼	○	△	○	○	○	□	△	□	○
	uneven-stand softwood	▼	□	▼	△	△	△	○	▼	▼	○	▼	▼	▼		○										○	○	○	○	□	△	○	○	▼	○	△	○	○	○	○	○	○	○
	hardwood	▼	□	▼	△	△	▼	△	▼	▼	○	▼	▼	△	○	○										○	○	○	○	□	△	△	○	○	▼	○	○	○	○	○	○	○	○
Recreation	saltwater oriented	□		▼	○		▼			▼			▼	○				○						○					△							○	○	○	○	○			
	freshwater oriented	□	○	▼	○	▼	▼	○	▼	○	○	□	△	△	△	○	○	○										△							○	○	○	○	○			
	wilderness	▼	▼	▼	□	▼	□	▼	○	▼	▼	○	□	▼	○	○	○		○									▼							○	▼	○	○	○	○	○	○	○
	general recreation	○	○	□	○	□	▼	○	△	○	△	○	□	□	△	○	○											△							○	▼	○	○	○	○	○	△	○
	cultural recreation	○	△	▼	○	▼	▼		▼	▼	□	△	□	△	△	○	○				○							○							○	□	○	○	○	○	△	○	
	driving for pleasure	▼	▼	▼	△	▼	▼	△	▼	□	△	○	□	□	○	○	○				○	○													○	▼	○	○	○	○			
Water management	reservoir	□	□	△	○	▼	▼	△	▼	▼	▼	○	▼	□	□	○	○			○	○	○	△	○	○	▼	▼	△	○	▼					▼	○	○	○	○		○		
	watershed management	▼	▼	▼	△	▼	▼	○	▼	▼	△	▼	□	△	▼	○	○			○	○	○	△	○	○				▼		○	○	○	○		○		○	○	○	○	○

▼ Incompatible   ▼ Incompatible   ▼ Bad
□ Low compatibility   □ Low compatibility   □ Poor
△ Medium compatibility   △ Medium compatibility   △ Fair
○ Full compatibility   ○ Full compatibility   ○ Good

FIGURE 8-6 *The McHarg matrix is actually three matrices combined. The items listed in the rows are the set of possible units. In the first part of the matrix, McHarg rates the compatability of each use with all other potential uses, using a scale of four alternatives (fully, moderate, low, and incompatible). In the second part of the matrix, McHarg extends the compatability analysis to the natural determinants of the landscape. Finally, he uses both preceding parts to estimate the impacts of each potential use on five environmental quality measures on a four-part scale (good to bad). For example, urban land-use is rated "bad" for soil erosion, flood/drought control, stream sedimentation, water pollution, and air pollution, while watershed management is rated as "good" for all five. (Redrawn from* Design with Nature, *copyright© 1969 by Ian McHarg. Used with permission of Doubleday & Co., Inc.)*

A typical example of the McHarg approach to land planning is the design for controlled urbanization of the Green Spring and Worthington valleys in Maryland. These valleys are adjacent to Baltimore but were relatively inaccessible until a new freeway was built. It was predicted that the now predominately rural and agricultural population would increase from 17,000 to 110,000 by the year 2000. Given that urbanization would occur inevitably, McHarg and his colleagues were asked to develop an ecological plan to minimize the environmental impacts of development in the valleys.

To accomplish the task, McHarg analyzed the physiognomic features of the region. He found that the region consisted of three major river valleys which were bounded by rather steep wooded slopes and adjacent to a single large plateau which was partially forested (Figure 8-7a). Furthermore, the landscape showed conventional drainage patterns: The plateau was well above the mean level of the area and so had little groundwater, while the valleys contained all the major rivers and aquifers. Analysis of soil types revealed that the slopes were subject to high erosion rates when disturbed. The valleys, collecting all the run-off, were subject to periodic flooding.

This information suggested that the three major physiographic regions could be classified readily by suitability for development. The valleys and wooded slopes were judged to be intolerant of urban development and best suited for recreation, open space, aesthetic values, and current levels of agriculture. On the other hand, the plateau was thought to be tolerant of development. Consequently, McHarg's plan (Figure 8-7b) emphasizes cluster housing on the plateau with sharply limited building on the forested slopes (the plan being to leave enough trees to prevent erosion). The valleys and unforested slopes would be zoned to exclude further development and prevent erosion, contamination of the aquifers, and probable flood damage.

The results can be summarized in a series of transparent overlays that collectively reveal the suitability of any one area for various modes of development. Formal development is outlined in a matrix developed along the ideas of Figure 8-6, in which different uses can be examined for compatibility with the characteristics of the area. For example, agriculture, forestry, recreation, mining, and urban building each can

FIGURE 8-7 (pages 196 and 197) *The McHarg plan for the Green Spring and Worthington valleys, Maryland. Compare the physiography of the region to the plan derived by ecological determinism. (Redrawn from* Design with Nature, *copyright© 1969 by Ian McHarg. Used with permission of Doubleday & Co., Inc.)*

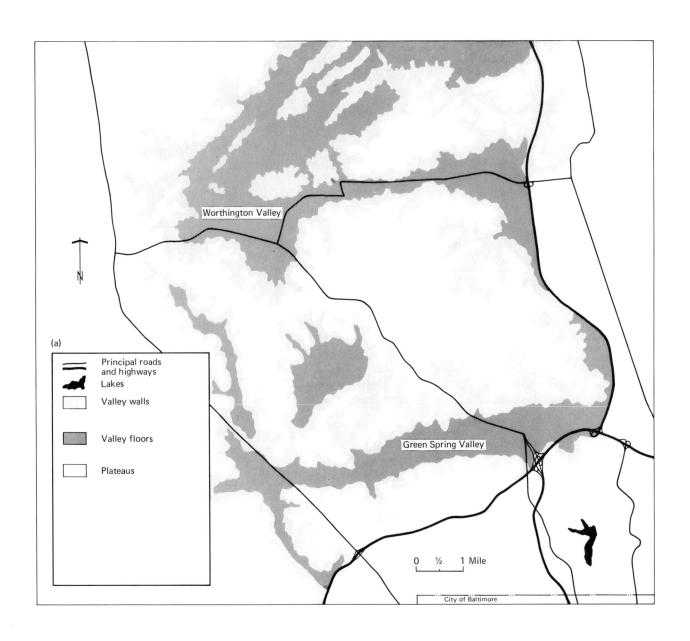

MECHANISMS FOR LAND-USE PLANNING AND CONTROL IN URBAN AND URBANIZING AREAS 197

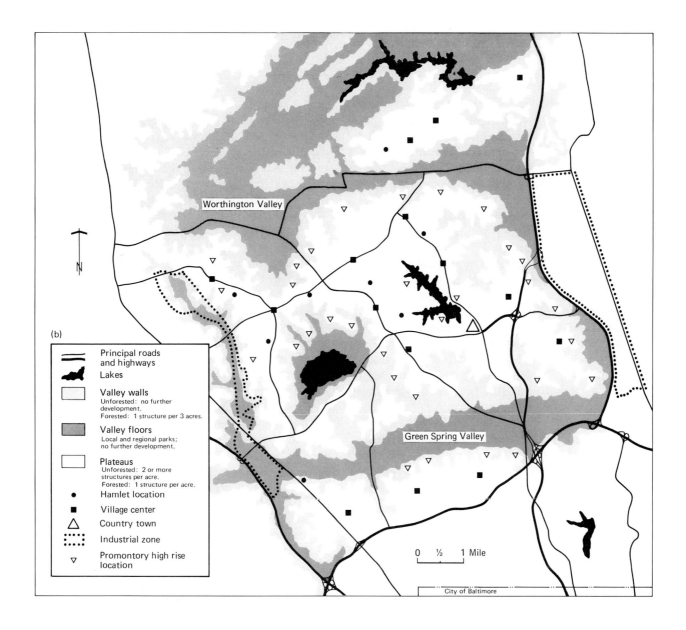

be assessed for compatibility with the wooded slopes of Green Spring Valley. All this information then may be made available for public use as a single map such as that illustrated in Figure 8-7b.

Ecological determinism clearly includes the first four steps of the generalized planning process, but here also step 2 is implicit because there are no formal requirements that the reasoning used to arrive at compatibility judgments be spelled out. Thus there is a strong dependence upon judgments of the staff. More serious, the prescriptive nature of the maps and the omission of step 5 create innumerable problems for implementation of such a plan. The Fourteenth Amendment guarantees that any landowner can challenge the plan, its boundaries, its judgments, and its conclusions; the burden of proof is on the planning staff. Hence unless provisions are made to deal with valid complaints, the whole plan could be endangered. Precisely this result may have happened to the comprehensive plan for the Brandywine basin in Pennsylvania, also developed by the McHarg group and reported by *Science* writer Peter Thompson (1969). The Brandywine plan envisioned protection of scenic features and of stream water quality under the conventional planning schema, with the addition of the use of such mechanisms as conservational easement and the power of eminent domain (that is, the power to ban development on a parcel) in cases of stubborn landowners. Thompson reported that the local citizenry regarded the plan as a suspicious incursion by outsiders, so that despite strong feelings in favor of the rural aspect of the valley and the desirability of clean streams, they rejected the plan in favor of developing their own.

Although Thompson favors the explanation that the Brandywine plan was rejected because it went against the self-interests of the landowners and did not allow for the vagaries of human nature, it is not clear that this is the only feasible explanation. We may argue that the inflexible prescription of the solution, without both local involvement and a detailed defense of the thinking that went into the decision process, was actually the mortal flaw. But whatever the reason, the history of the Brandywine plan shows how important the implementation process is to any social land-use plan. No matter how commendable the plan, it is worthless if it cannot be implemented.

The cause-condition-effect matrix and its application Jens Sorensen (1971) developed a method to expose the decisions that are made in step 2 of the generalized planning process. He wanted a method that could resolve resource-use conflicts clearly and prevent resource degradation without formulating a regional plan (this was left to others). Sorensen decided to accomplish these aims by specifying how each variable (the cause) would stimulate a set of processes (the conditions) which would generate another set of lasting changes (the effects). A particularly attractive feature of this method is that any effect can also be a cause, making the whole concept quite dynamic.

The entire process can be displayed in a stepped matrix (Figure 8-8). This particular matrix was constructed as part of the planning study for seven San Diego (California) County lagoons by the Laboratory for Experimental Design. There are 28 causes listed as inputs for matrix A. Let us choose "effluent output" as the cause we will follow through the matrix. If we trace the effects from this single variable as

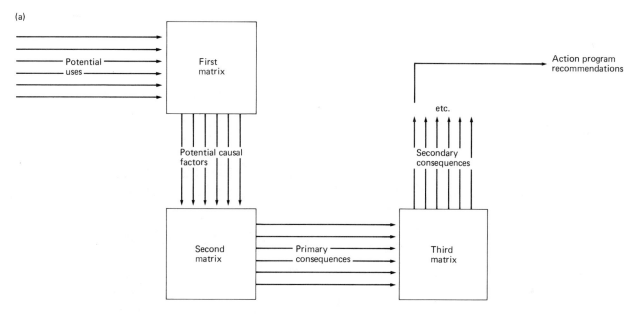

FIGURE 8-8 (a) *The Sorensen matrix as devised by the author. (b) (next page) The matrix as applied in the San Diego Lagoons planning study. The matrix is stepped, i.e., consists of a series of interrelated parts. Sorensen's concept starts with a row of potential uses (e.g., farming, recreation, housing). From each of these, one or more aspects can be identified as potential causal factors. In similar manner, the causal factors lead to initial conditions (i.e., intermediate change states), which lead to other conditions. At the end of the chain, documentation and plans for action may be attached. In the San Diego Lagoon study there are five levels of ramifying events, with the potential uses and action phases omitted. For example see Table 8-1. ((a) Reprinted by permission of Jens Sorensen. (b) From* The Coastal Lagoons of San Diego County, *plate 20, p. 52. Courtesy of The Laboratory for Experimental Design.)*

FIGURE 8-8b *Environmental impact matrix.*

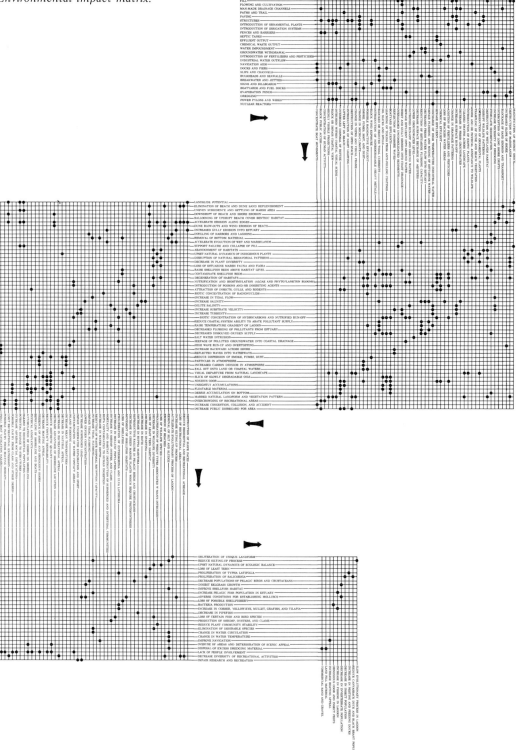

TABLE 8-1 *An example of ramifying consequences taken by following a single condition (effluent output) through the stepped matrix of Figure 8-8b.*

Condition	Effluent output
Primary effects:	sewage effluent produced nutrified runoff and groundwater sewage effluent into groundwater
Secondary effects:	contaminate shellfish keds degeneration of habitats nutrification and biostimulation (algae and phytoplankton blooms) attraction of insects, gulls, and rodents biotic concentration of hydrocarbons and nutrified runoff noxious odor dilute salinity increase turbidity seepage of polluted groundwater into coastal drainage
Tertiary effects:	loss of shellfish loss of wildlife species decrease in wildlife diversity proliferation of *Enteromorpha* and *Ulva latissima* contamination of water contact areas physical annoyance reproductive failure of pelagic birds and crustaceans reduce aesthetic quality decrease boating appeal decrease in *Ruppia maritima* increase in shrimp, oysters, and clams decrease in green marine plants which require photosynthesis increase suspended load impair underwater exploration and sport
Quanternary effects:	loss of possible shellfishery upset natural dynamics of ecologic balance loss of certain bird and fish species elimination of desirable species proliferation of *Typha latifolia* bacteria production increase in cobber, yellow-eye, mullet, grayfish, and *Tilapia* lack of people involvement decrease populations of pelagic birds and crustaceans overuse of areas and deterioration of scenic appeal decrease in pipefish production of shrimp, clams, and oysters adverse conditions for establishing molluscs impair research and recreation
Quintinary effects:	decrease in invertebrate population obnoxious odor and insect pests increase in fishing in lagoon

Source: Laboratory for Experimental Design, *The Coastal Lagoons of San Diego County*, 1971.

it passes through the five matrices (Table 8-1), we find that effects from the same factor sometimes appear in more than one matrix; some effects are beneficial, and some are deleterious. This is a graphic illustration of the snowballing effects that result from manipulating complex systems. Nutrification is cited as the cause of a change in species composition and stimulation of a possible shellfishery, and ultimately the original cause, "effluent output," accounts for 43 named items in the stepped matrix. On balance, this factor is supposed to increase fishing and the harvest of useful invertebrates, but at the cost of increased numbers of pest insects, noxious odors, and the creation of unsightly views. Ultimately the entire matrix becomes part of the information base for land planning and use in the areas that could affect the lagoons.

While its scope is quite limited, Sorensen's method has the important quality of fostering explicitly open decision-making, which is highly desirable.

Land-use capability ratings The next logical step in the development of ecological planning procedures is to combine explicit decision-making about important variables with flexible recommendations that are designed specifically to meet legal requirements. The new kind of land-use capability ratings (as opposed to older systems developed by the U.S. Soil Conservation Service for agricultural use) are still experimental and not well documented, but they attempt to do this. One excellent example is the land-capability map devised for land use in the Lake Tahoe basin, a region astride the California-Nevada line that is a favorite of gamblers, tourists, and recreationists and is now undergoing a surge of rapid urbanization (Figure 8-9).

Under contract with the Tahoe Regional Planning Agency, the planning team gathered data on the area both from surveying the region and from consulting published sources until they had formulated a list of 64 possible variables that might be important in determining the land capabilities of the region. From this long list only six variables were finally used—flood hazard, landslide potential, high water table, high soil erosability, poor soil drainage, and fragile fauna and flora. These variables were then combined into seven classes of land-development capability (Table 8-2). The importance of soil structure and drainage was recognized in the team's primary recommendation that building be restricted to a certain maximum percentage of land coverage by impervious surfaces (rooftops, parking lots, paving). Here the reasoning was that the concentration of water run-off by these impervious surfaces would accelerate the speed of the water in the run-offs, which in turn would accelerate erosion and increase the incidence of landslides and flooding.

Probably the outstanding quality of the land-capability plan for Lake Tahoe is the detailed attention it pays to implementation. The plan specifically notes that the various boundaries constructed are tentative; discusses the methodology used in the construction of the capability map; notes that such variables as social factors, fire, seismic activities, or scenic factors were not considered in the construction phase; and stresses that conditions in any parcel of land may warrant an exemption of that parcel from the overall plan. Although the Tahoe plan has not been implemented, this failure stems from the political organization (regional government at Lake Tahoe) rather than from a fault in the planning process.

MECHANISMS FOR LAND-USE PLANNING AND CONTROL IN URBAN AND URBANIZING AREAS 203

FIGURE 8-9 *View of the State Line gaming and entertainment center on the Nevada side of Lake Tahoe.* (Courtesy of Richard Weymouth Brooks.)

TABLE 8-2 *The land-capability rating system devised for Lake Tahoe (California-Nevada) uses slope, soil, and geology as primary factors determining the developmental capacities of the surrounding watershed.*

Tolerance for use	Slope (percent)	Relative erosion potential	Run-off potential	Disturbance hazards	Capability rating
Highest	0–5	slight	low to moderately low	low hazards	7
	0–16	slight	low to moderately low	low hazards	6
	0–16	slight	moderately high to high	low hazards	5
	9–30	moderate	low to moderately low	moderate hazard	4
	9–30	moderate	moderately high to high	moderate hazard	3
	30–50	high	low to moderately low	high hazard	2
	≥ 30	high	moderately high to high	high hazard	1a
	≥ 30	high	moderately high to high	high hazard, poor natural drainage	1b
Lowest	≥ 30	high	moderately high to high	high hazard, fragile flora and fauna	1c

Source: U.S. Forest Service, in conjunction with the Tahoe Regional Planning Agency.

Systems analysis and simulation in environmental planning During the time needed to properly execute a planning study, a region may change so that the study is no longer valid, or new information may turn up to reverse an earlier decision. These, of course, are usual problems with a dynamic system and a static model. One way to deal with unknowns is to incorporate flexibility for implementation, such as was done at Tahoe. Another possibility is to apply systems analysis and simulation techniques to the problem. Computers have been used a long time in regional planning, first to process data and, more recently, to produce maps directly. But the capability of a computer to build a model of a region is still in the formative stages.

Steinitz and his associates (1971) published a report on regional planning of lands around surface waters using systems analysis. Their data collection and interpretation techniques were pretty much like those discussed for others above, with the exception of a strong reliance on aerial photographs. The departure in technique was the use of a simulation model of the system. Since Steinitz and his coworkers were interested in recreation, they hypothesized that the system would consist of demand for recreation, the economies of meeting the demand, the ecological constraints upon recreational use, and the physical factors which would operate on both. Subsequently, systems analysis formed the framework for building reasonable models of each part; when linked together, these parts became the simulation model.

These investigators were interested in using the model to experiment with the plan they had developed. Since one property of a good model is that it represents a system adequately, it can be used as a surrogate; hence experimentation with the model suggests what similar action might accomplish on the real thing. After trying many alternative uses for a region, investigators might conclude what the "best" uses of the area would be under various assumptions. For example, two permitted levels of uses could be simulated to see which had the highest benefit in terms of recreational income generated and which had the greater lasting detrimental impact.

Simulation could be a powerful tool used in this way, but its potential remains to be explored.

Looking toward the Future

As we have seen, private ownership of land is not itself nearly as important in the area of land use as is the unrestricted freedom for owners to use land however they see fit. Land in a free-market economy has its highest value when it is urbanized, and so when in private hands and under the exclusive control of landowners, lands tend to be used only for this purpose. Furthermore, local governmental institutions tend to reinforce these patterns, and so de facto planning results. With no overwhelming evidence that anything was seriously wrong with this policy, and as long as the majority could profit, there was little public support for change. After all, if land is plentiful and cheap, if farmers can always go elsewhere, if human social and economic development are worth the sacrifice of some trees or a fossil bed or a rock arch or a naturally beautiful valley, why not make the trade?

In Europe, where countries are densely populated and lands for expansion really are scarce, there are strong motives to be careful about land use. In the United States, where land is and will remain plentiful for the near future, changes will come more slowly. But they are coming, if only because environmental protection is required by law and because much of the rest of the world may soon depend upon our agricultural productivity. The emergence of regional planning and governmental integration of planning, and public support for the protection of special environments like coastlines at the expense of property rights, may all be attributed to changing public attitudes about land tenure. As ecological planning at the regional level develops, the task of implementation undoubtedly will become smoother and facilitate more orderly use of land. However, it is by no means clear that land-use control will ever be as rigidly controlled by governmental agencies in the United States as it is in Europe.

It is in the further development of ecological land planning that hope seems brightest. Progress is being made, although most professionals admit the present state of the art to be embryonic and poorly documented. We need to publish these methods in a convenient form so that more people will be able to criticize, compare, combine, and improve them. Since the land is the focus of human activity, ecological land planning eventually will play a central role in our society. The planner must gather and evaluate data, assemble a useful scheme, interpret the data in terms of the planning goals, produce the plan, and see it through modification and implementation. Thus, the planner gets the opportunity to be a researcher, decision-maker, map-maker, and computer programmer, as well as an important voice in society. In a climate of increasing sympathy for land-use planning, there will be many opportunities for those who want to meet the challenge of an extremely broad but important occupation.

Questions

REVIEW QUESTIONS

1. Can land be a commodity that is most efficiently handled by market mechanisms? Your answer should
   a. contrast agricultural and wild open-space lands
   b. contrast urban vacant lands (inside city limits) and urban fringes
   c. compare two different blocks in the same suburban subdivision

2. Consider a group of farmers on land that promises to urbanize within 10 years. What do you think is likely to happen when
   a. the assessor values property strictly on current use
   b. an urban limit line is established short of the farmers' land
   c. the land has a conservation easement placed on it

3. Do you think the Hawaiian state-wide zoning law is good? Why or why not?
   Hint: This question is designed to force you to draw up a set of value criteria upon which you can make your judgments. For example, consider conservation of native species, preservation of tropical agriculture, orderly urban development, and the best possible pheasant hunting.

4. Why is it appropriate to call land use dictated by price "de facto land planning"?

ADVANCED QUESTIONS

1. Work through some of the pathways of Figure 8-6b or through Table 8-1 with the aid of 8-6b. Make a table listing the strengths and weaknesses of the cause-condition-effect matrix as
   a. a planning tool
   b. a data organizer
   c. a method for impact analysis

2. In the plan for the Green Spring and Worthington valleys, what modification would have to be made for
   a. unstable soils on the central plateau
   b. a significant increase in yearly rainfall
In more general terms, what limits must be placed on the scope of the plan?

3. Estimate the proportion of the valley and the county of Santa Clara that was urbanized for each decade. Use Figure 8-2 as your source and graphical estimation as your technique. What can you conclude from this analysis if you had no other information?

Further Readings

Note: Much of the material is either in planning documents or in contract reports, both of which are hard to obtain.

Babcock, R. F. *The Zoning Game.* University of Wisconsin, Madison, 1966.

   One of a series of books about the institution of zoning, especially its politics and subversion. For a contrasting view, see Malkieski (1966).

Belser, K. P. "The Making of Slurban America." *Cry California*, 5, No. 4 (Fall 1970), 1–21.

   Written by the former planning director for Santa Clara County, entertaining and profusely illustrated with photographs that show fascinating contrast over time. Substitute your county name, town names, and ordinances to get a reasonable picture of what has happened to *your* urbanizing county.

Costonis, J. J. *Space Adrift*. University of Illinois, Urbana, 1973.

Cullingworth, J. B. *Town and Country Planning in England and Wales*. University of Toronto, Toronto, 1964.

A detailed account of planning in England for those who want more than the brief mention of English land planning in this chapter.

Delafons, J. *Land Use Controls in the United States*. Publication of the Joint Center for Urban Studies of the Massachusetts Institute of Technology and Harvard University. Distributed by Harvard University Press, Cambridge, Mass., 1962.

Analysis by an English planner after a long tour of the United States. His perspective is valuable; he identifies the problems but is nonetheless sympathetic to the institution of zoning. A second edition was issued in 1969.

Ducsik, D. W. *Shoreline for the Public*. M.I.T., Cambridge, Mass., 1974.

Specializes on issues of marine-coast land use. See also J. Clark, *Coastal Ecosystems*. Conservation Foundation, Washington, 1974.

Eckbo, Dean, Austin, and Williams. *State of Hawaii Land Use Districts and Regulations Review*. Available from the firm in San Francisco, Los Angeles, and Honolulu; 1970.

Elliott, L. R., and R. O. Schultz. "Environmental Resource Data Analysis Techniques: A Compendium and Bibliography." In *A Critique of Water and Related Land Resources Planning*. Unpublished report to the California State Department of Water Resources.

Hagman, D. G. "Windfalls for Wipeouts." In *The Good Earth of America* (The American Assembly). Prentice Hall, Englewood Cliffs, N.J., 1974.

An excellent paper on the critical issue of socioeconomic equity as land rights are increasingly restricted.

Hawaii Department of Planning and Economic Development. *State of Hawaii General Plan Revision Program, part 5. Land Use, Transportation and Public Facilities*. Issued by the department, Honolulu, 1967.

Hills, G. A. "The Ecological Basis for Land Use Planning." Research report 46, Ontario Department of Lands and Forests, Toronto, 1961.

Holling, C. S. "Stability in Ecological and Social Systems." In Brookhaven Symposia in Biology, No. 22 (1969) 128-141.

An interesting simulation model based on predator-prey dynamics.

Kent, T. J. *Open Space for the San Francisco Bay Area*. Institute of Governmental Studies, University of California, Berkeley, 1970.

Laboratory for Experimental Design. *The Coastal Lagoons of San Diego County*. Available from the laboratory, California Polytechnic College, Pomona, 1971.

Leopold, L. B., F. E. Clark, B. B. Henshaw, and J. R. Balsley. *A Procedure for Evaluating Environmental Impact*. U.S. Geological Survey Circular, No. 645, 1971.

Lewis, P. H. *Regional Design for Human Impact*. Thomas Publications, Kaukauna, Wis., 1969.

MacArthur, R. H., and E. O. Wilson. *The Theory of Island Biogeography*. Princeton, Princeton, N.J., 1967.

Mace, R. L., and N. L. Wicker. "Do Single-family Homes Pay Their Way?" Urban Land Institute, Monograph 15, Washington, 1968.

Malkieski, S. J. *The Politics of Zoning*. Columbia, New York, 1966.

> The history of zoning and its improvement in New York City. See Babcock (1966).

McHarg, I. L. "Ecological Determinism." In *Future Environments of North America* (Darling and Milton, eds.). Doubleday, Natural History Press, Garden City, N.Y., 1966.

> The best succinct summary of McHarg's approach to ecological planning. This volume is generally valuable. Now that it is available in paperback, students of environmental issues should consider having it in their libraries.

McHarg, I. L. *Design with Nature*. Doubleday, Natural History Press, Garden City, N.Y., 1969.

> The full version, profusely illustrated and rich in worked examples, of the McHarg method. Novel and a sensation when first published, *Design with Nature* is now just one element in a rapidly proliferating set of methods.

Myrup, L. O. "A Numerical Model of the Urban Heat Island." *Journal of Applied Meteorology*, 8 (1969), 908-918.

Petersen, M. S. "Case Description: Morrison Creek Stream Group Basin, California." In *Environmental Quality and Water Development* (Goldman, McEvoy, and Richerson, eds.). Freeman, San Francisco, 1973.

Public Land Law Review Commission. *One Third of the Nation's Land*. A report to the President and the Congress. Washington, 1970.

> A review of the federal lands in the United States and their future uses.

Reilly, W. K. (ed.). *The Use of Land*. Crowell, New York, 1974.

Sorensen, J. *A Framework for Identification and Control of Resource Degradation and Conflict in the Multiple Use of the Coastal Zone*. M.S. thesis, Department of Landscape Architecture, University of California, Berkeley, 1971.

Steinitz, C., T. Murray, D. Sinton, and D. Way. *A Comparative Study of Resource Analysis Methods*. Department of Landscape Architecture Research Office, Graduate School of Design, Harvard, Cambridge, Mass., 1969.

> The authors cover 13 methods of ecological planning not covered in this chapter.

Steinitz, C., T. Murray, P. Rogers, D. Sinton, R. Toth, and D. Way. *Honey Hill: A Systems Analysis for Planning the Multiple Use of Controlled Water Areas*.

> A report to the U.S. Army Corps of Engineers (2 vols.). Available from the National Technical Information Service, Springfield, Va., 1971.

Sussna, S. "Land Use Control: More Effective Approaches." Urban Land Institute, Monograph 17, Washington, 1970.

> A short monograph of land planning in an American context. It is most valuable for a quick overview of land control mechanisms in open space, environmental, and urban contexts, and for its bibliographies.

Thomas, D. *London's Green Belts.* Faber, London, 1970.

Thompson, P. "Brandywine Basin: Defeat of an Almost Perfect Plan." *Science,* 163 (1969), 1180–1182.

   The original plan is available from the Institute of Environmental Studies, University of Pennsylvania, Philadelphia. Titles: *The Plan and Program for the Brandywine* (technical report, $3.50); *The Brandywine Plan* (popular, $2.00).

Thompson, W. R. *A Preface to Urban Economics.* Published for Resources for the Future by Johns Hopkins, Baltimore, 1965.

   Nothing has been said in this chapter about land economics. For those interested in this important topic, Thompson will serve as an excellent starting point.

U.S. Council on Environmental Quality. *The Quiet Revolution in Land Use Control.* Washington, 1972.

U.S. Department of Housing and Urban Development. *Urban Land Policy.* Division of International Affairs, U.S. H.U.D. Washington, 1969.

   Excellent bibliography and introduction to European methods of planning and control, not only for land but for entire regions. The social traditions of Europe contrast sharply to those of the United States.

Wagenseil, H. "Property Taxation of Agricultural and Open Space Land." *Harvard Journal on Legislation,* 8 (1970), 158–196.

   A survey of laws to relieve property-tax pressures on land that is not being employed for its "highest and best use."

Whitman, I. L., N. Dee, J. T. McGinnis, D. C. Fahringer, and J. K. Baker. *Final Report on Design of an Environmental Evaluation System.* Report to the Bureau of Reclamation, U.S. Department of Interior, from Battelle Laboratories, Columbus, Ohio, 1971.

Whyte, W. H. *The Last Landscape.* Doubleday, New York, 1968. (Paperback edition, Anchor, 1970.)

9. Sustained Yield, or What?

Sustained yield, the ideal objective of resource managers, refers to some size range of harvested yield that does not harm the capacity of the resource to keep producing the desired item(s). In previous chapters we have examined basic biological needs for human populations, all of which can be managed, but none of which really can be managed as a sustained-yield problem. Land can be managed, and land can be destroyed, but we have little capacity to build it; water can be used, and it can be converted into deadly solutions, but we do not create it to any significant extent. Food supplies can be thought of as sustained yield, but only if we can escape the desperate race to get enough food to feed the population at minimal levels; otherwise we are talking about the gross concept of global sustained yield. Certain parts of the food supply—for example, exploitable marine fisheries—are, however, excellent sustained-yield problems if we don't expect them to meet the demand. In fact, the concept of sustained yield first arose in fisheries science.

This chapter examines what usually are called *renewable resources*. Renewable resources do not form some well-defined block of substances separate from those discussed previously; rather, they refer to an indistinct group of resources that can be managed successfully for a sustained yield. Inherent in this definition is the requirement that the resource of interest have a generation time in terms of the human life span which is short relative to demand. This is strictly empirical and has little to do with biological or physical characteristics that could separate renewable and nonrenewable resources. For example, marine fish populations reproduce, some at extraordinary rates compared with human beings; in this sense they fit our definition. On the other hand, when faced with the expanding demand for fish protein, sooner or later even the most prolific species is hardly renewable. Other populations more or less fit our working definition, depending on their rates of reproduction and the demand for them. Whales or elephants, with their relatively slow reproductive rates and long life spans, are renewable only at low rates of demand; very few whales or elephants can be taken under a sustained-yield scheme. Redwoods, and especially giant sequoias, are at the low end of the production spectrum compared with southern yellow pines, which can attain harvestable size while sequoias are still saplings. Because they are derived from continuing biological processes, fossil fuels are really renewable resources, but their rates relative to demand are so slow that they are considered nonrenewable.

The arbitrary nature of these definitions should not obscure the general order that emerges from the binary classification. This will become clearer when we compare the differences between prevailing management practices and the alternatives that might be available for resources in each category.

Management Practices for Renewable Resources

We may begin with an enumeration of some basic principles of renewable-resource exploitation and management, especially as they have been practiced in American history. The underlying process that seems paramount is the preeminence of economic principles in the decision-making system governing the use of natural resources. So long as the economic model works, this approach is fully acceptable. But often enough the economic model indicates a course of action that is counter to the best long-term interests of the resource, and other action then becomes necessary.

PRACTICE 1: EXPLOIT IF YOU CAN

One of the most common practices is to exploit a natural system rather than attempting to manage it. This happens because it is cheaper per unit of material to do it this way; furthermore, less information is necessary to develop it as a resource. For example, systems such as estuaries and salmon populations can be managed, or they can be exploited. Historically the latter has been the usual choice, and even this is permissible if the exploitation process does not irreparably damage the resource, or if demand does not exceed supply, or if the resource is not vital to human societies. However, since demand has a record of outstripping supply, there should be increasing pressure to develop management practices the longer exploitation continues. Hence exploitation may be expedient and cheaper, but it tends to occur mostly in the pioneering stages of the development of a resource.

Unfortunately, however, this practice sometimes has persisted beyond the pioneering stages, and some resources have collapsed because the pressure to develop better management never appeared. Fin-whale populations were clearly overexploited, largely perhaps because they were not absolutely necessary to the societies exploiting them. Simple conservation proved to be an inadequate appeal. In cases like this it is not clear if successful management, which implies human control to some extent, could be developed without an outside authority that is willing and able to stop harmful overuse of a resource.

PRACTICE 2: OBTAIN YIELD AT MINIMUM COST

In practice 1, we already can see how important economic considerations are in management of natural resources. The simplest management approach emphasizes short-term efficiency and economic growth above other considerations and is stated here as practice 2. When an economic activity is dependent upon a resource, however, there should be strong pressure for its conservation. A farmer needs to worry about depleting the soil and thus should strongly support soil-conservation practices. A large logging firm that depends on its own large acreages of private holdings cannot afford to cut timber and then move on to land elsewhere so long as taxes on the land force it to strive for long-term income from its present holdings.

Historically, however, this has not been the case; the short-term view has predominated in resource development. Abundant resources have encouraged exploitation of private holdings and loosened governmental control over the exploitation of resources on the vast federal forest and range lands. However, the most flagrant cases have involved the truly international, free-good resources over which no one has proprietory interest or control. In certain cases, two, three, or even four nations have established workable management schemes with these resources; but rarely has more than that number been able to agree. Profits and responsibilities are spread too thinly, and the result frequently has been free-for-all competition until the resource has been exhausted.

What seems likely is the development of the sort of scarcity Whyte sees as the hope for remaining (if small) areas of open-space lands: The value of the resource rises, and pressure for its preservation increases as its scarcity becomes more and more pronounced. For biological resources, this probably will result in further deletion of species as difficult choices are made about what resources are essential to us. These resources then would replace other resources considered less vital and would receive intensive management attention similar to that presently being given to cattle and sheep. Of course, the cost per unit for these selected resources can be expected to rise as a consequence of the transition from exploitation to management.

PRACTICE 3: DO WHATEVER YOU CAN GET AWAY WITH

No society operates entirely on a free-market economy. In practice, societies set legal restraints and various incentives on the operation of the economy; at the extreme, the legal apparatus takes over the operation of the economy as well. In American society, the legal and political systems have had an increasingly powerful influence on renewable-resource management (see the discussion of forest practices later in the chapter). One well-known example has been the development of domestic petroleum reserves. With vast volumes of petroleum elsewhere in the world, a least-cost economic model would call for developing resources at known sites in the Middle East; few funds would be allocated to searching unknown areas of the United States for oil. Instead, by limiting foreign imports and giving the oil-depletion-allowance tax credit for oil pumped in this country, the federal government literally has driven the search for domestic reserves. Before the Wilderness Act of 1964 there were no incentives to preserve wild, free-flowing rivers in the face of pressures to dam them for multipurpose use because agencies could claim that the enhancement of lake and reservoir fisheries was a project having at least as much social value as conservation of extant stream ecosystems. Without a powerful control commission there was no incentive to cut back Antarctic fin-whale harvest rates; and without the development of limiting stipulations, unthinking forestry practices still predominate. No one could claim reasonably that all laws are beneficial or that any law is uniformly useful, but legal controls frequently default to operating guidelines based strictly on economic motives that worsen a situation.

PRACTICE 4: RUN THE RESOURCE DOWN AND...

In classic economic theory demand is suppressed by rising prices as supply declines. Theoretically this mechanism should work also for renewable resources, but as we have seen, this is not necessarily the case because other factors can become more important.

For example, if a resource is desirable but thought to be unessential, there may not be sufficient motive for its preservation, especially if prices and profits are high. If the resource is considered essential, declining supply may actually trigger intense competition and drive prices upward. And excess competition independent of supply may itself lead to the same result. Each of these three pathways can further exhaust resources, especially if they are biological and sensitive to exploitation.

The case of the Antarctic fin whales has been mentioned in several places. The passenger pigeon was formerly abundant in the United States (Figure 9-1),

FIGURE 9-1 *An old engraving showing the sport of passenger pigeon hunting in the late 19th Century. Shooting practices were such that the bird was virtually extinct before there was any indication of a problem.* (Courtesy of the Bettmann Archive, Inc.)

but shooting coupled with destruction of the socially driven reproductive behavior was so excessive that the species was extinct before market economics had any effect. Prices stayed low until the pigeon's extinction was certain. Fin whales cannot breed fast enough to keep up with hunting; passenger pigeons could not breed at all without large communal flocks, hence they were even more vulnerable to extinction. These are but two examples where market control has proved inadequate for resource management. To put much dependence on such mechanisms seems risky and insufficient unless the market is highly sensitive to the important features of the resource itself and then can match demand and supply indefinitely.

An alternative strategy is based on the idea of substitution of resources—if a resource gets scarce, switch to a substitute when it becomes economically feasible to do so. One argument put forward about future oil scarcity is that other oil-bearing sediments (tar sands and oil shales) are available to replace liquid reserves when prices rise enough to offset the higher costs of production.

Substitution of resources introduces a great deal of flexibility into management and exploitation decisions. By adjusting life styles, economies, and social habits, substitution can meet all but the few basic biological needs such as food, water, and space. Beyond these basic requirements individual philosophies are most important in determining which commodities are subject to substitution. At one extreme, conservationists would hold that all existing systems are important just because they are extant and because they have a right to continue; therefore none could be subject to substitution. They also might stress the possibility that each system might be useful in the future. Empiricists, however, would argue that existing systems are important only so long as they fulfill human needs, and, even then, become expendable if an equal or better substitute is available. Ecologists are somewhere in between. They can appreciate both the stability and diversity of nature derived from long periods of slowly changing conditions and the possible consequences of any disturbances to this stability. On the other hand, they are aware that change—even drastic change, whether produced by human beings or not—is a fact of nature and frequently is inevitable. They can, however, contribute only another voice to what is a political process, and perhaps they may provide information for better management if that is the objective of their society. Their contribution is to elucidate the mechanisms of change so that harmful uncertainties are minimized, even though politically determined goals may impose constraints.

PRACTICE 5: SCALE CONSUMER DEMAND TO SUPPLY

Instead of meeting or attempting to meet demand, another alternative is to scale demand to meet supply. The USSR did this for years with consumer goods—demand was allowed to go begging while the economy was diverted toward heavy industry; now the inequity is being corrected slowly. In market economies, this objective is often met more selectively. For example, game fish and wild waterfowl can no longer be purchased readily because laws forbid market hunting. But several decades ago market hunters operated freely in response to tremendous demand, especially to the public demand for passenger pigeon and waterfowl. Prohibition may act in conjunction with substitution to suppress a legitimate but overblown demand. It has never been effective to prohibit the shooting of tigers or leopards for their pelts, or the hunting of Everglades alligators for their hides, until the marketing of these products was also banned and the demand was satisfied by the substitution of alligator-grained cowhide and synthetic furs.

A major part of the "eco-activist" literature is concerned with scaling demand. For example, recycling paper and reusing grocery bags are designed to cut demand for pulp trees. The impact of such individual action is uncertain when compared with the decisions of a few large corporations and federal agencies, especially if they are trying to increase demand for overstocks of supply. In situations of real scarcity or emergency, such as has been experienced during wars and economic depressions, demand can be scaled down drastically; but its impact as a management tool in situations where the major concern is merely efficiency or lower cost per unit of yield is far less certain unless corporate and governmental decision-makers cooperate.

PRACTICE 6: INCREASE SUPPLY BY ECOLOGICAL ENGINEERING

"Ecological engineering" is a broad term that could be applied to all management activities of biological resources. Here we shall use it in a restricted context: It is the selective application of ecological theory to maximize yield. The two general methods that can be used to improve yields are designed (1) to improve the physical environment or other conditions important to production and (2) to foster biological relationships that benefit production and weaken relationships that detract from it. High-intensity agriculture provides numerous examples combining both methods, as does nearly any method of food production. In forest practices, redwoods benefit from monoculture and a natural understory of the nitrogen-fixing, soil-conditioning plants (the redwood-

FIGURE 9-2 *The growth habits of the Douglas fir are such that the seedlings will not thrive in the deep shade of older and larger trees. As a result, blocks of timber are clear-cut and intermittent islands of trees are left to protect against excessive erosion and to provide seeds.* (Courtesy of Weyerhaeuser Company.)

oxalis association). Periodic fire or flooding, whether natural or produced by human beings, help to increase production by suppressing competition, because redwood trees are admirably suited to withstanding fire, insects, and flood but do poorly in mixed culture. By taking advantage of the biological characteristics of this and other tree species, we can improve the timber yield. Douglas fir grows fastest in clear-cut areas (Figure 9-2), while jack pine requires

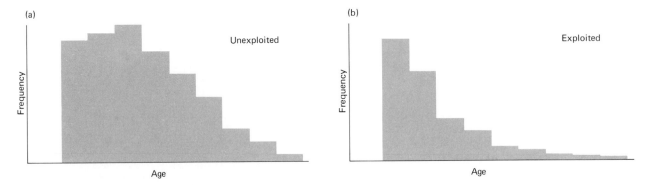

FIGURE 9-3 *Comparison of two age structures, one for an unexploited population and one for an exploited one. Note that the exploited population's shift in age structure can be the product of two different mechanisms (see text).*

periodic ground fires (to destroy the seedling hardwoods that otherwise eventually would eliminate them and to stimulate their own seed germination).

Establishing basic environmental conditions that insure long-term production is a simple first step toward the development of a management program. Current environmental conditions are not accepted without question, but instead considerable effort may be devoted to developing and maintaining another set of more favorable conditions. The development of a program to obtain the often minor gains in yield that are possible with finer adjustments of ecological systems may be more difficult to maintain and may have less predictable results, but such a program represents a possible second step in this approach. Obviously, there are objectives other than maximizing yield for developing ecological management schemes, but it remains a most important consideration.

PRACTICE 7: TRY TO OPERATE AT SUSTAINED YIELD

We have previously defined sustained yield, but there are many different ways to attain it or to fail to keep it. Renewable resources are variable systems by virtue of their responsiveness to the environment in terms of growth and reproduction. But in spite of these differences it should be possible at any point to stop and assess the resource, determine how fast it is reproduced, and from this information define the permissible take in terms of future population stability. This is routine management for fisheries, which must monitor catch statistics and other population indicators to determine population trends.

What would you expect to happen to an exploited fish population? Certainly the age structure of the population is one important indicator, because it reveals the dominant ages, which in turn reflect the impact of harvesting. In Figure 9-3, we see that the exploited population has proportionately more individuals in the youngest age classes. Certainly the harvesting method itself selects for larger and older fish (such things as size limits, gear that tends to let younger fish escape, and the greater yield from larger individuals are pertinent examples). This is differential mortality, and the age structure can show who is surviving. The same kind of age structure could be obtained, however, if just enough fish were harvested to decrease intraspecific competition for the food supply or for habitable space. The survivors then would have a sudden increase in resources, and a population explosion would result; then the age structure would be disproportionately young, more because of increased reproduction than because of differential mortality in older age classes.

The simplest way to differentiate these two alternative explanations is to examine trends in population size over a period of time or, more simply, to observe trends in the size of the catch in successive years. If fish get scarcer and catches decline, then overexploitation is suggested. If there are lots of young fish and catches do not decline, we might have the happy situation in which moderate exploitation has increased the catchable numbers of fish. In the usual situation, more than one indicator is available, and it will be consistent with others. The collapse of Antarctic baleen whale populations was signaled by several observations related to age structure. For instance, as time passed and overexploitation continued, it became obvious that larger, older individuals were no longer being sighted, and this was underscored by a decline in the size of individuals captured by whalers. In addition, despite the best equipment, the whalers were no longer able to meet their quotas, simply because there were not many whales left to take. Finally, what best signaled the severity of the situation was that newborn whales and young calves drastically declined in number, indicating the population had decreased to the point that it no longer was reproducing well and therefore was incapable of recovering from overexploitation.

In predicting a gloomy future for the Peruvian anchoveta fishery, Paulik (1971) used several such indicators: a sharp decline in the rate at which young fish were being added to the population, a shift in the age structure of the population toward younger age groups, declining catches, and the failure of the guano bird population to recover from an earlier decline—particularly noteworthy, since guano birds are dependent upon the anchoveta for food. In the past the anchoveta population was always large enough to support the fishery and the birds during most years. In certain years, however, the anchoveta disappear in response to a shift in offshore currents; the fishery suffers and the birds experience mass mortality from starvation. This had always been temporary in the past, and when the current shifted back to its original configuration, both the bird population and the fishery had recovered. The last time, however, the bird population did not recover, and Paulik suggested that the anchoveta population was no longer large enough to sustain both the fishery and the guano birds. Thus by the time fishermen had taken their "share," there were not enough anchoveta to sustain a growing bird population, and therefore the birds were not able to recover from this last population decline.

As we have mentioned, the predominant philosophy in marine fisheries has been exploitative, the aquatic equivalent of forestry's "cut and move" operations. With the development of an excessively large processing and fishing establishment in Peru, Paulik saw this history being repeated for the anchoveta in the anchoveta fishery. In 1972 the current patterns drifted, and the anchoveta declined, but there was a real question about the effect of heavy fishing. In 1974, the situation was still unclear. Paulik offered the hope that only one of several anchoveta populations would collapse and with its demise serve warning of the future for the others if rampant overexploitation continued. The fact that we have engaged in such practices of overexploitation does not undermine the concept of sustained yield, however. In fact, it should be clear that despite some operational difficulties (especially for stocks which are highly variable, hard to monitor, and susceptible to environmental fluctuation), the concept of sustained yield is a useful measure of the potential yield from a given aquatic population. Additionally, fisheries science already has the sophistication in population ecology and mathematics to determine sustained yields. The real problem here, as in other situations, is both political and economic. If calculated sustained yields conflict with economic goals and national, regional, or local interests, the concept is likely to be dismissed by decision-makers as overly theoretical and inapplicable to their particular situation. To judge from history, it would be surprising if such a conflict did not occur in more than a handful of cases, which is why so many populations suffer from overexploitation today.

PRACTICE 8: REACHING FOR SOPHISTICATION
IN RESOURCE MANAGEMENT WITH
NEW SCIENTIFIC TOOLS

One may calculate the maximum sustainable yield for a natural population at one time and then obtain widely different figures at another, even short, time later. Both the population and the environment change continuously, often in response to one another. The next step in renewable-resource management, then, is to extend our control over the environment and either raise productivity to an artificially higher level or at least insure a steady yield for any given time. This embodies the broader definition of ecological engineering (practice 6). Major advances in comprehensive ecosystem management are being made in aquaculture and mariculture in isolated or partly isolated systems. And technical developments will increase our understanding of ecological systems and, it is hoped, lead to further advances in comprehensive ecosystem management. Two examples are the development of remote sensing methods to study and monitor ecological units and the use of computer simulation to analyze complex systems.

It is difficult to keep abreast of the progress in instrumentation and analytical techniques used in remote sensing, and it is not within the scope of this text to cover them all. But the power and scope of some methods have reduced labor greatly and in many cases have accomplished things impossible to do otherwise. They especially have made the problems of comprehensive management less difficult to handle. In particular, radio transmitters attached to animals can trace their movements and activity patterns, which may yield valuable information for trapping or harvesting strategies. Aerial mapping, infrared films, satellite photography, and the like can be used to monitor large areas rapidly and cheaply, especially when analysis is done automatically by computer and results are printed out in various ways for direct use by resource managers.

The future of computer simulation for renewable-resource management also looks promising. The International Biological Program (IBP) has a number of comprehensive programs for ecosystem analysis taking place around the world, including a grassland study at Colorado State University. This program is attempting to unite all biological data on the grasslands ecosystem by using a comprehensive simulation model. Descriptive aspects of the grasslands and knowledge of the nature of interactions between its components then can be used to display some possible consequences of various management strategies before they are adopted. In simulating the system, modelers want to estimate the sustainable yield under various conditions and thus avoid the irreversible trial-and-error procedures that field workers often have resorted to in the past. A great deal of hope has been placed in this approach as a tool for management studies (see Watt, 1968).

Examples of Renewable-resource Systems Management

We now shall consider some current practices in three diverse renewable-resource systems: Sport fish and game, forests and ranges, and marine fisheries. The first is concerned with recreation, the second with the production of fiber and structural material, the third with food production. Yet these different objectives conceal what should emerge as basic similarities, insofar as each system is governed by the same practices. The object of these examples is to see how strong the similarities are.

SPORT FISHERIES AND GAME MANAGEMENT

The objective of the sport fishery or game manager is to maintain conditions that maximize the success of hunters and fishermen. Generally this means that the populations of interest should be as large and as susceptible to capture as possible. Given these objectives, it is not surprising that managers should become farmers who leave the harvest to others and that as farmers they should do all possible to promote their populations. Let us look at some of the methods used to accomplish these goals.

Manipulation of the ecosystem Manipulation of the physical habitats of desirable species has proved to be one effective means to insure the largest game population possible in a given area. For example, deer feed largely by browsing; that is, rather than eating grass, they prefer to feed on the tender sprouts and new leaves of shrubs and trees. It has been found that frequent controlled fires tend to stimulate production of this tender new growth, thus increasing the food supply as well as the deer population. Even more pervasive management is centered upon migratory ducks and geese. For instance, impoundments have been built at strategic points along the birds' migratory flyways in order to increase the suitable habitat at these points; moreover, many existing marshes have been preserved primarily because they provide a suitable habitat for game waterfowl. In addition, game management agencies may even grow grain stocks especially to provide food for the flocks and make them more accessible to hunters. While there is less control over other game populations, the basic approach is similar.

Some practices in sport fishery management provide even better examples of ecosystem manipulation.

Desirable species may be cultured intensively in hatcheries and released widely (see next section), and there may be open manipulation of the lakes and streams in which game species are present. At present the native fishes of New Zealand are being exterminated officially to be replaced by the familiar European game fishes that European settlers prefer. While this certainly is an extreme example, it is not uncommon for "undesirable" species to be poisoned or actively persecuted in order to increase the number of game fish.

Augmenting the stock A common method of increasing the catchable population is by augmenting the stock to supplement its natural reproduction (Figure 9-4). This method is particularly common for game fishes in heavily fished areas. State fish hatcheries are spread throughout the country and have regular stocking schedules. Particular attention is paid to stocking popular sites for the opening days of the season, a spectacle not unlike a circus (Figure 9-5). Sometimes the catches are made almost directly from the truck.

The rationale for hatchery operations is clear, but is it valid? Watt (1968) showed that where natural mortality rates are high or environmental carrying capacity is low, it makes little sense to stock additional individuals because they probably will be lost to fishermen anyway. However, Watt also noted that an obvious solution could be to plant the desired species solely to allow fishermen to catch them before they are eliminated naturally. For most situations, hatchery supplementation would have little long-term impact on the fishery, but it can have an overwhelmingly favorable short-term impact on recreation. Hence hatcheries make sense when they satisfy a demand that cannot be satisfied naturally. This implies that the

FIGURE 9-4 *Stocking trout in Brainard Lake, Colorado.* (Courtesy of Peter Menzel from Stock, Boston.)

FIGURE 9-5 *The hopefuls line up shoulder to shoulder on the opening day of trout season near Seattle, Washington.* (Courtesy of Doug Wilson from Black Star.)

circus aspect is exactly the proper role of hatcheries; it may not be natural, but it is good recreation management.

Regulations on harvesting practices While stocking may be a favorite method for increasing the chance of hunters' and fishermen's success, the unrestricted arsenal is not. Dynamite, hand grenades, land mines, and machine guns are not legal hunting gear; in fact, many state regulations prescribe which legal gear may be used rather than which may not. Attention is now aimed toward harmonizing legal restrictions on catch with the biological characteristics of the game population to obtain the greatest potential yield. For example, common regulations at present include prohibition of hunting one sex. "Bucks only" regulations on deer are still common, but more and more laws allow hunting of does; the older belief that hunting of does imperils the reproductive capacity of the herd is being replaced by the knowledge that moderate doe hunting does not do so. The improvement of many such regulations simply awaits greater knowledge and, perhaps more important, further evolution in human attitudes toward animals and nature.

Natural ecosystems have mechanisms that limit any population. As human populations continue to grow, fish and game managers may be forced to rely more and more on closed, artificial environments that eliminate these mechanisms to yield as large a game population as possible, though probably in a setting that is increasingly less natural. Currently, private clubs and enterprising individuals have pioneered in this area, and the growing industry indicates the likely future trend.

FOREST AND RANGE MANAGEMENT PRACTICES

For the limited number of game animals, exploitation yielded long ago to management and supplementation. For range and forest lands this is not yet true. The abuse of forest and range lands and the subsequent deterioration of the environment have long been favorite targets of conservationists, and justly so (Figure 9-6). Overgrazing, erosion, soil loss, drastic alteration of plant communities, and raging fires all have followed the lumberjacks and ranchers who were aided by various political and economic institutions. Some of these institutions are discussed in this section, but a thorough treatment would fill this book (see Dasmann, 1972, and section bibliography).

FIGURE 9-6 *Examples of forest management practices. (a) Clear-cutting Douglas firs. (b) Reseeding a clear-cut area by helicopter. (c) Hand-planting Douglas fir saplings. (Courtesy of Weyerhaeuser Company.)*

Nor are Americans unique in their treatment of forests and ranges. The cedars of Lebanon and the mountain forests of China are gone and have little chance of recovery, since the soil is gone too. Human activities have enlarged deserts, although authorities seem unable to agree on how important the human role has been in comparison to natural climatic change. In Europe, English forests nearly disappeared into wooden ships before contemporary England decided they should be encouraged as a national asset. On the European continent forested areas are limited but intensively managed, much as the pine plantations of the southeastern United States are.

The European situation signals the future of forestry in the United States. The transition from exploitation to management has already begun, but further progress depends not so much on technological innovation and further accumulation of scientific information as on changes in the political and economic conditions that shape forest practices. Forestry itself is an advanced science, facilitated by the ease of measuring and experimenting with trees. Hence the technical matters of measuring growth, the best conditions for various species, and the impacts of cutting are well known—so well known that foresters are well aware that simple, sweeping laws may be fine for one species but not for another. For example, a bill to ban clear-cutting would be excellent for hardwoods, which have limited seed dispersal, but disastrous for softwoods, which germinate and grow best on bare soil.

On small private plots and in the South, timber already is managed as a long-term agricultural crop. In the South this is possible because taxes are low relative to potential pulpwood profits; moreover, growth is so fast that small areas can be rotated for annual yields. Small wood lots in other areas can yield special hardwoods that are extremely valuable for making veneers to finish cheaper furniture woods. Again, the potential profit after taxes is adequate incentive.

That leaves the great western forests to supply basic building lumber. Most of these forests are on federal land, but that has not prevented small operators from using the most destructive cutting techniques. On federal lands the enforcement of stringent cutting regulations alone may be sufficient to improve the situation, but on private lands taxation is a more basic issue. At the University of California, Davis, Institute of Ecology a task force has assessed the situation and recommended changes in the political and economic system to alleviate tax pressures that promote destructive forest practices on California lands. Subsequently these recommendations were embodied into a new forest practices act proposed to the state assembly. In the state 9 million acres of private forest are subject to property taxes both for the land and for the trees. Fortunately, the state constitution bars taxation of trees less than 40 years old, but three factors counter this small advantage:

1. Assessments and tax rates have been rising with land and timber prices. With the increased timber demand resulting from the development of "second home" or "recreational" subdivisions, the effect of rapidly inflating timber prices on the landowner has been severe.

2. A lot of second-growth timberland has passed the 40-year mark.

3. Many countries, especially in extreme northern California, are heavily dependent on tree taxes for income; and so as their costs rise, they pressure private timberlands for more assets by raising taxes.

Obviously this system provides an overwhelming incentive to cut soon after the 40-year exemption, since taxes are collected every year and the land will provide no substantial income from the trees until they are cut. If the trees are not mature and are still growing fast at the 40-year mark, encouraging their logging is counterproductive; if, in addition, the trees are an even-age stand, it will be easiest to log the entire stand and save all the taxes on standing timber—in other words, this present policy promotes clear-cutting independent of the reforestation requirements of the particular species. Worse, tax policy encourages cutting *all* lands, including steep slopes where forests probably should be retained permanently to prevent severe erosion problems.

The solution proposed by the task force was simple: Reverse incentives by imposing a timber-yield tax and removing the standing-lumber property tax. The incentive to cut would remain, because the trees are potential income. But there would not be any incentive to cut all trees as soon as possible after taxes are billed on them. Damaging forest practices would still occur; for instance, owners still could cut 40-year-old trees and pay substantially the same taxes, or they could cut steep slopes without heeding the consequences. At least, however, such practices would no longer be mandatory as they are now. Obviously, the yield-tax system would be an improvement over standing-timber property taxation.

The study group also attempted to develop a framework for improved forestry practices, but not by specifying what is permissible. Rather, the group specified general objectives that could be met in a number of different ways, including such things as soil-erosion prevention, reseeding and reforestation, watershed protection, and environmental-quality improvement. To implement these procedures, the report specified that forest districts shall be established under the control of a Forest Resources Board. The board would then license foresters, who would adhere to a plan for logging an area that meets resource and conservation standards. The plan itself would be devised by a select group of certified forest planners and reviewed by the board. If the plan is not followed, the board would suspend operator licenses. This proposal offered two major advantages: It would end destructive forest practices that under current California law can legitimately ignore environmental quality and long-term forest production; and it would retain the flexibility to use local experience while state standards are met. In 1974, a modified version of the act sponsored in the Institute of Ecology report passed the legislature and became law, only to run into implementation problems (a court ruling that impact reports had to be filed for each and every lumbering operation).

Computer simulation in forestry and range management The role and scope of simulation studies in forestry and range management are both increasing rapidly. The grassland studies at Colorado State University are being synthesized now into a single grassland model called ELM (Ecosystem Level Model). This model is expected to represent grassland ecosystems from the Midwest to the Pacific Coast. The flow chart of the model produced at Colorado State (Figure 9-7) shows that the model emphasizes material and energy flows through several major compartments including both biotic and abiotic components. Its organization is relatively simple: weather, climatic, and water availability are calculated and used as inputs to determine biological events in the living

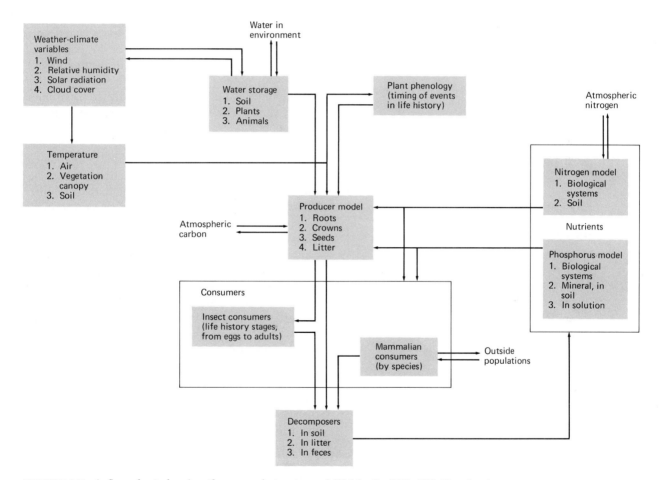

FIGURE 9-7 *A flow chart showing the general structure of ELM—the U.S. IBP Grasslands Ecosystem Model. This figure shows how the major submodels of ELM are linked together to mimic the behavior of the grassland ecosystem. Some of the flows have been inferred in this figure when no flow was indicated in the original but appears to be real.* (Redrawn from G. S. Innis, 1975, Grassland Biome Program, Colorado State University.)

past of the system. The producers are affected most directly by these factors, as well as by matters of biological timing (phenology) and nutrient availability. All these factors determine the biomass of plant produced, which subsequently is partitioned among various consumers (insects, mammals) and the decomposers.

ELM is a disaggregated model; that is, it specifies systems dynamics in great detail. As a consequence, either it requires much data to specify all these relationships or, if we cannot obtain the data, it requires that we make a number of assumptions—assumptions which may or may not prove correct. Naturally, because of the assumptions we are forced to make because of data limitations, questions about the reality of the final model will inevitably be raised. However, since these questions are difficult to evaluate without a detailed examination of each part of the model, we cannot deal with them here. If we are willing to assume that the model is realistic, we can proceed to use it to explore the effects of different policies—how, for example, repeated small fires might affect the grassland compared with less frequent but more intense fires or compared with regular mowing—and then proceed to devise recommendations for management of the grassland. This is the kind of procedure followed in simulation for land-use planning, and it is presently under way with ELM.

A management-oriented model for the Harvard Forest has been published by Gould and O'Regan (1965). Although less complex than ELM, this model also demonstrates the utility of the simulation approach for renewable-resource management. The model features routines that account for the influences of economics and policy-making on forest growth (Figure 9-8), including such factors as bank loans,

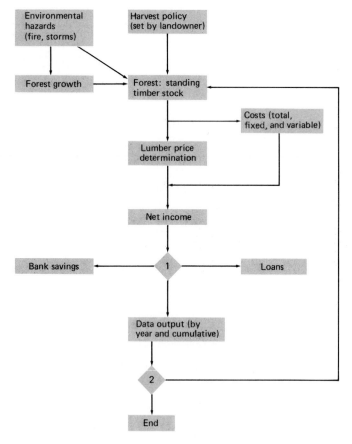

FIGURE 9-8 *Harvard Forest simulation model. Diamond 1 represents cash flow depending upon profit for the year. Diamond 2 keeps track of time. When the specified period is over, the program goes to "END" and stops.* (Adapted from Gould, E. M., Jr. and O'Regan, W. G., 1965, "Simulation: A Step Toward Better Forest Planning." Harvard Forest Papers, No. 13.)

capital reserves, operating costs, owners' decision-making, forest growth, and environmental variability. The model starts by specifying landowner policies that translate into a harvesting scheme. Then, depending on the policy and harvesting scheme, the normal (that is, completely unharvested) growth of the forest is reduced by the appropriate rate in each year. The volume of saw logs is multiplied by the current prices for lumber to obtain gross income, which is reduced by the operating costs (including taxes) to obtain net income. If the net income is negative, the loan routine may be called upon to supply credit; if it is positive, the net income is deposited to earn savings interest. Risk and uncertainty are simulated by the environmental hazard routine, which includes fire losses and storm damage.

One example constructed by the authors compared two strategies for cutting. One specified a harvest that would achieve a sustained yield, the other specified a harvest at rates commensurate with timber prices; all other factors were equal. The results are given in Table 9-1. Cutting and sustained yield gave 26,371 million board feet (mbf) over the 150-year period, compared to 24,058 mbf under the more opportunistic strategy. But the total volume of timber produced (including that still standing) was nearly equal for both strategies. The mean annual net income was, however, higher for the price-opportunity method, and so this strategy ultimately produced higher cumulative profit, measured here as total assets. Hence the strict sustained-yield strategy produces no more timber and definitely less profit; conversely, varying harvest in accordance with price did not substantially exceed sustained yield for the forest but took much better advantage of price fluctuations.

The Harvard Forest model raises some interesting points. Even though it is relatively simple, the model

Parameter	Case 1 Sustained yield	Case 2 Price opportunity
Total assets	$536,878.69	$709,935.15
capital	507,357.89	671,162.88
value of standing stock	29,520.80	38,772.27
Mean annual net income	21,074.69	25,621.60
Volume of standing timber (at end of run)	5,099.00	7,485.00
Total volume cut	26,371 mbf	24,058 mbf
Total cut and standing	32,070	31,543

TABLE 9-1 *Summaries of two test cases given by Gould and O'Regan (1965). For further explanation, see text.*

easily could be extended to examine the conclusions of the forest-practices study panel at the University of California, Davis, discussed earlier. Thus, it would provide additional evidence for or against current forest practices, and it could be used to explore still other policies. More abstractly, it also could be used to explore the use of simulation models for resource managers. Most interesting is the challenge of the Gould-O'Regan model to our cherished concept of sustained yield, because their model certainly does not provide a rationale for sustained-yield policy for the Harvard Forest. Whether or not their conclusion would stand for larger areas or longer times is problematical, but it is refreshing to see formal quantitative methods like simulation applied to complex problems, no matter what the results.

MARINE FISHERIES

The seas are vast and deep. Perhaps their illusion of infinity accounts for the optimistic hopes we hold for protein from the oceans. But as we have seen, Ryther's analysis deflates this optimism unless it is wildly erroneous. Furthermore, heavy and continued exploitation in international marine fisheries threatens future yields for all, as fishery after fishery collapses. Does this result from scientific ignorance? Despite the fact that fish populations are mobile, are found in an environment unlike ours, and are much more difficult to study than immobile trees, the problems again stem from the political and economic systems, not from lack of scientific knowledge.

In fisheries as in forestry, modeling approaches to resource management seem most promising where economic considerations are included in the dynamics of the system. Indeed, it is pointless to exclude them. The problem of salmon fishery management discussed by Paulik and Greenough (1966) is an excellent case in point. Pacific salmon are *anadromous*—they live in the sea but spawn in freshwater streams. They differ from Atlantic salmon in that the adults returning to fresh water die there after reproducing; they do not return to the sea. The life cycle of Pacific salmon and the nature of the harvest make the fishery unique in two ways:

1. The Pacific salmon migrate far out to sea, disperse widely, and are not sought except when they approach the spawning grounds. Each *year-class* (surviving adults hatched some number of years previously) is susceptible to harvest for only a short time, but that time is its reproductive period, a sensitive time for any species.

2. The fishery is not truly international, even though Pacific salmon spawn on both sides of the North Pacific, because the catch occurs mostly along the shorelines. This is because salmon reproduce only where they were hatched themselves.

The salmon-fishery model utilizes the following special features in its structure. Five sectors—migration, gear, gear decision-making, fishing, and management—simulate the passage of the salmon stocks through the coastal zone into fresh water (Figure 9-9), the period during which they are vulnerable to the commercial fishing fleet. The *migration* sector simulates the behavior of two species of fish, one of which migrates in such a dispersed and unpredictable manner that it can escape anywhere along the migratory pathway, while the other migrates in a tight school along a more predictable route so that they escape ready capture only at the end of the migratory gantlet. The authors call the former the *diffusion species* and the latter the *target species*. The *gear* sector has both fixed gear (such as fish traps and nets) and mobile gear (boats and seines). Decisions about gear involve its amount and its use in given areas at particular times. The closely related sector of *gear decision-making* applies only to mobile equipment and is the most comprehensive part of the model, for this sector represents fishermen's decisions concerning when, where, and how to use their gear. In order to decide where to fish, how long to fish, or whether to fish at all, the fishermen were hypothesized (1) to forecast future catches in each of three possible areas of Puget Sound to fish in, (2) to compare results from previous fishing experience with long-term predicted abundance of the fish in that area, and (3) to express the results in dollars of potential profit. With this information, derived from all possible sources, fishermen can stay in port and not waste their time in what they feel is fruitless effort.

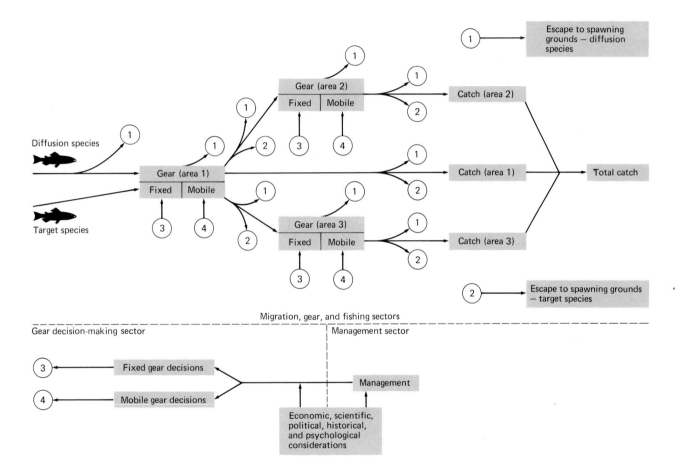

FIGURE 9-9 *The salmon fishery simulation model. The fishing and migration sectors are portrayed together for visual clarity.* (Redrawn from Paulik, G. J. and Greenough, J. W., "Management Analysis for a Salmon Resource System," in *Systems Analysis in Ecology*, K. E. F. Watt, ed., 1966, Academic Press.)

For those who decide to fish, the *fishing* sector is made a function of salmon-stock size, the amount of gear, and what proportion of the gear is being operated. In addition, the fishing success in any one area depends on competition among fishermen in the area and fishing success in other areas. Last, the *management* sector includes the political components. The managers can open or close fishing areas in a number of different ways and can respond to pressures to modify their decisions. Underlying each of these sectors are the data which the salmon model unifies: the scientific, managerial, and experience factors that interrelate all sectors in a realistic fashion.

The structure of this model is similar to that used for the Harvard Forest in that it depicts a process (the spawning runs for two salmon species) and the forces which manage those runs. For the Harvard Forest, the process was forest growth, and management was the force involved in production of timber. The model portrays both the salmon runs and the process of fishing in a way that can be experimented with. For example, fishing practices and management schemes at higher levels can be changed depending on the size of any one run or in response to demand or weather. Suppose a manager of salmon fishing wanted to know what the effects of a 10 percent larger catch would be on the diffusion species. He could increase the yield to this extent in the model and trace the effects on such things as salmon prices, the numbers of the target species, and the numbers of the diffusion species itself. He could look for unexpected feedback in fishermen's behavior or in public pressure on management. By designing a modeling program that carefully examines a range of alternative behavior, the manager can derive a series of impressions about how the system operates and how to manipulate it to achieve various objectives such as a larger sustained yield, or helping all fishermen make equal wages, or controlling gear. The net result is an improved salmon fishery—if the model was good in the first place. The promise of such a model is great, not only for salmon fisheries, but also for freshwater fish and for truly international marine resources.

Comparisons and Conclusions

The three resource-management areas considered in this chapter illustrate a continuum from exploitation to comprehensive management. For the most part, international marine fisheries have not progressed much beyond exploitation, and forestry practices currently are making the transition from exploitation to sustained yields with guarantees for environmental protection. Only sport fisheries and hunting have seen substantial progress toward management, if only because exploitation has proved hopelessly inadequate. However, even here management has focused on a small number of species which are advanced at the expense of other ecosystems. As we have seen, if the success of sport fisheries and hunting is to be measured in catches and bag limits, management will have to progress toward increasingly artificial ecosystems separated from natural ones.

But there is another choice. Some fishermen and hunters may rate kills as being less important than experiencing nature. This may move fish and game managers toward more comprehensive ecosystem management and away from becoming farmers of selected sectors of wildlife. If this happens, credit will be due to increased public understanding and valuation of complex natural environments.

Growing environmental awareness is a major force for change in managing renewable resources—a force that promises to be even more important than the concept of sustained yield. The forest-practices study panel report discussed above identified environmental awareness as the critical influence causing the public to seek comprehensive forestry practices. But while this is hopeful, it remains to be seen just how important environmental considerations will finally be, especially with demand increasing for all kinds of resources. If large sectors of society are in want of resources—renewable or not–environmental considerations could vanish.

Another major force, the conservation of resources and decrease in demand, also will have to assume greater importance. While each type of resource has a history replete with destruction and extinction, declining resources are a strong incentive to develop sophistication in resource management. The use of remote sensing, impact assessment, and simulation modeling seems likely to grow more important. This will require the expertise not only of scientists and economists but of politicians and diplomats as well. As we have seen, political considerations override all others.

While the development of sophisticated management tools and an expanded ecosystem viewpoint promise to emphasize commonalities among renewable-resource systems, no tool can reconcile larger demand and smaller supply indefinitely, or "do more with less," in Buckminster Fuller's words. Fish and game agents can retreat into an artificial ecosystem, a sort of mechanized provision for recreation; but the harvesting of timber and game at rates above sustainable yield auger for drastic future change. The change must be met by political, social, and economic forces, not only by technical ones.

Questions

REVIEW QUESTIONS

1. Is it true that the main effort in increasing sport fisheries is by scaling supply by ecological engineering? Why or why not? What are some biological factors that might limit the extent of this activity?

2. In which of the following situations would it be possible to augment stocks of game fish without disturbing native ecosystems? You may wish to review Chapters 1 and 2.
    a. a cold mountain trout stream in a recently disturbed area
    b. a warm-water reservoir
    c. a semi-isolated estuary
    d. an open ocean

3. What characteristics would you expect to find in trees that are favored by
    a. regular ground fires
    b. clear-cutting in small patches
    c. partial cutting of the tallest trees

4. What would be the effect of increased yield of a favored species on its ecosystem when
    a. the increase is accomplished by hatchery planting
    b. the increase is accomplished by nutrient fertilization

ADVANCED QUESTIONS

1. The ad valorum tax on standing timber in many states favors early cutting of trees. If such taxes were removed, do you think trees would be allowed to stand and grow longer? Why or why not?

2. Clear-cutting forests is a controversial practice, but it is better for some species and not good for others. Would clear-cutting be better or worse for species that
    a. require humid conditions
    b. require rich soil
    c. germinate seeds poorly in shade
    d. compete poorly for soil water with other species

3. Discuss the strengths and weaknesses of simulation as a resource management tool with respect to
   a. a heavily exploited population
   b. a population that is extremely valuable economically
   c. the ecosystem, including commercial and noncommercial species, in a large lake

Further Readings

Note: Renewable resources are a favorite subject in conservation texts. For entries into the literature see Dasmann (1972). See Chapter 6 for other pertinent references.

Boulding, K. E. "The Economics of the Coming Spaceship Earth." In *Environmental Quality in a Growing Economy* (H. Jarrett, ed.). Resources for the Future, Johns Hopkins, Baltimore, 1966.

> This paper is about the transition from expanding resource requirements to stable economy. It is well known and stimulated a number of commentaries on the future of the earth based on the analogy of a spaceship. This paper has been reprinted in a number of places. See also Boulding's contributions in *Future Environments of North America* (Darling and Milton, 1965), and in Murdoch (1971).

Cloud, P. E. (ed.). *Resources and Man.* Freeman, San Francisco, 1969.

Daly, H. E. "Toward a Stationary-state Economy." In *Patient Earth* (Hart and Socolow, eds.). Holt, New York, 1971.

> One of the best essays on the nature of economies and the problems of transition to a no-growth state. Highly recommended to all students.

Darling, F. F., and J. P. Milton (eds.). *Future Environments of North America.* Doubleday, Natural History Press, Garden City, N.Y., 1966.

Dasmann, R. F. *Environmental Conservation*, 3rd edition. Wiley, New York, 1972.

> A recent revision of a classic conservation text. Good treatments of forest and range management but, more important, entries into the literature beyond the scope of this book. See also Guy-Harold Smith, *Conservation of Natural Resources*, 4th edition, Wiley, New York, 1971.

Goodall, D. W. "Simulating the Grazing Situation." In *Concepts and Models of Biomathematics* (F. Heinments, ed.). Marcel Dekker, New York, 1969.

> A simple model with management implications, designed for heuristic purposes. Very much like the Gould and O'Regan model, but for range land.

Gould, E. M., Jr., and W. G. O'Regan. "Simulation: A Step toward Better Forest Planning." Harvard Forest Papers, No. 13. Harvard Forest, Petersham, Mass., 1965.

Grady, N. C. (ed.). *Agriculture and the Quality of Our Environment.* American Association for the Advancement of Science, Washington, 1967.

Institute of Ecology, University of California, Davis. *Public Policy for California Forest Lands.* Available from the institute on request; 1972.

> An excellent review and thoroughly researched proposal, it is a shame that it is not more widely available.

Johnson, P. L. (ed.). *Remote Sensing in Ecology.* University of Georgia, Athens, 1969.

> An introduction to the use of remote sensing techniques in ecological systems. See also National Research Council Report (1970) and Watt (1966).

Landsberg, H. H. *Natural Resources for U.S. Growth.* Resources for the Future, Johns Hopkins, Baltimore, 1964.

Landsberg, H. H., L. L. Fischmann, and J. L. Fisher. *Resources in America's Future.* Resources for the Future, Johns Hopkins, 1963.

> The second is the full version, the first a condensed edition, of a large study of resources both renewable and nonrenewable from an economic point of view. Most of the earthy facts and figures relating to resource adequacy are to be found here.

Lewis, P. H. *Regional Design for Human Impact.* Thomas Publications, Kaukauna, Wis. 1969.

Murdoch, W. W. (ed.). *Environment.* Sinauer Associates, Stamford, Conn. 1971.

National Research Council, Committee on Remote Sensing for Agricultural Purposes. *Remote Sensing with Special Reference to Agriculture and Forestry.* National Academy of Sciences, Washington, 1970.

> Compare with Johnson (1969).

Paulik, G. J. "Anchovies, Birds, and Fishermen in the Peru Current." In *Environment,* (W. W. Murdoch, ed.). Sinauer Associates, Stamford, Conn., 1971.

> A complete case study and excellent reading for anyone. In particular, note the similarities to the paper on salmon resources in the North Pacific.

Paulik, G. J., and J. W. Greenough. "Management Analysis for a Salmon Resource System." In *Systems Analysis in Ecology* (K. E. F. Watt, ed.). Academic, New York, 1966.

Ryther, J. H. "Photosynthesis and Fish Production in the Sea." *Science,* 166 (1969), 72–76.

van Dyne, G. M. "Implementing the Ecosystem Concept in Training in the Natural Resource Sciences." In *The Ecosystem Concept in Natural Resource Management* (G. M. van Dyne, ed.). Academic, New York, 1969.

> See also the papers by Spurr, Lewis, Bakuzis, and Wagner.

Watt, K. E. F. (ed.). *Systems Analysis in Ecology.* Academic, New York, 1966.

> See especially the paper by Savage, as well as by Paulik and Greenough.

Watt, K. E. F. *Ecology and Resource Management.* McGraw-Hill, New York, 1968.

> The first half of the book is very good for examples and principles; the second half (on mathematics and modeling) requires a much higher level of background.

Weiss, P. *Renewable Resources.* National Academy of Sciences/National Research Council, Publication 1000-A, Washington, 1962.

> Most interesting for individual comments on future research needs.

Whyte, W. H. *The Last Landscape.* Doubleday, New York, 1968. (Available in paperback), Anchor Books, 1970.

10. Conservation, Substitution, or Both?

Food, water, and land are important resources because in various ways they maintain life itself. But the nature of modern society is such that we really depend upon nonrenewable resources—energy and minerals, the raw materials of industry and commerce. *Nonrenewable resources* are those for which the rate of use far exceeds the rate of renewal. And since human populations and the rate of demand for all resources both continue to grow, one may argue that all resources except foodstuff and water are rapidly becoming nonrenewable. Thus, some of the resources we have considered (such as forests and oceanic fisheries) could be included here. Despite the imprecision of the definition, however, energy and minerals are so grossly nonrenewable that the concept of sustained yield clearly does not apply to them. We must choose between a program of conservation to insure that resources last as long as possible or blind faith that some abundant resource can serve as a substitute when the original resource is exhausted.

The choice between conservation and substitution is a major policy choice for society. If we emphasize conservation, we shall help insure minimum social disruption but possibly at the cost of increased unemployment and declining standards of living. But if we emphasize resource substitution, the national economy could continue to grow; and while the poor might benefit (if only because there is more economic "pie" to be divided), the potential costs include the ramifying impacts of resource shortages (such as the petroleum crisis of 1973–1974) and a social crisis that is only magnified by postponement.

Technology and resource economics are the two factors that will influence our decision-making most strongly. *Technological innovation* determines whether certain resources can be exploited, the environmental consequences of exploitation and production, and the potential for future substitution. *Resource economics* sets the price limitations within which technology must offer potential solutions and is a major determinant of the exploitation rate of existing resources. Together, these two factors can dictate conservation, substitution, or a combination of alternatives.

The problem for policy-makers is that in most cases they deal with situations for which there is little relevant information. For example, suppose we want to decide between conserving petroleum or assuming plentiful nuclear energy. We do not know how much petroleum is still to be found and exploited, how much it will cost to do so, how soon nuclear energy will be plentiful and inexpensive, or whether technology can control the potential environmental consequences of nuclear energy. As a result, policies regarding petroleum energy depend upon opinions, philosophies, and assumptions so different from each other that the policy itself is based on faith and preconception rather than on data.

The prospects for better information do not seem bright, and policy-makers have two unpalatable choices: if they choose conservation, they run into immediate reactions from a society suffering from rising prices and unemployment; if they choose substitution, the social costs may be displaced into the future but may become even more severe. These awesome consequences cannot be explored definitively in this chapter, but they are so important that they must be considered for each resource we examine. We shall begin with an illustration of the data problems which policy-makers face.

The Great Oil Controversy: Who Is to Be Believed?

The technology for locating and extracting petroleum liquids from reserves under ground is among the most advanced. Yet we have no reliable estimate of how much oil remains to be found or extracted. Hubbert (1971) estimated that 2.5×10^{12} (2.5 quadrillion) barrels of petroleum liquids plus 1.2×10^6 (1.2 million) cubic feet of natural gas existed before the discovery of petroleum. Since 1889, about 0.184×10^{12} (184 billion) barrels have been produced in the world, only a small fraction of Hubbert's estimate (excluding solid petroleum reserves, the tar sands, and oil shales). While this may seem to indicate plenty of reserves for the future, it must be remembered that worldwide petroleum consumption has been rising dramatically; if this continues, as it probably will, the reserves could drop rapidly. At present, geologists who specialize in forecasting petroleum yields do not agree either about the extent of remaining petroleum reserves or about the governmental policies that should be pursued. In fact, the controversy has been sharp and sometimes acrimonious.

Since crude oil tends to be buried far below ground and under the sea, and since not every geological formation with which oil is characteristically associated will necessarily contain oil reserves, some method must be adopted to make reliable forecasts about the size of future discoveries. One way is to use statistical data to extrapolate from past trends in discoveries of petroleum reserves, making allowance for new technology. Another method is to project only the statistics related to the abundance of reserves as found in new explorations. Those taking the first approach have produced a wide variety of estimates that often bear little resemblance to one another. For example, Moore (1966) used one equation to describe the curve of the rate of new discovery against the total area explored. By extrapolating the curve to the point where the rate of new discoveries fell to zero, he estimated that the ultimate magnitude of U.S. petroleum reserves would be 353 billion barrels (bbl). Using the same data but a different mathematical function, Hubbert (1966) estimated that only 190 bbl would be produced. Ryan (1966) used the same equation as Hubbert but estimated the parameters of the equation in a different way; he derived an estimate of 900 bbl, a figure he considered ridiculously high, but it certainly indicates the weaknesses of such projections.

A. D. Zapp (1962) estimated petroleum reserves from data of the actual volume of oil recently discovered per square foot of drilling in an area when the area had been adequately explored—about one well for every 2 square miles. His estimate for the United States—excluding the Alaskan Prudhoe Bay discoveries—was 590 bbl. Using the method of cumulative discoveries but including Prudhoe Bay, Hendricks (1966) and McKelvey and Duncan (1965) made estimates of 420 bbl and 650 bbl. The variation in estimates here is slightly less than that using production figures, but it still underscores the difficulty of making an accurate estimation by either method.

This lack of agreement leaves the agencies controlling national petroleum policies in a quandary. But as McKelvey (1972) has pointed out, it is far too much to expect unanimity from various authorities on the subject because there are bound to be weaknesses in any method that projects an unknown quantity; all we can hope for is improvement in the estimates and, consequently, greater confidence in them.

This issue is not merely academic; increased demand creates pressure to insure adequate future supplies of energy. Future pressures will be even greater, as more nations industrialize and increase energy consumption. Insofar as we use most of the world's annual energy production in the United States, and it is obvious that petroleum supplies are indeed quite limited, greater energy conservation is required. But that still does not tell us when or how urgently we must curtail our energy demand.

A Spectrum of Potential Energy Sources

In subsistence economies, wood is the predominant energy supply, as it once was in the early history of the United States. Wood is a ready source of heat for space heating and cooking, but it does not produce heat intense enough to fuel most industrial processes; thus more refined fuels are needed for industrial and economic development. This role once was played by coal, but for most of this century petroleum products have been dominant because they are compact, extremely energy-rich, versatile, and unmatched as sources of mobile energy. United States demand for energy has meant, in effect, demand for petroleum, and consumption has grown at a rate exceeding the demand for all other kinds of energy (Figure 10-1). It is clear to everyone, however, that the world's petroleum reserves are too limited to sustain the current level of U.S. demand, much less the future growth in world demand, especially now that some nations exceed the United States in energy-growth rate (Figure 10-2).

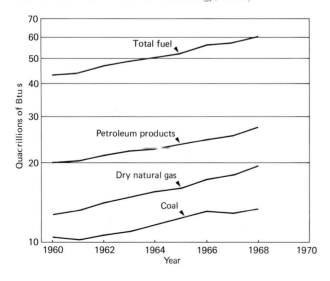

FIGURE 10-1 *United States energy consumption, 1960–1968, for all fuels and three fossil-fuel derivatives. Figures are in 10^{12} BTUs on a semilog plot, which makes an exponential growth curve linear.* (Redrawn from *Patterns of Energy Consumption in the United States*, U.S. Office of Science and Technology, 1972.)

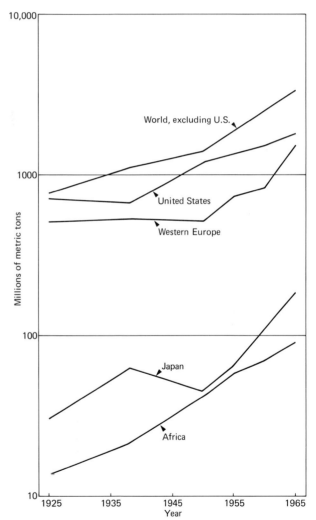

FIGURE 10-2 *World energy consumption, drawn on a semilog plot in 10^6 metric tons, with the U.S. included for comparison.* (Redrawn from *Energy in the World Economy* by J. F. Darmstadter. Published for Resources for the Future by the Johns Hopkins University Press, 1971.)

Derived from solar energy	Atomic power	Derived from lunar gravitation	Derived from earthly heat
Solar energy	nuclear fission	tidal	geothermal
Fossil fuels (and their derivatives)	nuclear fission		
Deep ocean currents			
Wind power			
Hydroelectric power			
Solid waste conversion			

TABLE 10-1 *The spectrum of energy resources, both proven and potential, classified by ultimate source.*

Thus the world faces a questionable future energy supply. Technological developments in energy are numerous, and events occur so rapidly that it is impossible to keep up with them. In this section we shall examine some major alternatives that could be operational by the year 2000. Readers seriously interested in energy resources, however, are urged to read the most recent publications in the field; the journal *Science* is especially comprehensive.

The spectrum of proven and potential energy sources is given in Table 10-1. All the primary energy sources can be placed into four categories according to their derivation: from solar energy, from nuclear energy, from gravitational effects of the moon, or from heat stored in the earth. Some of these primary sources also yield other energy sources. In the discussion that follows, we shall emphasize the economic and capability aspects of each energy source rather than dwelling on the technology of production. Technology, though certainly important, is not our primary concern here. Of course, various environmental consequences also are associated with each source, but discussions of these are grouped in Section D.

SOLAR ENERGY RESOURCES

The energy of the sun may be captured directly or indirectly, in chemical compounds and in fluid currents resulting from uneven heating by solar energy.

Direct capture of solar energy Solar energy may be captured by various methods and converted into heat or electricity. In one method, collector cells that contain water or air are specially coated to maximize their absorption of solar energy; the material absorbs the heat, which subsequently can heat buildings. Electric-power generation can be handled in the same way, albeit at a larger scale, because turbines to generate electricity can be run on steam generated by solar heat. Alternatively, silicon crystal photo cells (solar cells) can produce electricity on a limited scale by converting sunlight directly into electrical energy (Figure 10-3).

The attractiveness of solar energy lies first in its abundance and second in its small potential environmental impact. The amount of solar energy incident on the surface of the earth is estimated to be 3.56×10^{22} calories per year, more than will ever be needed. In spite of the apparent abundance of solar energy, however, it is an all-or-none situation; before we can depend on it, technology must master the problems of concentrating the sun's energy into useful forms and distributing energy from hot sunny areas to cooler, more cloudy areas. If these technological barriers could be mastered, we probably would have

FIGURE 10-3 (a) A home equipped with solar panels in the roof to provide for its own electrical and heating needs. (b) The solar generator at Odeillo in the French Pyrenees. ((a) Courtesy of John Messineo from Tom Stack and Associates. (b) Courtesy of Peter Menzel from Stock, Boston.)

all the energy we require. Until then, however, we cannot depend on the sun for any significant energy contribution. On balance, the potential of solar energy warrants further exploration, but at present its operational use remains unknown beyond application to space heating, which is technologically proven.

Fossil fuels Fossil fuels—petroleum, coal, and natural gas—are all products of the incomplete decomposition of ancient organisms, organisms whose livelihood depended either directly or indirectly on sunlight. The future holds declining prospects for petroleum products, since the supply is limited and exporters have been raising prices drastically. The combination of limited supply and skyrocketing prices is expected (1) to force a return to coal as the most important fossil fuel in the United States (coal supplies are large enough to last four to five centuries); (2) to stimulate research into other energy sources, particularly those derived from coal and nuclear energy; and (3) to make production of petroleum and its derivatives from petroleum solids (tar sands and oil shales) economically feasible. Petroleum liquids might last well into the twenty-first century, but natural gas reserves in the United States could be exhausted before the year 2000. Rather than continue to exhaust these reserves, however, it is desirable to conserve the remainder of our petroleum and natural gas as raw materials for industrial synthesis rather than use them as fuel.

The main thrust of technological developments in fossil-fuel production is the removal of sulfur from coal and the production of "natural" gas from coal (by a process called *coal gasification*). Probably 70 percent of U.S. coal reserves contain 1 to 3 percent sulfur, which upon combustion of the coal is converted to sulfur dioxide, an important air pollutant. Bulk coal is heavy and difficult and expensive to transport. These facts suggest that if we could remove sulfur and convert coal to a form that makes it cheaper to transport, it could replace petroleum with minimal radical technological and social change. Fortunately, we can attack both problems at once. Coal can be treated with solvents that remove the bulk of the sulfur and ash and simultaneously liquify the coal. Alternatively, coal subjected to high temperatures and pressures can be converted after several steps into synthetic natural gas by treatment with steam.

Magnetohydrodynamic power (MHD) is an entirely different solution to the sulfur problem and air pollution, but it would allow high-sulfur coal to be used in an unmodified form. MHD generators consist of a chamber in which gases resulting from the combustion of coal are heated to 2000°C and, as a result, flow from one end of the chamber to the other. The chamber has a magnetic field through which the expanding ionized gases must pass, thereby generating electric currents that then are captured by electrodes. Thus MHD generates electricity directly, and, moreover, its waste gases can be used to generate still more electricity because they can run conventional steam turbines. MHD is projected to be very efficient, with up to 60 percent of the fuel energy being converted to electric power compared to 40 percent for fossil fuel or nuclear plants. However, the primary advantage of MHD is that the dirtiest coals can be used, since once the coal is converted to hot ionized gases, the sulfur is also gasified and then converted to salts which are removed easily through precipitation.

The future picture for fossil fuels, then, includes a declining dependence on natural stocks of petroleum and natural gas, with a reawakened interest in coal and particularly its derivatives. Here again, however, the technology is not well developed, and therefore reliable predictions of eventual energy yield cannot be made.

Solid waste conversion   Organic wastes also represent an energy source derived directly and indirectly from the sun. For various reasons the U.S. economy produces prodigious volumes of solid wastes, which hitherto have been burned, buried, or otherwise disposed of without any energy return (Figure 10-4). The technology to convert these organic wastes into useful oil is now in the pilot-plant stage of development and presently involves two main processes: the production of heavy hydrocarbons by subjecting wastes and carbon monoxide to pressure and steam, and the anaerobic distillation of organic wastes to produce a mixture of low-energy oils and gases. Estimates of the volumes of fuel producible from solid wastes vary from 170 million to 2.5 billion barrels of oil, or from 3 to 50 percent of U.S. oil consumption in 1971.

The future of solid-waste conversion lies only partly in further technological progress. The energy content of oils produced so far is too low to be of much value, although conversion of these oils to methane is one promising alternative. More important, the effort involved in the collection, centralization, and separation of solid wastes is a socioeconomic problem that may have to await higher fuel prices and advances in technology before it becomes a worthwhile endeavor. Hence the conversion of solid wastes to an energy source may be more important from an environmental standpoint as a means of dealing with the solid-waste problem than it is as a fuel source.

Energy extraction from moving water and air Solar energy does not fall evenly on the planet. The consequences of this are large indeed, since weather and climate are determined by this uneven heating of the earth's surface. As air is heated, it expands and becomes lighter, and so it rises into the upper atmosphere until it cools to the point where it sinks back toward earth. This phenomenon creates high- and low-pressure centers, which in turn drive winds. In the earth's seas, tropical waters are heated to the greatest extent, making them less dense; the more poleward waters are cooler and heavier and consequently sink. The result is that poleward waters tend to flow deep in the oceans toward the equator, creating ocean currents. The warmer equatorial waters also evaporate to a greater extent, and when the water that evaporates is cooled as it rises in the atmosphere, it condenses to form clouds and then falls as precipitation, only to return ultimately to the seas as part of the hydrological cycle (Chapter 7).

Since these natural movements are all driven by the sun, utilizing them is still another means to solar energy. The extraction of energy from the deep ocean currents is based on the temperature gradients between the cold water in the currents and the warm water that surrounds it; the temperature gradient in subtropical oceans is enough to turn a submerged generator. One principal problem involves the transmission of power from the extraction facility, which must be located in the ocean, to the land surfaces for use; another problem is the rapid corrosion of the generators that result from salt water.

Wind power, which can be tapped by nothing more exotic than a windmill, holds more promise; but interest in large-scale projects is only beginning to develop. The main problem is that suitable locations having enough winds to produce power dependably are restricted mostly to offshore islands and coastlines where the cold sea and the warm land provide the gradients needed for the development of sufficient wind speeds.

FIGURE 10-4 *Each year millions of tons of solid waste are buried in landfill dumps such as this one.* (Courtesy of Daniel S. Brody from Stock, Boston.)

Despite the problems, the Joint Solar Energy Panel has estimated that 1.5 trillion kilowatt-hours (kwh) of electrical power could be produced if all suitable sites were utilized. Since this is far less than 1 percent of the energy consumed in the United States in 1970, wind power is likely to have a restricted impact, although it could be very important in places where it can be generated.

Hydroelectric power, the use of stream flow in rivers to turn generators, is well established and operational throughout the world. Cheaper than all other sources of energy, hydroelectric power can be produced in conjunction with other uses of fresh water, and it still has enormous development potential. Overall, perhaps 10.8 trillion kwh can be generated worldwide, with the greatest unrealized potential existing for the major rivers in developing regions of the world. Some, like the Mekong River in Asia, are being developed now, while the Amazon is one example of a river that remains totally unexploited in this regard. The United States, unfortunately, has already developed its hydroelectric power, and so little further potential remains.

ATOMIC ENERGY

Energy can be obtained by splitting radioactive elements apart in a process called *nuclear fission*. All presently planned and operating nuclear plants are of this type. Energy also can be obtained by reversing the process and fusing radioactive isotopes of hydrogen in a way that mimics the basic reaction of the sun; this is *nuclear fusion*.

Nuclear fission The simplest fission reactors utilize U-235, the highly radioactive isotope of uranium. When bombarded with neutrons, U-235 will readily undergo fission into other elements with the release of large quantities of energy. The generalized reaction is

$$\text{U-235} + \text{neutrons} \longrightarrow \text{fission products} + \text{neutrons} + \text{energy}$$

In conventional nuclear fission reactors, U-235 supplies its own neutrons, and it is necessary only to put enough of it together to start the fission reaction. Then the reaction will continue until the supply of U-235 falls below a critical minimum level. Apart from the process of fission itself and the safeguards involved in protecting against leakage and accidents, nuclear power plants using U-235 are the same as fossil-fuel plants. The energy released is used to heat some cooling solution, either molten metal or some other liquid, which in turn heats water into the steam needed to generate electricity. Since it is technologically quite simple, this type of nuclear plant has been springing up over much of the developed world. However, nuclear power plants based on the fission of U-235, despite their short-term promise, are no more than an interim solution to the energy problem because their fuel is scarcer than coal. Moreover, existing U-235 should be conserved, because it is vital as a starter for the nuclear fission process upon which the next generation of nuclear power plants (presumably, the breeder reactor) will be based.

More than 99 percent of the world's supply of uranium is U-238, a relatively stable isotope which cannot be used in conventional reactors. However, if U-238 is bombarded with enough neutrons, it will undergo fission in much the same way as U-235 and with the same energy release. The difference, besides the much larger stocks of potential fuel available, is that this fission process converts U-238 into other unstable isotopes which also can be used as nuclear fuel. As a result, this kind of reactor actually produces more fuel than the amount it initially uses; for this reason it is known as the *breeder reactor*. The reaction is

$$\text{Neutrons (from U-235)} + \text{U-238} \xrightarrow{\text{fission}} \text{products} + \text{neutrons} + \text{energy} \atop \text{(potential nuclear fuels)}$$

The reaction is conceptually simple, but because of technological problems, commercial breeder reactors are not yet available, though some pilot plants are being built in Europe. The main source of delay (besides lack of funding) has involved difficulties in building reactors rugged enough to withstand the longer time it takes for the fuel to burn down to critical level. In addition, there have been difficulties in developing the optimal cooling material (either molten metal or gas), as well as problems relating to provisions to safeguard against nuclear accidents. Despite these design problems, when breeders eventually become fully operable, they are expected to produce electricity quite inexpensively, to conserve nuclear fuels, and to replace the fossil fuels as the primary means of generating electricity.

Nuclear fusion The sun's heat is produced as the result of the fusion of hydrogen molecules to form helium. Incredible as it may seem, this reaction can be duplicated on earth. The reaction that has the most promise combines lithium (Li) with deuterium (H_2):

$$\text{Li} + \text{neutrons} \rightleftarrows \text{tritium (}H_3\text{)} + \text{energy}$$
$$H_3 + H_2 \rightarrow \text{He} + \text{energy} + \text{neutrons}$$

The most attractive feature of fusion is the abundance of fuel. Deuterium is so plentiful in sea water that the supply would last indefinitely. Lithium is rarer, but even this would last 1 million years. Hence commercial fusion would provide a cheap and practically limitless energy source.

Problems arise, however, concerning the possibility of achieving controlled fusion. (Uncontrolled fusion is the hydrogen bomb.) The prospect is so uncertain that we cannot predict whether it ever will be feasible, much less when. Because fusion produces so much heat, no physical structure can contain the reaction without itself being destroyed, and so ways must be found to keep the fusion process from melting the facility. Presently, two basic designs are being considered to overcome this problem. One, called *magnetic containment*, would use powerful magnetic fields to keep the fusion reaction from physically touching any wall of the reactor. The other, called *laser containment*, would use very small fuel pellets (of the order of 1 millimeter in diameter) and would heat them so quickly with enormous pulses of energy from laser guns that fusion would occur before the pellets could expand and come into contact with the vessel. Hence this method uses inertia to achieve reaction containment.

ENERGY FROM LUNAR GRAVITATION

As the moon rotates around the earth, its gravitational field interacts with the earth's to produce regular cycles in the gravitational force at any one spot. A result is that ocean waters regularly rise (high tides) and fall (low tides). Where both favorable coastal geography and tides of sufficient amplitude exist, it is

FIGURE 10-5 *Advancing tidal bore in the Bay of Fundy near Moncton, New Brunswick. The tidal bore frequently reaches heights of 50 feet or more. (Courtesy of Civic Relations Department, Moncton, New Brunswick.)*

possible to harness the tides to produce electrical power. For obvious reasons, the most favorable sites are large bays with narrow inlets and large tidal amplitude, where a dam can be erected across the mouth. Water surges in and out of the basins with a force that can be used to turn power generators. The Bay of Fundy (formed by Maine, New Brunswick, and Nova Scotia), is one possible site mentioned frequently (Figure 10-5), and plants now exist in France and the Soviet Union.

Tidal energy is clean and inexpensive, but experts agree that it is extremely limited. Tidal power worldwide probably would generate only 64 billion watts, even less than wind.

ENERGY FROM THE HEAT OF THE EARTH

At several places on the earth, the heat from the interior comes close enough to the surface that it can be used to drive turbine generators; the thermal features of the Yellowstone region are examples of geothermal activity with this sort of potential (Figure 10-6).

FIGURE 10-6 *Old Faithful in Yellowstone National Park is probably the most spectacular evidence of geothermic activity in the western United States, and is thus unlikely ever to be exploited for power generation. (Courtesy of Will Faller from Monkmeyer Press.)*

Where the heat is on or near the surface, the technology to convert geothermal energy to electric power already exists and is being applied to commercial use. If the geothermal resource is steam, it can be pumped to the surface and put through the generators with nothing more than filtering to remove particulate impurities. If it is hot water, the heat can be exchanged to another fluid with a lower boiling point, and its vapor will then drive the generators. For dry, "hot rock" geothermal resources, it is necessary to create an underground cavern (for example, by a small nuclear explosion) and then circulate water or another suitable fluid through this chamber to extract the heat. The problems are in finding new geothermal sites, developing the technology to tap deep sites, and disposing of the salt-laden water that comes from these geothermal sites. Despite these problems, the estimated 100,000 megawatts of electrical generating capacity make geothermal potential significant for local areas where conditions are favorable.

Energy and Environmental Considerations

The use of energy normally entails environmental costs. For convenience we shall identify three phases in which such costs can occur: during exploration for a resource, during production and processing of it, and during consumption. This arbitrary classification allows us to compare the impacts of different forms of energy. In this section we shall consider only the first two categories in detail; the impacts of consumption are reviewed in Section D.

IMPACTS OF EXPLORATION

The environmental impacts of every exploration depend on the location of the energy reserves, the nature of the exploration process, and the sensitivity of affected ecosystems to perturbation. Only energy resources that are buried within the earth or under the seas really require substantial exploration, and since there is a sophisticated technology in engineering geology that allows us to obtain a great deal of information from existing maps and carefully placed bore holes, we would expect the most serious environmental consequences of energy explorations to be found only where local ecosystems are especially sensitive to disturbance. The premiere example of this is exploration for petroleum in the Arctic regions of Alaska, where the tundra is disturbed easily and where scars of any disturbance may persist for decades (Figure 10-7). Even though the exploration bore holes themselves are small, the size of the equipment and the number of personnel needed to support exploration cause considerable land disturbance with largely unknown consequences.

IMPACTS OF PRODUCTION

In contrast to the relatively minor environmental effects of exploration, economically efficient production and processing techniques have large potential impact. For many kinds of energy having no substantial impact from exploration—solar, tidal, hydroelectric, wind—the impacts from production either are unknown presently or would probably be only aesthetic in nature. For others—geothermal power in particular—there is at least the speculative possibility that waste heat might escape to the environment and produce ecological damage. There also are possibilities

FIGURE 10-7 *The United States made extensive explorations for oil in the Alaskan tundra in the mid-1940s. Here we see the effect of scraping away the vegetation layer at Umiat, northern Alaska, nearly 30 years later. The vegetation insulated the underlying ice, causing the surface to subside as the ice melted. Consequently, the entire scraped trail has subsided below the relief of the undisturbed area. The characteristic microrelief in the trail (polygons) results from differential melting. (Courtesy of U.S. Army Cold Region Research and Engineering Laboratory.)*

of earth subsidence or enhanced earthquake hazard (see Chapter 14).

In short, the potential environmental impacts of these forms of power are largely unknown. All of them, insofar as they are cleaner than the fossil fuels today, could contribute to improved environmental quality. This is particularly true of MHD power generation from coal, and of solar, hydroelectric, and tidal power. However, because of limitations in the possible energy output from most of these sources, they could provide only small parts of our requirements. Therefore, unless a breakthrough is achieved in direct solar utilization for electrical power, we must depend upon technology to clean up fossil fuels and confine the nuclear reactions which are so important yet so potentially damaging.

Fossil fuels often lie deep within the earth or under the ocean floors. To get to solid veins of the highest quality low-sulfur anthracite coal, deep mine shafts must be excavated; but they are costly and dangerous and produce vast quantities of waste rock for a now scarce resource (less than 2 percent of U.S. coal reserves). The remaining stocks of coal mostly are shallow deposits, and so the prospect is that deep mining will disappear in favor of *strip mining*, which requires no excavation; the overlying rock (called the *overburden*) is removed entirely, and the exposed coal then may be broken up and removed. Stripping is cheaper and is likely to be used even more widely when low-quality coal deposits can be used more extensively (as with MHD power generation). Stripping is ugly, however. It chews up landscapes, leaving big pits and mountains of overburden, and it disrupts water tables. Recently, however, an effective federal bill to stop "cut and run" stripping was passed, and it should help prevent the worst abuses and even correct some of the old ones in Appalachia. Among other things, the new law requires licensing, bans stripping on alluvial valley floors, compensates owners of water rights if the water table sinks, and diverts outer coastal-shelf oil revenues to the restoration of the scarred areas in Appalachia. Most important, from now on, strip miners would be expected to restore the "approximate original contours of the landscape" before they abandon a site.

The special environmental circumstances of Alaskan North Slope oil and tidelands oil make production particularly hazardous. The 1969 Santa Barbara oil spill (see Steinhart and Steinhart, 1972) and a number of marine oil-tanker disasters have made the public aware of the dangers involved in the production and transportation of oil in aquatic environments. The controversy stemming from the Santa Barbara spill continues to this day as both federal and state governments slowly revert toward the granting of leases for offshore oil exploration and production in the Santa Barbara Channel. Proponents of renewed production cite the need for larger domestic stockpiles for national security, while opponents see more spills leading to polluted beaches, dead sea birds, and possible long-term ecological effects.

Alaskan oil poses special problems because transporting it is particularly difficult in that frigid environment. To be pumped south through the Trans-Alaska Pipeline, crude oil must be heated to 140°F: If the pipeline were on the ground, it would melt the permafrost; but if it were suspended in air, it might block caribou migration routes. Once the oil reaches the port of Valdez through the pipeline, it then must be transported by sea to its eventual destination, and the oil tankers would have to negotiate the narrow and dangerous Valdez Channel. Therefore, the danger of shipwrecks and massive oil spills in this area must be considered real possibilities. For better or worse, however, the Trans-Alaska Pipeline has been approved and is now under construction.

Despite such environmental dangers, it is nuclear power that potentially could pose the most severe hazards. Successful fission is accompanied by the escape into the environment of some tritium, which has a short radioactive life but is incorporated readily into biological systems. Theoretically, then, tritium could pose a genetic danger of unknown magnitude, though its emissions are not strong enough to produce direct radiation damage. In fact, tritium is used widely in research as a marker in living cells on the assumption that it is rather harmless, and there has been no evidence to the contrary. But the same cannot be said for other isotopes involved in nuclear fission. Basic fuels and fission products are much more dangerous radioactive substances both because of their longer half-lives (the time it takes for half the element to lose its radioactivity) and because of the strength of their emissions. Thus accidental explosions in nuclear power plants could spread radioactivity over large areas.

Some critics liken fission plants to time bombs and oppose their construction entirely. Even some sympathetic scientists question the wisdom of constructing a power source that always has the danger of producing radioactive contamination, especially since the process also uses an extremely rare and valuable fuel. Still another set of environmental problems arise from the very low efficiency of conventional fission plants—of the order of 1 percent. The other 99 percent of the heat energy is wasted and must be disposed of somewhere in the environment. So far no satisfactory alternative has been found. (This topic is discussed in Chapter 14.)

Finally, nuclear plants cannot consume their fuel completely, and even when the critical minimum density of U-235 in a fuel rod is too low to sustain fission, the material that remains is still dangerously radioactive. These fuel rods must be reprocessed, and waste products must be disposed of. Transportation of such highly radioactive materials to reprocessing plants always presents the possibility of accidental radiation exposure along the route. Disposal is yet another problem. At present, reactor wastes are disposed of on AEC sites, but as the number of nuclear plants continues to grow, it will be increasingly difficult to find sufficient disposal sites that possess all the critical features to minimize radioactive leakage into the environment.

Meeting Mineral Needs

Since several of the forms of energy we have discussed are minerals, it should not be surprising that the basic principles for exploration, exploitation, and use are the same for all. There is a need for the formulation of better inventories of our global mineral stocks as well as a need for better methods to get these mineral stores from the earth. This information then could be used in making decisions regarding domestic development of a mineral resource (if any is available domestically), the magnitude of the imports necessary, and the need for acceptable resource substitutes. The main difference between minerals and energy is that energy needs can be met (without regard for the limitless future) by dependence on a few fuels which can be provided by a few countries; complex industrial development, however, demands a diverse range of minerals that cannot be found within any one political entity and, in fact, must be imported from all parts of the world. This fosters mutual international dependencies that will continue to grow, within resource constraints, as more countries reach advanced industrial development.

Each continent is characterized by abundance in some groups of minerals and scarcity in others. Similarly, the continents as a whole are deficient in some minerals that are relatively abundant in the seas, such as manganese nodules in the deep sea. The known reserves of the United States and the world for the four major mineral groups is shown in Figure 10-8. Within the fuels group discussed above, coal is abundant worldwide, but the remaining minerals are not. In the United States, crude oil is actually scarce (a situation that has improved with the discovery of Alaskan oil) compared with other places. The United States has great resources in the ferrous (iron-type) metals, especially iron and molybdenum, but few in nonferrous metals—practically none in tin and aluminum. Of the precious metals, platinum is scarce throughout the world and practically nonexistent in the United States.

The total known reserves of some minerals are controlled by a few nations. For example, chromium stocks are mostly in South Africa and Rhodesia, nickel stocks are in Cuba and New Caledonia, and industrial diamonds are supplied by the Congo Republic. The patchiness of mineral distribution is a powerful political influence that shapes alliances, hostilities, and uneasy coexistences, simply because trade for all necessary industrial raw materials is a fact of life. No one recognizes this better than the isolated, mineral-poor Japanese, who are great traders and empirical politicians.

If minerals other than energy resources are unevenly distributed and subject to unstable political arrangements, can mineral-poor nations hold any hope for insuring sufficient mineral stocks? A means frequently suggested is resource substitution, whereby a rare resource is, with the aid of technology, replaced by a more abundant one. Indeed, many have such faith in technology that they believe its ability to rectify resource shortages through substitution or some other means is almost limitless. Preston Cloud (1971) attacked this approach and labeled it the "technological

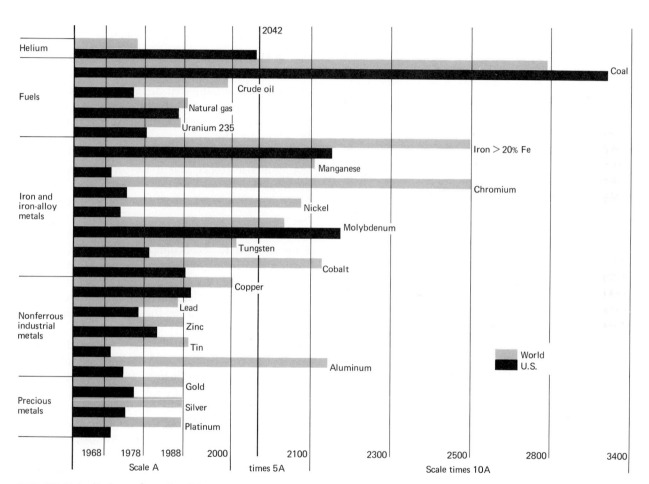

FIGURE 10-8 *Estimated stocks of four major mineral groups for the United States and the world. The bars indicate the extent of known reserves of each resource.* (Redrawn from Cloud (1971), after USGS data.)

fix," which is at once "full of hope and full of risk." The hope, he maintains, is false, but the risk is very real. Cloud contends that it is unrealistic to believe technology can compensate for every resource shortage, though certainly it can be very important in some cases. Limits in fossil fuels might be "fixed" by the development of nuclear or solar energy, for instance, and the present limits on various minerals might be "fixed" by the resources of the barely explored seas.

For marine mineral resources, at least, there is more hope than there is with Idyll's optimistic forecasts for marine fisheries. Offshore oil appears to be a major untapped resource with large potential. Deep-sea manganese nodules are numerous and fairly easily harvested, and there is always the protein potential of oceanic ecosystems. Nevertheless, the seas are not a panacea that is limitless and certain. If extensive mineral exploitation of the seas is begun, safeguards will be needed to prevent concomitant destruction of biotic systems from increased pollution, diversion of nutrients, or other changes having presently unknown effects. To be blunt about it, we know so little about ocean processes that extensive exploitation could start another series of unintentional experiments with consequences that are seen only in hindsight.

The biggest potential problem is, however, a continuation of the exploitation policies that have characterized our use of marine ecosystems. If 200-mile limits are adopted widely as an alternative to better management of common-property resources, individual nations can pillage or manage as they please without regard for the consequences that in-shore mismanagement may have for the deep oceans or for whatever remains of international seas. Our knowledge of systems, even though fragmentary, is sufficient to warn us that enlightened management of oceanic ecosystems is now more critical than ever as more eyes turn toward the seas for solutions.

Perspective on Mineral Policy

We shall close this chapter with an examination of the prospects for energy conservation. While there may be hope for unlimited energy from nuclear fusion or solar power, let us assume that we are likely to have shortages of energy for at least the near future. If we cannot afford to pay more and more for foreign oil, if we cannot be assured of abundant and inexpensive domestic stocks, and if we cannot count upon an immediate and effective "technological fix" to keep supplies abreast of demand, we must look to ways to conserve energy.

Although energy conservation is the surest long-term solution to reconciling demand and supply, it has received little detailed attention. Certain truisms are evident, for example, that more efficient energy use is possible and that energy conservation will become more important in the future; but when we begin to specify the details of a conservation program, things are more difficult. In some cases (such as with solid-waste conversion and MHD generation) technological progress may help us to save energy by increasing the efficiency of usage. In the past, technology often has had the opposite orientation; as Berg (1973) has pointed out, an environment of cheap energy has allowed us to lower developmental capital costs by designing equipment that requires more fuel but is cheaper to build. As energy costs continue to rise, it may be cheaper to reverse priorities and give energy efficiency the highest priority. Higher standards for residential insulation and development of home solar heating may help; estimates vary, but 25 to 30 percent savings over present electrical and gas consumption rates are possible.

As long as the population continues to grow, maximum expectations seem to center upon cutting the rate of growth of energy demand rather than upon stabilizing total demand. This is symptomatic of the whole problem with energy conservation: Our society has evolved in an environment of abundant energy, and so our institutions and social norms are governed by expectations of more of the same. This does not mean we are incapable of change, but it seems unlikely that such conservation can occur voluntarily and without the major social changes that would be undertaken only reluctantly. The energy-saving home may be a blockhouse—thick-walled and windowless—rather than the open, airy structure most of us prefer. Private transportation may be reduced severely unless new forms of fuel are available, and mass transit could become the primary means to move people and goods, as inconvenient as that might be. Multiple-family units, which share heating, cooling, and generating units, might become the dominant form of housing, at the cost of familial privacy (Figure 10-9). These all involve dramatic social change, making it obvious why apparently neutral technology is so much easier to support.

This perspective on energy conservation also makes it obvious why mineral policy will be biased toward technology and substitution rather than toward conservation. Continuing use of minerals to sustain economic growth, with its immediate benefits and only the possibility of long-term costs, will be favored over conservation, with its immediate costs and only the possibility of long-term benefits, by decision-makers whose welfare depends upon political support. Thus the burden of proof rests squarely upon those who favor conservation, and changes in this direction will occur in small increments.

FIGURE 10-9 *Co-op city in the Bronx, New York is virtually a centrally run city in itself. Apartments, schools, stores, and offices all receive services from a central point.* (Courtesy of I. A. Gonzalez from Black Star.)

Questions

REVIEW QUESTIONS

1. In the petroleum supply-demand equation, ultimate recoverable supply is certainly important, but demand is probably the key factor. Why might this be so?

2. McKelvey (1972) stated that "better methods need to be devised and applied more widely" in preparing comprehensive resource estimates. Assuming that such methods are devised, what are two probable effects on public policy?

3. Fossil fuels are the product of biological populations that are renewable; hence they also are renewable. Why are fossil fuels treated, then, as nonrenewable resources?

4. Illustrate the concept of the "technological fix."

ADVANCED QUESTION

1. Construct a flow diagram showing the systemwide impacts of strip mining or nuclear power upon the global ecosystem, emphasizing positive as well as negative impacts. You may use the Sorensen matrix or any other technique you choose.

Further Readings

Note: The issues of nonrenewable resource development can be covered in relatively few readings—in particular, see Hubbert (1971), Cloud (1971), and Holdren and Herrera (1972). Refer also to the readings of Chapter 9, especially Landsberg et al. (1963). It is an understatement to say that the literature in this area changes rapidly. Interested readers are urged to monitor The American Scientist, Science, Nature, *and other periodicals for recent developments.*

American Association for the Advancement of Science. *Science*, 184, No. 4134 (1974).

    A special issue devoted entirely to energy and the energy crisis.

Berg, C. A. "Energy Conservation through Effective Utilization." *Science*, 181 (1973), 128-138.

Brown, T. *Oil on Ice*. Sierra Club, San Francisco, 1968.

    An account of the problems of transporting crude oil south from Prudhoe Bay, Alaska. See also Chapter 11 for more oil-spill literature.

Chapman, D., T. Tyrell, and T. Mount "Electricity Demand Growth and the Energy Crisis." *Science*, 178 (1972), 703-708.

Cloud, P. E. "Realities of Mineral Distribution." *Texas Quarterly*, 11 (1968), 103-126.

    This paper presents the evidence on which Cloud bases his position on mineral policy.

Cloud, P. E. "Mineral Resources in Fact and Fancy." In *Environment* (W. W. Murdoch, ed.), pp. 71-88. Sinauer Associates, Stamford, Conn., 1971.

    Update and expansion of the 1968 paper.

Cornell University. *Summary Report of the Cornell Workshop on Energy and the Environment*. Ithaca, N.Y., 1972.

Daniels, F. "Direct Use of the Sun's Energy." *American Scientist*, 55 (1967), 15-47.

    Thorough review article of solar energy use and potential and fully referenced as well. Short version of a book published in 1964 by Yale University.

Gillete, R. "Oil: Did USGS Gush Too High?" *Science*, 185 (1974), 127-130.

    The latest word on the "great oil controversy."

Hammond, A., W. D. Metz, and T. H. Maugh. *Energy and the Future*. American Association for the Advancement of Science, 1973.

    An excellent single source that is sold inexpensively and available widely.

Hendricks, T. A. "Resources of Oil, Gas, and Natural Gas Liquids in the U.S. and the World." U.S. Geological Survey Circular, No. 522, 1965.

Hirst, E., and J. C. Moyers. "Efficiency of Energy Use in the United States." *Science*, 179 (1973), 1299-1304.

Holcomb, R. W. "Power Generation: The Next Thirty Years." *Science*, 167 (1970), 159-160.

    An assessment of coal-burning versus nuclear power generation.

Holden, J., and P. Herrera. *Energy*. Sierra Club, San Francisco, 1972.

An excellent sourcebook and a highly readable distillation of voluminous literature.

Hubbert, M. K. "Reply to J. M. Ryan." *Journal of Petroleum Technology*, 18 (1966), 284-287.

Hubbert, M. K. "Energy Resources." In *Resources and Man* (P. E. Cloud, ed.). Freeman, San Francisco, 1969.

This paper forms the basis for the 1971 paper listed here and in turn is derived from papers published in 1962 and 1967 but not as widely available. See Ryan (1966) for a critique of the method.

Hubbert, M. K. "Energy Resources." In *Environment* (W. W. Murdoch, ed.). Sinauer Associates, Stamford, Conn., 1971.

An update of Hubbert's 1969 paper.

Jones, L. W. "Liquid Hydrogen as a Fuel for the Future." *Science*, 174 (1971), 367-370.

Lincoln, G. A. "Energy Conservation." *Science*, 180 (1973), 155-162.

McKelvey, V. E. "Mineral Resource Estimates and Public Policy." *American Scientist*, 60 (1972), 32-40.

Recent and representative of the more optimistic and technology-confident point of view.

McKelvey, V. E., and D. C. Duncan. "United States and World Resources of Energy." In *Symposium on Fuel and Energy Economics*, 149th National Meeting. American Chemical Society, Division of Fuel Chemistry 9, No. 2 (1965), 1-17.

Moore, C. L. "Projections of U.S. Petroleum Supply to 1980." U.S. Department of the Interior, Office of Oil and Gas, Washington, 1966.

Osborn, E. F. "Coal and the Present Energy Situation." *Science*, 183 (1974), 477-481.

Ryan, J. M. "Limitations of Statistical Methods for Predicting Petroleum and Natural Gas Reserves and Availability." *Journal of Petroleum Technology*, 18 (1966), 281-284.

A very interesting paper that shows just how wide the ranges of predictions about ultimate petroleum production is, and therefore how weak any case built on them must be. McKelvey (1972) and Cloud (1971) call for better estimates, which is to be applauded, but it remains to be seen exactly how they are to be obtained. Ryan does have a point: The search is far from complete or reliable.

Scientific American, 224, No. 3 (August 1971). Special energy issue.

Steinhart, C. E., and J. S. Steinhart. *Blowout!* Duxbury, North Scituate, Mass., 1972.

U.S. Government, Joint Economic Committee. *The Economy, Energy, and the Environment.* Washington, 1970.

This one should be read by all interested students and is only 55 cents for 131 pages. It is full of data and free of bias on a very touchy subject.

Weinberg, A. M. "Raw Materials Unlimited." *Texas Quarterly*, 11 (1968), 90-102.

Weinberg, A. M. "Social Institutions and Nuclear Energy." *Science*, 177 (1972), 27-34.

White, D. E. "Geothermal Energy." U.S. Geological Survey Circular, No. 519, 1965.

Zapp, A. D. "Future Petroleum Producing Capacity in the United States." U.S. Geological Survey Bulletin 1142-H, 1962.

Section D
Environmental Quality: A Measure of Mankind

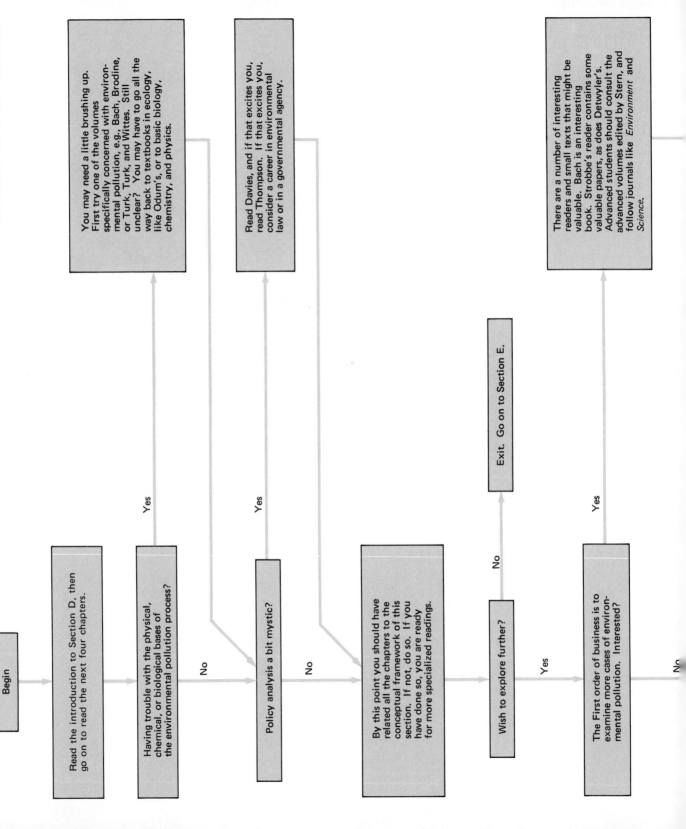

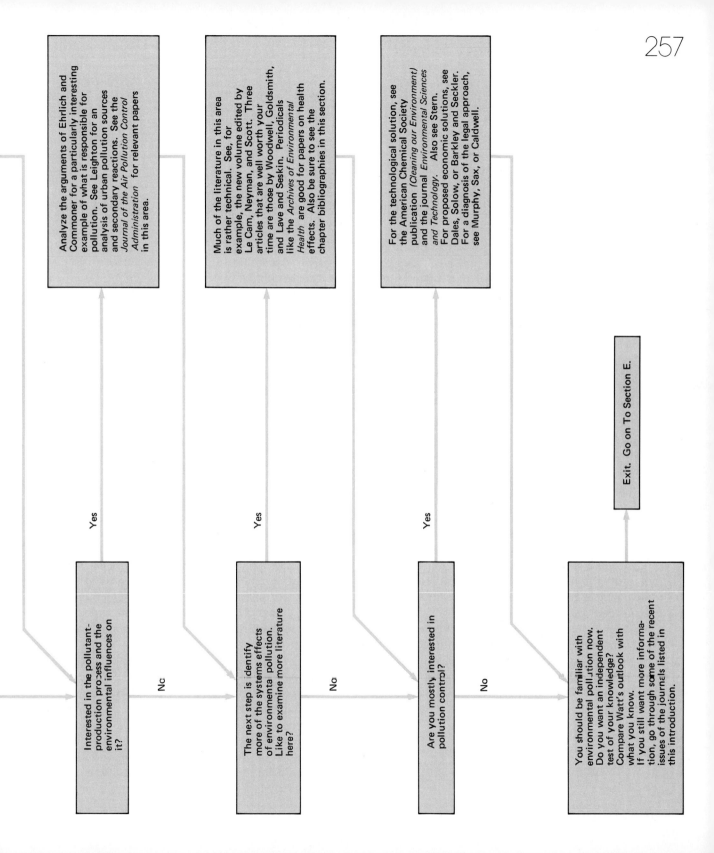

Is environmental quality a measure of human beings and their works? In at least a general sense this is true, because we have used the environment: We have exploited it for resources, dumped our refuse into it, and reshaped it toward our ends. No other species has ever had such control over so vast a domain. As we look around us, we see dirty water, foul air, refuse and litter strewn over the countryside. Insofar as these things are part and parcel of environmental quality, there is no denying that the maintenance of environmental quality has been sacrificed to our desire to obtain more conveniences and a better life for ourselves.

Now the situation has changed dramatically, and environmental quality has become more important than ever before. But when we try to apply operational definitions to "environmental quality," we find some new difficulties. The word "environment" itself is a semanticist's nightmare, since in its broadest context it is nothing less than everything perceived about and beyond one's person. Hence a uniform, operational definition of "environment" is impossible; any definition is specific to each observer and each particular situation. Similar remarks apply to "environmental quality," which also requires some arbitrary definition of the word "quality." The usual course has been to restrict the purview of the deterioration of environmental quality to detrimental changes that result from the degeneration of a resource such as air or water and, further, to include only that deterioration traceable to human activities. This is the classic meaning of environmental pollution.

As "environment" and "environmental quality" have become familiar concepts, however, we have begun to see the proliferation of additional meanings. Environmental quality can be more than simple physical, chemical, or biological changes detected by some sophisticated measuring device or by some index of species abundance; as people have become more concerned with their immediate surroundings, their perceptions (Section F) about acceptable lifestyles have crystallized into value standards concerning such things as noise, visual quality, and spatial arrangements. These are considered to be determinants of environmental quality, even though they have little to do with the more classical definitions of this term. Furthermore, there is no reason to believe that additional environmental parameters will not also be considered as part of environmental quality in the future.

As the varieties of interpretation included under the heading of environmental quality continue to proliferate, it may become increasingly difficult to discern the underlying similarities as they become even more obscured by the obvious differences. It is not a particularly startling observation to note that air pollution is not the same thing as excessive sound or that air and water pollution differ from each other in numerous ways. However, air pollution and excessive noise do share common properties, even if they often are not obvious.

In describing environmental pollution at the most general level, we may construct a model that describes the process of environmental pollution (Figure D-1). The model consists of two major levels called *compartments*. One of these, the "global dynamics" compartment, includes the entire array of elements that constitute the production, dispersion, and effects of various pollutants. (Note that these effects may be instrumental in producing still others.) The second, the "control system" compartment, includes elements that can modify one or more components of the global dynamics compartment.

Systems aggregated at such a high level are usually so general as to contain little useful information; this particular model is no exception, except for whatever conceptual help it provides. If we break down the pollution system in greater detail, we then may distinguish subsystems here labeled *"source," "potential target systems," "array of observed effects,"* and the elements of the control system, namely the *"policy process."* Some of the major interactions are shown, but this diagram assumes that readers acknowledge the additional interactions that must occur within each compartment. For example, "potential target systems" may include ecosystem interactions that either could negate or accentuate the effects of a particular pollutant.

Environmental Pollution: Sources

It is hardly necessary to point out that all environmental pollutants must have a source, but there is far more to it. Chemical and physical pollutants are the products of transformations such as those that result from a set of chemical reactions. Some reactions that result in pollution are found even in nature, but with a few exceptions (for example, eutrophication of lakes; see Chapter 11), they are both rare and unimportant compared with those that result directly and indirectly from various human activities. Even the nonclassical forms like "visual pollution" result from human activity and usually are conflicting structures or human rearrangements of natural environments that are not pleasing to one or more segments of society.

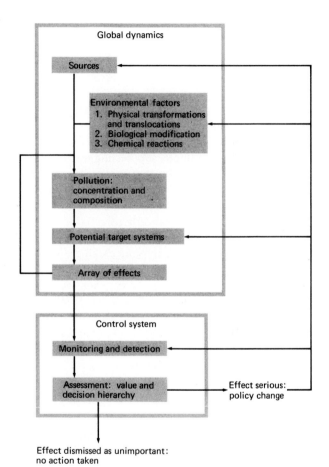

FIGURE D-1. *Environmental pollution: the process.*

An additional complication arises because most sources can produce a range of potential pollutants, not just one. And many pollutants can be derived from a large number of different sources. Because of all the possible interactions, it is important for policies dealing with particular pollutants to recognize that simple solutions applied to complex polluting systems are likely to produce complex and often unexpected results. For example, suppressing hydrocarbon output from internal combustion engines by combining more air with the fuel during combustion merely substitutes an increased nitrogen oxide output for the previous production of hydrocarbons.

Whatever the nature of the environmental pollutants derived from the primary sources, all are subject to the action of physical, chemical, and biological factors that can transform them into other secondary pollutants. The best examples of secondary pollutants are found in the atmosphere (photochemical smog; see Chapter 12). In Figure D-1, all these pathways are buried within the environmental factors that operate on pollutants between their sources and potential target systems.

Environmental Pollution: Potential Target Systems and the Array of Effects

In our general system, the spectrum of environmental pollutants becomes important only when the pollutants impinge on sensitive systems. For our purposes, the identification of these systems is not important—we are interested only in the fact that they react to the pollutants in ways we perceive as deleterious. The following list emphasizes the different kinds of harm such reactions can produce:

1. They can be harmful to human health.
2. They can attack large numbers of species and thereby disturb entire ecosystems.
3. They can be aesthetically disturbing.
4. Their range of physical effects, when taken together, can modify the physical environment.
5. They can disrupt or interfere with important economic activities.

Environmental Pollution: Policy Formulation

We have said that primary pollutants are subjected to environmental influences which may increase the pollution problem in various ways. Further, we have seen that this pollution becomes important when it enters potentially sensitive systems; once detected, the resulting array of effects is the important output of this whole process. Once the array of effects is considered important, the pollution problem enters the domain of the control system, which attempts to improve pollution control by reformulating policy in appropriate ways. Any person or responsible group ("decision-makers") must consider four elements in the process of policy-making:

1. Knowledge of the pollution generation system, with stress placed on the array of effects but which ideally also includes precise understanding of the generation mechanisms involved.
2. Value judgments assigned to the observed effects of the pollutants, which requires an assessment of the relative importance assigned to each by various interest groups.

3. Constraints placed on the solutions available.
4. The decision-makers' resources, which can be used to overcome some of the constraints.

All four of these elements are included in the component labeled "the value and decision hierarchy" in Figure D-1.

The limits that the control system imposes on the global dynamics sector depend on an indistinct boundary between "an acceptable effect" and an unacceptable one. There is no real issue either when it is impossible to demonstrate a harmful effect from a particular pollutant or when the effects are so obvious and dramatic that cause and effect are clear. However, for most of the complex chains of pollutants, sources, and effects, inadequate knowledge about the system may be a major problem for decision-makers. This is the case in handling problems of air pollution and solid wastes, for example. In other cases, the combination of agency resources available to the decision-maker and constraints on the potential solutions may be the important factors, especially when economic production and pollution generation are inseparable. In such cases, the value systems used may insure that agency resources are small and that constraints are huge and impossible to overcome.

This process of policy formulation is the critical part of the pollution process because none of the elements that play a part in it are completely objective. Scientists may gather objective data, but their use and interpretation can be highly subjective. (See the controversy over the climatic effects of air pollution in Chapter 12, for example.) As we have noted, the values, constraints, and resources of policy-makers are all interdependent and usually inseparable. In some cases constraints on certain solutions may be erected because they conflict with an underlying (and often hidden) set of values, while others may establish their value systems from the constraints they perceive. For example, the major U.S. automobile manufacturers may value the internal combustion engine because of its long history of use and the large investments they have put into it, and so they see major constraints in the technology and cost of replacing it. On the other hand, automobile drivers may value their cars because no suitable transportation alternatives are available, and they may value the internal combustion engine as a power plant because other power plants may be less convenient or more costly. Thus their familiarity with the internal combustion engine may condition drivers' values strongly.

Environmental pollution and its control promises to be an explosive and fascinating arena for policy formulation and analysis. The different value systems of groups such as industry and conservation organizations have been evident for some time, and conflict has resulted in many environmental areas. Often in the past, well-organized conservation organizations and citizens' groups have been the only ones concerned with pollution effects, and others have accepted the effects as an inevitable or reasonable cost of living in our present society. Now, however, the general public is being drawn into active participation as the processes of production are challenged in the interest of environmental quality. While the question of cost is still to be met by the public, there is heartening evidence that the public is willing to sacrifice some economic growth to gain environmental quality.

Selected General Sources on Environmental Pollution

American Chemical Society. *Cleaning Our Environment: The Chemical Basis for Action.* American Chemical Society, Washington, 1969.

Bach, W. *Atmospheric Pollution.* McGraw-Hill Problems Series in Geography, New York, 1972.

Barkley, P. W., and D. W. Seckler. *Economic Growth and Environmental Decay.* Harcourt, Brace, Jovanovich, New York, 1972.

Brodine, V. *Air Pollution.* Harcourt, Brace, Jovanovich, New York, 1973.

Caldwell, L. K. *Environment: A Challenge to Modern Society.* Doubleday; Natural History Press, Garden City, N.Y., 1970. Also in Anchor Paperback (1971).

Commoner, B. *The Closing Circle.* Knopf, New York, 1972.

Dales, J. H. *Pollution, Property, and Prices.* University of Toronto, Toronto, Ontario, 1968.

Davies, J. C. *The Politics of Pollution.* Pegasus, New York, 1970.

Detwyler, T. R. *Man's Impact on Environment.* McGraw-Hill, New York, 1971.

Ehrlich, P. R. *The Population Bomb.* Ballantine, New York, 1969.

Goldman, M. I. *The Spoils of Progress.* M.I.T., Cambridge, Mass., 1972.

Harte, J., and R. H. Socolow. *Patient Earth.* Holt, New York, 1971.

Lave, L. B., and E. P. Seskin. "Air Pollution and Human Health." *Science,* 169 (1970), 723–733.

LeCam, L., J. Neyman, and E. L. Scott (eds.). *Proceedings of the Sixth Berkeley Symposium on Mathematical Statistics and Probability, Vol. VI, Effects of Pollution on Health.* University of California, Berkeley, 1972.

Leighton, P. A. *Photochemistry of Air Pollution.* Academic, New York, 1961.

Murphy, E. F. *Governing Nature.* Quadrangle, Chicago, 1967.

Sax, J. *Defending the Environment.* Knopf, New York, 1971.

Solow, R. M. "The Economist's Approach to Pollution and Its Control." *Science,* 173 (1971), 498–503.

Stern, A. C. (ed.). *Air Pollution,* 2nd edition, 3 vols. Academic, New York, 1968.

Strobbe, M. A. (ed.). *Understanding Environmental Pollution.* Mosby, St. Louis, 1971.

Thompson, D. L. *Policies, Politics, and Natural Resources.* Free Press/MacMillan, New York, 1972.

Turk, A., J. Turk, and J. T. Wittes. *Ecology, Pollution, Environment.* Saunders, Philadelphia, 1972.

Watt, K. E. F. *Principles of Environmental Science.* McGraw-Hill, New York, 1973.

Woodwell, G. M. "Effects of Pollution on the Structure and Physiology of Ecosystems." *Science,* 168 (1970), 429–433.

Selected Periodicals

INDEXES AND ABSTRACTS

Bioresearch Today/Environmental Pollution
Current Contents
Environment Information Access
Pollution Abstracts
Reader's Guide to Periodical Literature

JOURNALS AND REPORTS

Air and Water Pollution
American Scientist
Clean Air and Water News
Council on Environmental Quality, Annual Reports
Ecology Law Quarterly
Environment
Environmental Law
Environmental Pollution
Environmental Reporter
Environmental Science and Technology
Environment Law Review
Environment Research
International Journal of Environmental Studies
Journal of the Air Pollution Control Association
Journal of Environmental Sciences
Journal of Environmental Systems
Journal of the Water Pollution Control Federation
Science

11. Ecology, Water Quality, and the Cesspool Mentality

In the broadest sense water can be described as "polluted" whenever it is enriched with any kind of nutrient. Some more common measures of water pollution include its clarity, the concentration of certain chemicals in it, and its frequency of algal blooms—all of which are interrelated. Note that this definition does not exclude many examples of natural enrichment, because pollution is very much a natural occurrence. For example, estuaries are biologically rich because they trap nutrients carried downstream. Without this enrichment, the fertile estuary would be little more than a brackish-water desert. In addition, many aquatic organisms secrete chemical substances that are serious "water pollution" to other species; these substances may be part of the defense against competitors and enemies. A famous example of such toxic substances is the "red tide," in which spectacular marine fish kills result from toxins secreted by periodically large populations of the alga *Gonyaulax*.

To most people, however, the term "water pollution" raises images of sewage flowing into a river, chemical wastes pouring into a lake, or closed beaches and massive fish kills; in short, water pollution is blamed on sophisticated industrial chemicals and vast volumes of sewage. At least we must concede that the scale and effects of human-related water pollution are overwhelming compared with all natural sources. In this chapter, we first shall examine the scope of human activities and their effect on water quality relative to these natural processes that may have similar effects. Second, we shall attempt to assess the reasons why water quality has hitherto been such a sensitive indicator of human presence. Third, and perhaps most important, we shall examine the possible corrective schemes that could restore water quality and the costs of doing so, including some schemes that take advantage of natural ecological mechanisms.

The sparkling, clear lake and the undisturbed, unpolluted stream are rare today when even moderate human activity occurs in the watershed. Why is this so? In the most simple terms, the reason is contained in the argument advanced by White (Chapter 7) and subsequently expanded by other authors from the viewpoint of economics. Manipulation of the environment to benefit human beings is at least as old as Judeo-Christian teachings and probably as old as human societies. In economic terms, the environment has been regarded as a free good; for most of our history we have used as much of it to absorb wastes as was necessary or desired.

In *The Closing Circle* (1971), Barry Commoner advances the argument that technology is the root of the crisis for human survival because it is the means for the manufacture of goods which bear pollution costs that lead to deterioration in environmental quality. This viewpoint contains an underlying assumption about industrial development—that human benefits from development have been more important than environmental costs. Under this hypothesis, the cost of keeping a river clean of industrial wastes would not be worth the resulting higher prices for manufactured goods. This does not mean that all technology must bring deterioration of the environment—only that it is cheapest to leave environmental quality out of consideration. I shall call the free-environment attitude toward water quality the *cesspool mentality*. When you dump wastes freely in the belief that they will no longer be your problem, you are guilty of the cesspool mentality. When your intentions are good but your performance goes unchanged, you are guilty of the cesspool mentality. Wherever there are large cities and massive industrial development, they are surrounded by the results of the cesspool mentality. Throughout industrial history the cesspool mentality has accompanied us. Can we change this in the future?

The Scope of the Problem

The term "cesspool mentality" is misleading in some ways because it implies a conscious choice of water pollution over the higher costs that would be required for environmental protection, something that certainly is not always true. Nevertheless, it *is* an appropriate description of the global extent of water pollution. The status of the Hudson River, the Great Lakes, and the Mississippi River are matched by the status of the Rhine River of Germany, Japan's Tokyo Bay, and the Seine River in France. Canada contributes to the deterioration of the Great Lakes, especially of Lake Erie. Similar problems are cropping up as industry develops on the shores of Lake Baikal in Siberia, and the Thames River in Great Britain formerly was famous for its unbelievable state of filth. In the oceans, oil slicks from ocean shipping mark the major commercial sea lanes, and sewer pipes belch directly from coastal cites into estuary and sea. Long-lived, biologically active pesticides run off the land into streams and rivers and finally pour into the sea, where they mostly remain until they break down. Sadly, it has been true that whenever there is significant human presence, there is water pollution. Besides all the waste deliberately dumped into water courses for disposal, some major problems are practically inescapable, if for no other reason than that water is such an excellent solvent for all sorts of substances. Salts, dyes, and acids all can be carried far from the actual site of disposal. A striking example is the discovery of DDT in penguins in Antarctica, where it has never been used.

In short, the variety of pollutants that enter natural waters worldwide is vast. Many of them, especially in more economically advanced countries, are the sorts of industrial and technological wastes that Commoner sees as symptomatic of a technological-environmental crisis. However, the problems of quantities of "natural" wastes are also important, especially for certain watersheds. Domestic sewage and garbage, for example, routinely enter natural waters. Treatment plants and septic tanks in developed countries are imperfect, and in developing countries untreated wastes routinely enter streams with no attempt to treat them.

In contrast to the quantities of all types of human pollutants entering natural waters, natural sources of pollution are less dramatic if no less widespread. In the terms of Barry Commoner, this is because natural systems have no advanced technological machinery producing exotic wastes, and natural ecosystems have mechanisms for dealing with the organic materials that regularly enter these systems. The natural recycling system—the decomposer element of the food web discussed in Chapter 1—automatically deals with the accumulation of organic materials in most ecosystems. For example, yearly leaf falls and annual mortality of all sorts of aquatic organisms will be dealt with by the decomposers.

It is important, however, to recognize the quantitative components in decomposition or in water pollution, whether natural or technological in origin. Large and concentrated human populations have demonstrated repeatedly the capacity to so thoroughly overload the environment that the decomposers cannot keep up with the influx of otherwise completely natural substances. Clearly, this type of effect does not require a sophisticated technology. We see, then, that the quantitative aspect may be very important to the system.

The Natural Purification Cycle: An Example

Consider any body of water. The members of the aquatic community eventually die, and run-off and rivers carry other organic materials into the system. A variety of decomposers—bacteria, fungi, aquatic insects—will consume the organic materials and break them down into the simple inorganic compounds that are essential nutrients for plant growth. Thus the decomposers can stimulate plant growth, although nutrients also can be washed into the system directly by run-off. In undisturbed lakes, organic material will accumulate, increase nutrient levels, and the lake will become biologically richer and richer (the *eutrophication* cycle). In rivers, downstream flow will carry organic matter and nutrients out of the system and eventually into the sea; here, then, eutrophication must await a slowdown in current velocities.

Like other systems, however, cycles of decomposition and nutrient release are sensitive to disturbance. If for any reason there is a large pulse of organic matter introduced into the system, decomposer activity may increase so dramatically that they use all available oxygen in the water; other organisms suffocate, and the whole system breaks down. In a lake, the first sign of increased decomposition may be an algal bloom from the increase in nutrients (Figure 11-1), followed by selective fish kills, then the appearance of species specialized to poorly oxygenated waters (*indicator species*), and finally a richly odoriferous reek of decomposition as nearly all organisms except the decomposers are eliminated. In rivers, this phenomenon is modified because of flow. The initial concentration of organic material may have to be greater to show a corresponding effect on the biota, and river flow inevitably will dilute organic matter downstream. Consequently, the biota will show a gradient downstream (see Figure 11-4 and its accompanying discussion).

Human organic wastes can be expected to have the same qualitative effects, and water pollution policy can be formulated to control inputs to a level of acceptable eutrophication. In contrast, those pollutants having no natural analogues require different policies,

FIGURE 11-1 *Algal bloom in a lake in Wisconsin.* (Courtesy of Daniel S. Brody from Stock, Boston.)

since there will be no natural system to deal with them. Substances like mercury and DDT—which are active biologically, are readily incorporated into food chains, and are dangerous poisons—require strict control to prevent their escape into the environment. Other substances, like insoluble inorganic salts, may be so chemically and biologically inert that they constitute no substantial danger and require little regulation. What we see, then, is a complete spectrum of effects in which we must carefully examine the sensitivity of potential target systems and the characteristics of potential pollutants.

A Conceptual Flow Chart of Water Pollution

A simple flow chart (Figure 11-2) summarizes the chain of events from the release of a pollutant to its final disappearance from natural waters. This flow chart is essentially the same as Figure D-1, but it is possible to specify more detail for the various important factors. For example, important physical factors include the current circulation system, possible temperature stratification, and wind mixing. Chemical factors include resident chemicals that change pollutants into more or less dangerous ones, and biological factors include decomposers that may remove substances. The result is rarely a uniform distribution of pollutants in the environment but rather a very patchy one. For example, sewage sludge is heavy, sinks fast, and is little affected by water movement. Furthermore, it is poorly assimilated by organisms other than decomposers and detritus feeders. Hence most of the sludge ends up on the bottom until it is broken down into lighter, more soluble substances in the

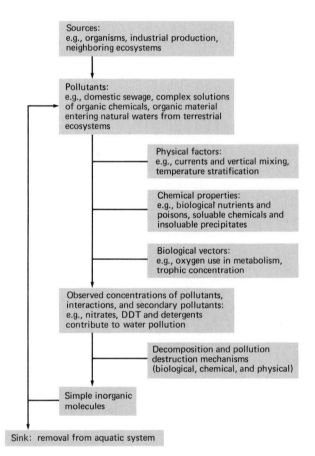

FIGURE 11-2. *A simplified flow chart for the process of water pollution from sources to sinks.*

degradation process. In contrast, some more uniformly distributed pollutants (such as nitrate salts, which are highly soluble in water) can be dispersed by currents, are absorbed by algae, and rapidly pass through aquatic food chains over and over again. It should not be surprising that smaller nitrates would be more widespread—they are less likely to be tied up relatively inertly in the bottom sediments. Once in the lake water, the nitrates will remain biologically active until they are washed out of the lake.

Although secondary water pollutants resulting from interactions among chemicals and organisms are poorly known in aquatic systems, there is no reason to believe that such interactions do not exist. One rather far-fetched example would be organisms with DDT concentrations—when eaten by other organisms, they could dubiously be called "secondary pollutants." Rather better examples of interaction may be found in the "destruction mechanisms" discussed later in this chapter.

Types of Water Pollutants

Water pollutants can be divided into a number of overlapping categories. The simplest classification recognizes organic and inorganic types. Inorganic pollutants are often small molecules whose effects are largely nutritive and which do little more than speed up the natural eutrophication process; but others, like various acids and radioactive compounds, may be highly toxic. Larger compounds can be assimilated into biological systems only after they are broken down into smaller molecules. Some larger molecules may be nutritive, but many are toxic or not easily decomposed, such as various fractions of petroleum and the "nonbiodegradable" detergents.

Still another way to categorize pollutants is to recognize their effects. Both organic and inorganic pollutants are chemicals, and so their effects are due to molecular properties. On the other hand, the release of heat ("thermal pollution"; see Chapter 14) will have physical effects that may lead to chemical changes, but it does not participate directly in reactions.

A third classification might emphasize the character of each pollutant—thermal, gaseous, liquid, solid, suspended, or biological (for example, pathogenic bacteria or viruses). But to continue to establish such classifications is pointless. Clearly there is a whole spectrum of pollutants, each of which has a series of properties and each of which can be placed in some arbitrary classification depending on the properties emphasized.

Now we shall examine how various pollutants affect the major categories of surface waters.

FLOWING WATERS

Flowing water is always a nice place to dump wastes—so long as one is upstream of everyone else. (Economist Ansley Coale has suggested that cities could solve their water pollution problems if they had to draw supplies downstream and deposit used water upstream!) As we have seen, rivers are nice places to locate cities, industrial plants, and sewage treatment facilities because they will dilute and carry away what is dumped into them. The proliferation of pollutant sources along major rivers has created large water-quality problems in Europe and the United States, mostly as a result of the Industrial Revolution and the resulting great increase in the production of toxic chemicals. Regular fish kills along the lower Mississippi River usually have been traced to specific releases of toxic chemicals, especially the pesticide Endrin (to which fish are highly susceptible). The load of human wastes, agricultural chemicals, and industrial output in the Sacramento–San Joaquin River delta is already so large that the proposed southward export of the relatively pure upstream water which normally would dilute these wastes could thoroughly poison San Francisco Bay. New York City damages water quality in the Hudson River so badly that pollutants from its outflow can be found far out to sea in the Atlantic. Recent newspaper reports claim that the Rhine River has become so loaded with chemicals that it could be used as one enormous tank for developing photographs.

It has been and still is the fate of many rivers that pass through highly developed areas to be used as a convenient dumping ground. While each contribution may be small, the total pollution load is cumulative, especially if toxic substances are involved. The farther one proceeds downstream, the worse the situation becomes. Dumping into a river only succeeds in exporting the problem to someone else. And because there are multiple sources and usually unnatural pollutants, natural purification is inadequate.

STANDING WATERS

Warren (1971) defines water pollution as "any impairment of the suitability of water for any of its beneficial uses, actual or potential, by man-caused changes in the quality of the water." To determine whether or not "any impairment" exists is a difficult question that depends on the economic impacts and the interest groups involved. If we accept this definition, we can see readily that lakes are sensitive targets, since many standing waters retain their water mass a long time. (A molecule of water entering Lake Tahoe in California and Nevada, for instance, takes 700 years to leave it.) Hence standing waters are very susceptible to eutrophication or poisoning, depending on the pollutants entering them. However, most lakes can withstand some eutrophication, and the exact amount which will be permitted must be determined by assessing who is harmed by how much and by considering the alternatives available. The only solution to prevent pollution of Lake Tahoe was the export of wastes into the rivers of other drainage basins; the solution for Lake Washington in Seattle (Edmondson, 1968, 1973) was to export the sewage of Seattle into Puget Sound, which is so big that the impact of the sewage was reduced below the impact threshold. To return briefly to flowing waters, disposal into these waters might, in some cases, be a less costly alternative than using a nearby lake, provided there were no other alternatives.

Perhaps the most famous and intensively studied case history is that of the Great Lakes and of Lake Erie in particular (Figure 11-3). It is fashionable to refer to Lake Erie as a "dying" or "dead" lake, but this is far from the truth. For reasons we have already given, it is more accurate to say that the Great Lakes, and particularly Lake Erie, have absorbed great quantities of pollutants, and the effects have been deleterious for the most part. Lake Erie is the smallest and shallowest of the five Great Lakes; it is ringed by more large cities—Cleveland, Toledo, Buffalo—than any of the others; and it receives the industrial wastes from the rivers of heavily industrialized northern Ohio. For all these reasons it is the most heavily polluted of the five lakes.

Originally, the Great Lakes were glaciated, oligotrophic (nutrient-poor) lakes. Beeton (1965) has shown how each of the lakes has eutrophied (Figures 11-4 and 11-5) as measured by the concentration of total dissolved solids. There seems to be little doubt that human activities have been at least partially responsible for the accelerated eutrophication of all the Great Lakes. Beeton also pointed out that chloride and sulfate ions have increased markedly (both can be traced to domestic and industrial waste), whereas magnesium concentrations (from the limestone beds around Lake Michigan) have been stable.

FIGURE 11-3 *View of Lake Erie over the Cleveland business district.* (Courtesy of the Cleveland Growth Association.)

The chemical changes in the waters of the Great Lakes are reflected in their biology, particularly in the most severely affected one, Lake Erie. Vast changes in the fauna and flora of Lake Erie have been recorded, although there actually has been an *increase* in total biological productivity measured as the weight of organic matter produced per unit time (Table 11-1). The difference is that certain species have either disappeared or become scarce, to be replaced by large numbers of other species which were not nearly as desirable. For example, various important commercial species of pike, lake trout, lake herring, and whitefish have nearly disappeared. By 1960 the whitefish catch had fallen to 36,000 pounds from a maximum of 4.1 million pounds prior to 1948. However, Lake Erie still produces a commercial catch of 50 million pounds annually, about half the total Great Lakes catch, because a few other species (perch, carp, smelt, and drum) have increased to compensate for the loss of the more desirable species.

The biotic changes seen in Lake Erie are true to some extent of the other lakes, especially Lake Ontario, which not only has its own pollution sources but in addition is downstream from Lake Erie. The fate of Lake Erie warns of the potential future for the other Great Lakes and for lakes in general. Big deep lakes, small shallow lakes, and mere ponds can all be eutrophied; and once the process occurs, it is not so easily reversed as it is in flowing waters. Hence we must pay particularly close attention to keeping these bodies of water as clean as they should be or need to be.

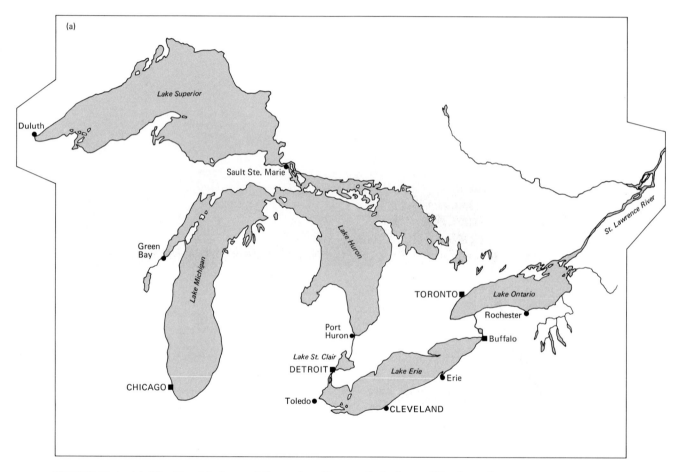

FIGURE 11-4 (a) *The Great Lakes and the major cities on their shores. The flow of water is eastward, through the St. Lawrence River into the Atlantic Ocean.* (b) *(top, next page) The development of nutrient loads in the Great Lakes, as indicated by total dissolved solids.* ((b) Redrawn from A. M. Beeton, 1965, "Eutrophication of the St. Lawrence Great Lakes," *Limnology and Oceanography* 10, pp. 240-254.)

TYPES OF WATER POLLUTANTS 273

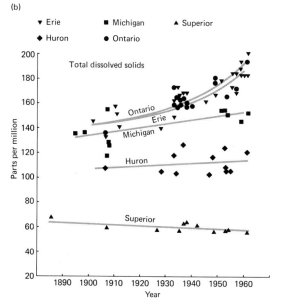

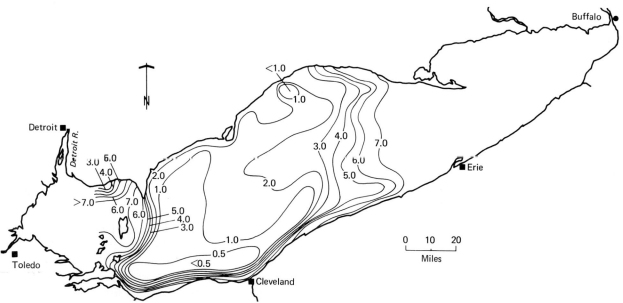

FIGURE 11-5 *The distribution of dissolved oxygen (DO) along the bottom (in the hypolimnion) of Lake Erie in the summer of 1959. O_2 concentration is measured in parts per million (ppm). Note that oxygen depletion is well correlated with depth.* (Redrawn from A. M. Beeton, 1963, Great Lakes Fisheries Commission, Technical Report 6.)

TABLE 11-1 *Changes in the fauna and flora of Lake Erie.*

Biological component	Change in abundance (Densities measured as number per square mile)	
Algae	shift from predominately green to blue-green species, but quantitatively unknown	
Plankton	probable increase, but not well documented	
Benthic fauna (1930–1961)		
burrowing mayflies (*Hexgenia*)	139: 1	
midges (can withstand low O_2 levels)	73: 322	
clams	221: 438	
snails[a]	40: 221	
sewage worms	677:5949	
Fishes	Maximum poundage landed (Year)	Poundage landed in 1960 (10^6 pounds)
lake herring	44.5 (1890)	0.018
whitefish	4.1 (1914)	0.036
walleye pike	13.3 (1957)	1.8
blue pike	26.8 (1936)	0.012
Total catch, all species	73.6 (1890)	50.5

Source: J. F. Carr and J. K. Hiltunen (*Limnology and Oceanography*, 10, 1965) and N. S. Baldwin and R. W. Saalfeld (Technical Report No. 3, Great Lakes Fishery Commission, Ann Arbor, Michigan, 1962).

[a] Some species have increased dramatically, while one has become nearly extinct. As might be expected, the latter is sensitive to eutrophication and decreased oxygen tensions. See Harman and Forney, *Limnology and Oceanography*, 15 (1970), 454.

OCEANS

The ultimate repositories of water pollutants are the oceans. They are vast and old, and they can absorb a lot of what first gets poured into fresh water. One would expect that unless there are reactions with salt that inactivate them, the pollutants would have the same effects in the seas as in the Great Lakes: reduction of dissolved oxygen concentration, changes in species diversity, and changes in the chemical composition of the water.

To a certain extent this is true, but no ocean has been studied as comprehensively or as thoroughly as Lake Erie. Despite warnings to the contrary, the oceans are still too vast for most pollutants to have much impact. Exceptions are DDT and chemically similar pesticides that are toxic at low concentration and persistent enough to be spread widely. Residues of DDT have been discovered not only in Antarctic penguins, but in marine algae, various invertebrates, seals, fish—in short, it is widespread in marine food chains.

There are also many examples of local pollution effects. Sewage poured into the Pacific off southern California was blamed tentatively for the decrease in the giant kelp weeds in several areas (see North, 1964). Further investigations revealed that the amino acids in the sewage were stimulating growth of the sea urchin population that feeds upon kelp. The increased number of urchins is thought to be capable of eating the kelp faster than it can grow. Here the effect of sewage is marked, but highly localized; once the sea urchin population is reduced (by whatever means), the kelp have been shown to recover their initial densities rapidly.

Most oil spills also have locally restricted effects. The effects of the Santa Barbara oil spill did not extend beyond the immediate area. When the *Torrey Canyon* was wrecked off the English coast, it released a load of oil over the English and French coasts, but the consensus of English scientists (Smith, 1968) was that effects were local and temporary (Figure 11-6). The investigation showed that most damage stemmed from detergents used to disperse the oil. The oil involved in both these spills was crude oil: Most of it was highly insoluble in water and simply floated until the lighter, more toxic fractions evaporated. Far more damage has been found to occur after spills of refined, more miscible petroleum fractions such as aviation gasoline and fuel oil.

Because of the immediate threat to human health, heavy metal pollution of water has attracted much attention. The mercury problem was first brought to public attention by the incident at Minamata Bay, Japan, in 1953. Metallic mercury itself is not very soluble or biologically active and tends to settle quickly. However, bacterial action converts metallic mercury to an organic derivative, methyl mercury, which is highly soluble, easily absorbed into food chains, and very much a human poison. At Minamata Bay, mercury from a plastics plant was released and absorbed into many marine organisms that formed a large part of the people's diet. Hence the fishermen's skills fed the wastes of the plastics plant right back to the people. Of those affected, the lucky ones died; many others were reduced to a state of mental incompetence. Fears of mercury poisoning also have been raised from time to time for San Francisco Bay, which accumulated tons of mercury from hydraulic gold mining in the nineteenth century. Undoubtedly, much of this mercury has been converted already into methyl mercury below the surface of the bay mud; it needs only to enter the human food chains to do its damage.

Mercury seems to be slightly less of a problem at present than DDT and its related compounds because it is not dispersed so readily or used so widely. As expected, most of the incidents and warnings of oceanic pollution occur in bays and estuaries where human impacts will always be the most severe. However, this is no solace. Those parts of the seas (continental shelves and estuaries) which Ryther (Chapter 6) has pointed out as being the most productive are precisely the places where oceanic pollution has been most significant. Thus pollution of marine environments threatens marine food production in a real way.

FIGURE 11-6 *Oil spilled on the Normandy coast was cleaned up with chalk powder, straw, and other chemically neutral absorbent substances. Detergent was used only in the worst areas. Consequently, recovery was relatively rapid.* (Courtesy of J. Pavlovsky from Rapho/Photo Researchers, Inc.)

Environmental Effects of Water Pollution: Physical and Chemical

Having discussed the kinds of water pollutants and their occurrence, we can deduce their impact on natural systems. Usually pollution induces physical and chemical changes in water. The chemical effects of water pollution usually have been evaluated in terms of their biological effects (for example, chemicals that are toxic or that merely stimulate plant growth). Sophisticated toxic chemicals seem to worry scientists less than the vast quantities of more natural pollutants, if only because really dangerous poisons (especially those affecting human health) cannot be disposed of freely without uproar. On the other hand, nutrients are more voluminous and can be more difficult to treat. In the 1950s, biodegradable detergents (that can be broken down by decomposers) were rare—most available types depended on an active ingredient that accumulated and formed a characteristic foam on streams. By the 1960s the biodegradable detergents not only dominated the market, but most contained little or no phosphorus, a basic nutrient for plant growth. This change will slow down the process of eutrophication in some lakes, but it would be a problem in lakes where algal growth is not limited by phosphorus concentration, especially if the ion replacing phosphorus in detergents (nitrogen) happens to be the real limiting factor. Ryther and Dunstan (1971) have shown that this is indeed true in the Atlantic Ocean off New York City. Thus phosphate-free detergents are not ubiquitous panaceas for abatement of water pollution.

One might also expect the addition of various pollutants to affect the physical characteristics of natural waters. The addition of very much suspended material can cut light penetration into the water and affect photosynthesis in streams and in lakes. Limnologists have documented many natural examples of this phenomenon, but unless the influx of pollutants were massive such an effect from sewage or industrial activity seems unlikely. Changes in heat balance are far more important. For example, a primary danger from exploitation of Arctic oil (Chapter 7) is the melting of pack ice if oil is spilled over a wide area. The dark oil would absorb solar energy far better than ice; the additional heat could melt existing ice and prevent refreezing. If a nuclear power plant is located on a lake, its waste heat can modify the internal circulation patterns and temperature stratification to the detriment of the lake.

Further work in this expanding area of interest should reveal a much larger catalogue of important effects, especially where organisms may interact with pollutants to create new physical and chemical conditions, which is covered in the next section.

Environmental Effects of Water Pollution: Biological

Some biological consequences of water pollution are well known. We have examined the case of Lake Erie and discussed how pollution affected the biota. But we did not mention that some organisms associated with the pollutants can themselves cause changes in the physical and chemical environment. For example, extensive decomposer activity accentuates the extent of the anoxic bottom waters of the lake. Flowing waters differ from lakes in that constant mixing of the water prevents the establishment of such surface-to-bottom gradients, but with severe pollution, streams can be changed completely below the pollution source; only with sufficient distance downstream does the stream gradually return to its normal condition. Streams and lakes share biological reactions which vary according to the concentration and composition of pollutants, precisely because living systems, as outlined above, react similarly. The difference is a result of the parameters of flow and constant mixing in streams, which makes them less susceptible to a given concentration of organic wastes.

Suppose a body of water is polluted with a quantity of sewage or other effluent that results in eutrophication. The pollutants are ecosystem operators because they change species diversity, one of the first indicators of ecosystem stress. Since species differ in their abilities to react to increased nutrients (some being favored and others disadvantaged), those adversely affected will decline in number or disappear, and consequently species diversity usually will decline, at least in the short term. In lakes, some species (for example, trout) will first show behavioral changes such as inhabiting shallower water because increased decomposition reduces oxygen concentration in the deeper water. Later their populations may decline, while those which feed on organic detritus and can withstand reduced oxygen tensions (for example, midge larvae) will increase. Ultimately, as more and more nutrients are released by the decomposers, spectacular algal blooms of a few species may appear with increasing frequency. In the most extreme cases, the lake biota may be reduced to a few hardy species, or new species characteristic of very unusual conditions may replace those which had an advantage at a less intense level of eutrophication. In rivers, the classic example is found in the succession of communities downstream from a sewer outfall. Directly downstream only a few bacteria live; slightly farther downstream are simple communities dominated by highly specialized sewage organisms like *Tubifex* worms. As the sewage is increasingly diluted, there is a gradual return to the community found upstream of the outfall (Figure 11-7). In this case diversity is an indicator that normal conditions impose less stress upon the biota than sewage.

It is important to note that pollution effects need not follow any set of rules. Large volumes of dilute sewage may be needed to stimulate an ecological change, but more often a small difference in the form or concentration of the pollutant can make a vast difference in results. Suppose the pollutant were a small inorganic nutrient (for example, phosphorus), rather than something as complex as sewage. Then

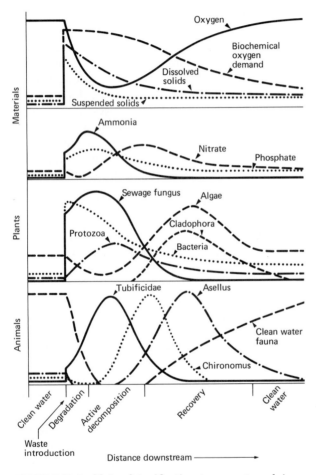

FIGURE 11-7 *Natural purification process at work in a stream. Note the changes in abundance of various species with distance from pollution source and more important, the concentration of oxygen and nutrients.* (Redrawn from Warren (1971) after Hynes (1960) *The Biology of Polluted Waters*, Liverpool University Press.)

the algae could use it directly without intervention of decomposers and the resulting lowered oxygen concentrations, and the increased productivity could mean more food for trout and then more trout. Therefore, though trout are usually very sensitive to pollution, in some cases pollution actually can benefit them. Oceanic examples of the same nature are areas of deep-water upwelling in which nutrient-rich water stimulates primary production and then the entire food chain, including the fish so useful to us. In freshwater lakes, there is no reason why carefully controlled rates of pollution could not achieve the same beneficial effects for freshwater fisheries. The dividing line between benefit and harm is a fine one, however: Too many nutrients, and generally undesirable blue-green algae will predominate instead of those algae useful in the fishes' food chains. These are the algae that usually are shunned as food by other organisms, the ones that accumulate until they rot in the water and on the beaches.

Some evidence has also appeared of insidious and subtle effects of pollution on organisms, effects of disorganization and disorientation at relatively low levels of pollution. An interesting case reported by Sprague et al. (1965) illustrates a nonlethal effect on salmon. They reported that water pollution was interfering with the spawning-stream homing response of Atlantic salmon (*Salmo salar*); the homing response is cued by chemical properties of the home stream. Other reports have suggested that oil spills have damaged the olfactory organs in some marine fishes. As we have discussed, other effects from pollution include changes in species diversity, and these changes indicate fundamental changes in the system. While the general theoretical expectations of the impact of pollution on ecological systems is known (see Woodwell, 1969), much more research is needed in order to predict effects of water pollution in specific cases.

Environmental Effects of Water Pollution: Human Health

One of the few subjects that receives unequivocal support is human health. No one is against better health and longer life for all, and so it is not surprising that where water pollution endangers public health, the standards set to control it are the most stringent. After an accidental escape of sewage, beaches surrounding lakes are closed quickly, and often bans on fishing are imposed. The behavior of public health authorities almost implies that human beings are a pollution indicator, a highly sensitive species whose existence is threatened by anything but the purest standards of water quality.

How dangerous is water pollution to human health? The answer depends on the kinds of pollutants and their concentration in the aquatic system of interest. The methyl mercury episode at Minamata Bay represents one case of a highly dangerous chemical that is effective at relatively low concentrations. On the other hand, many sewage components are no threat in themselves but are inseparable from dangerous ones (for example, hepatitis virus in human waste). Certainly in the public view if there is any question of a real hazard, prompt action may make the protection of public health one of the most powerful arguments to end water pollution, just as it is proving to be an important—if not the most important—rationale in air pollution control (Chapter 12). The agencies that control public health or water pollution are responsive to a constituency which believes human health is worth any cost, and they will act conservatively and react strongly to any supposed water pollution or health threat. For example, hepatitis, cholera, and typhoid fever are all serious diseases that have been and still are responsible for periodic epidemics; they all can be traced to water contaminated by human fecal wastes. Therefore, water found to be contaminated by human feces is quarantined promptly until the source is found and the problem corrected.

While it may seem irrational to question the supremacy of good health for all, we must consider possible future modification of this position for two reasons. Rene Dubos (1965) has argued that sickness is not a simple matter of exposure to disease but rather a complex interaction between the disease agent (in this case a component of the water pollutant), the condition of the potential target, and the genetic and physical environment of both the agent and the target. Policies designed exclusively to suppress the agent can have the unfortunate effect of making the target population more susceptible to a major epidemic in the future by permitting the susceptibles to survive. If there is a wide range of susceptibility (as, for instance, is known for lead poisoning), the trade-off between control costs and human health can be distorted seriously by marginal returns. Of course, this raises complex moral and ethical issues, but they are likely to arise in the future anyway.

A second problem is that to keep human health paramount may require policies which turn out to be environmentally damaging. Using DDT as a broadcast spray to suppress malaria-carrying mosquitoes is one example, but only one of many. If our aim is to keep every individual alive as long as possible, it is inevitable that other elements of the world ecosystem which either threaten the health status or just happen to be in the way will be jeopardized. Once again, this raises ethical and moral issues that need to be addressed now.

Despite questions of strategy, it seems inevitable that public health will continue to be a major force favoring progress in water pollution control.

Environmental Effects of Water Pollution: Aesthetics

The aesthetic issue contains elements of the entire spectrum of environmental effects of water pollution. For example, water quality in Lake Tahoe is important not because of increasing algal growth, or decreasing clarity, or the health hazard of raw sewage that escapes into the lake per se. These are simply supporting elements. Instead, water quality is important because it is inseparable from the population's desire for a lake that is a clean, unspoiled paradise in the midst of green timberlands and tall mountains.

FIGURE 11-8 *A sea bird being cleaned after an oil spill in San Francisco Bay.* (Courtesy of Georg Gerster from Rapho/Photo Researchers, Inc.)

Water quality is part of the image of unspoiled nature, where we can escape the problems and by-products of modern society and enjoy the idyllic setting. As might be expected, not all people, especially local landowners and casino operators, agree that this is what they want. Of course, the water quality of Lake Tahoe is only part of the entire environmental scene at Lake Tahoe, but it is the central element that provides the rationale for land-use control and sewage exportation in the lake basin.

Oceanic oil spills also have caused reactions derived from aesthetic considerations. As we have seen, potential damage to biological systems depends on the constitution of the spillage, and the most damaging components are those which mix readily with sea water. Still, components of oil that float have always received the most attention because they affect sea birds, are highly visible, and wash up on the beaches (Figure 11-8). Despite the fact that the net short- and long-term effects on the system affected may be exactly the same regardless of visibility, would you expect equal concern for the high mortality of barnacles or limpets as for an oil-blackened beach?

Inasmuch as aesthetic quality is an important part of the human environment, it should be considered. But as various conservation organizations have discovered, such concerns may not be shared widely by broad segments of society. Hence arguments based on aesthetics alone are not likely to be convincing, especially when major economic forces oppose them. There are signs, however, that aesthetics may be the only way that intangibles and unquantifiable elements can be included in broad environmental concerns. The reason: The mere acceptance of "environmental quality" as a desirable objective *requires* that unquantifiable variables be considered legitimate concerns, and these are often precisely the same variables previously described as "aesthetics." In short, while aesthetics may never prove to be as powerful an influence as human health, environmental awareness is causing greater consideration of aesthetics than ever before.

Approaches toward Pollution Control

The control of domestic sewage does not appear to be a major problem in the United States. The systematic collection of wastes and the technology of treatment are well known, although no one denies that there are problems of temporary overloads and leakages at treatment plants as well as problems in achieving especially high standards for waste release into highly sensitive environments. Moreover, technology continues to expand in this area of concern. There are opportunities to be explored in the use of sewage as fertilizer and in the recycling of domestic waste water reclaimed from treatment, to name only two topics in this rapidly expanding field.

The real control problems arise with sophisticated industrial wastes. For them the technology is not as simple; we cannot use amplified decomposition to break toxic compounds down into harmless substances. Instead we must find ways to isolate harmful pollutants, separate them from quantities of water, and recover them whenever possible. Despite the magnitude of the engineering challenge, however, the main problem is that the American economy has developed a style of operation which assumes the right to pollute water. The aquatic environment has been treated as an externality for so long that it is difficult to reverse the trend. Short of a complete revolution characterized by a return to small populations and a primitive level of technology, there are no solutions to water pollution control that would at the same time be easy to implement, be effective, and still be fair. A few suggestions that have been made follow.

CHANGE FROM PRIVATE ENTERPRISE TO STATE CONTROL OF PRODUCTION

Some hold that clean water is impossible in a free-enterprise system and that centralization of production is the only way to solve the water-quality problem. One might expect that since the cesspool mentality has characterized human societies in general, the nature of the political system in any country would have little effect on successful water pollution control. One obvious reason is the universal desire for industrialization, economic growth, and higher standards of living. One could hardly expect less developed countries, regardless of government type, to forgo development just to have clean water. Goldman's (1972) review of the situation in the USSR substantiates this hypothesis because hitherto the environment of a vast undeveloped countryside has meant little compared with economic development. However, if future attitudes change, the centralized decision-making of socialist societies would undoubtedly make it easier to control water pollution, whereas the decentralized nature of free enterprise would probably be incapable of concerted action except in an emergency. However, even if it were more effective, whether or not centralized decision-making is the best overall solution is another question.

PROHIBITION BY STATUTE

Another solution is simply to ban the dumping of wastes into natural waters. Where simple eutrophication is a serious problem or where highly toxic materials are involved, this is the only feasible solution. For example, the only way to keep Lake Tahoe clear has proved to be the export of sewage out of the lake basin, since it is not economically feasible to build treatment plants that are thorough enough to suppress eutrophication if sewage is retained in the basin. Similarly, no one seriously would consider the free release of radioactive ions into rivers. The dangers of radioactive escape warrant the inclusion of separate steam-generating and cooling circuits in nuclear reactors.

Prohibition fails, however, if the consequences of abrogation are not clearly deleterious, or if there are no viable alternatives for some form of pollutant release, or if prohibition is unenforceable. It is not difficult to detect large industrial sources of water pollution, but it may be an entirely different matter to enforce an existing statute or to institute a prohibition action if there is no known alternative method of waste disposal or if the violator is politically powerful. On the other hand, local pollution control boards could force the repair of faulty septic tanks if the time and work force required to check several thousand or million facilities were justified by the end-results.

The problems involved in prohibiting water-borne waste disposal are most severe for pollution involving simple molecules that are not amenable to further treatment and for substances that originate from many sources. In both cases, the costs of enforcement hardly ever justify the pollution control obtained, and so enforcement is rarely attempted. The laws stand as hollow mockeries of pollution control. In the words of noted Berkeley epidemiologist J. R. Goldsmith with respect to air statutes, this is environmental "Canutism"—named for King Canute, who commanded the sea to stop rolling in and succeeded only in getting his feet wet.

WATER POLLUTION STANDARDS

If outright prohibition does not work in many cases, a more flexible and perhaps more enforceable scheme is the establishment of some set of minimum standards for various pollutants. This mechanism still depends on a pollution monitoring system, but it allows for the economic hardships and dislocations that would accompany a complete ban on dumping wastes. The establishment of appropriate standards is a problem, but it is not as difficult as the problem of impartial administration. The parallels between enforcing pollution standards and applying land-use codes are distinct: Both are plagued by uneven application of the law. For example, larger firms have greater legal resources and are less likely to be prosecuted than small ones; standards can be so flexible that they are too responsive to the desires of the industries and municipalities that otherwise would be breaking the law; and regulatory agencies are often controlled by those whom they are supposed to regulate (some would say all regulatory agencies eventually suffer this fate).

These problems are certainly not unique to agencies controlling water pollution, nor are they necessarily more severe. Nor are the establishment of standards and their subsequent enforcement necessarily inequitable or useless—far from it. Nevertheless, the establishment and enforcement of standards do contain a real danger, namely, dependence upon the fiscal resources and ethical standards of the controlling agency. The enforcement of building codes and industrial safety standards have been used as political weapons in the past (for example, building-code inspections in Chicago). In water pollution control, Stan Benjamin of the Associated Press reported that the owner of a small textile firm in Grafton, Massachusetts, was prosecuted, by admission of the state, to act as an example for bigger firms. After the case was successfully prosecuted (the standards had been violated by all textile manufacturers in the area, large and small), the net outcome for Grafton was the closing of the plant and the loss of employment; for Grafton this was the crunch of improving environmental quality. In this case control agencies did not use their authority as a political weapon, but the sheer magnitude of monitoring pollution levels, assigning responsibility, and enforcing the law impartially can lead to the same result. No agency with limited resources can possibly handle the job well and impartially; the prospect of Canutism tinged with potential favoritism is inescapable even with the best of intentions.

POLLUTION TAXES AND EFFLUENT FEES

One way to control water pollution in the free-enterprise system is to place progressive taxes or effluent charges on the volume of water pollution generated. The greater the output, the higher the taxes, which then could be used to alleviate the effects of the pollution. This is the economist's market solution: the hope that high taxes would act to discourage high pollution output by supplying an economic motive for rethinking the cesspool mentality.

The main problem with the concept of pollution taxes is that they may not prove to be effective for pollution control unless the public responds to rising prices by cutting demand. If not, manufacturers could pollute all they wanted and pass the environmental costs to consumers as increases in price. Provided that such costs can be transferred indefinitely, raising the tax scale would just mean higher prices.

A second problem is that tax or fee levels would be difficult to set equitably. Government agencies would have to set the fees, and large polluters often have an important voice in these decisions. Consequently, the likely result would be low fee schedules with a minimal effect on pollution levels. If fees are too stiff, economic dislocation (lost jobs, reduced profits, lessened taxes) might result. In either case the effectiveness of the fee or tax program depends strongly on the establishment of appropriate charges.

POLLUTION RIGHTS AND PROPERTY RIGHTS

One of the more interesting proposals for the control of water pollution has been suggested by Dales (1968). Dales traces the cesspool mentality not to the development of free-market economics, but, more basically, to the failure of societies to extend the establishment of property rights to cover the environmental externalities. Once this supposition is accepted, his solution is deduced easily: establish property rights for environment by instituting *pollution rights*. Each unit of pollution rights entitles an industry or municipality to dispose of x units of waste. The total volume of pollution is limited because the number of pollution rights would be limited; they could not be increased without social approval, only transferred from one owner to another via a governmental broker. Finally, polluters would have to possess the proper number of rights for their pollution levels or else violate the law and be subject to prosecution.

Dales's proposal ultimately depends upon a set of socially determined standards limiting the extent of pollution, but once these are set up, pollution rights could automatically allocate the distribution of pollution production among industry and municipalities. If, for example, a particular industrial operation wished to start a new process that is more polluting or wished to increase production, it would have to buy a sufficient quantity of pollution rights on the open market. This is the important difference between the pollution-rights system and the tax system: Each producer would have to compete with others for a limited number of rights. Thus each town and industry must calculate how much it can afford to pay for pollution and bid for additional rights from the government accordingly. This shifts the responsibility to the manufacturer, who then must compete with others to find the "best" price. Conversely, those who sell pollution rights may calculate how much cutting their own pollution rates is worth to them economically. The net effect is to incorporate pollution into the price system, the lack of which is often cited as the primary reason why water pollution is a problem. If used in conjunction with effluent standards, a pollution-rights system could be an effective means to control water pollution.

On the other hand, to sell rights in advance instead of collecting fees or taxes after the fact does not escape the requirement for rigorous and costly enforcement, the same as for water quality standards. Without vigorous enforcement nothing would prevent a company or municipality from blatantly exceeding the effluent levels allowed by its purchased rights, especially if prosecution is unlikely or if penalties are light. Dales himself rules out pollution rights for air pollution by automobiles (and by implication other numerous, small sources of air pollution) because the administrative and regulatory problems would be overwhelming.

TOWARD THE FUTURE

We have seen two basic solutions: legal prohibition versus an economic blend of coercion and persuasion. The former can be effective, given sufficient enforcement; the latter is more efficient economically. Neither alternative is fully satisfactory, and so the search for better pollution control goes on, spurred by the desire for a solution that is efficient, effective, and fair. There are no easy answers on the horizon, only a widespread feeling that a better solution *must* exist. The 1971 report of the Council on Environmental Quality put strong emphasis on the need for better pollution control methods. Earlier, Professor Dales had emphasized the problems of standards, but in his book he went on to point out another scarcely noticed fact:

It seems very hard for people to learn that *everyone* must pay to reduce pollution, and that the important question is not *who* lays out the money in the first place, but *how much* is paid to achieve what benefits.[1]

Biological Approaches to Pollution Control

The use of biological systems for pollution treatment is natural and desirable, and we already have examined some specific cases. A few examples of wastes are, however, so poisonous that no biological system can handle them. As has been intimated, for such

---

[1] From J. H. Dales. *Pollution, Property, and Prices,* University of Toronto, 1968, p. 86.

pollutants there is no alternative to preventing their escape into the environment. The chlorinated hydrocarbons of the DDT group and the closely related PCB plastics (polychlorinated biphenyls, another group of chlorinated hydrocarbons) have been shown to be environmentally dangerous, and DDT at least has been largely removed from use.

Ecological alternatives exist for cases of organic wastes that are not alien to natural systems. Sewage treatment plants are basically places where amplified decomposition takes place in a specialized, isolated ecosystem. In the many cases where sewage treatment has been inadequate, usually the environment is depended upon to decompose the waste problem on the site. If this solution continues to be widespread, it can be done best in streams, as long as natural purification is adequate, and it is most appropriate where increased productivity both is desired and can be controlled. Clearly, all ecosystems reach a point beyond which increased productivity can be accommodated only by an entirely different ecosystem, one dominated by decomposers.

The implicit assumption of the cesspool mentality is either that ecosystems can handle water pollution without significant change or that whatever damage does occur is not significant to human interests. While this viewpoint no longer exists in its pure form in many places, greater effort must be given to delineating the conditions under which enhanced productivity would be desirable or permissible. For example, is it necessarily true that an oligotrophic lake is better than a eutrophic one? In Lake Erie, the main effect of pollution on the fisheries has been on the species composition, not on the total catch; and so if one were to agree that fisheries were harmed by the extent of water pollution, it would have to be because dietary tastes favored whitefish over perch and pike over carp. In an interesting and controversial paper, Hubschman (1971) claimed that it would be a waste of time and effort to clean up Lake Erie; we should use it as a dump instead in order to save the other Great Lakes. Regardless of its merits, Hubschman's arguments point to the very real problem of choosing between putting pollutants in one place or another, provided that the choice between pollution or no pollution is utopian and unattainable. Because natural waters vary in their ability to absorb pollutants, the choices concern *where* rather than *whether*.

Nevertheless, the choices now being made are not really choices between sewage treatment or no sewage treatment (Figure 11-9). Rather, they are choices relating to how much treatment the sewage should

FIGURE 11-9 *A secondary sewage treatment plant, showing tanks of activated sludge in operation.* (Courtesy of Daniel S. Brody from Stock, Boston.)

undergo before it is released. Primary sewage treatment is merely the removal of solid wastes and therefore is mostly physical treatment. Secondary treatment involves the removal of dissolved organic materials through their bacterial decomposition into simple salts. And tertiary treatment is the removal of these salts. Current American policy requires at least secondary treatment and, in certain special cases, either tertiary treatment or export of sewage wastes. Where eutrophication is a serious problem, tertiary treatment may be indicated, but many argue that the costs of tertiary treatment are too high to simply dump this purified effluent back into the environment. Dryden and Stern (1968) have reported that the effluent resulting from the tertiary treatment of the sewage of Lancaster, California, was pure enough to serve as the water supply for recreational lakes. Under existing circumstances the treatment was adequate for eutrophic conditions, but the authors claimed that the water could be refined further so that it would be suitable for oligotrophic lakes as well. It would seem that water of this quality could be recycled easily as drinking water, a suggestion which has been made in the past and which will be repeated in the future. Still another possibility is to grade water quality much like cattle and use the appropriate grade of water for different purposes. Perhaps less money could be spent on bigger plants and diverted instead into more advanced installations to produce small volumes of pure water.

An interesting suggestion made to improve the performance of sewage ecosystems has been to use the waste heat of power plants to maintain high-efficiency sewage treatment plants. This scheme recognizes that to a point biological reactions proceed faster at higher temperatures, and so more effluent can be processed per unit time without additional cost if the power plant can be located near the treatment plant or if the cooling water can be piped there.

Perhaps the best solution is translocation of pollutants not to other bodies of water but to terrestrial ecosystems. This is a logical consequence of tertiary treatment—the water is recycled or released, and the organic matter is used as fertilizer where possible. Analysis of the probable effects of these nutrients reveal some expected patterns. For example, Cole et al. (1969) reviewed the effects of various organic wastes on terrestrial plants. There is evidence that for some crop plants organic wastes can be a growth stimulant, while for others they can be as deadly as soil salinization from irrigation. Again, depending on the plant, certain diseases may be encouraged. Spraying sludge on young pine trees has been demonstrated to be an effective growth stimulant. A great deal of applied research is needed to clarify the conditions and procedures under which such wastes that otherwise would be dumped into natural waters can be utilized for beneficial means.

The use of sewage and other organic wastes as nutrients for terrestrial ecosystems is only one intermediate strategy between the release of untreated sewage into the environment and tertiary sewage treatment. Future water treatment procedures undoubtedly will develop engineering techniques that, in conjunction with an ecosystem approach, enable us to dispose of large volumes of dilute wastes by using the normal decomposition cycles of natural ecosystems with minimum perturbation of these ecosystems (Figures 11-10 and 11-11). The key requirement is to have enough area to dispose of wastes at a low

FIGURE 11-10 *Bermuda grass grown on waste water effluent near San Antonio, Texas. When photo was taken the operation had been going for seven years. (Photo courtesy of Professor G. Tchobansoglous, University of California, Davis)*

rate per unit area while still having sufficient assimilative total capacity. The designation of appropriate areas has only begun, but the economy and usefulness of this method are outstanding features that make it worthwhile to search further for appropriate specific applications.

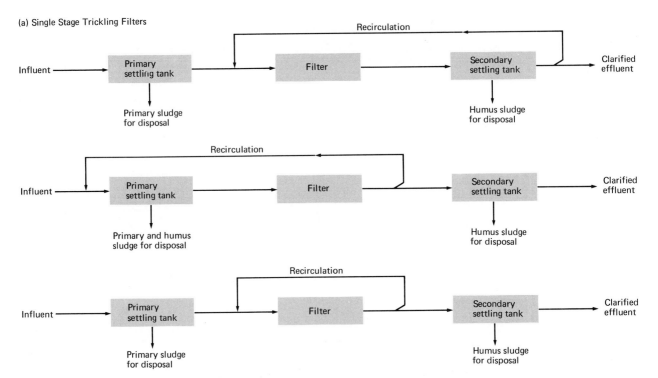

FIGURE 11-11 *Two processes important to sewage treatment: (a) the trickling filter system, and (b) (next page) the activated sludge method. Both of these methods can be varied and combined in innumerable ways, but both are very highly specialized decomposer-based ecosystems that seek to maximize the rate of decomposition. Note that the production of undigested sludge is a problem for both processes and the basis of other disposal schemes, such as conversion into agricultural fertilizers, a process that has been experimented with since the 1930s.* (Redrawn from C. E. Warren, *Biology and Water Pollution Control*, W. B. Saunders Company, 1971. (a) after Clark, Viessman, and Hammer, *Water Supply and Pollution Control*, Intext Educational Publishers, 1971, p. 490; (b) after Imhoff and Fair, *Sewage Treatment*, 2nd ed., Wiley, 1956, pp. 137, 154.)

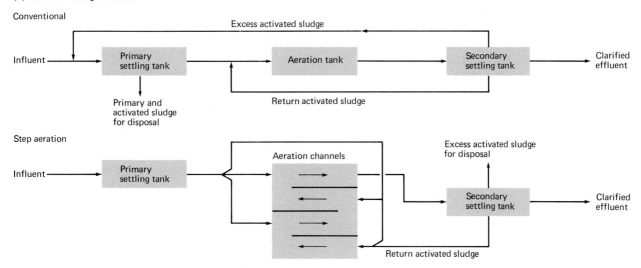

(b) Activated Sludge Methods

In Conclusion: Nature and Human Beings in Water Pollution Abatement

Three major points have emerged in this chapter.

First, water pollution of some degree is completely natural; except for certain highly sophisticated or toxic compounds, the impact of human sources of water pollutants is derived from their quantity rather than from their quality.

Second, because water pollution is natural, aquatic ecosystems possess mechanisms for dealing with certain kinds of materials. However, these ecosystems are variously susceptible to degradation and breakdown signaled by decreased diversity, reduction of oxygen tension, and accumulation of organic matter. A corollary is that accelerated eutrophication can be beneficial if done carefully.

Third, the development of environmental concern has centered on water quality as one important component. The proliferation of schemes to control water pollution signals the end of the rampant, condoned form of the cesspool mentality, even if the ideal solution still has not emerged. Increasing attention is being given to the development of various uses for the water pollutants as part of pollution abatement policy, perhaps another rediscovery of the wheel. For example, night soil (human fecal wastes) has been an important fertilizer in Asian rice paddies for centuries, when these wastes otherwise would have ended up in places where they would have been far less useful and, more often than not, even harmful.

The conclusion is that water pollution control seems less a matter of technique than of desire. Compared with air pollution (Chapter 12), the sources of water pollution are fewer and easier to control. Furthermore, the technology for doing so exists already and is being refined. The price is measured in terms of economic disruptions such as rising costs and prices and increased unemployment, factors that are always immediate human concerns. All of which can be summed up in just two questions: *Who* pays the costs? *Who* gets the benefits?

Questions

REVIEW QUESTIONS

1. Why is water pollution a surgical tool for natural ecosystems?

2. Why is it inappropriate to say that Lake Erie is dying? Is it more appropriate to claim that Lake Erie is aging or becoming senile? Why?

3. Why is Lake Erie actually more productive now than before pollution became a serious problem?

4. Compare the probable biological effects of
   a. a "new town" built on the Ohio side of Lake Erie
   b. a new petrochemical plant on Lake Erie
   c. a power-generating, coal-burning plant on the Ohio River
   d. a PCB plant on Lake Huron

ADVANCED QUESTIONS

1. Is it possible to reverse the impact of technology, as some authors have suggested, and completely suppress water pollution?

2. Is a subsidy for the research, development, installation, and use of control devices for industrial water pollution likely to work? Would it be an efficient solution? Hint: One author has claimed that an appropriate analogy would be a subsidy not to quit high school, but there is no unequivocal answer. This question would be an excellent focus for a student debate.

3. North and Hubbs (*Fish Bulletin, California Fish and Game*, No. 139, 1968) and MacIntyre and Holmes in Murdoch (1971) discuss the relationships among sewage and kelp, kelp and sea urchins, and sea urchins and sea otters. After examining the history of what has happened with these organisms, explain possible consequences of an introduction of sea otters on
   a. the abalone fishery
   b. the kelp (*Macrocystis sp.*)
   c. the fishes living in the kelp beds
   d. the sea urchins

Further Readings

Note: New information is appearing fast in the whole area of pollution control. Besides the general scientific journals, readers should turn to such a source as Environmental Science and Technology *for the latest information on technological developments and to legal and political science journals such as* Ecology Law Quarterly *for statements about socioeconomic control of water pollution. See Crocker and Rogers (1971, p. 147) for further annotation.*

Barkley, P. W., and D. W. Seckler. *Economic Growth and Environmental Decay.* Harcourt, Brace, Jovanovich, New York, 1972.

> An excellent, inexpensive, short text on the problems and hopes of economics and its environmental consequences. It is as cogent as Dales (1968) but considerably broader in its treatment.

Beeton, A. M. "Eutrophication of the St. Lawrence Great Lakes." *Limnology and Oceanography*, 10 (1965), 240-254.

Synopsis of the Great Lakes-Lake Erie pollution problem. See also Carr and Hiltunen, *Limnology and Oceanography*, 10 (1965) 551.

Cole, H., W. Merrill, F. L. Lukezie, and J. R. Bloom. "Effects on Vegetation of Irrigation with Waste Treatment Effluents and Possible Plant Pathogen-irrigation Interactions." *Phytopathology*, 59 (1969), 1181-1191.

Commoner, B. *The Closing Circle*. Knopf, New York, 1971.

Council on Environmental Quality. *Environmental Quality*, 1971. The 2nd annual report of CEQ. Washington, 1971.

For only two dollars, a nice summary of the problems and governmental response to environmental quality, particularly pollution.

Crocker, T. D., and A. J. Rogers. *Environmental Economics*. Dryden, New York, 1971.

One of a series of short texts in environmental economics. This one treats the issues, but in a rather light-hearted tone and at a fairly low level of detail.

Dales, J. H. *Pollution, Property, and Prices*. University of Toronto, Toronto, Ontario, 1968.

Davies, J. C. *The Politics of Pollution*. Pegasus, New York, 1970.

Good for describing current mechanisms of pollution control.

Dryden, F. D., and G. Stern. "Renovated Waste Water Creates Recreational Lake." *Environmental Science and Technology*, 2 (1968), 268-278.

Dubos, R. *Man Adapting*. Yale, New Haven, Conn., 1965.

Edmondson, W. T. "Lake Washington." In *Environmental Quality and Water Development* (Goldman et al., eds.). Freeman, San Francisco, 1973.

Edmondson, W. T. "Water Quality Management and Lake Eutrophication: The Lake Washington Case." In *Water Resources Management and Public Policy*. University of Washington, Seattle, 1968.

Goldman, M. I. *The Spoils of Progress: Environmental Pollution in the Soviet Union*. M.I.T., Cambridge, Mass., 1972.

Hawkes, H. A. *The Ecology of Waste Water Treatment*. Pergamon, New York, 1963.

Hubschman, J. W. "Pollution Abatement in Lake Erie: Then What?" *Science*, 171 (1971), 536-539.

See also the reply by Rosenblum and Hollocher, *Science*, 172 (1971), 1294.

Hynes, H. B. N. *The Biology of Polluted Waters*. University of Liverpool, Liverpool, England, 1963.

A faunistic approach to water pollution biology, more a catalogue and treatise than a position paper.

Keup, L. E., W. M. Ingram, and K. M. Mackenthun (eds.). *Biology of Water Pollution: A Collection of Selected Papers on Stream Pollution, Waste Water, and Water Treatment*. Federal Water Pollution Control Administration, Washington, 1967.

An excellent source to turn to as a first choice.

McGauhey, P. H. *Engineering Management of Water Quality.* McGraw-Hill, New York, 1968.

North, W. J. "An Investigation of the Effects of Discharged Wastes on Kelp." California State Water Quality Control Board Publication, No. 26, 1964.

Regier, H. A., and W. L. Hartman. "Lake Erie's Fish Community: 150 Years of Cultural Stress." *Science,* 180 (1973), 1248–1255.

A recent update on the saga of Lake Erie.

Ryther, J. H., and W. M. Dunstan. "Nitrogen, Phosphorus, and Eutrophication in the Coastal Marine Environment." *Science,* 171 (1971), 1008–1013.

An example of a uniform policy that affects different systems in opposite ways.

Schelske, C. L., and E. F. Stoermer. "Eutrophication, Silica Depletion, and Predicted Changes in Algal Quality in Lake Michigan." *Science,* 173 (1971), 423–424.

Compare with analysis by Beeton (1965).

Smith, J. E. *"Torrey Canyon" Pollution and Marine Life.* Cambridge University, Cambridge, England, 1968.

A report of the "Torrey Canyon" oil-spill incident and aftermath. There is no more complete and carefully researched account available.

Sprague, J. B., P. F. Elson, and R. L. Saunders. "Sublethal Copper-Zinc Pollution in a Salmon River: A Field and Laboratory Study." *International Journal of Air and Water Pollution,* 9 (1965), 531–543.

Strobbe, M. E. *Understanding Environmental Pollution.* Mosby, St. Louis, 1971.

A collection of readings in environmental pollution. Some of the papers are very useful; others either are highly technical or contain little more than rhetoric.

Warren, C. E. *Biology and Water Pollution Control.* Saunders, Philadelphia, 1971.

Actually a useful text in basic ecology, with some development of the role of biologists and ecological theory in water pollution control.

12. The Sewer in the Sky

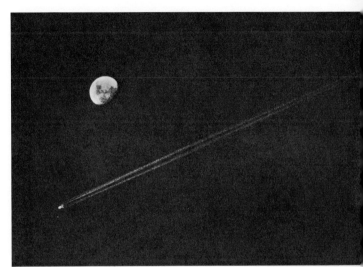

Some authors prefer not to separate air pollution and water pollution, and instead they emphasize the general properties of environmental pollution. This is justified for a number of reasons. Air and water are basically fluids, air being a particularly complex mixture. For another thing, mechanisms for air and water control are similar in principle and often in practice as well, because both readily accept the variety of pollutants discharged into them. In fact, since water and air normally interchange, they both can be contaminated from a single source of some pollutants, for example, heat. Finally, both air and water pollution are ubiquitous, since fluid media are highly mobile.

Nevertheless, the differences between air and water pollution seem more important than their similarities, if more in degree than in kind. For example, the interactions between physical factors and air pollution are better known and probably more important than corresponding processes in the aquatic environment.

Pollutants in air and water are often chemically dissimilar. Complex compounds are more likely to be flushed into aquatic systems, since many industrial processes use water to wash out and trap by-products. Many heavy and nonvolatile compounds are unlikely to get into the air, although there are exceptions such as metal that escapes up smelter smokestacks. The atmosphere is more likely than water to receive combustion products and materials of low molecular weight. Even if heavier compounds are ejected into the atmosphere, most are too heavy to stay there long and will quickly settle out, whereas even the heaviest materials may stay indefinitely in a lake basin.

There are other major differences. The role of natural ecosystems in the destruction of air pollutants is much smaller than we have seen for water, if only because the air is sparsely inhabited. The health hazards of polluted water are much more obvious than those for air; they also are easier to abate. Finally, the aesthetic impact of water pollution may be serious, but it is neither so pervasive nor so obvious to the public as fouled air.

In this chapter we shall examine the specific problem of air pollution: its origin, its distribution, its local effects and global implications, and some strategies for its control. Given numerous differences between air and water pollution, readers should compare Chapters 11 and 12 and attempt to relate both to the general topic of environmental pollution.

Air Pollution in the World and Nation

If one were to ask where air pollution is most likely to be found, major cities, urbanized regions, and major highway corridors certainly would be cited.

FIGURE 12-1 *Night or day? Actually this is London at 11 AM on a foggy day.* (Courtesy of Louis Goldman from Rapho/Photo Researchers, Inc.)

This impression is supported by fact (witness New York and Los Angeles), but aerial observation is required to reveal how widespread smog and haze are. Precise measurements are difficult to get from remote areas, but pollutants ejected high into the atmosphere are known to disperse far and wide. In populated areas locally high concentrations of air pollutants are a logical result of industrial emissions, internal combustion engines, and production economies in general. Tokyo, Paris, New York, Rotterdam, and many other cities share the problem and its sources. Even in rural areas, extensive agricultural burning creates dense smoke clouds that (at least visually) match the

TABLE 12-1 *U.S. Public Health Service air pollution rankings for 40 American cities. The figures are composite densities of dust, ashes, gasoline fumes, and sulfur dioxide. The data show an expected correlation of higher ratings in large, heavy industry cities and a trend toward an increasing pollutant load even in the short time data have been available. Compare with the data of Table 12-2 and Figure 12-1.*

1.	New York City	457.5	21. Paterson-Clifton-Passaic	304.0
2.	Chicago	422.0	22. Canton	302.0
3.	Philadelphia	404.5	23. Youngstown-Warren	294.5
4.	Los Angeles-Long Beach	393.5	24. Toledo	287.0
5.	Cleveland	390.5	25. Kansas City	285.5
6.	Pittsburgh	390.0	26. Dayton	280.0
7.	Boston	389.0	27. Denver	280.0
8.	Newark	376.5	28. Bridgeport	261.0
9.	Detroit	370.0	29. Providence-Pawtucket	261.0
10.	St. Louis	369.0	30. Buffalo	260.0
11.	Gary-Hammond-East Chicago	368.5	31. Birmingham	259.5
12.	Akron	367.5	32. Minneapolis-St. Paul	257.0
13.	Baltimore	355.0	33. Hartford	254.5
14.	Indianapolis	351.0	34. Nashville	253.0
15.	Wilmington	342.0	35. San Francisco-Oakland	253.0
16.	Louisville	338.0	36. Seattle	252.5
17.	Jersey City	333.5	37. Lawrence-Haverhill	a
18.	Washington	327.5	38. New Haven	246.0
19.	Cincinnati	325.5	39. York	246.0
20.	Milwaukee	301.5	40. Springfield-Chicopee-Holyoke	241.0

Source: U.S. Public Health Service.

[a]Tentative positioning.

more sophisticated industrial and automotive emissions. Dense smoke clouds have been reported over such unlikely places as East Africa and Southeast Asia. But if there is no such thing as unpolluted air anywhere, it is another matter to determine what degree and kind of pollution is "significant."

In the United States, the same trends seen elsewhere hold, with additional attention being paid to geographical influences on air pollution concentration, especially when the subject of interest is the *potential* pollution problem. It is well known that certain combinations of meteorological and topographical circumstances can concentrate daily outputs of pollutants to deadly levels. A number of incidents involving this kind of acute, concentrated air pollution have had dramatic health effects. Two more famous and oft-cited examples are London, England

(1952), and Donora, Pennsylvania (1948), where excess mortality was attributed to air pollution. London is characterized by heavy fogs that trap foul air (Figure 12-1), and Donora sits in a canyon that can serve the same purpose. Given the sources and the right conditions, other disasters could occur. The topography of Los Angeles, which is situated in a bowl-shaped valley, poses similar concerns.

Apart from the well-publicized cases in California and the industrial cities of the world, in areas that practice regular burning of agricultural stubble and the leftover slash of logging, there is a semipermanent or seasonal pall of smoke. In the southeastern United States, where industry is expanding rapidly, the topography is generally flat and unable to contribute to critical pollution concentrations. Summer high-pressure areas, however, tend to stagnate in the Southeast and to produce the same effects as mountains, canyons, or other land forms. This kind of stagnant pressure system precipitated such a severe air pollution crisis over Birmingham, Alabama, in the summer of 1973 that the Environmental Protection Agency had to enforce a shutdown of industry until meteorological conditions improved. Increasing numbers of acute air pollution incidents in the Southeast have caused some experts to predict that, as industrialization continues, the region will suffer from poor air quality unless appropriate preventive action is taken. However, the publicity attendant to previous air pollution crises such as in London and Donora may prove to be sufficient stimuli to prevent occurrence elsewhere. In contrast, less attention has been given to problems associated with the chronic incidence of lower concentrations of pollutants, which may prove more important to human health and society. In the Southeast and elsewhere there can be little satisfaction in merely preventing mortality that is related to acute air pollution.

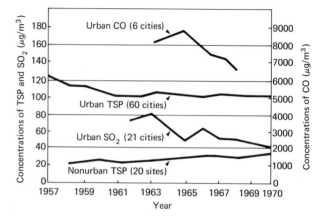

FIGURE 12-2 *Chronic levels of air pollution measured as indices SO_2 and TSP (total suspended particulates) by the Mitre Corporation. These are the most extensive data available, but should be accepted tentatively.* (Redrawn from *Environmental Quality,* Report 3 of the Council of Environmental Quality, 1972.)

In 1967, the most polluted air (measured as a composite index of particulates and sulfur dioxide) was found over larger cities with concentrations of industry and automobiles, although there were exceptions like Atlanta, Miami, Phoenix, and New Orleans —all less industrialized urban areas (Table 12-1). Since the Environmental Protection Agency established a network of air-quality monitoring stations for the nation, fragmentary data indicate that most of the country, including big cities like Chicago and New York City, have shown an irregular trend toward improved air quality, largely due to increased use of low-sulfur fuels and antipollution devices (Figure 12-2 and Table 12-2). No real change has occurred in the countryside, which shows that the improvement has been

TABLE 12-2 *Air-quality data for 10 cities, 1967–1972, expressed as index numbers based on the earliest reading. TSP: total suspended particulates; SO$_2$: sulfur dioxide, both measured as concentrations in micrograms per cubic meter of air. Some cautions must be observed with these data: Those marked with an asterisk are based on statistically insufficient data, and all cities have only one downtown monitoring station. Thus improvements downtown may simply represent a shift in the distribution of pollutants or quirks in the location of the station.*

			Year					
City	Pollutant	Value when first measured	1967	1968	1969	1970	1971	1972
Boston	SO$_2$	0.23*	1.00*	2.83	3.48*	2.57*	0.96*	0.70
	TSP	1.23		1.00	0.93	0.87*	0.92*	0.87
Chicago	SO$_2$	2.18		1.00	1.06	0.69	0.42	0.27
	TSP	1.49		1.00	1.21	1.00	1.03	0.87
Cincinnati	SO$_2$	0.36		1.00	0.92	0.39	0.58	0.81
	TSP	1.48	1.00	0.89	0.94	0.91	0.87	0.78
Denver	SO$_2$	0.22			1.00	0.77	0.45	0.41
	TSP	1.24	1.00	1.15	1.22	1.31	1.27	1.64
Los Angeles	SO$_2$	0.14*				1.00*	1.86	2.14
	TSP	1.22	1.00	1.41	1.02	1.37	1.45	1.29
New York City	SO$_2$	4.35	1.00	0.70*	0.39*	0.21*	0.20*	0.14
	TSP	2.18	1.00		0.65	0.75	0.65*	0.58
Philadelphia	SO$_2$	1.13		1.00	0.77	0.94	0.41*	0.50
	TSP	2.00	1.00	0.74	0.84	0.90	0.66	0.52
Pittsburgh	SO$_2$	0.89	1.00	1.06	1.07	0.81	0.70	0.89
	TSP	1.78	1.00	1.21	1.08	0.95	0.83*	1.01
St. Louis	SO$_2$	1.04	1.00	1.09	0.88	0.69*	0.12	0.23
	TSP	1.49	1.00		1.66	1.37*	0.79	0.83
Washington	SO$_2$	0.36			1.00	0.94*	0.72*	1.39
	TSP	1.13	1.00	1.01	0.87	0.89*	0.86	0.98

SOURCE: U.S. Environmental Protection Agency, National Air Sampling Network.

made in cities, where it is most needed. Nevertheless, some cities, notably in Los Angeles, have made little progress.

Sources of Air Pollutants

There are natural sources of air pollutants, just as there are natural sources that enter aquatic systems. One well-known natural source is found in conifer trees, which emit *terpenes*, volatile substances found in resin. On bright days in coniferous forests the air may be smoky blue with terpenes. Similarly, the smoke of natural forest fires may produce clouds of smoke stretching hundreds of miles downwind (Figure 12-3). Severe winds may erode soils and fill the air with particles, especially in barren places with no vegetation. In many cases human agents have caused the destruction of the vegetative cover. Finally, and most dramatically, major volcanic eruptions have been known to inject particulate matter in the atmosphere with effects similar to human-derived air pollutants (Mitchell, 1961).

Nevertheless, in air as in water, natural sources are neither so pervasive nor so troublesome as those of human origin. The internal combustion engine is most recognized by the public as the modern villain that has supplanted the smokestack and the steel mill. Additional sources are infinite and result from a wide range of human activities and technologies. An exact, exhaustive catalogue is unimportant, however, except to specialists interested in a specific group of pollutants; from the viewpoint of policy formulation, to categorize air pollutants by their sources seems far more useful. Viewed in this way, air pollution sources can be either *mobile* or *stationary*. The primary mobile source is the automobile (by virtue of its abundance), supplemented by other vehicles. The stationary sources are much more varied and numerous, including such diverse elements as power plants, industry, agricultural burning, garbage dumps, and heating plants. In general, these processes have a number of by-products, some of which are useful and worth saving, but others of which are not. Furthermore, by-products may be difficult to collect or refine, or if allowed to accumulate they may become a health hazard or a poison of the process that produces them. In general, then, synthesis and energy conversion can be expected to be major contributors to air pollution.

FIGURE 12-3 *The natural air pollution of a smoky forest fire in Missouri.* (Courtesy of Grant Heilman.)

The Chemical Nature of Air Pollutants

Air pollutants from incomplete combustion and a variety of other sources reflect the chemical nature of their parent substances. The burning of fossil fuels yields carbon, nitrogen, and sulfur compounds—all components of living organisms preserved during fossilization and released by combustion. Agricultural burning yields a high proportion of very heavy substances, mostly heavy carbon compounds only partly burned and collectively known as *particulates*. (To many the term "soot" may be more familiar.) Typically, pollution sources yield some mixture of elements—particulates, carbon monoxide, carbon dioxide, nitrogen oxides (designated as NO_x), and sulfur dioxide—in proportions that depend on the substance from which they are derived. For example, the fossil fuels vary widely in chemical constitution. The lighter, more volatile constituents of petroleum—natural gas, bottled gas, napthalenes—are highly flammable and burn readily with almost no residue. Gasoline burns well but tends to form NO_x compounds when burned with abundant air. The solid fossil fuels are much less combustible and present greater air pollution problems. For instance, coal was the fuel that caused the mortality episode from acute air pollution in London in 1952. When, in addition, the coal has a high sulfur content, sulfur dioxides are added to the already sizable output of particulates and carbon oxides. Anyone who has tried to burn asphalt, a solid derivative of petroleum, needs no reminder of its effects on local air quality.

Besides products produced by combustion, entirely different chemical elements may result from such processes as smelting and fabrication as the by-products of purification and synthesis respectively. Of particular concern are heavy metals (for example, lead, arsenic, cadmium, beryllium), many of which are cumulative poisons of biological systems at very low concentrations. Where common combustion products are measured in parts per million, the heavy metals often are detectable in parts per billion. When finely divided, these elements can easily become airborne and produce a severe health hazard, even if they are present only in a limited area for a short time. The danger posed by these heavy metals underscores the importance that toxic chemicals can assume when injected into the atmosphere, where most creatures may be unable to escape them.

Since air has such a low density, chemicals of high molecular weight cannot remain in the atmosphere very long, and unless very poisonous they will go undetected since most will soon settle out. Lighter materials may stay in the air for a long time since few biological mechanisms cause them to settle out; during the time they are airborne, they may interact with other chemicals and produce effects on organisms (via chemical reactions within living systems). The exact interactions depend to the greatest extent upon the chemical nature of each pollutant. For example, rubber degenerates quickly in the presence of ozone, and carbon monoxide is a deadly respiratory poison because it combines so strongly with hemoglobin in blood. This kind of information gives us some degree of prediction and warning about the consequences of air pollution if we know what chemical compounds are in the pollutant mixture. But because concentrations are critically important and difficult to measure, it is hard to predict exactly what day-to-day effects will be. This is yet another example of complex systems behavior.

FIGURE 12-4 Los Angeles city hall through a blanket of photochemical smog. (Courtesy of American Stock Photos/Tom Stack and Associates.)

Photochemical Smog: An Example of Interaction

An important consequence of mixing various pollutant chemicals together is that they may react with one another and with other natural components in the air to magnify the effects they might have had alone. The case of vehicular emissions illustrates this process well. The direct products of combustion, the *primary pollutants*, consist of oxides of carbon and nitrogen. They can react with water vapor in the air to form carbonic acid, which is no more potent than the fizz of carbonated water, and the more serious and caustic nitric acid. In the presence of sunlight, the primary pollutants also can react to form entirely new compounds (*secondary pollutants*) such as peroxylacylnitrate (PAN), sulfuric acid droplets, and ozone; these, along with other compounds, form the brownish haze called *photochemical smog* (Figure 12-4). Sunlight provides the energy needed for the formation of these secondary pollutants.

Geophysical Influences on Air Pollution

While knowing the chemical nature of air pollutants is important, it is insufficient to explain the distribution of air pollutants through any particular volume of atmosphere. To do this, it is necessary to examine the pollution sources in relation to the geophysical setting in which they occur.

Geography The best way to demonstrate the influence of topography on the distribution of air pollution is by analogy to the sea and the incoming tide. As the waves rush toward the land, they strike resistant rocks, then flow around them and, if possible, over them, pushing with particular force up surge channels and spreading out over flat slopes. Since air pollutants have mass and are carried passively about in the air (which, like water, is a fluid), the analogy is remarkably exact. Air moves over and around mountains, surges through passes and into canyons, and spreads out over valley floors, carrying air pollutants much as the sea carries salt and sand.

Using this analogy, it makes sense that a source located in a canyon is likely to cause air pollution problems in that canyon, just as a city built into a steep-walled valley is likely to wallow in its own pollution. An analysis of topography will yield useful information to planners who want to determine which sources have minimal impact on air quality.

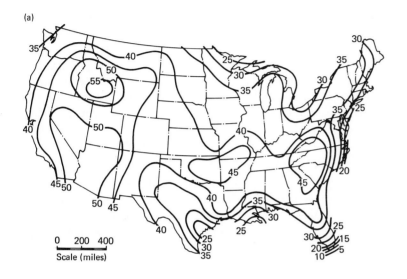

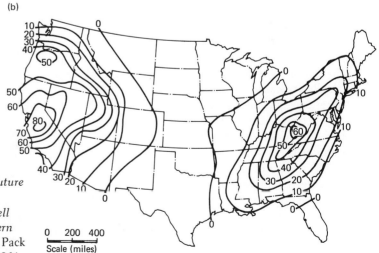

FIGURE 12-5 (a) The frequency of stagnating anticyclonic inversions. (b) Predicted isopleths for future air pollution problems. (Isopleths are lines of equal frequency of the variable in question.) Note how well correlated these two figures are, at least in the eastern part of the United States. ((a) Redrawn from D. H. Pack in *Science* 146, p. 1125, 27 Nov. 1964. Copyright © 1964 by the American Association for the Advancement of Science. Used with permission. (b) Redrawn from M. Smith in "Recommended Guide for the Prediction of the Dispersion of Airborne Effluents," 2nd edition, *Am. Soc. Mech. Eng.*, 1973. Used with permission.)

Climate Long-term climate patterns are a second determinant of air pollution distribution insofar as they determine the probabilities that conditions likely to favor buildup of pollutants will occur. The interaction of climate with the topography of the region is especially important in this regard. For example, areas that are often windy are not likely to have serious local air pollution problems, especially if the landscape is also flat. Agricultural burning in the Central Valley of California is permitted on windy days when the smoke is likely to be well dispersed throughout the atmosphere. Areas that have continental climates (characterized by hot summers and stagnant high-pressure systems) or areas that have little wind, are particularly susceptible to air pollution episodes; these conditions are dominant over most of the country. Some areas, such as the Southeast (Figure 12-5), frequently develop stagnant air conditions that particularly favor the buildup of air pollutants.

Weather The day-to-day expression of variation in climate is *weather*. Any region with a climate not usually favorable for the buildup of air pollution can experience an unusual weather pattern that is favorable. This was true of London and Donora; the fatalities that resulted were not less real, no matter how unusual the weather was.

The importance of wind, temperature, high-pressure systems, and the like is that each affects the effective volume of air into which pollutants can be diluted; hence, in conjunction with source-output rates, each determines the local concentration of air pollution. The conditions that most favor the buildup of air pollution are the result of global cycles of uneven heating and cooling. If one were to measure air temperature from the surface upward, one would ideally find that the temperature decreases smoothly the higher one goes into the atmosphere. This is what we expect to result from cooling of the earth, but we often find that at some point the air temperature will actually increase over a short distance (usually at about 1000 miles) before dropping again. For obvious reasons this phenomenon is called *temperature inversion* (Figure 12-6). Temperature inversions can arise in several ways:

1. Heated tropical air rises, disperses poleward, and cools until it sinks. If it encounters air rising from local solar heating, inversions will occur when the temperature of the subsiding air and the temperature of

FIGURE 12-6 *A graphic illustration of a temperature inversion. The plumes of water vapor from the cooling towers of this nuclear power plant in Pennsylvania condense into clouds, which rise until they encounter the inversion layer. The clouds are contained by the inversion layer and therefore mark it perfectly.* (Courtesy of Dr. C. L. Hosler, College of Earth and Mineral Sciences, Pennsylvania State University.)

the rising air are similar enough to prevent further movement in either direction. This area of stable air is the inversion.

2. A cool air mass may move into an area, displacing warmer air over it and preventing vertical mixing much as described in item 1. When the cool air mass is very low in the atmosphere, the potential for dangerous pollutant concentrations is particularly large.

3. Local solar heating on sunny days causes surface air to rise. As it rises, it cools; and as it cools, it eventually starts to sink again. Then the inversion develops as in item 1. However, if the air is heated greatly, it can get turbulent enough to break up an inversion layer.

4. At night solar radiation absorbed by the earth is reradiated as heat back into the atmosphere, forming a very shallow mass of warm air near the ground with cooler air above. These conditions often produce dense fog and inversion layers similar to those from frontal intrusions (item 2).

Temperature inversions create conditions favorable to pollutant buildup, especially when they last for more than a day. The effect of a temperature inversion is to "clamp a lid" on the atmosphere. The inversion prevents dispersion of pollutants through itself and limits the vertical mixing of the air beneath. The simple physical truth—that the more air there is, the more dilute a given volume of pollutants will be—depends upon the action of weather factors that control the size of this air volume.

Factors that break up temperature inversions include the day-night cycle and strong winds. Of course, any inversion produced by passing frontal systems or local heating tends to disappear with the coming of night, and strong winds can make the air turbulent enough to destroy any inversion. Inversions due to stagnant high-pressure systems (especially of tropical origin, which are constantly reinforced by an inflow of tropical air) may persist until the primary system moves off. The interaction of air pollution and atmospheric dynamics is sufficiently well understood to form a useful part of the air-quality monitoring system and to be an important consideration in the formulation of air pollution policy.

Biological Influences on Air Pollution

In our discussion on water pollution, we found more evidence of the importance of biological systems on water quality than of the physical and chemical properties of water on pollution dispersion. In the atmosphere the opposite is true: Compared with the knowledge of physical and chemical interactions, almost nothing is known about the influence of organisms on air quality, apart from the small contribution of natural air pollution made by pine trees and other plants. Part of the reason for this can be laid to the vastness of the atmosphere compared with the volumes involved in most natural waters, plus the small size of the pollutants (mostly the result of metabolism) produced by organisms. In addition, while aquatic organisms form an important part of the water pollution system, terrestrial organisms have only a negligible effect on absorbing and breaking down air pollutants compared with the physical and chemical influences.

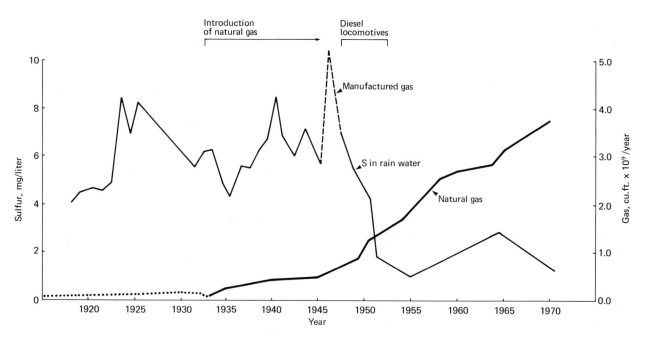

FIGURE 12-7 *The history of acid rain in western New York State (Ithaca), 1915–1970. When natural gas replaced other fossil fuels in 1933 and diesels replaced steam locomotives in the early 1950s, the sulfur content of the precipitation fell accordingly. In general, the acidity of the rain reflects human activities (largely the use of fuels containing sulfur) quite accurately.* (Redrawn from G. E. Likens in Cornell University Water Resources and Marine Sciences Technical Report 50, October 1972.)

We are the only organisms that can improve air quality significantly, and then only through the use of technology and policy to suppress undesirable emissions. However, we are no better equipped to remove pollutants from the air than any other organism is, whereas many aquatic organisms carry out a cleansing function in the process of taking oxygen from the water; they inadvertently or deliberately remove a variety of chemicals and particles from the water. And many organisms depend on a high level of "organically polluted" water to get their food.

Natural Pollution-destruction Mechanisms

Fortunately, chemical contamination of the atmosphere is a steady-state phenomenon that would decline naturally if rates of pollution generation were decreased. We already have noted that most of the heavier particulates settle out pretty rapidly and hence do not persist for more than a few weeks. Some pollutants are unstable and degenerate into simpler compounds rapidly or are absorbed into water vapor. Once hydrated, many pollutants can then be washed out of the atmosphere during rain and snow storms. Reports of "acid rain" (sulfuric acid from hydrated sulfur dioxide) over Europe and North America show this mechanism in action (Figure 12-7).

Besides rapid removal of pollutants from the air, the sheer size of the atmosphere is an important factor in preventing important concentration of air pollution. Many primary pollutants are chemically stable and impossible to destroy, but when thoroughly mixed with the atmosphere they are diluted to very low (and presumably insignificant) concentrations. At one time, there were fears that CO_2 emissions resulting from the combustion of various fuels might significantly modify the gaseous composition of the atmosphere, leading to an increase in the earth's temperature (the greenhouse effect, discussed below). However, there is reasonable doubt that the rate of increase of CO_2 in the atmosphere is rapid enough to increase global temperature, especially since the recent trend has been one of cooling.

Our saving grace is that many problems we identify as pollution depend on conditions which favor locally high concentration of pollutants; without such conditions, dispersion of pollutants throughout the atmosphere would occur naturally. However, the slow accumulation of pollutants throughout the atmosphere can lead to significant problems. Many atmospheric physicists believe that the diluting capacity of the atmosphere is more limited than is commonly believed, and there is always the possibility that organisms can concentrate pollutants. If we depend on dilution, we may simply be exchanging one set of problems for another.

Climatic Effects of Air Pollution

We have seen how climate and weather can play important roles in the distribution of air pollution. The physical properties of certain pollutants can have even more important effects on weather and climate. On the local scale, Changnon (1968) has presented evidence that abnormally high rainfall at LaPorte, Indiana, is a product of steel-making activities in northwestern Indiana (Table 12-3). The excessive rainfall

TABLE 12-3 *Differences between average precipitation at LaPorte, Indiana, 1951–1965, and mean precipitation at surrounding weather stations expressed as percentage of the mean values.*

Weather parameter	Percent above average
Annual precipitation	31
Warm-season precipitation	28
Number of days with precipitation exceeding 0.25 inches annually	34
Number of days with thunderstorms annually	38
Number of days with hail annually	246

SOURCE: Changnon (1968), by permission.

is restricted to a narrow corridor downwind from the steel mills; the excess precipitation is thought to be a product of the heavy output of particulate materials from the smokestacks of the mills. The particulates act as *condensation nuclei,* that is, as centers about which water vapor can accumulate. On the average, by the time the nuclei reach LaPorte, they are heavy enough to fall out of the atmosphere as rain. Clearly the LaPorte experience is an example of great, if unintentional, success in cloud seeding. If only such success could be transferred to places where it would be appreciated!

In places where condensation nuclei are not so densely produced or where atmospheric water vapor is not so common, the result can still be extensive cloud formation or lower atmospheric haze, either of which can scatter incident sunlight. In areas where clouds or haze persist for long periods of time, reduced light penetration can affect crop growth or create damp conditions unfavorable for many plants. One special case of such cloud formation results from the contrails of jet aircraft, which produce water vapor via fuel combustion (Figure 12-8). R. A. Bryson of the University of Wisconsin has predicted that jet contrails along certain heavily traveled routes could produce complete cloud covers along these routes 75 percent of the time.

Bryson's special concern about high-flying supersonic aircraft and other pollution processes that inject particulates high in the atmosphere is their effect on climate. In the stratosphere, even large particulates may have a long residence time since they are not likely to settle out naturally by weakened gravitational forces. As long as they stay in the atmosphere, these particulates blanket the earth and reduce incident sunlight at the surface by reflecting the light back into space exactly as the low altitude contrails do, but on a global basis instead. An appropriate analogy is a ball completely surrounded by a spherical mirror; the ball is in the dark. Hence, all other factors being equal, the effect of high-altitude particulates would be to cool the earth and change climate and weather patterns globally.

In a number of papers, Bryson and his colleagues have developed this argument as evidence that human activities have led to the observed decrease in global temperature (Figure 12-9), whereas others have predicted that global temperatures should rise because of the burning of fossil fuels. This argument is based on the fact that one combustion product of these fuels (CO_2) has thermal properties which affect the radiation of heat from earth back into space.

FIGURE 12-8 *Jet contrails over Colorado Springs, Colorado.* (Courtesy of Myron Wood from Photo Researchers, Inc.)

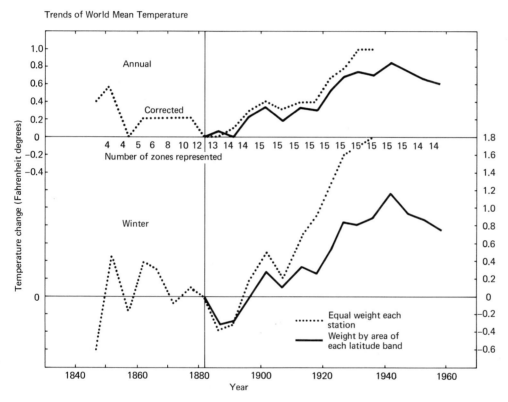

FIGURE 12-9 *Global temperature trends, 1840–1960, measured as mean annual temperature changes. The long-term warming trend, 1890–1940, has slowed in the last three decades, but continues at this reduced pace.* (Redrawn from J. M. Mitchell, "Recent Secular Changes of Global Temperatures," in *Annals of the New York Academy of Sciences* 95, p. 237, 1961. Used with permission.)

(This is called the *greenhouse effect,* because greenhouses do the same thing: They admit sunlight but slow the reradiation of heat accumulated during the day out of the greenhouse). Although CO_2 does not absorb the short-wave sunlight entering the atmosphere nor scatter it like particulates would, it does absorb long-wave heat being reradiated from the earth. Thus CO_2 promotes heat retention in the atmosphere, which should produce higher temperatures.

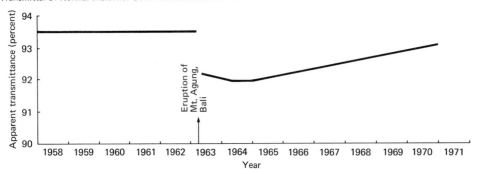

FIGURE 12-10 *Mauna Loa, Hawaii is well isolated from most air pollution sources, and as a result, the clarity of the atmosphere has been reported to be relatively unchanging (measured as light transmission through the atmosphere). With the eruption of Mt. Agung in Bali in 1963, a great quantity of dust was thrown into the upper atmosphere, markedly reducing the clarity of the air. In the decade following the eruption, the clarity of the atmosphere has gradually increased back to normal levels.* (Redrawn from H. T. Ellis and R. F. Pueschel in *Science* 172, p. 846, 21 May 1971. Copyright © 1971 by the American Association for the Advancement of Science. Used with permission.)

Of course, global heat balance is a very important matter, and so it is of more than casual interest to know what is happening. The evidence (Figure 12-5) is not very clear, for the very recent cooling trend has followed a long-term period of gradual warming. If Bryson is correct, the eventual outcome will be increased ice and snow, expansion of the polar regions, dropping sea levels, and eventually another period of glaciation. Conversely, if the earth warms up, sea levels would rise from melting ice and flood coastal cities. In either case, there would be real changes in wind patterns, temperature regimes, seasonal changes, precipitation, and other characteristics that depend on uneven heating and cooling of the earth's surface. The consequences of such change would be dramatic at least, and catastrophic in all likelihood.

National Oceanic and Atmospheric Administration data show a dramatic effect of dust thrown into the atmosphere by the 1963 eruption of Mt. Agung in Bali upon incident light received at Mauna Loa, an isolated extinct volcano in the Hawaiian Islands. The data (Figure 12-10) show a dramatic drop in light transmission, followed by a gradual return to the normal value of 93.5 percent. These data support Mitchell's (1961) contention that the recent global temperature depression is a short-term aftermath of several major volcanic eruptions; hence it is not to be regarded as a serious consequence of air pollution. Other meteorologists agree at least that the observed temperature depression has been far too small compared with known, natural variation in global temperature patterns to draw any supportable conclusions about pollution effects. But it is clear that such effects could occur in the future, and it is possible that they are in progress now.

Remember that water and air are both fluids that obey various physical laws and react to heat flux. But a major difference between them is that most natural waters are complex mixtures of chemicals which are in a sense naturally polluted, and so aquatic organisms have evolved mechanisms for using these solutions for nutrients; furthermore, water is slow to heat and cool, and it reacts slowly to changes in heat flux. Water buffers environmental changes for aquatic organisms in many ways.

The atmosphere, however, is mostly devoid of foreign chemical elements in its natural state and reacts quickly to heat-flux changes; it is a more sensitive barometer than water to changes in environmental conditions that affect heat flow in general, and more specifically to changes in the concentration of those pollutants that can change heat balance, such as particulates and CO_2.

Corrosive and Soiling Effects of Air Pollution

Reference has been made to "acid rains" which result when sulfur dioxide is washed out of the atmosphere. This is one of the more spectacular effects of immersing the planet in an atmosphere that closely resembles a bath of corrosive (if dilute) solutions. For example, both natural and synthetic rubber are broken down by ozone: tire side walls, windshield-wiper blades, rubber gaskets, and similar items crack and fail more quickly when immersed in photochemical smog. In Europe, famous sculptures have been eroded by air rich in sulfur dioxide; this is particularly true of marble statues and bronze carvings in England and Italy (Figure 12-11).

Heavy particulate materials, particularly carbon soots, may not be very active chemically, but their presence has been the subject of numerous complaints because of soiled laundry and the general griminess that settles on windows, automobiles, and

FIGURE 12-11 *Bronze statue in Rome, showing the corrosive effects of pollution in the air and in rainfall.* (Courtesy of Dr. Eugene A. Eisner from Photo Researchers, Inc.)

even oneself. Families living near busy landing strips have had to adapt to a continuous black rain of carbon, generated by the inefficient combustion of jet engines as heavy airliners struggle to leave the ground. Residents can easily attest to the dirtiness of their cities, even if they cannot accurately measure the fallout rate of particulate matter. There seems little doubt that any person could rapidly differentiate a heavy industrial city from one that depended solely on service industries with nothing more than a superficial inspection. For better or for worse, these heavy, highly conspicuous particulates command public attention and are identified most readily with the air pollution problem. Some agencies still measure pollution by estimating particulate concentration, as crude and ultimately meaningless as it is.

Biological Effects of Air Pollution

Evidence is accumulating that biological systems can be affected radically by air quality. Both human and various plant tissues are known to be sensitive in different degrees to atmospheric pollutants. Because California is so diverse and important agriculturally and because air pollution has a long history there, more is known about crop damage there than elsewhere. Some crop plants are extremely sensitive to air quality, particularly to the products of automobile exhaust. Commercial crops of spinach and lettuce, two leafy vegetables whose marketability depends heavily on their appearance, can be ruined. In the famous smog belt of southern California, spinach has been eliminated by smog in the Los Angeles–Riverside basins, and citrus trees that have survived residential development have been partly defoliated and have suffered reduced productivity. In all these cases, smog damages the leaves and impairs photosynthesis. Consequently growth is reduced, and in extreme cases the plant is killed.

Considerable national publicity also has been devoted to the effect of air pollutants on pine trees in the mountains east of San Bernardino. Here, where topography and meteorology combine to produce the worst possible conditions, ponderosa pines have been defoliated in large numbers, largely by ozone. When weakened by air pollution, the trees are all the more susceptible to other enemies such as bark beetles and various diseases or to environmental stresses such as drought.

There is, of course, considerable variation in individual susceptibility, just as there is between different, even closely related, varieties. A ponderosa pine may be badly damaged, but an adjacent tree of the same species may be virtually untouched. Members of the cabbage family are generally resistant, as are melons, wheat, barley, and apple trees. While ponderosa and Jeffrey pines are highly susceptible to damage, Monterey pines are much less so, while palms and redwoods are quite resistant.

The evidence for smog damage in animals is much less plentiful. Experiments with laboratory mice have been inconclusive compared with similar experiments using cigarette smoke, and the evidence for direct physiological effects at observed environmental pollution levels remains weak. However, an interesting indirect effect has been reported from England, where heavy soots formerly so blanketed the trees that moths protectively colored to blend in with the light-colored bark of trees were exposed to predators as bold white patches on the blackened tree trunks. As long as the situation persisted, the soot promoted the selection of a dark-colored variant of the same species whose color was much closer to the color of the sooty trunks; the lighter, more easily seen moths were eaten, and therefore fewer lived to produce more light-colored moths. This particular example is known as *industrial melanism*.

The scant evidence for direct pollution effects on animals should not be taken to mean that such effects do not occur. In contrast to most plants, animals do not have high surface-volume ratios of exposed tissues except for the lungs. Skin surfaces are usually quite impermeable and insensitive to air pollutants; hence effects logically would appear first in the lung tissues, where they would be more difficult to detect and to definitely link to pollution. Respiratory systems also would have the ability to filter out these particulates and various chemical substances since they do so normally. Unlike plants, then, these organisms might be protected at least partially from pollution damage. However, this does not deny that many animals may be staggering about with diseased lungs.

Air Pollution and Human Health

Enormous effort has been expended on the search for air pollution effects on human health, centering on various lung and respiratory diseases. Health studies have taken three forms: studies of specific cases, often via autopsy; laboratory experiments with animals; and correlation analysis, which uses statistical methods to link known levels of pollution and known health problems.

It is not unusual to find highly abnormal respiratory systems in autopsies. For some special cases of pollution—chronic heavy cigarette smokers, coal miners' "black lung"—lung degeneration may be so severe that the lung is no longer functional. Studies on laboratory animals, especially mice and guinea pigs, show how a respiratory tract will respond to chronic exposure to various pollutants. Prolonged exposure can cause lesions or stimulate abnormal growth, or it can produce emphysema (the replacement of the elastic tissues of the lung by more rigid networks of fibers so that the affected parts of the lung become nonfunctional). This sort of laboratory research, however, is not evaluated easily. Mice especially have been inbred so thoroughly that by careful selection of the proper strain one could show almost anything. And there is always the problem that rodents and human beings are different, and so results obtained from the former might not apply to the latter. Correlation analysis, on the other hand, attempts to show that there is a statistically significant association between degree of pollution concentration and some standard of health effect. For example, one could attempt to correlate mortality from lung cancer with changing levels of pollution over a number of years, or with differences in pollution levels in various places.

The strength of evidence discovered thus far depends on the type of effect observed. There has been a history of mortality associated with acute episodes of air pollution, as we have already seen (Table 12-4). These dramatic incidents are unequivocal but have become rare; hence they are important mostly in a historical sense. Far more difficult to deal with are potential chronic health effects, which are easily confounded with other environmental hazards and with aging. These are difficult to trace unequivocally to air pollution per se.

TABLE 12-4 *Recorded episodes of mortalities related to acute air pollution. The deaths are recorded as excess deaths, which are those remaining after the normal deaths are subtracted from the total observed.*

Year of episode	Location	Excess deaths recorded
1880	London	1000
1930	Meuse Valley, Belgium	63
1948	Donora, Pennsylvania	20
1950	Poca Rica, Mexico	22
1952	London	4000
1953	New York City	250
1956	London	1000
1957	London	700–800
1962	London	700
1963	New York City	200–400
1966	New York City	168

SOURCE: From *Atmospheric Pollution* by W. Bach. Copyright © 1972 by McGraw-Hill, Inc. Used with permission of McGraw-Hill Book Company.

One source of uncertainty involves the mechanisms by which air pollution can affect health. The possible mechanisms are not well known, although it seems probable that air pollution acts in a manner like cigarette smoking. Under this hypothesis, one might expect that air pollution effects could be seen first as an increase in the incidence of such lung diseases as lung cancer, asthma, pneumonia, and emphysema. Secondary effects of lung degeneration might include increase in the incidence of heart disease as falling lung capacity increased demands on the heart. Available statistics from the U.S. Public Health Service (plotted in Figure 12-12a) do not specify emphysema as a separate disease and show increases only for "broncho-pulmonic disease" and respiratory cancer. California data (Figure 12-12b) list emphysema separately and show rapid rises in its incidence and in respiratory cancer.

A second problem concerns the meaning of "correlation" and puts us into a circular causal pathway in reasoning. For example, the mortality data suggest trends, but they do not necessarily imply that air pollution is the causative agent. Let us assume the data are gathered and analyzed carefully. Unless there is solid information linking air pollution and one of the diseases *causally*, an observer ultimately must make an arbitrary judgment about the meaning of the trend with respect to air pollution—and this is an implicit hypothesis about causation. Then to claim an important correlation is circular reasoning and may prove to be spurious without other supporting evidence (that is, evidence that other factors are responsible for the observed effects). In our specific example we could hypothesize that the emphysema death rate is a function of air pollution concentration. But we would have to show that higher emphysema death rates in the industrial areas of the city where a higher level of air pollutants would be expected are (1) not the result of higher concentrations of people who might be too poor to afford adequate health care, (2) not because of greater exposure to environmental hazards than in the suburbs, or (3) not a change in recognition of the disease.

Strong emphasis on the correlation-causation problem is necessary because the desire to prove a connection between air pollution and human health is great.

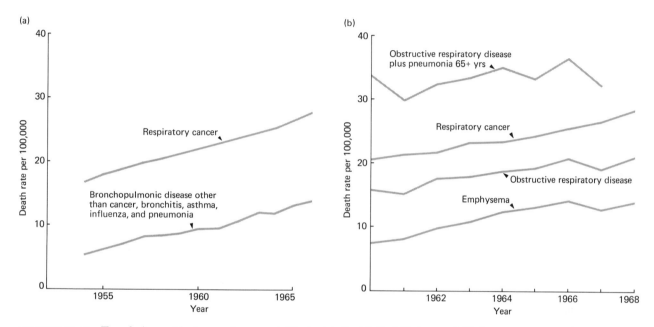

FIGURE 12-12 *Trends for respiratory system mortality in (a) the United States, and (b) in California.* (Data from (a) U.S. Public Health Service, and (b) California State Department of Public Health.)

Even with these strict cautions, however, some studies have suggested that at least a segment of the population is susceptible to pollution damage. Winkelstein et al. (1967) evaluated air pollution and health effects in Buffalo, New York, paying specific attention to the confounding effects of economic status. They divided Buffalo into sections by average air pollution concentration (measured as suspended particulate matter) and by median family income; they found (Table 12-5) that within each economic level average annual death rates for some lung diseases rose with increasing air pollution. This suggests that air pollution is a possible cause of elevated death rates, but in common with all statistical approaches it does not rule out other causes, nor does it mean that median family income is an adequate correction factor for health risks from other hazards.

In their review of existing literature, Lave and Seskin (1970) attempted to deal with the causality problem. They concluded that there is "an important association between [chronic sustained] level of air pollution and various morbidity and mortality rates." They emphasized that little evidence implicates other

TABLE 12-5 *Average annual mortality rates per 100,000 population for white males 50 to 69 years old in Buffalo, New York, and environs. The data are only for cases in which chronic respiratory disease was specified as cause of death, 1959–1961. Note that the mortality rates rise as air pollution level does for all economic levels, if not always smoothly. The important point is that the air pollution effect is more pronounced than the economic one.*

Economic level	Air pollution level			
	1	2	3	4
1	NA[a]	0	126	188
2	64	75	96	105
3	NA	65	51	103
4	35	47	114	NA
5	42	63	0	NA

SOURCE: Winkelstein et al., *Archives of Environmental Health*, Vol. 14 (1967), 162–171. Copyright 1967, American Medical Association.

[a]NA = not applicable.

possible causes, especially when all evidence is taken into account. Because they recognized the basic weakness of correlation analysis for specifying causal factors, they claimed only a significant association; but they still estimated the direct cost of air pollution (both in premature deaths and lost working time) to be the staggering figure of 2080 million dollars if air pollution were only half as severe as it is now.

On balance, there seems little doubt that air pollution is a health hazard. Adequate laboratory evidence shows that air pollutants of various kinds can damage tissues and cause their degeneration. This is especially true for the heavy metals, the more complex secondary components of photochemical smog, and sulfur dioxide. Hence one might expect some health effects when these substances are released into the air, but it is quite another matter to conclude either that certain health problems contain pollution effects or that pollution accounts for a certain percent in a world of multiple causality. Air pollution may contribute to a shortened life or select against those predisposed to respiratory diseases, but whether or not it is a significant contributor to chronic-disease mortality remains controversial.

Lave and Seskin believe that the evidence linking health and air pollution is adequate and that air pollution *is* a significant environmental hazard. Is the controversy to be solved unequivocally? Is better evidence likely to be available soon? What are the risks of a wrong policy decision? Without much ability to experiment with human subjects, significant improvement in the evidence does not seem likely. This leaves us to make the best decision we can on the basis of present evidence. To act slowly might risk human health, which, as we have seen, is not popular; to act quickly could result in a great socioeconomic disruption without any producing immediate health benefits.

Air Pollution and Aesthetics

Granted that the health issue is important—we are all quite aware of our mortality—the aesthetic issues may prove to be the catalysts that finally provoke action, so long as the public measures the severity of pollution by its visibility. As superficial as it may seem, much money has been spent to reduce the industrial output of airborne particulates and black smoke. Industry has its electronic precipitators to collect dust particles, and airlines have fan-jet engines to cut down on smoke at take off; both proudly publicize their concern in advertisements that proclaim how much action is being taken to fight pollution. This does not mean, however, that reducing the concentration of suspended particulates is a worthless objective *unless* the only thing accomplished is the replacement of a visible pollutant with an invisible one.

Of course, it is often much cheaper to change the composition of the pollutant output than it is to suppress it. For example, the reduction of particulate output in a badly functioning internal combustion engine can be accomplished by tuning, but at the expense of increased output of lighter-weight carbon oxides. The addition of pollution control devices to internal combustion engines reduces carbon oxides, but with a commensurate increase in nitrogen oxides. Ironically, in most cases the less visible products (especially the nitrogen oxides) are by far more serious, and so we have actually worsened the situation by reducing visible components. This certainly underscores the danger of allowing aesthetic considerations to be the most important criteria in establishing standards for pollution control.

A Case Study: Air Quality in the Los Angeles Basin

When the conversation turns to Los Angeles, which do you think of first: Disneyland, Hollywood, or smog? Los Angeles is as famous for its smog as for the other two. Los Angeles has a highly visible photochemical smog problem, but visibility has one important ancillary benefit. Since its smog problem developed long ago, Los Angeles has had an air-quality monitoring system since the early 1960s, longer than most other places. In addition, the air pollution problem has been studied more intensively in Los Angeles than anywhere else. Hence this locality is a logical place to examine the observed interactions of air pollutants with environmental variables and to relate them to public health.

Data from the Los Angeles County Air Pollution Control District (LACAPCD) do not pinpoint individual sources, but they give a general picture of the air over the urban area. Observed concentrations are higher in the winter than in the summer, but because the weather conditions that favor both temperature inversions and photochemical smog are more frequent in summer, the effects are likely to be more severe then. The combined effects of weather, topography, and source strength have the effects expected from previous discussion (Figures 12-13a, b). The two areas of greatest source strength are downtown Los Angeles to the north and the Long Beach-Signal Hill industrial areas to the south, with both concentrations being displaced downwind as much as the hilly topography allows. For the Los Angeles basin this means that the Long Beach pollutants drift northward, while the downtown production is funneled eastward into the San Gabriel Valley.[1]

The northern concentration is mostly the result of automobile exhaust, while the southern concentration consists mainly of industrial effluents, especially from coal-burning power plants and oil refineries. While industrial sources are easier to control, the county also must cope successfully with the automobile problem merely to maintain present levels of air

---

[1] Recent evidence reported in *The Los Angeles Times* (28 August 1974) indicates significant leakage north into the San Fernando Valley and east through the Santa Ana canyon into the Riverside area.

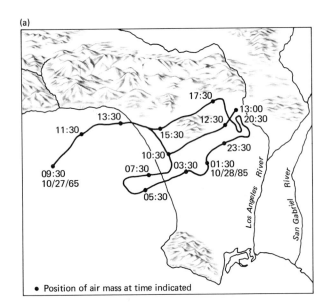 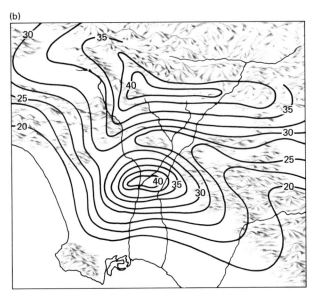

FIGURE 12-13 *The air pollution basin, Los Angeles–Long Beach. (a) The topography and wind circulation pattern for the basin for October 27–28, 1965. The air mass moves back and forth with land and sea breezes, but the general trend is east-northeast through the pass in the mountains leading to Riverside and San Bernardino. (b) The observed concentrations of oxidant, September 1960–64. The northernmost center of air pollution concentration drifts eastward; the southernmost center is blown northeastward.* (Redrawn from W. J. Hamming, W. G. MacBeth, and R. L. Chass in *Archives of Environmental Health* 14, pp. 137–149, 1967. Copyright © 1967 by the American Medical Association.)

quality. The predicted course of events from the statewide Air Resources Board (Figure 12-14) assumes that adequate control of automobile emissions will occur, an assumption that the board itself admits is not very likely. Even then, in the face of growing numbers of people and automobiles, and taking into account present and probable future technology, the Air Resources Board has warned that present air-quality standards in the Los Angeles basin will not be reached by programs designed solely to limit emissions. To meet these standards, drastic changes are required, such as a program that limits population or the various programs proposed by the Environmental Protection Agency to limit automobiles. For example, von Wodtke (1970) has used Air Resources

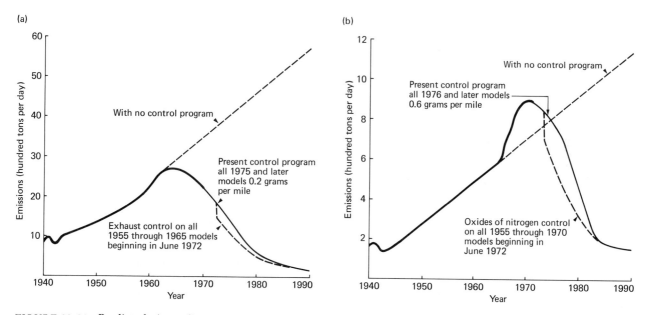

FIGURE 12-14 *Predicted air quality in the Los Angeles basin using three different assumptions. The air quality is measured as emissions in hundred tons per day for (a) hydrocarbons (HCO) and (b) nitrogen oxides (NO_x).*

Board data to calculate that the carrying capacity of the Los Angeles basin is only about 14 million, about 2 million more than the 1970 population. He calculated this figure by using current air-quality standards, probable future technological developments, and an estimate of minimum air volume required per capita. Of course, this is only a crude indicator because it is difficult to see how any volume of air per capita could limit a population.

Hamming and associates (1967) used data from the routine monitoring program of the LACAPCD[2] to estimate the public health consequences of the air pollution problem. They estimated that at the present rate of increase in air pollution (measured as the concentration of the sulfate ion), dramatic increases in mortalities related to air pollution could be expected in the near future; they predicted that pollution deaths would occur approximately 9 percent of the

---

[2] Los Angeles County Air Pollution Control District

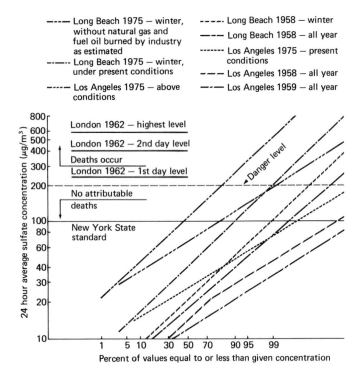

FIGURE 12-15 *Prediction of air pollution related mortality in the Los Angeles Basin.* (Redrawn from W. J. Hamming, W. G. MacBeth, and R. L. Chass in *Archives of Environmental Health* 14, p. 139, 1967. Copyright © 1967 by the American Medical Association.)

days (Figure 12-15). Their estimate makes an important assumption in that it did not predicate any specific relationship between mortality and air pollution but was based only on the supposition that sulfate concentrations known to have killed people in London in 1952 would prove to be equally effective in Los Angeles.

Faced by this prospect, it is easy to understand the pessimistic tone of the Air Resources Board: Without major reductions in source strengths, regional growth will likely increase the risk of disease related to air pollution in the basin. Some even advocate a health crisis as the most dramatic and effective way to expose the potential dangers of the present situation. Perhaps the publicity given to Los Angeles's smog has been effective, for in recent months the city has experienced a decreasing population.

Air Pollution Abatement: The Special Problem of Air Quality

Considered all together, air pollution has a pervasive, ramifying effect on ecological systems (portrayed as a block diagram in Figure 12-16). Perhaps the human health effects are not in themselves an imperative to reduce air pollution, but combined with potential climate effects, evidence of plant damage, corrosion, soot, and material destruction, a strong effort to reduce this fouling of our air is certainly prudent.

No element is more closely identified with the environmental crisis or more needs governmental action than air pollution. However clear the desires for clean-up may be, however, it is another matter to find an effective and equitable solution. Air pollution may share certain similarities with water pollution, but control is not one of them. The most troublesome difference is the mobility of air sources. Water borne organic wastes are routinely collected and treated to some degree before release into the environment, but it is impractical to collect or even monitor the output of the many millions of individual farmers, bonfires, and automobiles. The dispersion of many small sources is one serious challenge to any technological solution for air pollution.

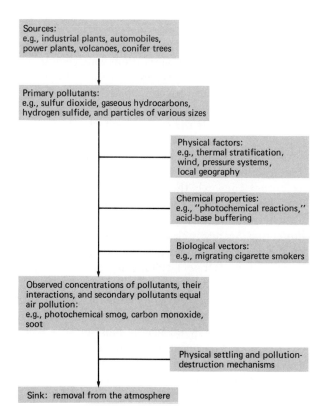

FIGURE 12-16 *Block diagram for air pollution, sources and effects.*

The Technological Situation

"Technological optimism" holds that pollution abatement can be achieved if sufficient scientific and engineering talent is applied to the problem. The reasoning is as follows: Since technology is responsible for pollution, suitable reversal of that technology, modification of production procedures, or if necessary, cessation of an unrectifiable process should provide an appropriate solution.

The technological solution is always the answer, provided we restrict our viewpoint to the productive process alone. But there is a serious flaw in this reasoning because

1. Technological solutions are always limited and directed by the prevailing social, economic, and political climate.
2. Technological solutions can provide imperfect panaceas if the criteria of solution are themselves imperfect.

A good example of the first limitation is provided by the saga of the internal combustion engine, detailed later in this chapter. The second problem with technological solutions can be seen in the aftermath of the London smog situation. Following two health episodes in 1952 and 1956, the London authorities dictated two changes: the replacement of coal furnaces by fuel oil for home heating and an increase in the height of smokestacks to suppress low-level inversions. This policy resulted in cleaner air over London and ended the health problem. However, Swedish scientists who have been measuring the acidity of their rainfall have claimed that the London solution has exported the problem downwind over the North Sea to Scandinavia. If this proves true, the pollution problem for London has been solved only if Scandinavia is not considered. Viewed in a slightly different way, Scandinavia now pays the costs of the London pollution abatement program.

A technological solution to pollution control can entail either new solutions or modifications of existing processes. Of course, modification is preferred by powerful and vested economic interests concerned with protecting an investment, and it may prove easier to implement when the novel solution is clouded in uncertainty. But modification is by no means the certain or best possible solution for all cases. For example, electrostatic precipitators are successful in removing particulates, and pollution control valves reduce vehicular hydrocarbon emissions, but these kinds of solutions can create other air pollution problems. The growing literature on acid rains suggests that the removal of particulate materials has increased the acidity of rain by removing elements which can neutralize the acids and thus are potential buffering agents, while the reduction of hydrocarbon output in automobiles (by more complete combustion of petroleum) has increased the emission of nitrogen oxides.

We conclude this section with the developing crisis of air quality in Los Angeles. The California Air Resources Board (ARB) believes that incremental technological solutions to improve the performance of the internal combustion engine are inadequate to meet 1975 standards; at best, they can prevent worsening of the situation, because more people and more automobiles threaten to overcome pollution reductions achieved by technology. The ARB suggested that three other solutions would prove necessary:

1. Limit the number of internal combustion engines.
2. Develop a comprehensive, nonpolluting urban transport system.
3. Limit population and industrial growth.

These suggestions outline a policy that does not depend upon incremental technological improvements; they strive instead for a more sweeping and challenging shift in transportation technology coupled to some basic socioeconomic changes.

The technical advisory committee of the ARB that made these recommendations realized it was proposing radical and challenging solutions which would not be implemented easily because they called for an unprecedented degree of governmental authority. Nevertheless, the three sweeping proposals above merit serious attention. If the strategy of technological improvement is favored by existing institutions and by our uncertainty about the future, we must be all the more careful to consider broader policy changes, for we do not know whether existing technology can be developed or whether it can be refined in time to prevent serious consequences. The Environmental Protection Agency is clearly pessimistic that air-quality standards can be met in Los Angeles by 1975. Its proposal for gasoline rationing in the Los Angeles basin reflects this attitude.

It is not popular either to doubt the potential of technology or to be overly pessimistic about it, since improved technology ultimately must make a major contribution to regaining air quality. Each of us must decide whether excessive pessimism about the potential of technology is a worse alternative than blind belief in its infallibility.

Social Mechanisms for Air Pollution Abatement

Different countries vary considerably in their approaches to air pollution control. Lund (1968) was singularly impressed with what he termed the "velvet glove approach" of European officials. He noted their dependence upon education, persuasion, and voluntary cooperation to reduce emissions from stationary sources. Although this approach had proved successful often, there were exceptions such as the Ruhr Valley of Germany, where air-quality problems remained serious despite such efforts. In Europe, pollution control functions usually are vested at the national level, but much of the responsibility actually resides with local inspectors; hence there is a piecemeal pattern of methods, approaches, successes, and failures.

The situation in Australia resembles that in Europe but with a much lower level of achievement. Cleary (1969) reported that authority was vested in each of the states, with wide variation in the results. The net impact has been scant with the possible exception of Victoria, and the federal government is only beginning to show signs of interest in air quality. Perhaps the situation reflects the present undeveloped and sparsely populated character of the country.

The United States has a strong local tradition in air pollution control. Municipalities and counties may form local authorities. Occasionally these may be combined with regional agencies if natural topographical or political unity favors them. The local agencies are the source of air-quality data, and they are empowered to take action against violators and impose bans under conditions of high pollutant concentration. The Air Quality Act of 1967 recognized local authority by restricting the federal role to research and the establishment of ambient air-quality standards, while local authorities were expected to find their own means to implement them.

With the passage of the 1970 Clean Air Act, which set ambient air-quality standards for 1975, the federal government entered air-quality management on a large scale. The Environmental Protection Agency is charged with enforcement of the standards, but EPA's strategy has been to require each municipality and region to implement programs of its own, stepping in only when a local plan is inadequate. The proposal for gasoline rationing is one example of EPA action; in other cases, the EPA has threatened to preempt local authority to control land use (zoning and building permits) if such developments contribute to worsened air quality.

The requirements of the Clean Air Act can be met in a number of different ways. Five general alternatives are outlined below.

AMBIENT AIR QUALITY STANDARDS

One of the easiest ways to devise an enforcement system is simply to adopt the same requirements that have been imposed by the Clean Air Act. This is a cheap alternative because monitoring costs are low: Only a few stations, or even one, may be needed to measure ambient air quality. But this method is usually ineffective because there is no sure way to determine which sources violated the standards. This simple procedure could be effective only when ambient air measurements really reflect the production of one or, at most, two stationary sources in an isolated air basin. Otherwise, ambient air standards imply no real control procedure, except for a lax or heavy-handed application of bans on classes of pollution sources.

EMISSION OR EFFLUENT STANDARDS

An alternative regulatory approach is to abandon dependence on ambient air standards (except in a general way) and switch to some sort of emission standards, such as those for vehicles. The regulatory agency would attempt to set emission standards at some level that would not exceed ambient air standards under average conditions. Once the standards are established, regulation depends on routine monitoring of sources (for example, an automobile inspection program) in order to detect and convict violators.

The problem resulting from applying effluent standards revolves around the same equity issues we discussed for water pollution in Grafton, Massachusetts (Chapter 11). With limited enforcement resources, who is monitored? Who is prosecuted when all are violators? Sources that are both numerous and highly mobile (such as the automobile) are particularly difficult to monitor, especially if the enforcement agency is short of both manpower and funds. If the agency decides to prosecute, say, 10 percent of the worst offenders, it may succeed only in finding the poorest owners who can afford nothing else.

POLLUTION FEES, EFFLUENT TAXES, AND
PURCHASE OF POLLUTION RIGHTS

These three examples represent "market" solutions to the air pollution problem and were discussed as potential methods of water pollution control (Chapter 11). While these proposals could work when coupled with a strong source-monitoring program, they are impractical for mobile sources that collectively are important but individually are insignificant.

Presumably appropriate fee schedules could reduce atmospheric pollution within the classic market mechanism of supply and demand. Many economists favor this type of solution for its self-adjusting efficiency that requires no external intervention, but other authorities contend that efficiency is not an important system characteristic. They point out that economic efficiency is likely to discriminate against the same economically weak, politically powerless, and legally resourceless people as any system of government intervention.

SUBSIDIES AND TAX INCENTIVES

Financial incentives represent the obverse of systems that use supply and demand to adjust fees. They may take the form of public bonds or subsidies to be used to develop pollution control devices, or they might appear as tax breaks for the installation or development of such devices. Practices like these are used widely in Europe, but at present subsidies are not popular with the American public; here there are widespread feelings that pollution control costs should be borne by industry.

A number of opportunities remain for some form of public assistance in the development of pollution control technology, and it seems likely that tax breaks or direct subsidies may yet become an important part of American programs, as they presently are in Europe. Direct subsidies and tax incentives could be made openly if favored recipients have friends in powerful governmental positions or if demand for rapid progress is urgent. However, covert subsidies and hidden deals are not unknown, and there is little reason to suspect they are absent from the environmental area.

TECHNOLOGICAL SUBSTITUTION

While subsidies for the modification of existing technology may not be favored in the United States, there is considerable support for the development of technological alternatives. Two examples, however indirect, include the development of mass public transit systems and of nuclear power plants. While the former may be built primarily for transportation benefits and the latter as a dependable future source of electric power, both have definite ancillary benefits for air pollution control. Public transit systems that offer publicly acceptable alternatives to the automobile can greatly reduce chronic levels of air pollution, and nuclear power plants would markedly reduce the demand for high-sulfur coals. Of course, such substitutions could produce a different set of environmental effects as deleterious as air pollution—high-speed trains could produce unbelievable urban sprawl, and nuclear plants would replace SO_2 with heat—but this is a characteristic of complex systems and is not unique to these cases.

Major technological innovations ordinarily require large initial investments for research and development, sums which are no smaller when the responsibility for development rests with public agencies. Underestimation and inflated costs are not solely found in the aerospace industry. The new San Francisco Bay Area Rapid Transit District (SFBARTD, or BART for short) has already cost 1.5 billion dollars, more than 2.3 times the initial estimate for only about half the originally envisioned project; it still is plagued by implementation problems that both raise costs and damage its public image. Since San Francisco is connected to the north and east only by bridges and it had become apparent some time ago that bridge construction could not keep pace with the ever-increasing vehicular demand, the question is not whether the investment was wise, but whether it was wisely spent.

Ideally, the best approach would be to combine public resources and private ingenuity in the European style, while emphasizing substitute technologies and improvement of existing production methods. If some proportion of public funds could be used by private enterprise with appropriate guarantees that any results would be accessible to the public and to other companies, specialized private companies with the most expertise could develop technological innovations at the lowest possible cost. Such an approach does not assure us automatically of a solution to the air pollution problem, and it must be watched carefully to prevent waste, but it does provide a mode of operation that combines the social objectives of government and the resourcefulness of private enterprise. This is subsidy, to be sure; but it is already in use as funds are diverted from highways to mass transit research.

Another Case Study: The Internal Combustion Engine

Highly integrated industrial societies depend on specialized production of goods and exchange of services, and these depend in turn upon a sophisticated and flexible mode of transportation. The development of the petroleum-burning internal combustion engine combined speed and flexibility to meet individual needs and to connect almost any two points. It is apparent that the automobile transformed the shape of cities, stimulated new industries, and changed American society.

Nearly all facets of American society are dependent upon the internal combustion engine. If we were deprived of it suddenly, the disruption would be enormous. Yet, the demands of the internal combustion engine for petroleum and the effects it has on air

FIGURE 12-17 *Poorly maintained automobiles can be significant contributors to air pollution; the harmful components, however, are invisible. Most of what you see here is condensing steam.* (Courtesy of Rohn Engh from Tom Stack and Associates.)

quality have made it the target of much abuse (Figure 12-17). Can the internal combustion engine be modified to ameliorate these effects, or must a substitute power plant be found?

The standard reciprocating-piston internal combustion engine is "dirty" because, no matter how it is tuned, it disperses a range of atmospheric pollutants. Rich fuel mixtures produce many particulates and hydrocarbons, while leaner fuel mixtures (containing lots of air) produce nitrogen oxides.

Engines that use very clean fuels (such as propane and butane) are not permanent solutions, since these fuels are not common. Variants of the internal combustion engine, such as the rotary or gas turbine engine, are not cleaner than the reciprocating engine.

In essence, then, the requirement for cleaner air is to intercept the combustion products other than CO_2 and H_2O and prevent their release.

Current solutions feature "hang-on" systems that depend on tuning adjustments and accessory equipment to recycle exhaust emissions. A promising development in the near future is the catalytic afterburner, which would complete the combustion process after the emissions leave the engine. This approach has at least three faults, however; these systems require regular maintenance, they increase fuel consumption, and they contribute to lessened engine performance. If possible, a maintenance-free system would be preferable, if for no other reason than to eliminate inspection programs that are both expensive to run and often subject to favoritism, patronage, and corruption.[3]

Quite apart from the technical considerations is the question of how technological improvements are to be implemented. The usual practice is to phase in new pollution control devices by requiring them on new automobiles. Implementation then is completed as older automobiles gradually disappear. In California, however, there also has been some pressure to require installation of newer devices on older vehicles. Policies retroactive to older vehicles contain, however, the same inequities that we have discussed for pollution fees and effluent standards.

Public disillusionment with the ability of the internal combustion engine to serve transportation needs without destroying air quality has created growing pressure for potential alternatives like public mass transit, steam and electric vehicles, and bicycles. All

---

[3] In 1975 catalytic converters for hydrocarbons were required for new cars. When it was discovered that quantities of sulfuric acid resulted, the requirement was suspended.

of these undoubtedly will play roles in future transportation schemes, but none is a specific substitute for the privately owned internal combustion vehicle. Public transit is the most efficient mode between specific points, but it cannot function effectively for individual needs. On the other hand, electric and steam power do not possess the range of gasoline engines. Perhaps one answer lies in emphasizing each mode in areas where they are most suited (for example, mass transit between and within city cores; individualized vehicles in suburbia and rural areas). The diesel engine is another possible alternative, and some domestic manufacturers favor the rotary engine as being most easy to change to. Furthermore, the sole rotary engine available in 1973 was one of two to meet 1975 Clean Air Act Amendment emission standards; but this is true not because the rotary engine is cleaner, but because its high power-to-weight ratio means a smaller engine and more room for "hang-on" antipollution equipment. It would be well to remember that this solution is paid for in increased fuel requirements, and only a comparison of the systemwide effects of more fuel versus dirtier air can tell us which alternative is less attractive.

The other engine to meet 1975 standards is the Honda *stratified combustion* engine, which is a reciprocating engine with redesigned two-part cylinders; it gets its name from the fact that the fuel mixture is not uniform but is richer in the prechamber containing the spark plug and leaner in the main body of the cylinder. The stratified engine has been shown to produce much smaller volumes of air pollutants because combustion in the prechamber produces hydrocarbons that subsequently are burned in the main chamber with practically no nitrogen oxides. Another possibility is a modification being developed at Cornell University. Appealing in its simplicity, this engine uses six cylinders in a standard V-8 as the main power plant and the last two as an afterburner. Its inventor claims the engine also will meet 1975 standards easily. It remains to be seen whether such modifications will be developed successfully. Further in the future, awaiting technological breakthrough, are hydrogen power and fuel cells. Both of these energy sources would be highly mobile, but both depend upon lower costs for hydrogen and the development of nuclear or solar power.

A Final Prospectus on Air Pollution

Growing public awareness of air pollution is a major element of environmental concern. Air pollution also exemplifies the problems that face policy-makers when the causes are deeply ingrained in economics. Many difficult choices between personal economic gain and environmental quality remain unrecognized by the general public. A steadily growing catalogue of pollution effects adds a sense of urgency to the search for appropriate solutions, but it does not necessarily suggest one.

The ideal situation is one in which air pollution reduction can be effected without the threat of serious economic dislocation, but this is surely a rare situation. Transportation, power generation, industrial production, and even agriculture contribute air pollutants while they supply goods and services. The painful choice is one of priorities: Is the environment worth economic sacrifices? When does the refusal to make a sacrifice now lead to even higher costs later? It is useless to pretend that such basic conflicts will be resolved easily, and there are signs of growing polarity. Environmentalist organizations have enjoyed an all too brief period of growth; in recent months opposing organizations have been founded. How to combat pollution is one permanent bone of contention between the two forces.

We shall close with an intriguing suggestion made in a paper by Bohn and Cauthorn (1972). They argue that air pollution can be used constructively to stimulate plant growth in the same way that has been suggested by others for waterborne wastes. They reasoned that normal air discharges could be dissolved in water instead and diverted into the soil, where microorganisms would break complex substances down into plant nutrients. Clearly, this suggestion is not applicable to mobile sources, but searching for constructive uses of airborne effluents is imaginative and could supplement other pollution control programs.

Questions

REVIEW QUESTIONS

1. What are temperature inversions? Why do they act as "lids" on pollution? How can inversions be broken up?

2. Even if producing constantly, particular pollution sources can create smog problems for only one day. By the next morning the problem is alleviated. Why? What circumstances might cause a smog problem to persist longer?

3. Some authorities have criticized mortalities from emphysema as a relevant measure of air pollution health effects on the grounds that emphysema is a popular cause of death today. How could one refute this argument?

4. Measurements taken along median strips of freeways often show that carbon monoxide levels frequently exceed the lethal dose, yet few people die from this problem. Why might this be so?

5. We noted in this chapter that ambient air standards could work under some restricted circumstances. What are some examples?

ADVANCED QUESTIONS

1. Construct a flow chart showing all the processes involved in the formation and destruction of temperature inversions.

2. If plants differ in their susceptibility to air pollution damage, can genetic selection be used to alleviate the problem?

3. Lichens are sometimes used as a biological indicator for air pollution. What precautions must be observed before attaching significance to the abundance of lichens in any one place?

4. List the assumptions of von Wodtke's calculations on the carrying capacity of the Los Angeles basin. If his model ultimately proves to be wrong, where are its weaknesses likely to be found?

5. Recently evidence has been presented that fluorocarbons from spray cans can escape to the stratosphere and break down ozone which screens ultraviolet rays. Explain (1) why the fluorocarbons can get into the stratosphere and (2) why increased incidence of skin cancer is expected.

Further Readings

Note: See also references on the economics of pollution and pollution control in Chapter 11.

Bach, W. *Atmospheric Pollution.* McGraw-Hill Series in Geography, New York, 1972.
    A short but comprehensive volume by a single author presenting his own views on health effects, meteo-

rological interactions, sources, and the other topics in this chapter, all thoroughly supplemented by references to the recent literature. This text is one of the finest summaries available; its level is only slightly higher than the one presented here, and I highly recommend it. Advanced students should see Stern (1968).

Bascunana, J. L. "Divided Combustion Chamber Gasoline Engines: A Review for Emissions and Efficiency." *Journal of the Air Pollution Control Association,* 24 (1974), 647-679.

A thorough review of the technology and air-quality potential of the stratified charge engine.

Bohn, H. L., and R. C. Cauthorn. "Pollution: The Problem of Misplaced Waste." *American Scientist,* 60 (1972), 561-565.

The authors' thesis is that the soil is the natural and most beneficial place to dispose of wastes.

Bryson, R. A. "All Other Factors Being Constant: A Reconciliation of Several Theories of Climatic Change." *Weatherwise,* 21 (1968), 56-61, 94.

Bryson, R. A., and J. E. Kutzbach. "Air Pollution." Association of American Geographers Resource Paper, No. 2, 1968.

These two papers present the views of R. A. Bryson and his colleagues and their evidence that air pollution is having a significant effect upon global climate.

California Air Resources Board. "Recommended Ambient Air Quality Standards." California Air Resources Board, Report of the Technical Advisory Committee, September 1970.

California Air Resources Board. "Air Pollution Control in California, 1970." 1970 Annual Report, January 1971.

Changnon, S. A. "The LaPorte Weather Anomaly: Fact or Fiction?" *Bulletin of the American Meteorological Society,* 49 (1968), 4-11.

This paper stimulated a flood of letters. See those by Landsberg, *Science,* 170 (1970), 1265; Holzman and Thom, *Bulletin of the American Meteorological Society,* 51 (1970), 335; Hidore, *Bulletin of the American Meteorological Society,* 52 (1971), 99; and the subsequent reply by Changnon, *Science,* 172 (1971), 987.

Cleary, G. J. "A Status Report on Air Pollution Control in Australia." *Journal of the Air Pollution Control Association,* 19 (1969), 490-496.

de Nevers, N. "Enforcing the Clean Air Act of 1970." *Scientific American,* 228(b, 1973), 14-21.

Hamming, W. J., W. G. MacBeth, and R. L. Chass. "The Photochemical Air Pollution Syndrome." *Archives of Environmental Health,* 14 (1967), 137-149.

Lave, L. B., and E. P. Seskin. "Air Pollution and Human Health." *Science,* 169 (1970), 723-733.

Lund, H. F. "An Evaluation of European Industrial Air Pollution Control Practice." *Journal of the Air Pollution Control Association,* 18 (1968), 586-589.

Mitchell, J. M. "Recent Secular Changes of Global Temperature." *Annals of the New York Academy of Sciences,* 95 (1961), 235-250.

National Academy of Sciences. *NAS Report on Technological Feasibility of 1975-1976 Motor Vehicle Emission Standards. A Critique of the 1975-1976 Federal Automobile Emission Standards for Hydrocarbons and Oxides of Nitrogen.* National Technical Information Service, 1973.

Stern, A. C. (ed.). *Air Pollution,* 2nd edition, 3 vols. *Air Pollution and Its Effects. Analysis, Monitoring and Surveying. Sources of Air Pollution and Their Control.* Academic, New York, 1968.

The outstanding technical compilation for specialists and research libraries. Also useful for more general readers for some of the papers and for the extensive bibliographies provided by the many authors.

U.S. Department of Health, Education, and Welfare. *A Digest of State Air Pollution Laws.* Washington.

Issued yearly, this is a summary of state laws regarding air pollution abatement, reprinted without comment. As such the series is a good source of raw materials for a project comparing different approaches to air pollution control.

U.S. Public Health Service. *Smoking and Health.* USPHS Publication 1103, Washington, 1964.

Winkelstein, W., S. Kantor, E. W. Davis, C. S. Maneri, and W. E. Mosher. "The Relationship of Air Pollution and Economic Status to Total Mortality and Selected Respiratory System Mortality in Men." *Archives of Environmental Health,* 14 (1967), 162–171.

13. A Question of Mounting Imbalance

Marx (1971) estimated that 7 million automobile bodies, 48 billion cans, 26 billion bottles, and 25 million pounds of toothpaste tubes find their way to the trash heap every year in the United States. On average, each person generates 5 pounds of solid wastes every day; thus about 190 million tons of solid waste are produced in the United States every year. This includes not only domestic wastes but also wastes generated by all industrial processes. To this volume of municipal wastes we must add 1.7 billion tons of mining wastes and 2 billion tons from agricultural sources (cattle feed lots, slaughterhouses, and such). These large volumes of solid wastes are highly visible (and often very pungent) reminders that human society must find ways to handle this ever-growing volume, in one way or another.

Despite these impressive statistics and the fact that some municipalities are running out of places to put their solid wastes, this area of environmental pollution has a number of promising developments which brighten the prospects for viable solutions considerably. Solid wastes are generally easy to contain; they often possess valuable substances that can be reclaimed; and they are, in general, particularly amenable to technological solutions. In the United States we may be proud that the whole problem of solid waste no longer is confined merely to its disposal. We are entering a transitional phase in which new technologies and social conventions are being explored, open dumps are being closed, and the comprehensive planning of waste management practice is moving ahead rapidly. In the future we may see the volume of waste decrease rapidly as most of it is used as a resource rather than as a waste; in this way the problem may practically evaporate. This chapter examines the reasons for such optimism.

Solid Waste Disposal in the Here and Now

Let us assume that we wish to manage solid wastes so that their environmental effects are minimized. What general choices are available? We could dump the wastes where their impact on us would be minimal, such as on ocean floors. We could bury our solid wastes and hope they stay out of sight; in combination, we could use various schemes that minimize the bulk of these wastes. Finally, we could try to reclaim part or all of the supposed "wastes." All these general methods are currently in use for various substances.

The predominant American practice in solid waste management is to use virgin materials and then to dispose of them without further reuse. In its most primitive—and unfortunately predominant—form, disposal is simple dumping. This method is simple and effortless and offers the least individual inconvenience, but its costs are infamous. One can find trash dumped along roadsides, on river banks, and in ravines in almost any rural area. Then there are the rusting automobile bodies collecting in small farm lots and junk yards. Beer cans turn up in the most unlikely places; in fact, the accumulation of aluminum cans along roadsides was a major reason for Oregon's "bottle law," discussed below. Casual litter accumulates in city streets, making city agencies responsible for picking it up. Municipal garbage dumps differ from casual disposal only in their systematic collection of the wastes and in their use of more formal management procedures. If we include variants such as burning, open dumps are the sum total of solid waste management for probably 95 percent or more of the country.

Today the drawbacks of open piles of garbage and trash are perfectly obvious, and this method of disposal is clearly undesirable; the low cost of collection and hauling is not enough to cancel out the undesirable environmental effects. For communities, the cheapest alternative to open dumps is to use landfill, in which solid wastes are compacted and buried in deep pits with alternating layers of soil, with or without prior incineration to reduce the wastes to the smallest possible volume before burial (Figure 13-1). Certainly landfill is far more sanitary than open dumps. However, it shares some of the disadvantages that open dumping has, and it brings at least one additional drawback. Landfill requires much larger areas of land than open dumps in exchange for the improved sanitation, and large areas such as these will

FIGURE 13-1 *The two most predominant disposal techniques: (a) the open dump, and (b) a landfill operation.* ((a) Courtesy of Lynn McLaren from Rapho/Photo Researchers, Inc. (b) Courtesy of Ray Ellis from Rapho/Photo Researchers, Inc.)

be limited in crowded urban areas. Moreover, just as in open dumps, dissolved wastes can enter the groundwater, and if incineration is used to reduce bulk, air pollution is exacerbated. For these reasons, even sanitary landfill is not a long-term solution.

If land-based disposal has no long-term future, why not dump our collected wastes in the oceans, which seemingly are unlimited? Unregulated ocean dumping from ships and private citizens may increase no matter what laws exist, yet the Council on Environmental Quality has concluded that the dangers of ocean dumping are so severe that it should be stopped wherever possible. In reaching this conclusion, CEQ (1970) cited the unknown effects of ocean dumping on marine ecosystems and pointed out that the possible consequences of this practice on these systems could be great. Of special concern are toxic chemicals and radioactive wastes and the possibility of massive deoxygenation of deeper waters should there be a general increase in ocean dumping. It is clear that if we permit ocean dumping to increase, we shall have done nothing more than transfer our disposal problem from one site to another.

To sum up, present practices of solid waste management that emphasize disposal can be no more than transient solutions and ultimately give way to better ones. Environmental consequences of our present disposal policies are known well enough to be considered real; moreover, as resources become more obviously limited, it becomes increasingly urgent that we find constructive uses for solid wastes.

Transitional Solutions in Solid Waste Management

The 1965 Solid Waste Disposal Act may be credited as the first major step toward improved solid waste management. Including the 1968 amendments, the act made solid waste a federal responsibility as well as a local one, increased the role of environmental protection as a factor in solid waste management, increased the funding for planning programs and technological improvements, and made federal agencies responsible for program development.

The immediate consequence of the Solid Waste Disposal Act was to stimulate state and regional solid waste planning under federal funding. At present, solid waste management authority is centralized in the Environmental Protection Agency, which has provided funds to communities, regions, and states in order to encourage the development of solid waste disposal plans. Such funds can be used to build facilities; to develop collection, disposal, and management procedures; or to engage in other activities relevant to solid waste management, especially methods that would minimize environmental impacts. These particular EPA funds favor programs that emphasize practical management plans rather than those that stress more innovative research, but they are a logical step in solid waste management and are aimed at all aspects of waste management, from the private dump to the most sophisticated recycling program.

By 1972, almost every state had participated in developing better solid waste disposal programs under the provisions of the 1965 congressional act, and 32 states had actually completed plans for solid waste management (Table 13-1). The Oregon plan is a typical example of the designs being proposed by the states. Basically, this plan is an attempt to systematize solid waste disposal statewide. It encourages uniform practices throughout the state and contains mechanisms to make new technological developments available quickly to city and county authorities.

The Oregon plan calls for an integrated process to cover all phases of solid waste management: collection, storage, transportation, and disposal, preferably handled as a public utility in the same manner as water, light, or power. As part of this process, the plan calls for the coordination of plans developed by local, county, state, and even interstate agencies, and for the provision of a sufficiently long lead-in time to allow for acquisition of future disposal sites. The plan emphasizes the need for a stronger role for county governments in planning and enforcing solid waste regulations, primarily because county governments must deal with volumes of solid waste that originate in cities, where the agencies have no real authority.

The Oregon plan relies heavily upon current solid waste practices in the state. It calls for the continuation of private operations and suggests some revisions that either use new developments in disposal technology or curb the worst current abuses. For example,

TABLE 13-1 *The status of solid waste facility planning in the United States, 1972.*

State	Solid waste laws	Rules and regulations	Disposal permit required	Political subdivisions		State (cont.)	Solid waste laws	Rules and regulations	Disposal permit required	Political subdivisions	
				Technical assistance	Financial assistance					Technical assistance	Financial assistance
Alabama	×	×				Nevada	×				
Alaska						New Hampshire	×	×	×	×	
Arizona	×	×				New Jersey	×	×	×	×	
Arkansas	×					New Mexico		×	×		
California						New York	×	×		×	×
Colorado	×	×	×	×		North Carolina	×	×			
Connecticut	×	×	×	×	×	North Dakota		×		×	
Delaware		×	×	×		Ohio	×	×	×		
District of Columbia		×				Oklahoma	×	×			
Florida		×				Oregon	×	×	×	×	
Georgia		×				Pennsylvania	×	×	×	×	×
Hawaii	×					Rhode Island	×	×		×	×
Idaho		×				South Carolina		×	×		
Illinois	×	×	×	×		South Dakota	×	×		×	
Indiana	×					Tennessee	×	×	×		
Iowa	×	×	×			Texas	×	×	×	×	×
Kansas	×	×	×	×	×	Utah		×		×	
Kentucky	×	×	×	×		Vermont		×	×	×	×
Louisiana		×				Virginia		×	×	×	×
Maine				×		Washington	×	×	×	×	×
Maryland	×	×	×	×	×	West Virginia		×		×	
Massachusetts	×	×	×	×	×	Wisconsin	×	×	×	×	
Michigan	×	×	×			Wyoming	×			×	
Minnesota	×	×	×	×		American Samoa					
Mississippi		×				Guam					
Missouri		×				Puerto Rico					
Montana	×	×	×	×		Trust Territory					
Nebraska						Virgin Islands					

Source: U.S. Council on Environmental Quality, Third Annual Report, 1972.

one recommendation of the plan suggests that compaction and volume reduction be prerequisites for disposal at public sites, while another calls for development of a network of collection stations to serve as convenient transfer points between sources and collectors in rural areas where individuals usually have been responsible for their own waste disposal. The first suggestion is aimed at minimizing requirements for fill land, while the second is considered to be the least expensive way to completely centralize public costs and requirements for fill land, and to control all waste collection and disposal. As such, the plan provides a disposal program aimed at eliminating roadside dumps while minimizing its cost to the public.

In its present form, the Oregon plan recognizes the limits of current technology, economics, and public attitudes. It emphasizes improvements that are implemented easily with little social or technological change. Nevertheless, the Oregon plan does provide for the use of new technological developments as soon as they are available, and even more important, it stipulates that the state plan shall be updated regularly as additional information on waste management becomes available. Some of the mechanisms to improve waste management practices which the plan suggests should be evaluated include: the formation of a single solid waste control agency, the use of state power to prosecute violators of acceptable solid waste practices, and the designation of specific disposal areas in the development of new planning codes.

Why is the term "transitional" appropriate for developments in solid waste management such as the Oregon plan? Because though the future appears to contain some innovative developments that might make the dump an obsolete institution, all of these require further development to be useful on a large scale. Therefore, current solid waste planning cannot rely heavily on future potential, and for this reason the state plans must propose refinements and improvements on what is basically an old solution. Thus systematic planning for appropriate dump sites and for ways to minimize their impacts is both appropriate and wise until better methods are proven. The Oregon plan contains the commitment to change as developments occur; in this sense, it is not only transitional but also innovative.

A plan developed by the Office of Solid Waste Management Programs (OSWMP) for four counties in the Appalachian Mountains of North Carolina is typical of a regional approach that has more specific implementation plans than state-level planning. Before the plan of disposal was developed, 168,000 people used 24 authorized sites (23 of which were for open-air incineration only), 325 scattered dumps, and most often any convenient spot to dispose of their wastes. There was little organized collection and no regional authority, a large area to be served (1679 square miles), and extremely dispersed waste generation. OSWMP decided that plenty of sites were suitable for landfill disposal, and following this basic decision, they delineated solid waste disposal zones of various sizes on the basis of population and the road network. They then did a cost analysis of these different-sized zones in order to determine which size would be operated at the least cost. They did this by using costs associated with collection and with landfill administration to calculate the cost per ton of waste disposal, and since it turned out that the size of the dump dominated all other considerations, OSWMP selected large zones with big dumps as the best alternative. Next OSWMP set about to determine the cheapest collection scheme from a set of three choices. The first alternative made individuals responsible for carrying wastes to the landfill site. The second, in essence, involved maintaining the existing pattern: In some places individuals were responsible for their own disposal, and in others organized collection and disposal already existed. The third alternative, and the one finally chosen, involved placing large bulk containers at convenient spots. Individuals would dump their wastes into these, and the wastes would be collected regularly and transported to landfill sites. The next step was to locate landfill sites; this was done by selecting the point within each zone that would minimize how far collection vehicles had to travel. Finally OSWMP proposed implementation of the plan by a new regional commission set up to be responsible for providing and placing the bulk containers and for funding and operating the entire system.

Innovations in Solid Waste Management

The major innovative research in solid waste management has been aimed toward recovering the tons of valuable materials that are disposed of yearly. If we could achieve success to this end, we would be reducing our dependence upon declining resources and simultaneously solving the solid waste problem. The U.S. Bureau of Mines is one agency that is actively working in this area; its theme is that solid wastes are really "resources out of place," and they should be put to constructive use. Despite some problems of practical application, innovation is moving forward rapidly in several areas. The ones we shall consider are recycling, composting, using wastes for the construction of environmental structures, and fuel production.

SALVAGE, REUSE, AND RECYCLING

Contrary to popular belief, recycling, salvage, and reuse are all well-established practices—in fact, about 25 percent of all municipal solid wastes is recovered. Most of this, however, does not result from public participation in recycling programs. Much is recovered by the manufacturers who produce the waste. At the end of many manufacturing processes, there are wastes that can be recovered and reused to produce additional quantities of the product. This, plus wastes salvaged by industries that specialize in waste recovery, make up the majority of recycled material at present. Although the public could help to improve recycling of wastes, it is obvious that recycling has not been widely successful. This is due to several factors:

1. Generally the cost of virgin materials is at least as low as for recycled materials.
2. Past abundance of raw materials has favored technological developments that are specialized toward the use of raw materials.
3. Recycled materials are more difficult to collect than virgin materials, and moreover are more expensive to transport since wastes are more dispersed than most raw materials.
4. Recycled materials require labor-intensive sorting, whereas raw materials do not. This, too, adds to their expense.
5. Many modern products are complex mixtures that are not reusable and are nearly impossible to recycle.

TABLE 13-2 *Recycling of major categories of materials commonly found in municipal wastes.*

Material	Percent municipal waste	Percent recycled	Date of estimate	Recycling trend direction	Reason
Paper	40–50% by weight = 53 million tons	19	1969	downward	Virgin materials are cheaper, and technology favors virgin materials.
Ferrous metals	7% = 9 million tons; 106 million tons used each year	20	1967	downward	Publicly consumed scraps are mostly tin cans which cannot be used.
Nonferrous metals	<1% = 9.8 million tons	30.8	1967	upward	These metals are valuable and worth scavenging.
Glass	6% = 12.8 million tons	4.2	1967	downward	Most glass recycled from manufacturing waste; municipal glass is more expensive, much more than virgin materials.
Textiles	0.6% = 5.7 million tons	4.3	1968	stable	Textiles are extremely difficult to recover.
Rubber	1% = 3.9 million tons	26.1	1969	downward	Virgin materials are cheaper and more attractive; trend may change if tires are used in pyrolytic energy recovery.
Plastics	1.1% = 2.1 million tons and growing rapidly	unknown			

SOURCE: Environmental Protection Agency Report SW-29c (1972).

The factors set forth above account for the variable rates of recycling for various materials (Table 13-2). Metals, especially ferrous metals, have always been the most heavily recycled materials, because demand has been relatively high (Figure 13-2). Demand has been falling, however, since the newer steel-making processes can accept less scrap. In many industries, it simply is not economically feasible to use recycled materials. The paper industry is one in which new materials are cheaper, and this is true for practically every potentially recyclable material. Glass, textiles, rubber, and plastics present additional difficulties because they are difficult to separate from refuse mixtures. And plastics are recycled least successfully of all because most cannot be reformed for other uses. Technology does not hold much promise in this area at present.

Despite these difficulties, there clearly are opportunities for recycling to increase in importance. Economic conditions can change, and in fact it seems inevitable that raw resource supplies must get more expensive. Concomitantly, rising prices for resources can stimulate industrial processes designed to use recycled materials. Thus a chain of events favoring recycling is dependent at each point on shifts in prices that can occur at any time.

If changes favoring recycling are to occur, social changes may well lead the way. For example, federal agencies now use recycled paper whenever possible. The 1970 Resource Recovery Act provides subsidies for research into recycling, and Oregon has a landmark bill that requires beverage containers to be recyclable. The so-called "bottle bill," effective as of October 1972, provides that nonrecyclable beverage bottles and pull-top cans cannot be sold in the state. The bill was passed in response to a united electorate proud of their natural environment and intolerant of a growing problem of roadside litter (62 percent of which was nonreturnable beverage containers). Predictably, since the bill became effective, the roadside litter problem has practically evaporated, with litter down by 41 percent in late 1973, and with nonreturnable bottles almost completely gone. Economic impacts, however, are mixed. Since returnable bottles are cheaper to fill than new bottles are, both local industries such as breweries and consumers of such products have paid lower costs per unit of beverage; but distant beverage industries are at a competitive disadvantage since their transportation costs have risen sharply because of the added costs of moving empty bottles back to the factory for refilling. In addition, bottle manufacturers have suffered declining sales, although the conversion of their facilities to making refillable bottles may help, and grocers have complained loudly about the higher costs of handling returned containers.

FIGURE 13-2 *A metal reclamation facility in operation in Madison, Wisconsin.* (Courtesy of Daniel S. Brody from Stock, Boston.)

FIGURE 13-3 *A composting bin such as might be maintained by an organic gardener.* (Courtesy of Philip Jon Bailey.)

Oregon's bottle bill recognizes, although not in a quantitative way, the environmental quality is not merely an economic external. As we progress toward recycling more wastes, it will be important to remember that it may be bothersome to sort household refuse before collection, and recycling certainly will be more expensive as long as raw materials cost less, as environmental quality remains a common property resource and as industry uses methods that favor new resources. However, as recycled materials become more widely used, prices may actually drop, even while we reap environmental and resource benefits.

COMPOSTING

Composting is the degradation of organic material. It is practiced commercially on a limited scale in Europe and in India, and it is popular among organic gardeners in the United States. At the very least, composting could help reduce the volume of solid wastes that have to be disposed of; at best, it could be a significant source of soil conditioner and fertilizer, a useful recycling of organic wastes.

Unfortunately, composting has not fared well in any large-scale application. After the decomposition process is completed, the compost is a very low-grade fertilizer and is most useful as a moisture-holding soil amendment. Composting has not proved successful commercially, primarily because not only are the initial costs for a processing plant high, but the market price per ton is low, and markets for the product are almost nonexistent. While there are composting plants overseas, this method does not seem commercially feasible in the United States. In fact, even where composting has been commercially successful, whenever commercial fertilizer has become widely available at a reasonable cost, composting has fallen into disuse. This, of course, would be true only of areas where the compost had been used primarily as fertilizer. Many plants currently operate not because of the commercial success of the product but instead because in these particular situations composting has been found to be a suitable method for the disposal of solid wastes.

These conclusions do not necessarily apply to small-scale operations in which individuals do the composting with no large investment in equipment. In such cases, individuals provide their own guaranteed market and supply their labor free (Figure 13-3). Moreover, though composting is not commercially feasible in this country at present, McGauhey (1971) has made an ingenious suggestion that would allow us to readily develop a compost-production industry at a later date if it became more feasible. He suggests that at this time we adopt sanitary landfill as the preferred disposal method, since sanitary landfills are cheaper than composting. Because this method in effect begins the composting process anyway, we can afford to table the question of composting until some future time when conditions may favor commercial production. McGauhey suggests that these landfill sites may well be future mines for potential compost, which, when needed, can be unearthed and finished off aerobically in a short time (commercial aerobic composting can be accomplished in as little as two or three days).

Building Environmental Structures from Solid Wastes

Let us consider a different approach to solid waste management that does not depend upon appropriate economic conditions to make it work or upon a well-developed technology to separate complex solid wastes mixtures. Instead of speeding decomposition of the waste, and burning it or compacting it, we could devise methods that actually take advantage of the slow decomposition of many solid wastes; by putting them in the proper places, solid wastes can be used to construct special environments. The advantages are evident immediately. There are no problems of individual incentive, since the refuse collection systems already exist and only need to be coordinated. No new technology is required: The machinery to handle the wastes already exists. Finally, this may result in minimal environmental quality problems as long as the selection and placement of wastes is carried out carefully.

Constructive use of solid wastes is not a new idea. In the past, automobile bodies and the like were used to stabilize river banks and to serve as a framework for temporary dams. Each year there are more automobiles on the roads, and more automobile bodies end up rusting in junk yards, farm lots, and gullies. Why is it not possible, with only a little less convenience and a bit more cost, to use these hulks, along with old washing machines, refrigerators, and other large chunks of metal, to slow stream erosion, to provide cover for fishes in lakes, and even to improve potential trout habitats? Is concrete actually more dependable and longer lasting, or is it just our aesthetic sense or old habits that cause us to prefer poured concrete?

Besides their uses in waterways, solid wastes have been used to build special recreation environments. The city of Munich, Germany, built a mountain of trash and World War II rubble and then turned it into a gigantic landfill by covering the whole pile with earth. The small mountain they created is now a park and monument to war victims. Virginia Beach, Virginia, used an EPA grant to support a similar effort; the city has built a landfill mountain and appropriately named it "Mt. Trashmore." Perhaps with the use of a little imagination, new uses can be found even for old landfill sites. Moreover, if cities are to use landfill as a disposal method in the near future, we should allow our imaginations to run freely and attempt to devise even more novel uses for landfills, perhaps on a still larger scale. Consider this scenario: The State of Wisconsin contracts with Chicago, Detroit, Toronto, Cleveland, and Milwaukee for their solid wastes for 20 years, hires a fleet of barges to collect it from these cities, and transports it to Green Bay, from where it is hauled north of Rhinelander. There it is built up into Wisconsin's tallest mountain and leased out for the development of ski resorts to serve the entire North Central region of the United States. Besides the profit gleaned from fees cities pay for garbage removal, there are also the potential profits from the mountain resorts that would be constructed on the mountain. Certainly, this plan could have its drawbacks, not the least of which might be the potentially great costs involved in transporting wastes on such a grand scale for great distances; but despite the problems in such a scheme, regional development based on garbage is at least an intriguing idea.

In a more realistic context, cities running out of sites for landfill operations are looking for innovative ways to use their solid wastes. San Francisco is investigating a plan to transport its solid wastes to the Sacramento–San Joaquin River delta, where it would be used to rebuild islands that are steadily subsiding. The subsidence problem results from the highly organic nature of the alluvial soils of these islands, which are easily compacted when plowed and worked for agriculture. As a result, entire islands are settling at the rapid rate of 5 centimeters per year. Landfill on the delta islands that uses organic solid waste would compensate for subsidence. Without some compensation, continued subsidence will decrease the safety margin provided by the levee systems that presently are the only defense against submergence of these islands, many of which are already below water level. If massive flooding occurs or a more extensive levee system must be built to prevent this from occurring, the higher costs resulting from either could reduce the attractiveness of continuing what has been a highly productive agriculture in the delta. The economic damages that would result are a compelling reason to seriously consider using organic wastes for delta landfill. This plan constructively uses solid wastes, and collections require no more than rough sorting. Whether or not the delta islands should even be preserved for intensive, specialized agriculture or allowed instead to revert to wildlands and wildlife habitat that is sometimes flooded and sometimes dry is important, but quite another issue.

Other possibilities for intelligent placement of solid wastes have barely been explored. The Japanese are using it to build artificial reefs for fish cover in their Inland Sea, and other biological uses should be easy to discover. Indeed, other uses for various materials are appearing all the time. Instead of reclaiming such items as bottles for the same uses or reprocessing them as a source of reclaimed raw materials, waste glass can be ground up and used in various ways. For example, it can replace asphalt in road surfaces, it can be pressed into glass bricks, or it can be blown into glass wool. As oil prices continue to skyrocket and with them the price of petroleum derivatives such as asphalt, enormous volumes of "waste" glass may suddenly become an economically valuable substance. Sawmill wastes and other wood scraps seem to be limited in their potential only by imagination, research, and economic conditions. Thus though it may seem fanciful for Wisconsin to consider building a ski industry on trash, ideas of this sort undoubtedly will be considered more seriously when we can regard solid wastes as a potentially valuable resource rather than merely as a disposal problem.

Energy Derived from Solid Wastes

One of the most promising uses of organic solid wastes is to convert them into energy. Two EPA-sponsored demonstration plants in Baltimore and in St. Louis use proven processes to convert solid wastes into fuel.

In Baltimore, a Monsanto-designed plant is being built to handle up to 1000 tons of waste daily; acceptable materials include nearly anything organic, including scrap tires and plastics. The process used is *pyrolysis* (destructive distillation in the absence of oxygen), which produces a mixture of low-grade petroleum-type gas and oil. Since these products are not very valuable, it is not economically feasible to transport them elsewhere, and therefore the Baltimore facility is designed to burn its own pyrolytic products in order to produce steam, which can then be piped to other facilities. Though the pyrolytic process itself is technologically proven, it is still not known whether there will be a ready market for the steam produced by the plant. But even in the absence of insured markets for the steam, the Baltimore plant offers the advantage of reducing the volume of solid wastes that must be disposed of. The St. Louis facility is a joint venture of the city and the Union Electric Company and is already in operation. The plant presently accepts only household wastes that have been screened for engine blocks, old stoves, and similar heavy metals. Thus here, too, the wastes are mostly organic. These wastes are shredded into small particles and separated gravitationally. The heavy fraction is then passed through a magnetic separation system, thereby allowing recovery of additional ferrous metal scrap; the remainder is discarded at present. The light fraction is fed into the boilers as fuel at the ratio of 1 part wastes to 4 parts coal, whence it is no different from any other coal-fired electric power plant. Though this plant might be considered by some to be technologically less innovative, the advantage of the St. Louis facility is that, unlike the Baltimore plant, its operation is proven; its success does not depend upon future events.

The Bureau of Mines is the federal agency most active in solid waste research, and much of its effort has gone into energy production from solid wastes. Some research has gone into pyrolysis, but the major attention has gone into the *hydrogenation process*, which chemically reduces cellulose to produce hydrocarbons—in effect, crude oil. The process requires high pressures (100-25 atmospheres) and temperatures (24-380°C.), carbon monoxide, and steam, as well as solid wastes. Yields have reached up to two barrels of crude oil per ton of solid waste, with a low sulfur content as an added bonus. Where solid waste can be sorted more carefully (so that it is predominately made up of such things as manure or animal waste), yields are correspondingly higher. The fuel produced by this method, unlike that from pyrolysis, is useful for a variety of purposes and is practically indistinguishable from fuel oil derived from natural crude. The principal problem to be resolved is that it is not yet known whether the fuel produced from solid waste will be economically competitive with natural fuel oil.

There appear to be three general alternatives in solid waste management:

1. Some form of disposal, without further use
2. Some form of reuse, resynthesis, or structural use, collectively termed *recycling*
3. Energy production from solid wastes

While open dumps still predominate in the United States, there are strong pressures to replace them with sanitary landfills, and eventually there will be pressures to reduce what ends up in sanitary landfills to an absolute minimum. Thus alternative 1 would seem to have a short future, with its actual time remaining being most dependent on the rate of progress toward alternatives 2 and 3. While economic competitiveness usually is cited as the controlling factor, it is clear that the factors which control these competitive prices are more important than the prices themselves. Thus such factors as resource supplies, environmental protection, public opinion, legal requirements, and new technology are all major determinants of prices, though they certainly are not the entire set of important influences. All the possible unknown influences, of course, make it difficult to say exactly what will happen in the future except to make the general observation that we certainly shall be less wasteful.

Earlier it was observed that solid waste management was an easier problem to solve than other forms of environmental pollution, primarily because the materials being dealt with are solids. As such, solid wastes are inherently easier to move about, to handle, and to shape into useful things. This property also makes the development of a truly comprehensive waste management program possible for solid wastes. In fact, some attempts are being pursued already in this regard. Systems analysts have been working on regional analysis and planning of solid waste management, of which the study of Appalachian North Carolina landfill locations is one example. Other studies have been conducted on finding the most efficient allocation of labor in refuse collection (Marks and Liebman, 1972) and on the various effects of new types of collection vehicles (Chaucey and Pinnell, 1972). It seems likely that systematic analyses of collection and disposal will become sophisticated for large regions. While it is less likely that comprehensive models will be constructed that integrate the various processes of collection and disposal with the recycling and reclamation of wastes, at least it now appears possible. A development such as this would be a major accomplishment for environmental problem-solving because it would include the use of solid wastes as a resource. Solid waste management then would become a question of resource use.

Questions

REVIEW QUESTIONS

1. If so much solid waste is produced annually, why has it not inundated us?

2. What is recycling? Can you think of conditions that might make recycling less beneficial than disposal?

3. How can solid wastes be used to produce energy? What are the limitations of each method?

ADVANCED QUESTIONS

1. Chicago is located in a much colder climate than Atlanta. Would this make any difference in composting? How might we go about finding an answer?

2. Using organic garbage to counteract settling of soils could be considered a natural thing. Why?

3. We noted that organic wastes would likely have one kind of effect on ecosystem dynamics. What might be the effect of inorganic substances? Could you generalize validly or would you have to consider each material separately?

Further Readings

Note: Most literature dealing with solid waste management appears in government documents, principally under the sponsorship of the Solid Waste Management Office and its successor, the Environmental Protection Agency. Other significant sources include the journal Environmental Science and Technology *(for digests of industrial efforts and views) and the U.S. Bureau of Mines.*

Chaucey, J., and C. Pinnell. In *Solid Waste Demonstration Projects* (U.S. EPA, SW-4p, 1972, see below).

Environmental Science and Technology. "Solid Waste Utilities: Ready, Set...". *Environmental Science and Technology,* 5 (1971), 752-753.

An optimistic perspective on future governmental-industrial cooperation to use solid wastes for electric power generation.

Kenahan, C. B. "Solid Wastes: Resources Out of Place." *Environmental Science and Technology,* 5 (1971), 594-600.

A survey of the activities of the U.S. Bureau of Mines in innovative solid waste technology.

Kenahan, C. B., R. S. Kaplan, J. T. Dunham, and D. G. Linnehan. "Bureau of Mines Research Programs on Recycling and Disposal of Mineral-, Metal-, and Energy-based Wastes." U.S. Bureau of Mines Information Circular (IC) 8595, 1973.

The latest update of the Bureau of Mines research announcements. The bibliography includes all pertinent references for Bureau of Mines literature: Reports of Investigations (RI series), Information Circulars (IC series), and Technical Progress Reports (TPR series); it is therefore an excellent key into the literature.

Marks, D. H., and J. C. Liebman. *Mathematical Analysis of a Solid Waste Collection System.* U.S. EPA, 1972.

Marx, W. *Man and His Environment: Waste.* Harper & Row, New York, 1971.

Turk, A., J. Turk, and J. T. Wittes. *Ecology, Pollution, Environment.* Saunders, Philadelphia, 1972.

Includes a chapter on solid wastes that is short but good.

U.S. Council on Environmental Quality. *Environmental Quality: The Third Report of the C.E.Q.* Washington, 1972.

Devotes considerable space to solid waste and includes some pertinent statistics.

Ocean Dumping. 1970.

U.S. Environmental Protection Agency

Composting of Municipal Solid Wastes in the United States. Report SW- 47r, 1971.

Computer Planning for Efficient Solid Waste Collection. Report SW-5rg. 1, 1972.

Energy Recovery from Waste. Report SW-36d, ii, 1973.

Initiating a National Effort to Improve Solid Waste Management. Report SW-14, 1971.

Oregon's Bottle Bill: The First Six Months. Report SW-109, 1973.

Oregon Solid Waste Management Plan: Status Report 1969. 1971.

Recycling and the Consumer. Report SW-117, 1974.
Short condensation of the two salvage industry documents listed below.

Regional Management of Solid Wastes. Report SW-80, 1, 1973.
Details of the Appalachian plan.

The Salvage Industry. Report SW-29c, 1, 1973.
Summary of the next citation.

Salvage Markets for Materials in Solid Wastes. Report SW-29c, 1972.
Excellent, technically detailed analysis with economic analysis.

Solid Waste Demonstration Projects. Report SW-4p, 1972.
See, in particular, the paper by Burch: "The Systems Approach to Solid Waste Management Planning."

U.S. Office of Solid Waste Management

American Composting Concepts. Report SW-2r, 1971.
Those interested in organic gardening and composting should read this short analysis.

Closing Open Dumps. Report SW-61+s, 1971.
A "cookbook" for local agencies and municipalities.

Recovery and Utilization of Municipal Solid Waste. Report SW-10c, 1971.

14. New Problems of a Powerful Technology

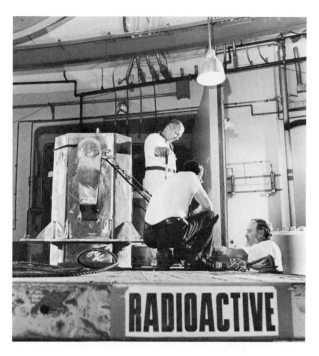

In this section we examined three important categories of environmental pollution, classified by their site of primary disposal. Like any classification, this one is imperfect insofar as some items do not fit neatly into any category, for one reason or another. For convenience and clarity, we have taken the reductionist view, while acknowledging the comprehensive nature of environmental pollution.

In this chapter we shall examine some kinds of pollution that do not fit readily into the original classification scheme but that share many of the properties we have discussed. The four treated below—radiation, waste heat, biocidal chemicals, and altering fluid balances deep in the earth—will serve as primary examples of problems that are not logically air, water, or solid pollutants. Radiation is particularly pervasive,

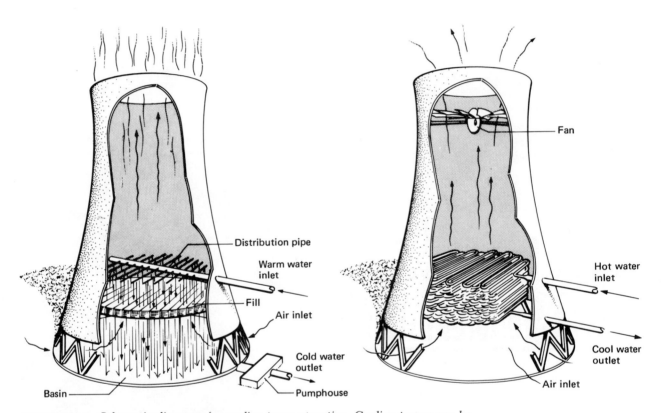

FIGURE 14-1 *Schematic diagram of a cooling tower operation. Cooling towers can be "wet" or "dry." Wet towers lose more water by evapotranspiration, while dry towers cost more per unit of cooling. The principle is the same for both: heat is exchanged to air by spreading out the water such that a high surface per volume ratio is attained.*

being found in air, water, and on the land; and it can arise from solids like fuel rods, from cooling water into which it has escaped, and in air as radioactive gases. The potential genetic and physiological hazards of radioactive compounds, combined with their high and pervasive mobility, make them particularly dangerous. Waste heat is interesting because the proliferation of nuclear power plants means there will be a lot more of it to deal with in the future. Despite its purely physical nature, quantities of heat are environmentally dangerous and must be dispersed with care into water or air or both. The choice of a disposal medium is particularly vexing and may have effects not yet known. Biocides are treated here because they also can be very mobile at the same time they are needed for disease and pest control. Altering fluid balances in the earth is a particularly good example of the systemwide consequences of human manipulation of the environment.

This "grocery list" of nonconforming pollutants does not include all potential candidates, yet it is enough to reemphasize the systemwide nature of environmental pollution. The sites of environmental effects are manifold, as are the number of effects that may be produced primarily, secondarily, and so on. This is precisely what we should expect from complex systems, and by now this effect should no longer be a surprise.

Waste Heat: Thermal Pollution

Among all environmental factors, temperature is one of the most important influences on biological processes. Organisms can be found active over an enormous range of temperatures, from freezing to boiling; but the range of temperatures over which any particular species can exist is normally rather narrow. Few organisms can exist at temperatures over 40°C (104°F), and so it is not difficult to imagine what the effects of hot water (up to 90°C, 194°F) would be on any ecosystems involved. There are almost no natural waters (except for a few hot springs) where this hot water could be dumped without severe damage to living systems. The advent of nuclear power generators threatens to create an enormous demand of cooling water (estimated up to half of the freshwater runoff of the United States per year), as well as the problem of what to do with all this heated water.

Is it best to put it back into cooler natural waters, or into air, or into the soil?

The utility engineers who must design and construct nuclear power plants to meet increasingly tough environmental standards now favor the cooling tower (See Figures 14-1 and 14-2) because all the heat is discharged into the air, where it can disperse readily (along with a volume of water vapor; see Figure 12-6), although power plants occasionally may discharge this heat directly into adjacent bodies of water. For exam-

FIGURE 14-2 *An operational example of cooling towers. Three 437-foot-high cooling towers of the TVA's Paradise Steam Plant in western Kentucky supplement the cooling capacity of the Green River. The base of each tower is large enough to hold a football field.* (Courtesy of Tennessee Valley Authority.)

ple, one major reason for locating plants along the coastline is because oceans are enormous heat sinks. If one uses equipment that will not be corroded by salt water, a lot of sea water can be withdrawn from the sea, used for cooling, and then returned without significantly changing ocean temperatures, provided the heated water is not released into a confined area, such as a bay. Plants located inland but along rivers can pump cooling water from the rivers into large impoundments from which the water either is not released back into the river until it cools or is repeatedly returned to the plant for additional use.

Nuclear power plants with conventional reactors are now operational. All that is needed for the construction of many more are appropriate sites and ways to meet various standards of environmental quality. Biologists and engineers have been stimulated to suggest a number of innovative schemes to use this waste heat in the same spirit as those for the utilization of solid wastes. These plans would allow us to use the waste heat rather than to disperse it wastefully. Bohn and Cauthorn (Chapter 11) included waste heat in their scheme to increase crop productivity (warm soil stimulates plant growth). Another possibility is the coupling of waste heat with the operation of water-waste treatment plants; again, the objective would be to accelerate desirable biological processes, in this case more rapid decomposition. Some authorities also have suggested placing nuclear power plants on deep ocean floors, where their waste heat would stimulate convection currents in the water; these currents would bring nutrient-rich water to the surface, where organisms that can utilize them are located.

This in turn could increase the productivity of the ocean in that area and subsequently might result indirectly in an improvement in the fisheries. In a somewhat more conventional variation of this idea, the U.S. Atomic Energy Commission is testing the feasibility of offshore nuclear power plants.

CAYUGA LAKE: HARBINGER OF THE FUTURE?

A short while ago nuclear power seemed the solution for abundant electrical power. With the development of working reactors, utility companies began to design power plants to such an extent that they threatened to proliferate across the landscape like weeds in a field. Yet many of these proposed plants remain unbuilt, their construction either postponed indefinitely or canceled altogether. Why? The safest assessment is that many factors are involved, including economic conditions, technological problems, and rising labor costs. Yet real environmental concerns certainly would have to be included in the list of reasons.

A case in point is the nuclear power plant proposed for Cayuga Lake, near Ithaca, New York. The history of the developments in this case has been described in remarkable detail by Eipper (1971), a participant in the power-plant controversy there. In 1967, the local utility company proposed that a new nuclear plant (to be named Bell Station) be built adjacent to an existing coal-powered facility on Cayuga Lake, a deep, narrow glaciated lake and one of the famous Finger Lakes of New York. To promote good public relations and demonstrate their environmental awareness, the utility company also announced that it was prepared to expend funds to spell out the possible ecological impacts of this new facility.

What the company did not anticipate was the response of some faculty members of nearby Cornell University who were dissatisfied with the company's promises. Far from being mollified with the official answers from the utility company to some pointed questions, these citizens were alarmed enough to develop an adversary position. The main point of contention was the belief that the company had underestimated the ecological effects of the plant's thermal output. By early 1968 it had become evident that the utility company planned to build the plant so that it met only the minimum legal requirements; this crystallized an organized opposition, whose first move was to produce a position paper articulating arguments in favor of delaying construction and providing additional environmental protection. Attempts to compromise failed, and so the remainder of 1968 saw each side plowing resolutely toward an inevitable confrontation. The utility company spent money on site construction, apparently confident that the plant would be built, while the organized opposition broadened its base of support into the state legislature. The efforts of the citizens' committee culminated in three legislative measures requiring, for the first time, some standards for heat discharge into the surface water of lakes. Faced with escalating requirements and considerable uncertainty about the future, the utility company finally suspended construction in April 1969.

The potential thermal modification of Cayuga Lake proved to be the principal issue of contention. Company plans called for cooling water to be drawn from deep within the lake and redeposited at the surface. While the company was emphasizing the beneficial effects of the waste heat for a salmon hatchery, the opposition was alarmed at the possibly disastrous consequences of the added heat on the ecology of the lake. They concluded that the company's plans would result in an extension of the growing season, an increase in the supply of available nutrients in surface waters, and an increase in biological activity within the lake—in sum, an acceleration of eutrophication in Cayuga Lake.

Some simple calculations will show how great the potential effects could be. By their own admission, the utility company estimated that throughout the year 1100 cubic feet *per second* of water would be taken from near the bottom of the lake, heated 20 to 25°F, and returned at the surface. Since the lake volume is estimated to be 3.31×10^9 (3.31 billion) cubic feet, the equivalent of the entire volume of the lake would be cycled through the plant 10 times per year! What is the environmental impact of such heating? A Cornell study group financed by a grant from the utility company calculated that the upper layer of the stratified lake would be approximately 1 foot thicker and that the lake would be stratified 10 days longer each year. (See Figure 14-3 for an explanation of the stratification process.) However, these are only estimates and cannot be considered reliable because a lake system is far too complex and there are too many interacting factors that we know nothing about to make a reliable calculation possible. A reliable estimate of the change in temperature (*heat balance*) and its consequences would have to consider various physical factors (for example, size of the lake, turbidity, lake depth, current circulation pattern) and the exact nature of thermal input, especially where the water is taken from and how it is returned. Given these considerations, the fact that the utility-sponsored study team would find some change of the kind predicted by the opposition is highly significant, and these predicted changes can be interpreted to be the *minimum* to be expected. The report does minimize the ecological consequences of the predicted shift in

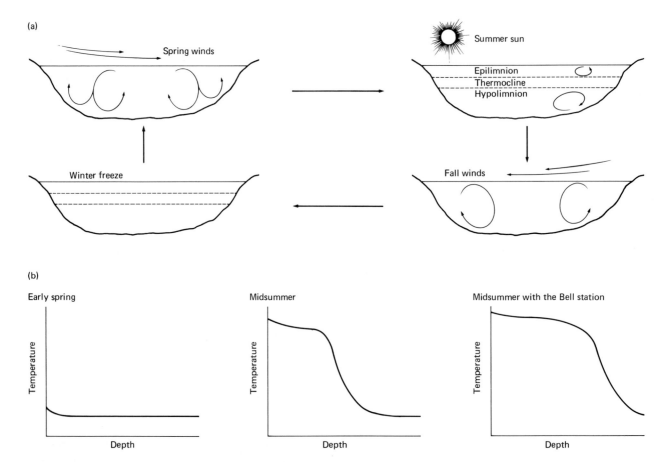

FIGURE 14-3 *The effect of a proposed nuclear power plant on Cayuga Lake, assuming direct thermal discharge into the lake. (a) The normal course of stratification results from differential heating by the sun with consequent formation of thermally different layers. The colder, heavier water at the bottom is called the hypolimnion; the warmed upper layer, the epilimnion; and the zone of rapid temperature change in between, the thermocline. Differential heating and uniform cooling alternate in a seasonal cycle. (b) Temperature profiles for the hypothetical lake in two seasons with the nuclear power plant. The thermal discharge would increase the volume of the epilimnion and allow it to persist for a longer time.*

heat balance. In contrast, the citizens' committee was very concerned with the consequences of a change in heat balance. For example, extended stratification might result in severe oxygen depletion in the deep waters of the lake, which in turn would be a severe blow to the trout fishery.

The volume of water needed by the proposed nuclear plant for cooling is so enormous that it creates unresolved environmental questions too serious to permit the construction of the facility and thus risk the future of Cayuga Lake. If the effects were large, the lake would be so changed that it would be impossible to restore it to its former condition. Hence to defer the construction of the plant was certainly the wisest choice under the circumstances. This delay should continue either until better information is available or until an alternative cooling scheme is instituted. One is entitled to wonder why the utility company, even after it had invested so much time and money in site construction, refused to consider cooling towers (Figure 14-1) as a compromise solution. Was the company so confident its interests would prevail in the end that it felt compromise was unnecessary? Whatever the reasons, the Bell Station remains in limbo as various government agencies continue to struggle with the problem.

Thus the dangers in using freshwater sources for cooling nuclear power plants are not precisely known, but major ecological changes are possible. More important, public opinion apparently favors putting the burden of proof upon the plant builder, at least for the present. In fact, even seaside plants that would have minimal thermal impact may well all end up offshore to minimize thermal effects.

Can it be any wonder that releasing heat into the atmosphere is so attractive to the designers of nuclear power plants? But there are problems even here. Wet cooling towers cannot transfer heat directly to the atmosphere; instead, water must be passed over the circulating pipes in order to transfer heat. In practice, then, much of the water passes through the cooling tower as steam and thus rises out of the tower. However, since many natural waters are hard, the water pipes rapidly become encrusted with mineral deposits. To counteract this, many plants would use sulfuric acid to dissolve and prevent such deposits. Unfortunately, some of this sulfuric acid is carried out with the steam and deposited downwind—constituting a problem of "acid rains" (see Chapter 12). Dry cooling towers are more economical with water but much more expensive to operate.

Biocides

We use an impressive array of chemicals to combat pest and disease organisms. These chemicals, collectively called *biocides*, are products of a pressing need to defend us, our beasts, and our crops against the "ravages of nature" and of the intensive research put into meeting these needs. In one sense, it is a credit to human ingenuity that we have created this bewildering variety of chemicals which are lethal for insects, molds, vegetation, and even vertebrates. Biocide research has identified the compounds that are lethal, identified the susceptible populations, estimated the dosages required to produce the desired effects, and even produced some combinations that increase the total effectiveness (for example, two pesticides together or a biocide with a wetting agent to increase surface penetration). Since this kind of information is needed to develop biocides, there is sufficient incentive to get it.

However, we often do not know what the system-wide effects of a biocidal treatment will be. Two cases within the past decade—the large-scale use of herbicides and defoliants in Southeast Asia and the discovery of the insidious effects of the DDT group of insecticides—have made us aware that such effects are real. Whether we shall ever know much more *before* a biocide is applied, however, is open to question. Such information is expensive: It costs a lot of money to monitor natural systems, especially when we do not always know where to look. For example, defoliation in Vietnam was intended to destroy crops and to expose the forest floor and thereby hamper enemy movement, but it succeeded mostly in changing the nature of the forest (perhaps permanently) and in showing how inordinately sensitive mangrove trees were to the defoliants. (For more information, see the summary report of the AAAS Herbicide Assessment Panel, 1974). There is great need for more research into the total impacts of biocides on natural ecosystems.

DDT AND ITS DERIVATIVES:
A POST HOC ANALYSIS OF IMPACTS

Some of the best documentation of pervasive environmental effects of biocides exists for the class of chemicals known as *chlorinated hydrocarbons*, representatives of which include DDT, Lindane, Chlordane, Endrin, and Aldrin. There is abundant evidence of DDT accumulation in a whole variety of animals living in even the most remote locations, and equally good evidence documents the causal pathways by which this situation arose (see Rudd, 1964). In 1972, the Environmental Protection Agency banned the use of DDT in the United States except for a few special cases. In underdeveloped countries, however, DDT is still used by the United Nations to combat malaria, the rationale being that the ecological consequences of DDT are preferable to possible health effects resulting from the discontinuation of the anti-mosquito programs. Despite claims to the contrary, the long-term effectiveness of DDT spraying has never been evaluated for these programs, nor is it clear that other, less persistent but more dangerous pesticides would not serve equally well.

The properties of DDT and their effects are well known.

1. *Application and persistence*   DDT is a potent insect killer that works well against populations which have not become resistant to it. But the stability of the DDT molecule has proved to be its most important property, because this has negated what little specificity it may have possessed originally. The property of persistence has meant that DDT can affect organisms other than the original target species, and this effectiveness can persist over a long period; it can be effective years later. And because it lasts so long, it can be dispersed great distances from the location of its original application.

2. *Fat solubility*   DDT is highly dispersible not only because it is persistent but also because it is extremely fat soluble; organisms readily absorb and store DDT in their fat deposits, and with their muscles, fins, and feathers take this DDT with them wherever they go. A migrating tern may transport its collection of DDT over 10,000 miles in a single trip. The rapid dispersal of organisms supplements the physical currents in the water and air that can also transport DDT.

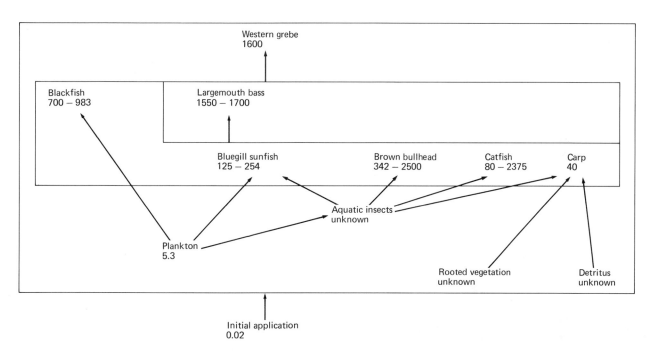

FIGURE 14-4 *An example of trophic web concentration of DDD. (DDD is a stable derivative of DDT.) The estimated concentrations are for visceral fat of each type. The diagram assumes an initial application of 0.02 ppm DDD. Note how important species differences are in determining the degree of concentration.* (Data from Rudd, 1964.)

3. *Trophic chain concentration* Another observed property of DDT in natural ecosystems is that it is concentrated at progressively higher trophic levels; since organisms selectively concentrate DDT from the environment in their fat tissues, as one organism eats another it is tapping some of the richest sources of DDT available. One would expect progressively higher concentrations of DDT in animals that occupy higher positions in the food chain, and this is indeed the case. It follows that the top trophic levels are the most severely affected (Figure 14-4).

4. *Physiological action* The final link in this chain of bizarre events is related to the varied manner in which DDT affects different taxonomic groups of organisms. Thus DDT is a direct and lethal poison for insects, but in birds it has the less direct effect of interfering with calcium metabolism and thence with

eggshell formation. Some evidence indicates that in human beings it even may confer protection from some forms of cancer, although this is certainly not a recommended prophylactic procedure.

Since the role of DDT in eggshell thinning has been most thoroughly documented, we shall study this phenomenon. It had been noted for some time that brown pelican and peregrine falcon populations were declining (Figure 14-5). Armed with the proper techniques for analysis and enough time to document their cases, investigators have successfully traced the decline of these species to reproductive failure. Specifically, the reproductive failure was found to result from DDT's interference with calcium metabolism, which led to thin eggshells that could not bear the weight of the parents during the incubation period. With the discontinuation of the widespread use of DDT in the United States, there have been indications that this eggshell thinning and its population decline may be reversed in some species in which it had been found. For example, there is evidence that the brown pelican population is recovering; and in another well-documented case, Rudd and his colleagues (1964) have discovered the same recovery for the Western Grebe at Clear Lake, California. The grebe population crashed as a result of the repeated application of DDT to Clear Lake in an attempt to control the Clear Lake gnat; but now that these applications are no longer being made, the DDT concentration in the grebes has declined, and the population is recovering.

The populations that are still endangered and that may never recover are the great birds of prey, such as the peregrine falcon and the bald eagle.[1] These species are at the top of the trophic pyramid and because of this are likely to have accumulated the largest DDT dosages. Moreover, like all top carnivores, they are rare species to begin with, and they may have already been reduced below the recovery level. If this is true, then no matter what future action we may take in biocide control, it will have come too late for them.

FIGURE 14-5 *Peregrine falcon at the nest. The effects of DDT in eggshell thinning are now known to be a major cause of declining peregrine falcon populations.* (Courtesy of G. Ronald Austing from National Audubon Society.)

---

[1] One great bird of prey, the osprey, has recovered since the cessation of DDT use. Perhaps this augurs well for the bald eagle and the peregrine.

Discoveries of DDT accumulation in birds and of the effects of this phenomenon led to further investigations of other species in various parts of the world. These investigations have led to the discovery of residues in penguins in far Antarctica and of potentially dangerous concentrations of residues in coho salmon in Lake Michigan and in marine phytoplankton in the Atlantic Ocean; and not surprisingly, residues have been found in the human body.

Though the effects of DDT on individual species certainly can be serious, especially to the species in question, of far greater importance are the ramifying ecological effects throughout an entire ecosystem. Such an effect is more serious because it can determine the fates of many species. For example, we do not know exactly what role the peregrine falcon plays in human affairs, although we can hazard some guesses. Peregrines once were thought to be important waterfowl predators and hence competitors with human beings for game birds; the name "duck hawk" lingers, even if it is not appropriate. On the other hand, peregrine falcons also used to live in cities and were important in preying on pigeons in such places. Perhaps peregrines were important in controlling bird populations that become pests when their numbers are not controlled by predation, or perhaps they are nothing more than the esoteric concerns of conservationists. If, indeed, the peregrine population already has been reduced to a size beyond what is necessary for recovery, we may learn the importance of this particular species sooner than we might like. However, even if the extinction of peregrine falcons has no apparent ramifying effects, this does not mean that the next system we tamper with might not have. By the time we know the truth, it may be too late to rectify our mistakes.

We have seen that it is always easier to understand why an undesirable reaction occurred after the fact than it is to predict it in advance. When we find ourselves unable to anticipate these effects, either we must be quite conservative about the actions we take, or we must be prepared to accept the undesirable consequences of our actions. The possible extinction of peregrine falcons is an instance of the latter alternative. Surely the former alternative would be preferable. Perhaps the lesson learned from DDT will result in our placing a greater emphasis on this approach.

BIOCIDES: A POLICY PERSPECTIVE

Because we are so ignorant about the effects of biocides on ecosystems, in effect we are running a series of experiments on nature with post hoc observations as our reward. This raises two questions of vital importance for biocide policy formulation:

1. If controlling pests with biocides is as risky as it seems, why do it in the first place?
2. If we must have some form of pest control, what are the alternatives to the use of biocides?

The answer to the first question is the simpler one. Nature does not necessarily obey our wishes. Diseases do not avoid us when we are susceptible, nor do plant disease and animal pests avoid energetically rich plants just because they bear a label like "Crops—Stay away!" Thus something is needed to destroy the diseases and agricultural pest organisms before they do too much damage; and the most easily handled, neatly packaged, and dependable solution has always been a lethal chemical. As evidence accumulates, however,

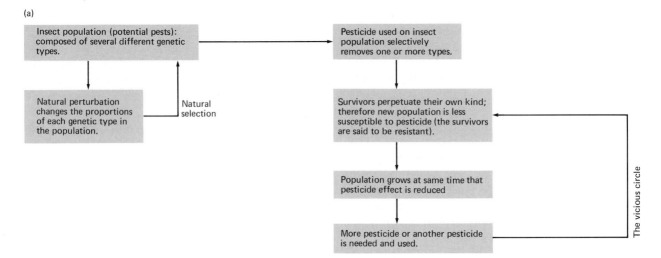

FIGURE 14-6 (a) *The mechanism of pest resistance: a notable case of destabilization and the vicious circle of chemical pesticides. (b) N. W. Moore's summation of the history of agricultural production emphasizing the effects of pesticides. The curve embodies the idea that greater production can be bought only with a reduction in ecological diversity, which in turn increases fluctuations in pest populations. Do you see why? If not, go back to Chapters 1 through 3. Recent increases in fluctuations are due to increasing pest resistance to pesticides. The three future alternatives given by Moore are based on the same ecological ideas.* ((b) Redrawn from N. W. Moore in *Advances in Ecological Research* 4, p. 124, 1967, Academic Press, Inc.)

the chemical alternatives lose their simple attractiveness. The catalogue of undesirable effects comprises three main items:

1. There is the possibility of significant long-term effects on nontarget systems, as has been demonstrated for the DDT group, especially with compounds having the otherwise desirable property of persistent toxicity.

2. The use of biocides is often accompanied by the destruction of all natural control mechanisms, because biocides often are not specific enough to avoid destroying the target's natural enemies as well. This property has a more detrimental effect on systems that are largely natural (forests rather than agricultural crops) and in situations where the pest can recover more quickly than its biotic control agents. The effect is greater in natural systems because they are characterized by more widespread control of pests by natural enemies. In these cases, if the pest population recovers from decimation before the population of its natural enemy does (as is frequent; see item 3), then it can reproduce and grow at an even greater rate.

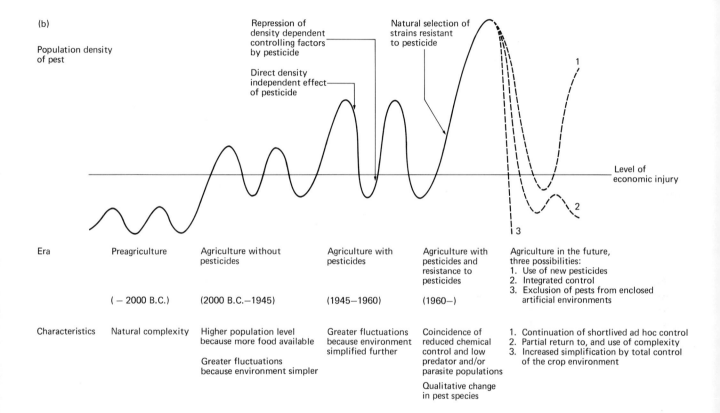

3. The widespread use of biocides has demonstrated the phenomenon of *pest resistance*, in which a target population is less and less affected by a biocide (Figure 14-6a). Natural populations are not made up of many identical individuals but rather of organisms having a range of variability that allows them, by natural selection, to adapt to changing environments, whether this be climatic differences or susceptibility to pesticides. Mosquitoes, houseflies, and other insects have been sprayed for generations with these pesticides and have evolved the ability to resist their effects. Since pesticides promote the resistance, the more of a pesticide one uses, the more one selects for greater resistance to the pesticide. Changing pesticides has proved to be no escape from the dilemma. Public health authorities in central California are predicting that it is just a short matter of time until an epidemic of mosquito-borne equine encephalitis occurs, since the carrier is resistant to practically all present insecticides and is in the process of spreading over the state. We have been hoist upon our own petard.

Moore (1967) concluded that pesticides, whatever the reason, have been counterproductive in reducing the number of pests (Figure 14-6b). He offered three possibilities for the future: continuation of our present course, which only would worsen the situation; pursuit of our present course but to such an extreme that we could exercise total environmental control (which, no matter the relative advantages and disadvantages, is an impossibility now and probably will remain highly impractical at least); or finally, acceptance of our limitations, which would allow a partial return to natural complexity. As evidence mounts, it is obvious that our "grand experiment with chemical control of insect pests" has been less than a spectacular success.

When we turn to the second question, that of alternatives to chemical control, we find techniques that remain proven. (For a detailed discussion of these alternatives, see Chapter 6.) Though all possible alternatives show some promise and certainly are worth pursuing, it is important to remember that they all require a great deal of expensive information to be implemented effectively. Moreover, none of them carries any guarantee of uniform applicability to all control problems. For better or worse, chemical control probably will continue to predominate for years to come, until more comprehensive systems-oriented methods can be developed.

THE WISCONSIN HEARINGS AND THEIR IMPACT

In 1969, the state of Wisconsin held hearings to determine whether or not DDT was an environmental pollutant under Wisconsin state law. This hearing ultimately produced a decision that changed national pesticide policy. Although the "DDT trial" was simply a hearing and its outcome could affect only Wisconsin, it attracted nationwide attention, witnesses of national prominence, broad financial support, and a lawyer from the Environmental Defense Fund (EDF). In fact, the Madison hearings were a showcase that turned into a focal point of national significance in the handling of the DDT problem. After nearly six months of testimony, the decision of the state examiner to remove DDT and its analogues from use in Wisconsin actually proved to be anticlimatic, for in the year between the testimony and the publication of the decision, the federal government had not ignored the proceedings. Pressure had been put on the U.S. Department of Agriculture and on the Department of the Interior for a national ban on DDT. In 1971 EPA, which has direct national responsibility for the environmental impacts of pesticides, initiated registration cancellation proceedings for DDT and some other biocides. (All pesticides must be registered with the federal government before they can be used.)

Finally, on 14 June 1972, EPA banned DDT entirely except for use on three minor crops. The ban was effective 1 January 1973 and has been only the first of a series of actions initiated by EPA. Investigations and proceedings similar to those taken on DDT have been started for other poisons. The ant poison Mirex was banned except for restricted use on fire ants. When mercury was found to be an ecological contaminant in circumstances much like those of DDT, over 780 mercury compounds and substances containing mercury were banned as well. Perhaps most importantly, EPA has proposed new rules for pesticide registration. Previously, only pesticide manufacturers could initiate review procedures for registration; the EPA proposal would give the general public the same right. All these actions were started by a set of hearings in Madison, Wisconsin.

At the hearings, the pelican was one important piece of evidence; its decline in abundance was instrumental to the ban on DDT. Interestingly, the brown pelican once again is reproducing successfully, and its populations are recovering their former abundance. But what of the peregrine falcon? It was not a common species in the first place, and now it is extinct over most of the continent. Its future remains uncertain, and it may be a fatality of the DDT "experiment." And—in case some of us regard the DDT ban as a shining example of human wisdom and become complacent—consider the possibility that the victory in the DDT battle was hollow. Perhaps DDT was allowed to disappear from the scene simply because it no longer was effective or because a more effective substitute was available for its big users. Nor is there a guarantee that governmental action on DDT signals fundamental attitudinal change on pesticides. With human populations continuing to grow, a solution that simultaneously preserves natural diversity and provides enough foodstuffs is necessary. While such a solution is sought, what will happen to DDT? It is still used in the world, and in the United States there has been some small-scale resumption of its use in response to increasing economic damage by the tussock moth in northwestern forests. Moreover, eastern forests are similarly threatened by the gypsy moth. One wonders how long it can be in cases like this of severe economic damage and consequent social disruption before the widespread use of DDT is resumed.

The reinstitution of DDT to suppress tussock moth epidemics in the Northwest draws attention to the basic dilemma of pest control. The tradeoff ultimately involves getting enough food from the global ecosystem every year to meet the demands of a growing population versus the ever-increasing ecological simplification that is a necessary consequence of meeting the demand for food. For the specific example of the tussock moth, the property of persistence that is such a problem for peregrine falcons is precisely the property so valuable for killing tussock moths. If the evidence is sound, we in effect are trading peregrine falcons for economic welfare in the Northwest; at least we are setting the precedent for resuming acceptance of this trade.

Deep Disposal: Fluid Injection

Deep disposal of wastes involves depositing wastes in locations so remote that it is assumed they will never be seen or heard from again. Since this method of waste disposal does at least imply the isolation of the wastes from living systems, it is strongly favored for highly toxic substances. There are several forms of deep disposal; one form involves dumping wastes into the deep ocean, but this method currently is out of favor, since it offers few guarantees that the wastes so treated will not come welling back up into shallow water someday. Since ocean disposal has been discarded as unsatisfactory, the remaining alternative is to dispose of wastes at sites deep within the earth's crust. One method involves the potential use of old salt mines and salt beds for the disposal of nuclear wastes. Another variant, called *deep-well fluid injection*, involves the injection of fluid waste into appropriate

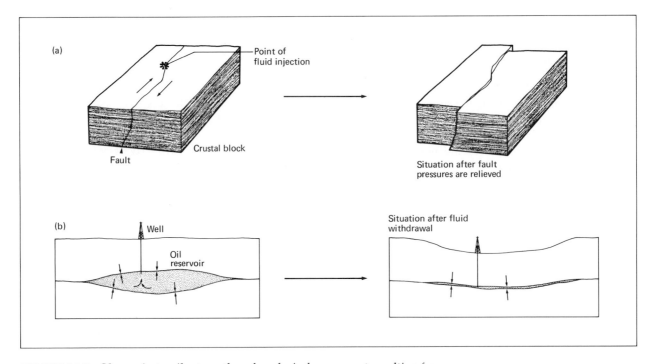

FIGURE 14-7 *Change in tensile strength and geological movement resulting from alterations of fluid balance. (a) Movement along a fault released by fluid injection. (b) Subsidence due to fluid withdrawal.*

spots in the earth's crust, the rationale being that certain porous rock layers can absorb great quantities of these wastes (very much like a sponge) and hold them indefinitely. A natural example of such a rock matrix is the oil shale of Colorado, which is saturated with oil but which holds it too tightly for the oil to be extracted by normal pumping procedures.

Deep-well fluid injection has worked reasonably well in many cases, but there are problems. First, the number of suitable sites is limited. Moreover, it can work only when the rock matrix is sufficiently porous to permit waste fluids to flow away from the point of deposition and also when the entire rock layer is deep within the earth; if any portion is exposed, it can provide the injected fluids a means of returning to the earth's surface. Finally, it also has been found that this technique can produce unanticipated geological repercussions. These geological problems will be our concern here.

The rocks of the earth are unchanging and enduring only in the short time perspective of human beings. To the contrary, abundant evidence shows that the earth is quite plastic. Continents move about, and rocks bend, break, shatter, and fracture in the processes of mountain building and subsidence. The analogy of a thin, brittle crust under continuous stress from the gigantic, restless forces underneath is not inappropriate. The evidence for plasticity of the earth's crust is impressive, but we are still far from a complete understanding of all the geological movements. Meanwhile, we are learning from experience some of the consequences of changes in fluid balances within the earth, and fluid injection has given us insight into some geological repercussions that can result. For example, it has been found that fluid injection can reduce friction between rock layers, enabling them to slide over one another more readily when stress is applied, thus producing the earth movement better known to us as an earthquake (Figure 14-7).

We now shall devote attention to two examples where fluid balances deep in the earth have been altered and were later implicated in earth movement.

SHIFTS IN THE EARTH: THE DENVER EARTHQUAKES

During the years from 1962 to 1966, an active program of deep-well injection was carried on at the U.S. Army Rocky Mountain Arsenal near Denver. The injection well was drilled very deep (12,000 feet) and was heavily used (receiving up to 6 million gallons per month) for disposal of certain unspecified dangerous wastes. Almost from the start the Denver area experienced a series of earthquakes ranging up to 5.5 on the Richter scale, creating public alarm and a search for the cause of the problem.

Healy and associates (1968) have documented the Denver episode thoroughly. They showed that the probability that so many earthquakes should occur in a short time coincidental with fluid injection was vanishingly small. Particularly convincing were these three lines of evidence:

1. Geological evidence indicated that very few earthquakes had occurred in this area since early in the history of life on earth (Precambrian time). This makes it unlikely that the 28 earthquakes around Denver from 1962 to 1966 were due to chance events.

2. Fluid injection was stopped from October 1963 to September 1964, during which time the earthquakes practically ceased (Figure 14-8).

3. Healy and his associates were able to show that there was a definite pattern in how the earthquakes spread from the site of the Arsenal well (Figure 14-9). They suggested that the spread of fluids in the rock released earthquakes farther and farther from the well as the fluids penetrated rock layers away from the injection site. All in all, the evidence was conclusive enough to halt the injection procedure and to initiate discussion about the possibility of withdrawing the fluid to reduce the potential earthquake hazard.

SHIFTS IN THE EARTH: A RESERVOIR FAILS

Our second example of fluid balances and earth shifts is not related directly to deep-well fluid injection. Instead, it concerns the more general problem of unanticipated changes that result from shifts in the fluid balance in the earth's crust. In this second example,

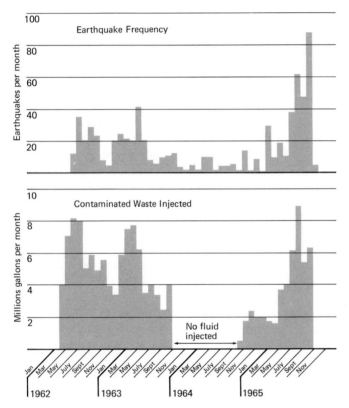

FIGURE 14-8 *These histograms illustrate the remarkable correlation between the frequency of earthquakes in the Denver area and fluid injection at the U.S. Rocky Mountain Arsenal, especially when injection was temporarily suspended.* (Redrawn from "The Denver Earthquakes" by J. H. Healy et al., *Science* 161, pp. 1301–1310, 27 September 1968. Copyright © 1968 by the American Association for the Advancement of Science.)

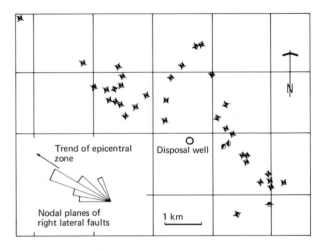

FIGURE 14-9 *The spatial pattern of earthquake epicenters in relation to the RMA well. Note the pattern of dispersal NW-SE along a suspected fault zone.* (Redrawn from Healy et al., *Science* 161, 1968. Copyright © 1968 by the American Association for the Advancement of Science.)

the shift in fluid balance resulted not from injection of fluids but from their withdrawal. On 14 December 1963, the Baldwin Hills Reservoir in Los Angeles was breached as a result of an incredible series of interlocking circumstances. The reservoir emptied itself into the city, resulting in more than 15 million dollars property damage and five fatalities. Since its failure, the reservoir exists no more; it was closed when subsequent investigation showed that even if it were reconstructed, another failure would occur eventually.

The investigation, published by the California Department of Water Resources in 1964, outlined the chain of circumstances that led to the failure (Figure 14-10). The Baldwin Hills area is characterized

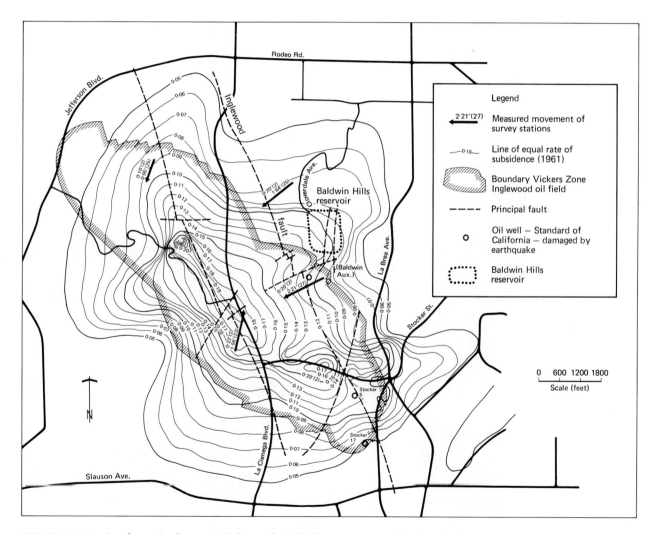

FIGURE 14-10 *A schematic diagram of the geological phenomena contributing to the failure of the Baldwin Hills reservoir.* (Redrawn from DWR Report, 1964.)

364 NEW PROBLEMS OF A POWERFUL TECHNOLOGY

by general land subsidence, particularly in the immediate area of the reservoir. And this subsidence process caused vertical displacement of crustal blocks along the Newport-Inglewood fault system. Because of this combination of forces, the site of the reservoir was subjected to stresses on its floor. When the reservoir was built, engineers attempted to compensate for the stresses related to the subsidence process in the region by lining the reservoir with asphalt; but unfortunately, the lining could not withstand the differential stress, and eventually that portion overlying the vertical displacement site broke, allowing water to seep downward and completely undermine the earthen foundation of the reservoir. Once water was admitted to the foundation under the reservoir, it was just a matter of time until failure was complete. Water seepage eroded the foundation on the downhill side, a process that accelerated as erosion continued. Finally, as water began to seep out of the hill, the foundation was breached so badly that the reservoir wall broke (Figure 14-11).

FIGURE 14-11 *The Baldwin Hills reservoir after the failure of the wall. The events of the day of failure were as follows: At 11:15 AM the caretaker noticed an unusual sound of water at the spillway intake. Later leakage from the reservoir was noticed at about the level of the second roadway in front of the dam. At 2:20 PM the first surface evidence of a possible breach occurred; a crack appeared across the main dam. Sandbagging was attempted to plug the crack. But the opening widened rapidly, and by 3:38 PM the breach was complete. By 5 PM the reservoir was virtually empty.* (Courtesy of State of California Department of Water Resources.)

Why is subsidence occurring in the Baldwin Hills area? Certainly natural forces are involved, but the nearby Inglewood oil fields may be an important influence. Subsidence is known to be associated with oil-drilling operations elsewhere, and it seems to be more than simple coincidence that the reservoir was located only a few hundred feet from the Inglewood oil field that underlies the Baldwin Hills. From the start of production at this oil field in 1924 to the day of the reservoir failure, oil was removed from the rock strata. The removal of the oil resulted in a substantial lowering of fluid pressures in the Inglewood field and thence to subsidence in the area. The investigating board confirmed this by noting that one oil field nearby had subsided 28 feet in only 26 years and by pointing to adequate evidence that changes in fluid balance could lead to such compaction and subsidence. Hence removal of oil from the nearby field accentuated subsidence in the area, which caused displacement along the fault underlying the dam, which caused the failure.

The failure of the Baldwin Hills reservoir shows that balances in fractured rocks are an important factor in determining whether an area will experience stability or earth movement, and this is true whether fluids are injected or withdrawn. The underlying rocks are a dynamic part of the environment, no matter how inert we may think they are.

Radioactive Wastes

The use of nuclear weapons at Hiroshima and Nagasaki, Japan, in 1945 indelibly demonstrated the dangers of radioactivity to the public. Consequently, the public has demanded always that radioactive wastes be handled carefully and disposed of in ways that insure their permanent removal from circulation. This requires that we recognize the special dangers of radioactive wastes, to wit:

1. Radioactive wastes can be dangerous at very low concentrations, down to one-billionth of the concentration of nonradioactive wastes.

2. Radioactivity is possibly *mutagenic*—it could cause genetic changes that can be passed on to offspring. This danger is in addition to the damage that can be done to individuals but that has few consequences for descendants.

3. Radioactive wastes cannot be destroyed. They remain radioactive for as long as their atomic structure permits.

4. Radioactive wastes can contaminate equipment and solutions, which then can be transported widely and allow the radioactivity to be involved in everyday chemical reactions, rendering the most harmless chemical compounds dangerous. Thus all equipment contacting radioactive wastes also must be treated as radioactive material.

This combination of strong biological activity and high potential mobility makes radioactive wastes a particularly difficult problem to handle.

The choices of radioactive waste managers fall into three basic categories: Dilute the wastes and disperse them widely; hold them until the radioactivity decays to a safe level and then disperse them; or store concentrated, highly radioactive wastes indefinitely. The alternative chosen depends mostly upon the form and radioactivity of the material to be disposed of (Figure 14-12). Presently the Atomic Energy Commission (AEC) controls most of the stages of radioactive waste disposal; before 1962 AEC put most of the liquid and solid wastes of low-level radioactivity (measured in microcuries = curies/10^6) into concrete-filled steel drums and dumped them in the deep oceans. Since then ocean dumping has ceased almost entirely, and these wastes now are being buried mostly in selected landfill sites on government reservations, although AEC does allow some low-level gases and liquids to escape into the environment if the levels of radioactivity are no higher than the background radiation of the planet. Ocean dumping was halted because, though in theory it should not result in any radioactive hazard, once the wastes are dumped, there is no way to monitor them. Since adequate landfill sites are available, this has become the preferred disposal method not only for low-level wastes, but also for medium-level (measured in millicuries = curies/10^3) and high-level (measured in curies) liquid and solid wastes. Radioactive gases emitted through smokestacks are captured on air filters, and these are buried, too. In addition, some high-level liquid wastes have been injected into rock, and intermediate-level wastes are also sometimes stored to permit natural decay if the half-life of the isotope involved is sufficiently short.

Presumably, if nuclear wastes were not to increase much above current rates of production, present methods of disposal would remain adequate. However, AEC estimates that the volume of liquid wastes will go from 100,000 gallons in 1970 to 6 million gallons by the year 2000, and that solid wastes will climb from the present 1 million cubic feet to 3 million cubic feet by 1980. The bulk of the increase is predicated upon the construction of nuclear power plants. The routine wastes of such plants include liquids and solids (usually high level) contaminated during fuel-rod reprocessing, contaminated cooling fluids (usually

FIGURE 14-12 *Two methods of disposal of radioactive waste. (a) Steel lined concrete tanks for the storage of liquid radioactive wastes. These million-gallon-capacity tanks will be covered with more than 10 feet of earth and will be monitored by leak detection equipment in the columns in the background. (b) Radiation contaminated equipment is sealed in plastic, crated up, and buried in special sites. This is the Atomic Energy Commission's Hanford Project in the state of Washington. (Photos courtesy of Battelle-Northwest Photography Unit, Richland, Washington.)*

low level), tritium escaping from cooling fluids (usually low level), and intermediate-level liquids resulting from the initial synthesis of fuel rods. Apart from this normal waste production, there are possibly nonroutine large solid wastes like reactor vessels or even large parts of power stations that might become contaminated, should a major leak occur. And there is the mundane but real possibility of truck, rail, or airplane accidents during transport of radioactive materials, and all the bulky items resulting from such an event that must be dealt with. Presently liquid wastes usually are reduced in volume through evaporation before disposal by injection or burial; but as wastes become more voluminous, AEC plans call for the precipitation of radioactive compounds to convert them into less bulky solids before burial. The most difficult problem is associated with high-level wastes, and these are mostly from fuel-rod reprocessing; they contain a mixture of long-lived radioactive elements in a bath of highly corrosive acids. These presently are being injected into deep rock, but AEC has been very interested in the possibility of their disposal in old salt mines. These would offer insured permanent storage, because the heat from radioactive decay would fuse the salt into an impenetrable coffin long before the containers corrode and leak their contents.

So far, the handling of nuclear wastes seems rational and reasonable, although many might disagree violently with this conclusion. There is much uncertainty about such possible problems as how much radioactive waste may leak eventually into general circulation, the magnitude and frequency of accidents involving radioactive materials, the possibility of future nuclear blackmail, biological effects of chronic exposure to leaking wastes, and the probability of and possible effects from the incorporation of radioactive elements into the global system. But radioactive waste disposal, indeed the whole range of nuclear development, is enshrouded so heavily in fear that any action is certain to be controversial.

Questions

REVIEW QUESTIONS

1. Based on the information in this chapter, sketch out the ecological effects of DDT that occur as the consequences of its trophic chain concentration and fat solubility.

2. Why did the Baldwin Hills reservoir fail? Is its failure conceptually related to the Denver earthquakes?

3. Compare the similarities of persistent pesticides with radioactive compounds or materials with long half-lives.

4. You know about the ecological effects of DDT; now sketch out the anticipated effects of a short-lived but broad-range pesticide.

ADVANCED QUESTIONS

1. Read Montague and Montague (*Mercury*, Sierra Club, San Francisco, 1971) and draw the parallels between it and DDT. Can you see similarities in the biological activity of small molecule, persistent, toxic chemicals?

2. Make a list of population properties that would make it difficult to find a biological control agent for a hypothetical pest population.

3. The success of the screwworm control method using sterile males depended upon a few key properties of the screwworm itself. Read R. C. Bushland, *Advances in Pest Control Research*, 3 (1960), 1-25, for a synopsis of the method on Curacao and identify these properties.

Further Readings

Cairns, J. "Coping with Heated Waste Water Discharges from Steam-Electric Power Plants." *Bioscience*, 22 (1972), 411-419, 423.

> A recent review paper on the innovative use of waste heat, due to be one of the most challenging areas of research accompanying construction of nuclear power plants. Includes a large bibliography.

California Department of Water Resources. *Investigation of Failure: Baldwin Hills Reservoir*. Sacramento, 1964.

> The investigation of the Baldwin Hills reservoir failure. The report assigns no blame directly, although readers are free to read between the lines.

Egler, F. E. "Pesticides—In Our Ecosystem." *American Scientist*, 52 (1964), 110-136.

> Egler was one of the first to frankly assess the problems of sophisticated pest control. Contrast to Stern et al., *Hilgardia*, 29 (1959), 81-101.

Eipper, A. W. "Pollution Problems, Resource Policy, and the Scientist." *Science*, 169 (1970), 11-15.

Eipper, A. W. "Nuclear Power on Cayuga Lake." In *Patient Earth* (J. Harte and R. H. Socolow, eds.). Holt, New York, 1971.

> These two references supply the details on the fight over the future of Cayuga Lake and provide entries into the thermal pollution literature.

Healy, J. H., W. W. Rubey, D. T. Griggs, and C. B. Raleigh. "The Denver Earthquakes." *Science*, 161 (1968), 1301-1310.

Henkin, H., M. Merta, and J. Staples. *The Environment, the Establishment, and the Law*. Houghton Mifflin, Boston, 1971.

> An annotated official record (transcript) of the hearings at Madison, Wisconsin. Not unique for its coverage of the hearing itself, but more importantly an example of the political process at work in environmental issues.

Hickey, J. J. (ed.). *Peregrine Falcon Populations*. University of Wisconsin, Madison, 1969.

> An edited collection of papers detailing the discovery and documentation of the decline of the peregrine. See also the special issue of the *Canadian Field Naturalist*, 84, No. 3, 1971.

Jahns, R. H. "Geological Jeopardy." *Texas Quarterly*, 11 (1968), 69-83.

> One of a group of papers in a special issue on environment. Several have proved famous enough to have been reprinted several times. Jahns stops short of providing the tectonic basis for geological hazards. For a general inexpensive reader see J. T. Wilson (ed.). *Continents Adrift*. Freeman, San Francisco, 1972. These papers are reprints from *Scientific American*.

McKenzie, G. D., and R. O. Utgard. *Man and His Physical Environment*. Burgess, Minneapolis, 1972.

> Contains several valuable papers, especially on nuclear wastes and salt beds and geological hazards from development and fluid injection.

Moore, N. W. "A Synopsis of the Pesticide Problem." *Advances in Ecological Research*, 4 (1967), 75-129.

> One of the most succinct summaries; far better than some much longer.

Rabb, R. L., and F. E. Guthrie (eds.). *Concepts of Pest Management*. Proceedings of a conference at North Carolina State University, Raleigh, and presumably available through the University. 1970.

> This volume is devoted to the ecology of pest control and features papers by major figures in the field. A very practical volume that is duplicated nowhere else, and a logical development from Egler's paper.

Rudd, R. L. *Pesticides and the Living Landscape*. University of Wisconsin, Madison, 1964.

Rudd, R. L. "Pesticides." In *Environment* (W. W. Murdoch, ed.). Sinauer Associates, Stamford, Conn., 1971.

> The 1964 book was the landmark, an important first when it was lonely to be against the free use of agricultural pesticides. It is still a useful volume. The 1971 paper is a condensed, updated version. See also the paper by G. R. Conway.

Turnbull, A. L., and D. A. Chant. "The Practice and Theory of Biological Control on Insects in Canada." *Canadian Journal of Zoology*, 39 (1961), 697-753.

U.S. Congress. "Herbicide Assessment Studies." *U.S. Congressional Record: Senate* (S 3226), 3 March 1972.

Wurster, C. F. "DDT Reduces Photosynthesis by Marine Phytoplankton." *Science*, 158 (1968), 1474-1475.

Section E
Relicts and Islands Along the Roads of Progress

SECTION E / SELF-STUDY ORGANIZER

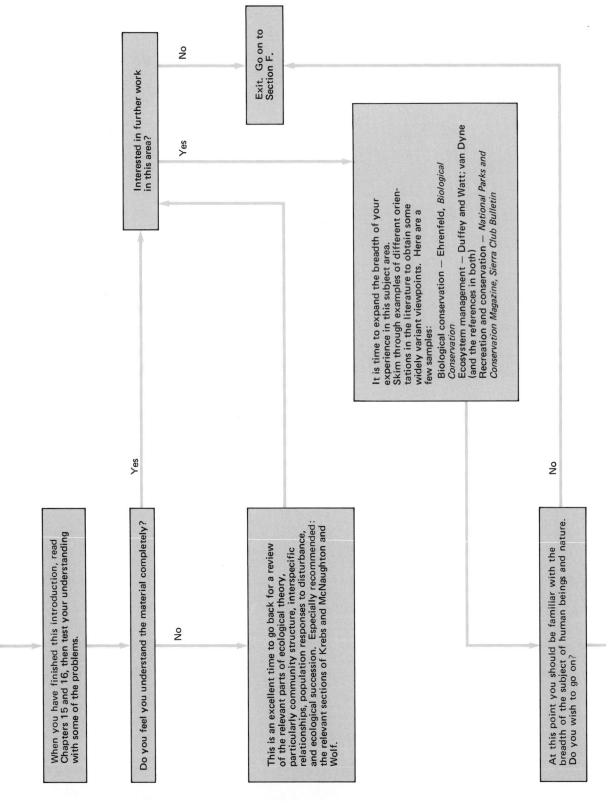

It is time to examine a few subjects in depth. We can organize our search in three different ways:

1. Emphasis on organisms (e.g., California condor, whooping crane, tropical forest ecosystems)
2. Emphasis on areas (e.g., Tsavo National Park, Mineral King (see Chapter 15 for a start in both), Everglades National Park/Big Cypress Swamp, Redwoods National Park).
3. Pot luck-serendipity of search (whatever the literature abstracts and journals might lead you to).

Examine as many examples as you have time and interest for, then go on. These studies can be very involved technically, so bear in mind that sometimes the time spent in pursuit of more detail than is readily available might be better spent tracking down additional, closely related cases.

Having digested this sort of information, you should be prepared to return to the literature on ecological theory with the purpose of ordering your information into an ecological framework. You will obtain the simultaneous benefit of a check on your understanding of ecological theory and applied ecology. For example, suppose that you chose to study the problems of Redwoods National Park. You should seek information on the ecology of redwood trees, especially their environmental requirements and their role in the forest community. See Drury for a start.

The next logical activity is to carry out some comparative analyses of different examples. Paradoxically, this could increase the power of your understanding and at the same time it would reveal the costs and uncertainties of science. Interested?

Yes →

Some sample exercises:

1. What are the commonalities of the endangered species discussed by Fisher et al.?
2. Compare the management practice in four or five national parks with respect to visitation.

No ↓

The last area to explore is the reaction of man to nature. For a good summary, read Nash or Leopold. The former is also a key to the early writings of Olmstead, Muir, Thoreau, and other pioneers of American conservation; should you wish to expand your efforts in this area. Also see Section F, Chapter 18.

It is a characteristic of this area that despite all of your work, you have probably only scratched the surface. Exit and go on to Section F.

During the past few years, environmental pollution has occupied the public consciousness as the foremost of environmental concerns, especially as an increasingly wider segment of American society has accepted the hazard it presents to public health. Even more recently, the specter of worldwide starvation and the "energy crisis" have forcibly underlined the dependence of complex and internationally interdependent societies such as ours upon reliable resource supplies. These events have focused attention upon the nature of different societies and economies which, while seemingly independent, have been shown to be just the opposite. American and European societies are so dependent on Middle Eastern oil that they scarcely can avoid involvement in many aspects of Middle Eastern society, economies, politics, and even warfare. The outcomes of these involvements can have large impacts on internal policies in turn.

While these problems are large, demand enormous expenditures of money and time, and at times even seem insolvable, there is still a larger problem. This is the nature of the present and future roles of human beings and natural systems on this planet. As human populations have continued to expand over the earth, it has been at the expense of natural systems that have evolved without us and are usually ill equipped to withstand our influence. This is, of course, a very general statement that most people take for granted and dismiss with a casual statement that this is the unavoidable cost of progress. Most people have an anthropocentric orientation (that is, human life has the highest priority) and are only too willing to accept the modification of the planet.

A few individuals, usually called *conservationists*, do not accept this premise. Their position may be called *biocentric*; it holds that human beings and natural systems must coexist harmoniously without the former's attempting to dominate or overcome the latter. The rationale for conservation can be summarized in four arguments:

1. The scientific argument. Natural systems with their long history on this planet have become integrated by the process of selection. As a result, understanding these mechanisms of integration might suggest how we could improve our own societies. Furthermore, these systems might possess many unexplored and unexploited resources we shall never have if they are destroyed before they are understood.

2. The need argument. Human biological roots lie in the natural systems we have come to dominate; these roots cannot be severed completely. Natural systems yield irreplaceable material resources; in addition, they provide aesthetic, psychological, and recreational benefits that are often immeasurable but invaluable.

3. The "insurance" argument. Some authors maintain that natural systems provide a measure of stability to human affairs because they absorb waste products and ameliorate human activities that could be extremely harmful on a large scale. This argument holds that keeping human societies surrounded by adequate "buffering systems" promotes our continued survival.

4. The morality argument. All life is sacred; the power we have to destroy natural systems demands a correspondingly high level of morality. With the power should come the responsibility to use it wisely.

Perhaps the major reason offered for public apathy toward the conservation of nature is the lack of data that would allow us to examine and prove or disprove the validity of each of these arguments. Such data would permit us to document the effects of specific past projects of major environmental impact so that it no longer would be necessary to accept on faith that these impacts are, in fact, real. This void of information has two principal causes: The extreme breadth of the problem, encompassing as it does practically all human activities, makes it very difficult even to know where to start collecting data; and the lack of funds further hinders our ability to carry out the necessary research. The latter in turn is a consequence of the more pressing and obvious needs for research in such areas as pollution, health, and resource development, subjects that presently can be tied more concretely to our continued well-being.

The ecology movement in the last decade has produced a growing awareness of the conservation problem. Wider understanding of ecology has improved public sympathy to the point where several million dollars in research funds are being spent specifically to improve our understanding of the interface between human beings and nature. The National Science Foundation has a program devoted to the broadest possible analysis of human impact on natural systems; a decade ago the political support did not exist. Increased consideration of the classic concerns of conservation is encouraging but is so recent that a severe shortage of high-quality information remains, and we lack a framework to help us determine priorities in the search for information.

Human Beings and Nature: Current Programs and Future Challenges

A flow chart representation of human beings and nature must be cast in such general terms that it would offer little benefit to readers. The only aspect of the problem that requires further attention is the choice of solution—and that, only so long as the proposition that natural systems should continue to exist is accepted. The two chapters that make up this section deal with two major issues: the known interactions between human beings and ecological systems and the conservation of space and habitat. In this section we shall use an operational definition of "nature" as follows: Nature is synonymous with the living systems with which human beings must interact. In short, nature is composed of the ecosystems that evolved before human beings and that must now compete with them for survival. It does *not* exclude us as part of the total system.

We shall examine the processes of extinction, the survival of species, and a new focus on the conservation of ecosystems. The universal solution offered to prevent extinction and to promote the survival of ecological systems is the setting aside of natural reserves, for which there are really no substitutes. We shall examine the problems of existing reserves, especially the national parks, as well as a few alternatives such as the role of zoos and the potentials of multiple-use concepts for reserves. Finally, we also shall examine the use of open spaces that are set aside specifically for human use, and the relationship between the use of urban open space and the use of open spaces of parks.

To establish more reserves may seem difficult and low in priority compared with other things, but this is easy compared with the process of protecting their integrity once they are established. This is the main challenge of the future, for we must know what our goals are, must have the results of costly research and the public support that this implies, and must be willing to manage our preserves by whatever means are necessary. We cannot expect reserves to care for themselves and to be free of outside interference. For if we do make this assumption, then it simply means that de facto management will be accomplished by the circumstances of the moment without reference to desired goals.

Selected General Sources on Man and Nature

Note: This list contains some unusual documents, particularly the Drury interview (done for the Bancroft Library at the University of California, Berkeley, and perhaps not widely available) and the Curtis book, which ought to keep the soapbox we all find ourselves on from time to time well greased.

Advance, A. B. (ed.). *Proceedings of the First World Conference on National Parks.* U.S. National Park Service, Washington, 1962.

Curry-Lindahl, K. *Conservation for Survival.* Morrow, New York, 1972.

Curtis, R. *The Case for Extinction.* Dial, New York, 1970.

Dasmann, R. F. *Environmental Conservation,* 3rd edition. Wiley, New York, 1972.

Drury, N. B. *Parks and Redwoods, 1919–1971.* An interview conducted by A. R. Fry and S. Schnepter, for the Bancroft Library/Regional Oral History Library, University of California, Berkeley, 1972.

Duffey, E., and A. S. Watt (eds.). *The Scientific Management of Animal and Plant Communities for Conservation.* Blackwell, Oxford, England, 1971.

Ehrenfeld, D. W. *Biological Conservation.* Holt, New York, 1970.

Fisher, J. L., N. Simon, and J. Vincent. *Wildlife in Danger.* Viking, New York, 1969.

International Union for the Conservation of Nature and Natural Resources. *Conservation of Nature and Natural Resources in Modern African States.* Morges, Switzerland, 1963.

Krebs, C. J. *Ecology.* Harper & Row, New York, 1972.

Leopold, Aldo. *A Sand Country Almanac.* Oxford, Fair Lawn, N.J., 1949.

Marine, G. *America the Raped.* Simon & Schuster, New York, 1969.

McNaughton, S. L., and L. L. Wolf. *General Ecology.* Holt, New York, 1973.

Nash, R. *Wilderness and the American Mind,* revised edition. Yale, New Haven, Conn., 1973.

Nicholson, M. *The Environmental Revolution.* McGraw-Hill, New York, 1970.

Stamp, D. *Nature Conservation in Britain.* Collins, London, 1969.

U.S. Council on Environmental Quality. *Environmental Quality.* Washington, 1972.

van Dyne, G. M. (ed.). *The Ecosystem Concept in Natural Resource Management.* Academic, New York, 1969.

Warren, A., and F. B. Goldsmith (eds.). *Conservation in Practice.* Wiley, New York, 1974.

Selected Periodicals

INDICES AND ABSTRACTS

Biological Abstracts
IUCN Bulletin Supplement
Zoological Record

JOURNALS AND REPORTS

Note: Three categories are represented here. Conservation organization publications, most useful for current events but usually requiring scientific analysis from other sources; general scientific journals which contain articles in subjects with at least peripheral interests in human beings and nature; and government publications.

Audubon
Biological Conservation
Bioscience
"bird journals" (*The Auk, Wilson Bulletin, etc.*)
Conservation Foundation Report
Ecology
IUCN Bulletin
Journal of Environmental Management
Journal of Wildlife Management
Living Wilderness
Nature and Resources
Sierra Club Bulletin
Transactions of the North American Wildlife Conference
U.S. Bureau of Land Management, occasional publications
U.S. Forest Service, unnumbered publications
U.S. National Park Service

15. The Retreat of Natural Systems

Natural systems are conservative. They evolve and change in ways that promote complex biological structure. Such systems tend to be composed of tightly interwoven and interdependent populations and are resistant to change; that is, they persist and preserve their biological history. Furthermore, they modify their environments in order to buffer external perturbations. But radical changes are seen, for when large climatic shifts sufficient to overwhelm system resistance have occurred, whole groups of organisms have disappeared in a mere geological instant. Dinosaurs are a famous and still mysterious example of this kind of disappearance.

For millions of years the dinosaurs dominated the earth. They radiated into many forms, from walking to swimming to soaring, and evolved to huge size in warm, wet, and favorable climates. However, early in their history a branch of the dinosaur family evolved; even though it remained inconspicuous and was suppressed by the bigger forms, this branch was to be the ancestor of the mammals that thrived when the earth turned colder and drier and when the great mountain ranges were thrown up. So even though they dominated the earth for so long, the dinosaurs eventually disappeared, leaving only a few reptiles to remind us of their heritage (Figure 15-1).

The history of human civilization has been characterized by great internecine strife, wars, revolutions, and other social and political upheavals of a scope that, to our knowledge, has been unprecedented in any other organism. Nevertheless, there is a familiar pattern: Civilizations have risen, spread, and declined much as species have. Like complex biological structure, human economic and political systems promote gradual change (and thereby discourage more radical forms) and maximum stability of environmental and social conditions. In short, whether intentional or not, both human and nonhuman systems pursue maximum stability for themselves involving interlocking structures and highly predictable future conditions. This is a consequence of the biological history of all organisms, including *Homo sapiens*.

Human Beings and Nature: A Brief History

The obvious difficulty with the arrangement described above is that both human and nonhuman systems are themselves interwoven. Human societies could not exist without nonhuman systems, and more important, they influence these ecosystems in numerous important ways. We already have examined a number of these influences—including such problems as pollution effects and the utilization of land for urbanization and agriculture—all of which have had dramatic impact. Even when change has been unintentional, our history is littered with a record of extinctions and population reductions in other species. The current rate is estimated at one species per year. There even is a plausible theory that implicates our Pleistocene ancestors as prime contributors to the mass extinction of the large mammals of that time (Martin, 1967). Martin argues that the fossil record supports the idea that primitive hunting groups, recklessly killing animals whose ranges had been reduced severely by glaciation, could have slaughtered enough large mammals to drive them to extinction. Martin calls this hunting excess "Pleistocene overkill" to denote the central importance human progenitors may have had on large mammal species then sharing the land.

Many contemporary examples illustrate the awesome capacity we have for the destruction of other species. Prominent cases include the American bison, the passenger pigeon, and the American alligator. In these cases each species was reduced from an enormous population either to carefully protected and "managed" remnants a fraction of their original size (the bison), to the finality of extinction which occurred when the last known passenger pigeon died in the Cincinnati Zoo in 1914, or to eventual recovery when persecution was relaxed (the alligator). These examples of our influence, stretching from early history up to recent times, share one common feature: They all stem from the direct exploitation of animal populations for food, hides, or sport. All attest to the power and efficiency with which we can exterminate life.

380 THE RETREAT OF NATURAL SYSTEMS

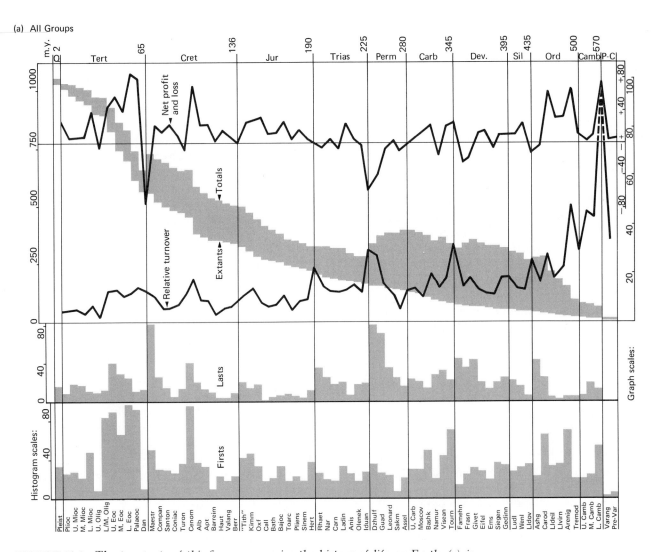

FIGURE 15-1 *The two parts of this figure summarize the history of life on Earth: (a) is the numerical analysis of all species, and (b) is just the vertebrates. These graphs illustrate the dynamic nature of evolution, speciation, and extinction. The number of species is a steady-state phenomenon controlled by a continuous process of loss (extinction) and gain (speciation). The diversity of living species has been increasing steadily through organic history, particularly since the emergence of terrestrial vertebrates. There appears to be a*

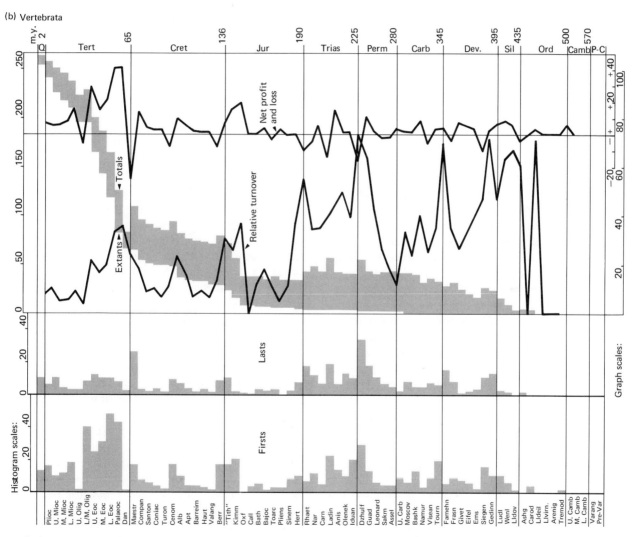

marked rise in extinction at the end of each geological era, and a less well-marked opposite trend in speciation. The vertebrate graph shows that speciation of mammals and birds has been the important event in the increase in species diversity. (Redrawn from J. L. Cutbill and B. M. Funnell in *The Fossil Record. Part III: Computer Analysis*, W. B. Harland et al., eds., The Geological Society of London, 1967.)

Habitat destruction is more insidious than direct extermination of a species. Some methods of habitat destruction are quite obvious, such as the conversion of forest and plain to agricultural land or clear-cut logging practices. Other, less dramatic changes (in human terms) may be just as important to a species' survival. For example, dams on rivers convert flowing water habitats to standing water with dramatic changes in the biotic communities found there. Below the dam, with flowing water conditions restored, the biota often will not be the same as that which existed before the dam was built; small changes in water quality are enough to promote permanent change.

The fate of the ivory-billed woodpecker is the result of just such a habitat change. As the remaining virgin timberlands of southern Louisiana are cut and replaced with second-growth forest (as good forest management dictates for highest timber yields), the inefficient, dead standing timber that foresters invariably wish to eliminate goes with the rest of the forest. As these disappear, faint hope for the survival of the ivory-billed woodpecker fades. In this case, even the preservation of the forest ecosystem would not be enough, for the ivory-billed woodpecker requires old, insect-riddled timber to survive. Nothing else will do.

There is no distinct boundary between species extinction by deliberate habitat destruction and the elimination of a species as an unintentional by-product of habitat modification; nor is there any compelling reason to separate them, for the end result is the same. As long as we must compete with and use other organisms to meet our needs, the choice will be in favor of sacrificing natural order to meet human demands for material and energy. This tendency is driven by worldwide human population growth, and as a result, the rate of destruction of wildlands and wildlife is always accelerating. Now, as our knowledge of ecological interactions continues to grow, concern is building not only for rare or attractive individual species, but also for the more serious matter of the survival of entire ecosystems.

Is such concern sufficiently widespread or will it occur in time to save much of nature from extinction? We must insure that the conservation of nature is not solely the province of a small part of the population. Although vigorous conservation campaigns are being conducted in some parts of the world, in many other areas society feels it cannot afford the "luxury" of protecting the native biotas at the expense of economic development—the two are usually in direct conflict. Many experts predict that the end of the large mammal herds of Africa is only a short time away. In recent times, Australia has seen 34 mammal and 13 bird species disappear, and the various lists of endangered and extinct species kept by different organizations throughout the world continue to grow.

Do these isolated examples imply a deliberate strategy of selective extinction, do they result from unforeseen accidents, or are they the unavoidable cost of human expansion across the planet? All these have been true in the past (the history of wolves, passenger pigeons, and kangaroos provide examples). Yet this fact is of historical interest only—it does not tell us much about the course of future events. With human demands on the biosphere continuing to increase, an examination of the history of our impact on living systems coupled with applied ecological research is needed urgently. From such a program we first need a checklist of the characteristics that make an ecological system susceptible to extinction so that we can determine which systems need special attention. Second, we need an understanding of ecosystems sufficient to establish comprehensive management procedures.

Susceptibility to Extinction

Some well-known characteristics that make species susceptible to extinction are described below.

Isolation and the breakdown of isolation   Populations that evolve in isolation may be so protected from competition and predation that they are extremely vulnerable to an ecological change. The best examples are the biotas of islands, which often display exactly these characteristics. For example, the famous dodo of Mauritius is one case where a flightless bird was easy prey for visiting sailors. The larger but also flightless moas of New Zealand were equally helpless before the Maori hunters. Living examples of isolated animals that could be destroyed easily are the giant tortoises and the Komodo dragon. For the marsupial mammals of Australia, the arrival of English settlers and their animals has proved to be a grim example of the consequences of the termination of millions of years of protection from more advanced mammals.

Small population size and low rates of reproduction   These two ecological characteristics indicate a restricted ability to recover from accidental disasters and thus a greater probability of extinction should an episode of catastrophic mortality occur. That such species could exist at all implies they are not normally subject to such drastic mortality; under such protected circumstances smaller populations and restricted reproduction can be important adaptations to prevent resource overexploitation. But depredations by human beings can change this advantage to a very definite disadvantage, since such a population lacks the capacity to recoup its losses. The blue whale is a prime example, with its large size, small numbers, and very low rate of reproduction.

Extreme specialization in feeding   A species that has become specialized to the point where it can feed on only one or two other kinds of organisms is in considerable danger if these resources are imperiled. The ivory-billed woodpecker is one example; another is the Everglades kite, which feeds solely on the apple snail of southern Florida. Species with narrow feeding preferences are greatly dependent on a stable food supply; any serious disturbance in this supply is likely to be threatening unless specific adaptations (such as resistant spores or resting stages) allow a species to survive periods of scarcity.

Specialization in habitat   Many organisms live in habitats that either are vulnerable to destruction or limited in extent, while others return to reproduce in such places. For example, many waterfowl species depend on marshes for food and nesting. Marine turtles in particular may require very specialized reproductive habitats. In fact they may lay their eggs on only a few sand beaches, where they are easily attacked; and many (if not most) of their offspring are eaten before they even reach the relatively safer sea. Conservation programs for the highly palatable green turtle recognize this problem by concentrating their efforts on the protection of the beaches the turtles visit to lay their eggs as well as by protecting the nests themselves from poaching. At this stage in their life history green turtles are most vulnerable.

Of course, some species are specialized in both food and habitat (an inescapable combination when the food resource is a major element of the habitat). An excellent example of this dual vulnerability is the blue whale, the world's largest living creature. Blue whales

FIGURE 15-2 *Timber wolves in natural habitat.* (Courtesy of Hugh M. Halliday from National Audubon Society.)

can support their bulk only with concentrated food resources, namely the shrimplike crustacean known as *krill*; these concentrations occur only in the polar upwelling zones where there are enough nutrients to support large populations. The restriction of blue whale habitat to that of its prey, in combination with its very slow rate of reproduction, have made the blue whale highly susceptible to extermination.

Species of transitory habitats Some species depend on fleeting combinations of environmental conditions that are created by such disturbances as fire, drought, or flood. Though this may seem to be a tenuous set of conditions on which to base perpetuation, some species are very successful in this regard.

Species that do well in transitory habitats are termed *opportunistic species* because they can reach widely separated and fleeting habitats and can utilize them fully once they do reach them. Depending upon the situation and the species, there is a range of situations from species that are perpetual pests to a desirable species in danger of extinction. Crabgrass is one example of an opportunistic species that thrives on plowed or open ground; hence it occurs widely in fields and lawns. It is a pest familiar to most homeowners. Kirtland's warbler of central Michigan is an example of an opportunistic species that is more desirable, but it also is less successful as an opportunistic species. This species breeds only in jack pines and only in those pines lying within a small range in height. As the pines continue to grow, survival of Kirtland's warbler will depend upon a management program that allows for a continuing supply of jack pines of the size required by the warblers. The specific program needed is one of controlled fire: Jack pines are adapted to forest fires, both because their seeds germinate only after fires open the cones and allow the seeds to be released and because the forest fires reduce the abundance of competing trees. A burning program must be used to insure that there will always be a sufficient number of proper-sized pines.

Habitat destruction Certain species have been nearly eliminated because they have suffered extensive range reductions. Some species, such as wolves (Figure 15-2), are so intolerant of human life that range reductions have been accomplished when little more than a few people moved into their range. Others have been unfortunate enough to occupy habitats that we want. For example, the tall grass communities in the Midwestern prairies of the United States nearly disappeared as the plains were plowed for agriculture. Habitat destruction is one of the most obvious threats to the survival of any species. But as simple as it may seem, habitat destruction can occur in so many different ways that one could build an extensive catalogue of variations illustrating the influences of species characteristics, ecosystem complexity, and the environmental impacts of a wide range of human activities.

Ecosystem consequences of the elimination of a species on another The removal of a species for any of the above reasons may seem a rather straightforward event, but frequently its consequences are by no means simple nor necessarily predictable. The ramifying effects of the removal of a species from an ecosystem illustrate the importance of the interrelationships that exist between various species. Suppose that a species were eliminated from an ecosystem. If it were an "important" species, its loss could irreparably damage the ecosystem of which it was a part. Such an effect could result from the extermination of a species that is (1) an important link in the trophic web or (2) an important agent in such biological processes as insect pollination of plants or one of many kinds of symbiosis.

Many examples of the first kind of relationship can be found among studies of predator-prey relationships. One is the case of the crown-of-thorns starfish, which in recent years has apparently undergone population expansions in many parts of the Indo-Pacific Ocean. Since the starfish is a voracious feeder on coral, the effect of its feeding on some heavily grazed coral reefs has been a denudation of the reef, followed by a rise in algal populations and a change in the nature of the coral ecosystem. For example, the fish that feed on coral are forced to leave and are replaced by algae eaters. The success of the crown-of-thorns in causing local extinction of some coral-reef species has led some scientists to believe that all reefs are in danger (although others disagree strongly). This would bring severe consequences for organisms dependent on coral for food and shelter, and for people who depend on corals for livelihood and protection from storms.

Despite obvious damage in areas of the Pacific and in the Red Sea, we do not know what caused the outbreak, and so we cannot determine what to do about it. Some scientists believe that dredging was responsible, because it opened up new places where starfish larvae could survive. Others believe shell collecting reduced natural predators, and still others believe it was a natural event or even an artifact of observational error (we did not look so hard for the starfish before the supposed outbreak). Consequently, we do not know whether we should persecute the starfish or simply leave it alone.

An example of the second kind of relationship has been given by Fittkau (1970). For many years, several species of Amazonian crocodiles (caimans) have been the major source of alligator skin and "stuffed alligators" proffered to tourists. The wholesale slaughter of the caimans has eliminated them over large parts of the Amazon basin. Fittkau showed that after the caimans were eliminated, the abundance of all fish species declined. He attributed this to the importance of caiman in mineral cycling in the waters of the Amazon basin. By eating fish and terrestrial animals, and, under unfavorable conditions, even subsisting upon their own stored energy, the caimans were the major elements in providing a stable source of nutrients (by their defecation) for the aquatic plants of their habitat. Hence the unique role of the caimans in making nutrients available (from the fish passing through the area that they consumed but also from various terrestrial sources and from their own stored energy) overshadows their role as predators of the Amazon River ecosystem. Consequently, where we might have expected reduced predation to have increased the various populations of fish, the elimination of caimans has had the surprising effect of actually decreasing fish production.

The Comprehensive View of Susceptibility to Extinction

An analysis of factors contributing to species extinction is useful because it can help focus attention upon the ecological and evolutionary processes that lead to it. However, remember that the comprehensive view which includes all important factors in a particular case is the only realistic one, because all the factors are interrelated; that is, all interact to affect a particular species. For example, one species may possess a strategy of high reproductive potential, high dispersal ability, and an environmentally resistant resting stage; another may have low reproductive potential and low dispersal, but it dominates its habitat and lives a long time, and its young survive well. The former will do well in unstable, changing environments, and the latter will do better in more stable conditions.

Despite the lack of clear distinctions among the responsible causative agents, three general conclusions are possible. The first is that in principle we can define a set of specific characteristics from which we can infer at least a crude indicator for the probability of extinction for a species. We have examined some of the possible criteria above—many other permutations could be specified as well. But the least obvious characteristic is that the most highly evolved ecosystems may prove to be the most susceptible to disturbance *if* community stability turns out to be a product of environmental stability more than of any biological property. If the survival of the whole coevolved complex depends on the stability of future environmental conditions, we could find that the stability we assumed to be a property of complexity will turn out to be a mirage. Then we might have the curious situation in which species of complex ecosystems could be more susceptible to extinction than species of simpler ones. This conclusion is contradictory to the conclusion of Odum (1969; see also pages 387–388).

The second conclusion is that biological change is inevitable. Sooner or later all species lose their adaptability as their environments change, and extinction is then inevitable. However, the world has not become an impoverished place, because new species have evolved continuously and even have contributed to the extinction of their predecessors. The impact of human beings, our technology, and our numbers upon the planet is simply a new factor in a theater of endless change. In some cases, we have had no effect; in others, we simply have accelerated the inevitable process of extinction. The eventual outcome, barring a moral revolution that confers greater respect

TABLE 15-1 *The strategy of ecosystem development.*

Ecosystem attribute	Developmental stages	Mature states
1. Energy accumulated as biomass	relatively large	relatively small
2. Biomass supported per unit energy	lower	higher
3. Nature of food chain	simple, tend toward linear	development of food webs
4. Species diversity	lower	higher
5. Degree of ecosystem organization	lower	higher
6. Specialization of populations and individuals	lower, more generalists	higher, more specialists
7. Rate of exchange of materials between community with environment	higher	lower
8. Importance of dead organic matter in ecosystem mineral cycles	lesser	greater
9. Selection for:	rapid response to stress, higher growth potential	ability to resist stress, lower growth potential
10. Community stability	lower	higher

SOURCE: E. P. Odum, "The Strategy of Ecosystem Development," *Science*, 164 (April 1969), 262–270. Copyright 1969 by the American Association for the Advancement of Science.

for nonhuman life, is a biologically poor planet of little diversity.

The third conclusion is familiar: In the final analysis, all species must reproduce and disperse faster than they can be eliminated, or else they will go to extinction. The process is a direct and logical application of ecological theory, even if it is difficult to measure dispersal and reproduction in ways that measure evolutionary success. But if we are interested in a program of dynamic conservation of ecosystems, comprehensive management programs will have to be developed, whether our information is adequate or not.

Responses of Natural Systems

In the previous section we examined some characteristics that could put a species in danger of extinction from the point of view of an environmental force which might impose a change on the system. Now let us examine the consequences of extinction for us and the adjustments made by natural systems in response to changes imposed upon them.

Ecological theory suggests that given enough time and environmental stability, ecosystems should evolve toward greater complexity. Odum (1969) has summarized the trends expected in ecosystem evolution (Table 15-1). He suggests that the direction of evolution is toward the maintenance of an extensive ecosystem with a definite capacity to resist externally imposed change. The evolutionary trends also require increasing specialization of ecological units (primarily in populations) and a consequent need for increased interdependencies among these various units. For instance, to return to the example of the coral reefs, one of the most highly evolved interactions is called the *cleaning symbiosis*. This behavior involves stereotyped behavior from large fish and very small ones (plus a few species of shrimp), in which the latter pick parasites from the larger ones, even from their mouths, without danger of being eaten (for details, see Feder, 1966). In this example, the small cleaners play an important but very specialized role in the coral system. In return for "services rendered" to many larger species, the cleaners have a dependable source of food.

The antithesis to a strategy of increasing specialization may be found in areas that are disturbed frequently and in harsh environments like the polar zones. In such places, fewer species can survive the vicissitudes of harsh climate or uncertain food supplies; hence only those few species that can tolerate such conditions are to be found. They can successfully dominate more kinds of food and breeding habitat because they lack competition from other species. Such species are generalists; they do well only where other species that could outcompete them are excluded by unfavorable environmental conditions.

Human beings occupy a unique position in this scheme. We are the only generalist species that can outcompete and replace more complex and highly evolved communities. Long ago, natural systems of temperate climates in the Northern Hemisphere disappeared as the land was taken for agriculture; now, ecosystems of the tropics are receiving increasing scrutiny as a source of new lands for expanded agriculture, and much destruction of African and South American tropical forests has occurred already.

Soon, perhaps, as human resource requirements continue to grow, even the deserts and arctic tundras will be invaded and their biotas replaced by more agricultural crops if technology permits (for example, desert agriculture that uses the Puerto Penasco method described in Chapter 6). The development of complexity and integration in an ecosystem requires time and the process of trial-and-error in which many different species may attempt to become a part of the ecosystem. They may persist if they successfully integrate into the community; if they cannot, they will disappear quickly. Clearly, we cannot synthesize complex communities or even develop the intricate relationships that compose them, although we sometimes can create conditions which encourage their formation. It is far easier, unfortunately, and even inevitable that some human activities will destroy these results of the evolution of countless eons. Our history is rife with familiar examples.

The total destruction of ecosystems by their displacement from the land is an extreme example, but far less interference is needed to produce serious disruptions in an ecosystem. Direct exploitation of various species by harvest or control programs can strongly modify energy flow to alter the biological structure of the ecosystem. Suppose, for example, that we wished to exploit a single species in a hypothetical ecosystem with a known trophic web (for example, plant D in Figure 15-3). As the harvested proportion of the plant D population increases, less energy is available to species which cannot feed on that plant. Consequently their populations will decrease, and the ecosystem will become less diverse and be dominated increasingly by fewer species. Note that the results

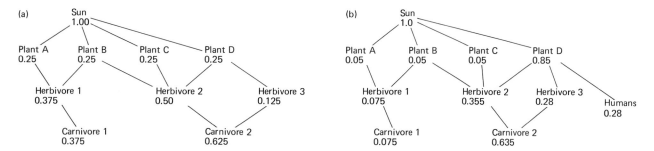

FIGURE 15-3 *(a) Hypothetical trophic web unharvested by human beings. (b) Web with one plant species harvested by human beings. Both these diagrams assume that the total energy available to each species is divided equally without any energy loss in proportion to the number of trophic links. Of course, these assumptions are ecologically unrealistic, but they do not affect our conclusions. In (b) consider the addition of an agricultural system in which the land is cleared and planted with Plant D such that it receives 85 percent of the available sun energy. Without any additional assumptions, this act shifts the balance of energy flow dramatically. The reduction of available energy to Plants A, B, and C ramify through the food chain, resulting in reduction of Herbivore 1 and Carnivore 1. These reductions in energy flow imply reductions in population sizes and therefore changes in diversity of the community. In turn, these changes imply changes in community stability and effects on parts of the food web not represented here.*

of the harvest lead to an altered ecosystem that, while it may retain a few characteristic species, is not the same as before it was exploited. The changes imposed on interspecific relationships bear profound implications for the future of that ecosystem.

Just how important the destruction of naturally evolved relationships is depends on the nature of the control systems and human dependence upon particular "balances of nature" (a popular term for stability of the populations that characterize the ecosystem). All too often we have found that the result is a kind of "revenge of nature," in the form of the release of the natural controls that had operated on some species in the ecosystem. Crabgrass is one example;

it thrives in disturbed areas and on bare ground but is rapidly eliminated by the taller plants that follow it in the process of natural ecological succession. Crabgrass is adapted specifically (by its preference for bare ground, its prolific seed production, and its wide dispersal of seeds) to take advantage of disturbed ground. Plowing the soil eliminates some weeds but favors crabgrass, which then presents a weed problem.

Agricultural crops (with current production practice) offer an enormous stock of potential food energy to species that can utilize them. We label as pests those species which are fortunate (and adaptable) enough to be able to utilize this store of energy. Many vertebrates and insects fit this description and thus may be called pests at one or more stages of their life cycle. Viewed from this perspective, pests should be expected as a perfectly natural development resulting from human tampering with natural systems. Furthermore, pests that can resist or recover from insecticides may be at even more of an advantage, since the insecticides may keep the ecosystems at their most simplified and highly disturbed state. The pesticide destroys natural control mechanisms and ends up increasing the number of pests. (This is often referred to as the "pesticide treadmill.") This observation is the rationale for reducing pest populations by encouraging the reinstitution of natural control mechanisms. Such biological control agents at least hold the promise of being effective control measures—that is, their actions are proportional to the size of the pest population—instead of merely being a mortality to which the pests are adapted already by virtue of their ecological characteristics.

The basic problem is that nature has a strategy of maximum dispersion—many species, many energy-flow pathways—while we pursue exactly the opposite strategy. In a world dominated by human beings, natural ecosystems have increasingly fewer refuges in which they may develop or survive, except under conditions imposed by us. The changes that accompany the decline of natural complexity will not be the same in every circumstance, and in some cases they may be ameliorated effectively. However, there is no way to recover a species once lost or coevolved relationships once destroyed, and unless human population growth is halted and our demands on the planet are stabilized, the destruction of nature will continue.

Do Human Beings Need Nature?

One might argue that natural ecosystems are important only if we can demonstrate that their continuing existence is vital to us. If this condition is not met, then the only argument left is that a particular ecosystem must be conserved solely for its own sake; otherwise we could destroy it and replace it with a system with a known human benefit. We might preserve some undisturbed specimen areas of each of these "useless" ecosystems for science and for those who have an interest in them. It is obvious, therefore, that the question of human need for natural systems is important in the formulation of broad social and environmental policies that, in the absence of technical information, involve basic values and moral judgments.

In fact, little information is available to support the contention that we need nature. Given this lack of data, judgments have been made on the basis of personal moralities and values. On what is presently the majority side of the issue is the argument that the ecological risks of development of the landscape (in the sense of the destruction, modification, and replacement of natural ecosystems) are small compared with the benefits that we will reap from them. Basically, variants of this argument constitute an anthropocentric position, in which the value of human life and the attainment of a better life for all individual human beings have the highest priority.

The essence of the minority argument is that human well-being is bound inextricably into all the world's biological systems and that we cannot separate the well-being of one from the other. Therefore, that which promotes natural systems will also benefit us. Differences between these positions are summarized in Table 15-2, from which it will be evident that there is superficially little in common between the two positions. However, a closer examination of Table 15-2

TABLE 15-2 *Some of the main arguments advanced for and against comprehensive biological conservation. The pro position implies a comprehensive view of nature, "biocentric" and perhaps somewhat misanthropic, and a belief that the risks from development that are imposed on nature are sometimes not worthwhile. The con position is just the reverse. These arguments are central to the entire man-environment relationship (see Chapter 20).*

Pro arguments	Con arguments
1. All life is sacred; by an extension in morality, all species have an inherent right to survive, and we must act as protectors for all species.	1. All species have a right to survive only so long as they earn it. Human lives are supremely important; if necessary, other species may have to be sacrificed for this cause.
2. Human beings make up only one part of the global ecosystem.	2. We are more than just a part of the global ecosystem: We dominate and control it for our ends.
3. If nature is destroyed, there are intolerable risks for human survival.	3. The unthinking conservation of nature is an esoteric matter unless documented proof is presented to show that human survival is bound closely to that of some other species or ecosystem.
4. It is unwise and wrong to undertake environmental modification if the consequences are unknown—and the consequences are generally unknown.	4. Viewed from the human perspective environmental modifications are necessary. If there are unfortunate consequences, they can be countered by further modifications or by provisions of adequate reserves of nature.
5. Once unique living systems are lost, they are lost for all time. Science should have the chance to study them.	5. Science for the sake of science is too expensive and esoteric to be justified in the face of more pressing needs.

shows that some differences are just a matter of degree and that there are some obvious intermediate positions between the two extremes. The intermediate positions arise from the political process. For example, suppose that an agency proposes to cut down a forest to build a recreation village. One side of the argument might require proof that such an action would harm public interests to a greater extent than the benefits generated by the village. At the other extreme, advocates might claim that no development could exceed the value of the undisturbed forest itself. Those taking a centrist position could support all or part of the development if certain conditions were met. And there starts the political bargaining process.

To date the burden of proof has rested on those who would stop human encroachment on nature and prevent the destruction of natural systems. Despite growing public willingness to value some natural systems having no direct human benefit solely for esoteric and aesthetic considerations, human nature supports continuation of an attitude that favors our welfare above all else. This is certainly the easiest position to take, because it avoids the moral dilemma of choosing a course of action that does not place human life and welfare first; the costs of such a policy are borne by the natural environment.

All these considerations still do not tell us whether or not we "need" nature, nor do they give us hints about exactly which natural elements we do need. A proponent of one view might argue that it will be possible to control environmental elements so well that we shall need no natural buffers to ameliorate our mistakes. Another, however, might argue that no system (however carefully designed) can guarantee everything will happen exactly as planned. For this reason, if no other, extreme caution in changing natural ecosystems always will be required to prevent unanticipated and irreversible consequences. The simple truth is that we do not know the answer, and different groups are proceeding on very disparate assumptions.

The Present: Contrasting Examples

There are innumerable examples of human impact on biological systems—species extinction is only one. However, only a few examples are needed to illustrate the kinds of impacts that have occurred in the past and that continue today, including some in the planning stage that dwarf anything ever attempted. Several of these have been discussed previously, such as the North American Water and Power Alliance (NAWAPA).

THE PREVENTION OF EXTINCTION: THE ZOO ALTERNATIVE

The danger of extinction facing many species today has spurred the search for alternatives other than the maintenance of large segments of natural systems. One alternative is to create small but critical reserves (in terms of species survival) more or less free from development, such as regional, state, and national parks; wildlife refuges; and national recreation areas. These multiple-use reserves require extensive discussion and are examined in Chapter 16.

A second alternative is that taken by zoological societies as part of species-survival programs. In recognition of the rapid disappearance of many animals (particularly large, conspicuous ones), many zoos have undertaken exchange and breeding programs in the hope of saving some vanishing species from extinction. But will the final result be the same as the stated objective? That depends upon one's personal values. If the desired end is rare species that can be seen and experienced in confined (if "natural") settings, the stated and attained objectives would be the same; but if the goal is to save a natural population of a rare species from extinction, the efficacy of the chosen method is open to debate.

Zoological gardens are logical means to acquaint the public with the diversity of living organisms and also increasingly with some principles of ecology and evolution that apply to the adaptations of organisms to their environments. But zoos usually cannot duplicate the natural environments of these organisms (Figure 15-4). Hence zoos are in reality museums with living collections of animals that cannot be used invariably as stocks for the reestablishment of wild populations in ecologically suitable areas. At least, the burden of proof must be upon the zoo to demonstrate that its breeding programs can produce animals which can survive in the wild. For this reason, saving species from extinction means little more than saving some individual museum curios.

Hediger's (1968) classic studies on the psychology of zoo animals in Europe show that animals which are removed from their natural habitats behave abnormally in the confining and unnatural habitat of small cages. This is certainly an extreme set of circumstances, but one may wonder if a population bred and raised in zoos and therefore accustomed to captivity would be prepared with the appropriate adaptations and behavioral responses to survive in a natural environment if an attempt were made to use it in reintroducing the species to its natural habitat.

Another problem, associated with zoological collections to some degree but very much more prominent with trade in exotic animal pets, is the high mortality rate of specimens during capture and shipment. This problem has often been presented in an extremely emotional light evoking an image of callous cruelty, an image that usually is exaggerated. Even so, this can be a critical problem for extremely rare animals, which are precisely the ones desired most highly by zoos and wealthy pet fanciers. The looting of the psittaciform birds (parrots and their allies) in Australia, prized for their bright plumage and friendly dispositions, resulted in an absolute ban on their export without special permit. (Now nearby New Guinea is raided instead.) Another widely known example of this problem results from the profitable trade in tropical fish from South America, a trade in which even more catastrophic losses are known than for birds.

However, we cannot dismiss zoos as frivolous sideshows for curious people, because they underline a critical policy question: Is the extent to which fragments of natural systems are permitted to survive actually determined by public wishes? In other words, does nature exist only for our benefit? Martin Krieger (1973), in a paper called "What's Wrong with Plastic Trees?" has raised precisely this point with regard to the utility of natural environments versus environments constructed by human beings. In some real ways zoological collections meet public needs, and it is clear that zoos will continue to perform a real and valuable function in our society; but it is equally clear that they do not provide the best way to preserve species from extinction. The question is the same as that raised by the beggar bears of Yellowstone National Park. In such circumstances what kind of "nature" is being saved?

FIGURE 15-4 *A Sitalunga bull in Cascade Canyon of the San Diego Zoo. The San Diego Zoo is one of the better examples of a synthetic natural habitat.* (Courtesy of F. D. Schmidt from the San Diego Zoo.)

THE DYNAMICS OF EXOTIC SPECIES AND THE EXTINCTION OF NATIVE BIOTAS

One elementary principle of biogeography is that species have been established by isolation. As we noted above, isolation is also an important factor in protecting native species from superior competitors; hence isolation confers some protection from extinction. If we reverse this reasoning process, we would conclude that the control of species introductions is one way to lessen the dangers to existing systems. One application of this idea is the ban on importation of a variety of plants and animals into the United States to prevent the entrance of agricultural pests and disease. Fresh Argentine beef has been prohibited from importation to the United States for many years to prevent the accidental importation of hoof-and-mouth disease. But many such importations have occurred. One of the most famous has been the worldwide introduction of rats from ships into the ports of many countries. Rats are extremely aggressive and have proved to be ruthless competitors with other rodents and efficient predators on other animals.

The previous discussion emphasized the role of the breakdown of this isolation in the extinction process and only implied the human involvement. However, a brief consideration will show that it is impossible to take such a cavalier attitude because human agents have been important in breaking down this isolation between biotas. Whether there is human involvement in the introduction of the exotic species or not, the principles are the same. The difference is that human agents can abrogate barriers more easily than anything else can. One example will illustrate these points. Before 1932, the marine lamprey (*Petromyzon marinus*) had not been found in any of the Great Lakes above Lake Ontario. The fishermen of the Great Lakes were able to exploit a rich lake trout fishery as a result. After 1932, however, when the Welland Ship Canal was improved, the lamprey began appearing in the rest of the Great Lakes. By 1949, they had virtually destroyed the lake trout fishery in the lower Great Lakes. Why? The Welland Ship Canal opened up the Great Lakes to the lamprey: It simultaneously provided a migratory pathway around Niagara Falls and a free ride (via ship traffic, which the lamprey could use as transportation by clinging to ship hulls). In the Great Lakes the lamprey found an ecosystem unadapted to its depredations, with the result being the near extinction of the lake trout population.

The ecological lesson is familiar. Ecosystems tend to develop some degree of resistance to change, but human interference can introduce transplanted species about the planet and upset the most ancient and complex ecosystems. Successful introductions, whether accidental or not, have left our influence deeply impressed upon essentially all natural systems.

POPULATION AUGMENTATION: THE LOGIC OF HATCHERY OPERATIONS

State fish and game departments are subject to considerable pressure from organized groups of sport fishermen and hunters who expect them to help increase the potential take. This pressure tends to limit the ability of fish and game agencies to undertake missions other than those having a direct and demonstrable effect on game animals (especially when budgets are chronically too small). This mission orientation is not necessarily too narrow except when it fails to accomplish its goals or when, in the process, there are wide-ranging effects on the ecosystem of which the game species is a part.

Watt (1968) has argued at length that the success of hatchery operations, as measured by an increased yield of game fish to fishermen, depends on when the supplementation is made, the size of the fish population in question relative to the size of populations that can be sustained in the environment, the kinds of mortality the population is subject to, and how long the extra individuals supplied by the hatcheries must endure before being removed by fishermen. Using these factors, Watt inferred that the highest success would be expected under these conditions:

1. When the time that hatchery fish must survive in nature before being caught is short
2. When the hatchery fish are not eliminated quickly by strong competition with wild fish
3. When weather-related or other catastrophic mortality has reduced the wild population to a point below the supportable level

Watt concluded that hatchery augmentation was likely to be most effective in situations where the fish would be in the water just long enough to be caught. For example, trout stocking in popular streams near urban areas historically has been a fruitful policy for fish and game operations, both in building public support for future operations and in serving the recreational needs of the public.

Watt's conclusions imply that the aims and methods of hatchery supplementation resemble those of agriculture and forestry. The logic behind hatchery operations embraces the conscious manipulation of ecosystems to favor particular species (Figure 15-5). In effect, fish and game management is actually "herd management" directly analogous to that undertaken by cattle ranchers. The "put-and-take" operation outlined above is a strategy for supplying meat to a special kind of consumer, namely, fishermen. However,

FIGURE 15-5 *Stripping eggs from female Chinook salmon at the Washington State Fish Hatchery. The eggs will be fertilized and hatched in tanks for stock supplementation.* (Courtesy of Keith Gunnar from National Audubon Society.)

this operation is necessary in order to meet the demand, whereas associated operations such as the elimination of predators and competing species seem far less beneficial. In New Zealand, for example, the development of game fisheries has included extensive destruction of native fishes as imported trout (familiar to and desired by the settlers of European stock) have displaced them. Perhaps the native salmonids (also trout) were not suitable game fish, and perhaps they had no obvious value to human beings. But the story of the replacement of native species has been replayed on island after island and continent after continent with man as the judge, jury, and executioner. Can we hold this kind of power responsibly and manipulate natural systems whenever we cannot immediately define the purpose of an existing ecosystem?

It would be unfair to infer that the sole purpose of hatchery operations is to gratify fishermen. For migratory fish cut off from their breeding streams by dam construction, hatcheries may represent the chief hope for survival. For example, the king salmon of the Pacific returns to the rivers of California to spawn, but many of the natural spawning beds no longer are accessible to the fish because of the many dams constructed on these rivers. Fortunately, the female king salmon can be stripped of her eggs fairly easily. The eggs then can be fertilized artificially, and the fingerlings can be raised in the hatchery; mortality rates are lower than those associated with natural spawning. In California, most spawning runs of king salmon are natural only to the point where the salmon encounter the dams; they remain there until hatchery personnel catch and strip them. Without the hatcheries, the southern stock of king salmon would be endangered because their reproduction is threatened. It is ironic that salmon survival depends upon human efforts to ameliorate human activities (the process of water impoundment).

POTENTIAL FOR BIOLOGICAL DISASTER: THE PANAMA SEA-LEVEL CANAL

What would happen if two entirely different ecosystems were allowed to mix? Imagine that instead of only one exotic species being introduced into a new ecosystem, two different ecosystems encounter one another for the first time. What might the effects be? Hopefully, we will never know for sure, for if this does occur, the effects are likely to be dramatic and irreversible. A project that would produce just such an encounter is being considered, however: the Panama sea-level canal. Plans now are being formulated for the construction of this new "Panama canal" (a possible misnomer, since the canal still could be built across another Central American country). The canal would be entirely at sea level; its construction would be a very large-scale project that would involve nuclear detonations to dig the canal. The project would be an engineering feat comparable to the massive transfers of water that have been proposed to satisfy the needs of North America (Chapter 7).

The present Panama Canal was completed in 1914 and was an immediate success because it connected two major oceans and shortened ship travel times by months. Fortunately, it has not allowed free mixing of the biotas of the two oceans it connects because the canal had to be built over a mountain ridge. In order for ships to pass through the canal, they must enter a series of locks and then pass through Gatun Lake, which is entirely freshwater. This effectively destroys all the potential migrants between the Atlantic and Pacific oceans, just as oceans prevent migration of freshwater organisms from the continents to isolated islands. Of course, a sea-level canal would not deter the free exchange of organisms between the Pacific and the Atlantic, whatever consequences this might have (Figure 15-6).

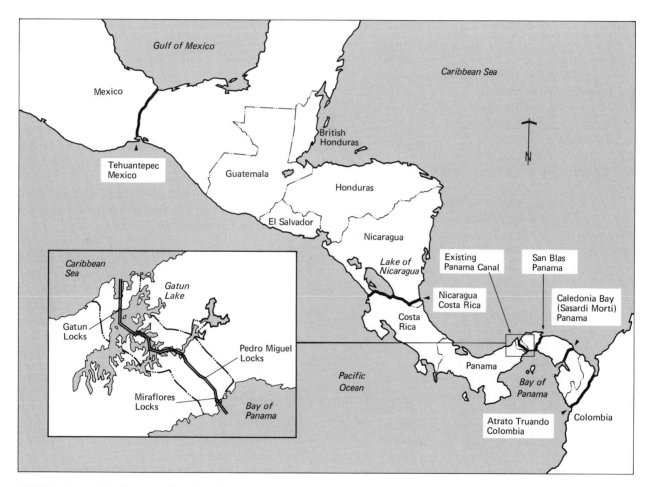

FIGURE 15-6 *The Panama Canal and proposed routes for the new sea-level canal. The inset shows detail of the present Panama Canal. Arrows on the large map indicate proposed routes for the new canal.* (Adapted from Ira Rubinoff, "The sea-level canal controversy," *Biological Conservation* 3, p. 33, 1970, Applied Science Publishers, Ltd. and "Central American sea-level canal: possible biological effects," *Science* 161, p. 857, 30 August 1968, copyright © 1968 by the American Association for the Advancement of Science.)

Why, since we have the present Panama Canal, is there pressure to build the sea-level canal? The most obvious reason is that the new generation of big ships, especially the supertankers, are too big for the present canal. The canal is too shallow, and the locks are not big enough; and so large ships now must go around Cape Horn. Moreover, in the absence of information about ecological consequences, the obvious economic savings to shippers generate considerable pressure to construct the canal as soon as possible.

Nonetheless, the new canal may never be built. The political problems involved in reaching a satisfactory agreement between Panama and the canal company are formidable. In addition, the economic problems of compensation for the displacement of people, for the construction costs themselves, and for an equitable distribution of costs among all interested parties are equally formidable. Finally, there are fears and dangers associated with the use of the nuclear devices that would be detonated to excavate the canal. These must be powerful enough to crumble mountains and blow the debris clear from the canal site; yet they must not be powerful enough to spread radioactive wastes throughout a wide inhabited area.

Besides these effects, the possible construction of the sea-level canal has raised concern among scientists such as Ira Rubinoff that catastrophic biological effects could occur. Rubinoff has been studying the effects of mixing certain Atlantic and Pacific species and has concluded that the Isthmus of Panama was submerged recently enough in geological time so that many related species presently separated by Central America have not yet developed complete reproductive isolation. Rubinoff presented four possible alternative consequences of renewed contact as a result of the sea-level canal:

1. The development of many fertile hybrids that might exist quite distinct from the parent populations. If the hybrids were successful and numerous, they could obscure the differences between the parents, and the whole complex would look like one highly variable species.
2. The development of many hybrids that are adaptively inferior but that would cause extinction of both parent populations.
3. The production of sterile hybrids that are eliminated rapidly and do not inferfere with the course of evolution of the parent populations.
4. No hybridization of any sort between the parent populations, but the distinct possibility of competition with one another. Competition could, in turn, lead either to coexistence or to the elimination of one or both parent populations.

The information needed to evaluate Rubinoff's hypotheses remains to be gathered, although it is possible to make some informed guesses about their validity. Hypothesis 2 seems unlikely, since adaptively inferior hybrids could hardly cause extinction of the parent populations without complete hybridization, in which case neither parent would produce its own kind. The remaining hypotheses represent the possible outcomes of the hybridization of not-quite-distinct species. Which hypothesis holds true for any particular case depends on the species in question as well as on various environmental factors. Evidence gathered by the scientists of the University of Miami's program in biological oceanography supports the contention that most of the Pacific fauna off Panama is indeed derived from Caribbean-Western Atlantic ancestors, which increases the likelihood of hybridization; but it is impossible to predict which species will hybridize until a species-by-species comparison has been made.

Rubinoff concluded that the possibility of a new sea-level canal provides a priceless opportunity for an ecological and evolutionary experiment using "before and after" comparisons to study processes of ecological interaction. Rubinoff emphasized the need for broad multidisciplinary teams of scientists and for studies lasting long enough to establish baseline information about seasonal and long-term population fluctuations. It is easy to see that the new canal could trigger major biological changes, but it is quite another matter to plan and carry out the studies to predict these changes. Provided that there are no insurmountable obstacles to overcome, we would need to study both marine ecosystems comprehensively, a task requiring many people and billions of dollars to provide the kind of information Rubinoff desires. In other words, the information cost for proper study of the sea-level canal is very high. Rubinoff was well aware of the problem of information cost, but he felt it was "not an impossible task" to do all the critical experiments.

Rubinoff's solution would be to create a control commission to study environmental manipulation and give it powers that include "adequate funding to support its own investigations." This is a naive suggestion. Experience indicates that the problems of research funding are not necessarily solved by the creation of new superagencies, no matter how important their mission. All too frequently, most agencies are chronically short of money, so that they are often unable to accomplish even their highest priorities.

More important is the fickle nature of public support for any project requiring long-term financial aid. Because of this tendency and other pressing demands, it is unlikely that the biological consequences of the sea-level canal can attract the support needed to carry out the necessary baseline studies unless the forecasted effects of such intermingling of faunas can be shown to have a dramatic and major impact on human welfare. Even if funds were unlimited, the number of possibilities that require investigation would take years to complete, precious time during which the project would remain stalled under steadily increasing pressure from groups interested in the fastest possible completion of the canal and everballooning costs. Finally, many critical relationships are likely to remain unknown despite our best efforts. For example, how well can we predict even the results of competition between only two species when the conditions vary with time and environmental conditions? (The ecology literature indicates that even small differences can drastically change the outcome of competition.) About the only alternative we have is to make well-informed guesses about the possible effects while waiting for the experiment itself, that is, the completion of the canal.

Despite all these practical difficulties and limitations, if the sea-level canal is built, it will be an experiment in combining entirely separate ecosystems unless an effective migration barrier is found, such as Gatun Lake has proved to be for the present Panama Canal. That we even have the power to consider changes of such magnitude in natural systems is awesome and frightening, especially since the major benefits of the new canal would go to such a small sector of the total human population. The situation demands a reevaluation of the careless modification of existing ecosystems that has prevailed so often up to now (even if reevaluation shows us how difficult it will be to improve our evaluation and planning of projects with major environmental implications).

A GLIMPSE AT A PROMISING FUTURE: THE WOLVES AND MOOSE OF ISLE ROYALE

Isle Royale, Michigan, is an island in Lake Superior near the Minnesota shore. Preserved as a national park in 1940, Isle Royale has been the site, under the direction of Durward Allen of Purdue University, of extensive studies of the relationship between the moose and wolf populations living there. These studies have shown that we can learn enough about natural ecosystems to institute intelligent management procedures and that such intervention is sometimes necessary, since the complete isolation of ecosystems from outside human interference is, for all practical purposes, impossible, and therefore is not a viable alternative.

Early in the twentieth century Isle Royale was covered with spruce and fir trees, which created an environment ideal for the luxuriant growth of lichens. The lichens supported large populations of woodland caribou, snowshoe hares, and lynx—typical mammals of northern coniferous forests. However, logging and repeated fires reduced the density of spruce and fir trees, and as a consequence the populations of animals that characterize this ecosystem also decreased. The reduction of spruce and fir simultaneously favored moose, beaver, and red fox—animals associated with the aspen and birch trees that thrive in burned-over areas. When the moose arrived (by swimming from the mainland) sometime before 1915, they found conditions to their liking and underwent a population explosion that crashed dramatically when a burn in 1936 destroyed a large part of the suitable browse (aspens, birch, and some shrubs). However, new browse growth was stimulated by the fire, and the moose population again grew, only to stabilize after wolves arrived over the ice in the winter of 1948 (Table 15-3).

Many fascinating aspects of the wolf-moose relationship have implications for ecological theory, but the management implications are derived from two goals:

1. Isle Royale has a wolf population that is worth conserving.
2. The wolf-moose interaction should be maintained on Isle Royale.

Research on Isle Royale has shown that the moose population is sensitive to the quantity of high-quality browse. If the moose population gets too large, the browse species are overutilized, and the moose population must decline. Since the wolves depend upon the moose as a major source of food, their numbers are sensitive to the number of moose. The situation for both species is further complicated because the birches and aspens are a fleeting source of moose browse. The first manifestation of this problem is that they soon grow too tall for the moose to reach; second, since they are not shade tolerant, new seedlings do not appear to provide a continuing supply of moose food, and they eventually are replaced by the spruce and fir. In addition, neither wolves nor moose can freely emigrate to escape food shortages that might occur. On the mainland, moose will leave an area long before it is overgrazed; whereas on Isle Royale, an excessively large population is doomed to stay and worsen its food situation until that is corrected when part of the population dies.

TABLE 15-3 *The history of the wolves and moose of Isle Royale.*

Year	Estimated number of moose	Estimated number of wolves	Comment
Prior to 1900	0	0	community dominated by woodland caribou and coyote
1915	200	0	this and entries through 1961 from Mech (1966)
1920	300	0	
1925	2000	0	
1930	1000–3000	0	
1936	400–500	0	crash in moose population due to over browsing and fire in 1936
1943	171	occasional stray	
1948	800	first permanent pack	
1953	?	5	
1956	300	14	
1959	194	19	
1960	600	20	
1961	600	21–22	
1967	1000	25	moose population estimate higher here probably because of a more thorough method of aerial census used. Wolf population increase apparently real

SOURCE: Mech (1966) and Jordan et al., *American Zoologist*, 7 (1967), 233–254.

These results suggest one possible management solution, given the two goals specified above. One could design, for example, a controlled burning program to be rotated regularly around the island. Such a program would burn out the spruces, firs, and whatever is left of the birches and aspens, thereby insuring a supply of suitable moose browse. It is clear that a fire suppression program would favor ecological succession toward spruce and fir and eventually would depress the moose population, with possible harmful consequences to the wolves, depending on how well they switch to caribou. The wolves and moose of Isle Royale underscore the point that a program of active management based on detailed knowledge of the system of interest is most desirable. However, the definition of that program depends on goals such as those given above. For example, suppose we wanted to promote a wolf-caribou predator-prey system: A zealously enforced fire suppression program might be the best course.

The explicit definition of system goals in ecosystem management and the development of a management plan based on these objectives heralds the advent of the ecosystem approach to the formation of carefully planned relationships between us and nature. Comprehensive and usually expensive, the ecosystem approach will require much refinement before we can begin applying it widely. On one hand, we are learning more about the systemwide effects of large engineering projects and the problems of well-meaning but intuitive management policies. On the other, the high cost of information discourages the will to invest the capital needed. Whether or not the momentum currently building for comprehensive ecosystem management can be sustained remains to be seen.

What Might the Future Hold?

Despite the extensive information requirements, more comprehensive ecosystem management will have to be the goal of future efforts to control the interrelationships between us and nature. While species (especially familiar, spectacular, cuddly ones) will continue to be the prime objects of public concern, ecosystem management is far more important in the long run both for the betterment of science and for the most comprehensive form of conservation. In any case, meeting the ecological requirements of a particular species requires a management scheme that deals with the ecosystem of which it is a part, whether the management policy adopted recognizes that fact explicitly or not.

The literature in this area of applied ecology reflects its relative infancy in the field of community ecology; though we now recognize the need, we have little of the information required to develop substantive management planning. Nevertheless, several major concepts are well-recognized parts of any plan. One is the concept of ecological succession, so important for the wolves and moose of Isle Royale; less well-documented facets include mechanisms that promote species diversity, the concepts of population and community stability, and the dynamic relationships among communities. These are major topics of interest in community ecology and are vital to the development of better management practices.

The excellent volume edited by Duffey and Watt (1971) contains a number of papers that repeatedly stress the need for aggressive but informed management of complex natural systems. For example, Naveh discusses the management difficulties involved in maintaining the ecosystems of the mountainous Mediterranean areas of Israel, the region where most of the urban and other intensive use of the country is centered. One problem is that the lush forests (now represented by tiny fragments of such famous trees as the Cedars of Lebanon) disappeared under ruthless exploitation many years ago. Scrub forests, shrub land, and grasslands succeeded the forests because the soil loss and the change to drier climatic conditions that resulted from deforestation excluded the possibility of the return of the original flora. In recent years, the modernization of the State of Israel (characterized by intensive grazing of livestock, planting of dense single-species pine plantations, and the spread of urbanization into the mountains) has destroyed all but small remnants of the wild vegetation. Here, in Naveh's understated terms, "the need for sound ecological management is very acute."

Naveh's problem was not only to maintain the long-term viability of the floral preserves that had been established but to encourage the floral structure which would be most diverse and interesting to potential visitors. He knew that the flora could be protected by preventing overexploitation, but he realized also that the diversity problem was more complex, because the plant community evolves toward *reduced* diversity. In ways reminiscent of California chaparral development, the scrub, shrub, and grassland part of Israel eventually is dominated by taller trees that shade the shorter vegetation, eventually resulting in a dense woody thicket.

Naveh concluded that human intervention would be necessary if both goals were to be accomplished. First, he would allow fires and grazing by goats, both of which reduce the competitive advantage of tall perennial thistles and woody vegetation, but only at a strictly controlled level to prevent the soil loss that accompanies overgrazing. Second, he would set aside large areas that would be protected as completely as possible from any form of human interference; these would be subject solely to natural influences. Third, Naveh would devote still other areas to experimental manipulation in order to obtain the data needed for a more informed management policy that would be more closely aligned to the dynamics of the ecosystem. In short, Naveh proposed immediate implementation of what is presently the best possible management program, but he coupled with it a large program of applied research specifically designed to improve management.

The problem of the Israeli highlands illustrates a highly desirable emphasis on ecosystems. Further progress depends on the gathering of more information and what problems are raised in obtaining these data. In an introduction to their volume, Duffey and Watt focus upon the critical lack of good data to support the new ecosystem emphasis of conservation:

> In recent years much has been written and discussed about the subject of conservation... but nearly all of it stops short at the point where technical knowledge is necessary to translate ideas into practice. Indeed, until recently, there was little awareness of this gap in our knowledge and, in fact, it was often assumed that the necessary scientific information was available.[1]

Furthermore, Duffey and Watt set the tone of the volume by defending the need to manage all ecosystems because, quite simply, there are no places where our influence is lacking. Given that all ecosystems—from those completely wild to those thoroughly artificial—must be managed on the basis of biological knowledge, Duffey and Watt concluded: "There is no reason to suppose that... the problems to be faced [in obtaining the necessary biological knowledge from these various ecosystems] are inherently different."

If this conclusion is correct, it follows that basic ecosystem-oriented research programs will be part of

---

[1]From E. Duffey and A. S. Watt (eds.). *The Scientific Management of Animal and Plant Communities for Conservation.* Blackwell, Oxford, England, 1971.

the solution to comprehensive management problems. Duffey and Watt suggest that the solution involves more and better ecosystem research, higher levels of integration between workers in the conservation field, and a more perfect translation of scientific discoveries into management policy. They term this next step forward an "ecologist's dream," and so it is. It is the orderly and uninterrupted development of science. However, it does not necessarily recognize that as many people are unaware of the difficulty, the costs, and the time involved in obtaining information as there are people who think it all is available already.

FIGURE 15-7 *Plastic trees such as this one at Disneyland in Anaheim, California can be part of convincing artificial environments. These trees feature steel roots down to 40 feet below the surface and literally hundreds of thousands of handmade leaves and blossoms.* (Courtesy of Christopher W. Bowen.)

The problem of cost and the question of need continue to be major barriers to changing our attempts to manage our relationships with natural systems. Information is needed to determine what the human costs of the destruction and disruption of natural systems are; but this information requires money, and that, in turn, demands public support. This system of positive feedback is reinforced by the misconceptions and erroneous assumptions Duffey and Watt point to. This realization casts a pessimistic light upon the prospects of future progress.

What if it proves difficult to stimulate much progress toward comprehensive management policy? We can turn to the extensive use of artificial environments. Plastic trees in Disneyland (Figure 15-7), natural settings for zoo animals, and even multispecies groups and plant associations in zoos and botanical gardens respectively offer alternative artificial environments. Why not, then, impose human will and order upon all the world? To create and maintain systems that mimic parts of nature avoids all the problems that accompany the preservation of natural systems, especially when carried out in conjunction with the preservation of those parts of nature's living order that are closely tied to the well-being of society. The creation of plastic trees and Disneyland fantasies is not inconsistent with the establishment of waterfowl refuges, because both serve different needs. The waterfowl refuge is a direct subsidy to hunters, while Disneyland may serve a useful function by absorbing recreation demand that otherwise would go to more fragile national parks. However, this strategy is workable only if we know enough to anticipate all we need from natural ecosystems before we inadvertently destroy them. No one now knows whether this can be done.

The final choice is to allow the course of events to determine the future. If we continue to impose human order over the earth and to disregard natural elements that have no direct relationship to human welfare, we may find out whether or not we need nature. If we are part of a fabric of nature and we cannot substitute for all the relationships we have sundered, the final revenge of nature could be our extinction. But perhaps our ingenuity is great enough, and natural systems are superfluous after all. Which carries the greater risk?

Questions

REVIEW QUESTIONS

1. We have examined a list of characteristics that make a species vulnerable to extinction. Construct an "ecological profile" of a hypothetical species you think would be extremely resistant to extinction, then compare your profile with the evolutionary history of modern man. Does man conform to your profile? If not, what special qualities of man do you think are responsible for the differences?

2. In this chapter, one of the main points made was that the ecosystem orientation is the current trend in the management and preservation of biological resources. Name three specific ways in which the ecosystem orientation differs from the consideration of individual species.

3. Consider four ecosystems:
   a. a biologically rich lake
   b. a biologically poor, deep, cold lake
   c. a tropical rain forest
   d. a patch of Arctic tundra

with respect to their degrees of complexity, their number of interdependencies, and their rates of recovery from disturbance. Now suppose that human beings harvest some part of the biomass of each ecosystem. Which is likely to be most severely affected, and which least? Why?

Hint: See Table 15-1 and Odum's original paper.

ADVANCED QUESTIONS

1. Rubinoff (1970) suggested that in the absence of adequate data, prudence dictated that some impenetrable biological barrier be constructed in a future sea-level canal to prevent migration of elements between oceans. Duke (*Biological Conservation*, 4, 1971, 7) suggested that Panama simply channel its solid wastes, copper mining wastes, and thermal effluents into the middle of the new canal to create "a useful localized River Styx." Evaluate his proposal: Is it likely to work or not, and for what reasons?

2. The Suez Canal is essentially a sea-level canal, but not very many species have crossed it. Of those that have, most have come from the Red Sea into the Mediterranean, rather than vice versa. Can you explain this? See Ben-Tuvia (*Copeia*, 2, 1966, 254) and a good atlas of the Nile delta area.

3. Compare the natural history and dynamics of two avian species, one of which (the passenger pigeon) is extinct and the other of which (the California condor) is nearly so. Identify the population characteristics of each that promoted decline.

For the condor, see Mertz (*American Naturalist*, 105, 1971, 437, and references therein).

For the pigeon, see Schorger (*The Passenger Pigeon*. University of Wisconsin, Madison, 1955).

Further Readings

Note: The conservation of biological systems occupies a great many pages of writing, much of it rather repetitive. The journal literature is especially full of studies of individual species. Keys to this literature will be found in the references below and in the section bibliography. Those citations in the following list that were not given in the text of this chapter represent sources of particular value, because they are thorough, indicate new lines of investigation, are the best summaries, or are readily available.

Cox, G. W. (ed.). *Readings in Conservation Ecology.* Appleton Century Crofts, New York, 1969.

This reader is considerably broader in scope than this chapter. It includes many classic papers and treats topics that have been covered in other chapters in this book. There are, however, some excellent selections useful for this chapter, especially Odum, van Dyne, Taber and Dasmann, and Leopold. Unfortunately, there is little editorial comment on these papers, and students may wish to obtain the papers from original sources.

Curry-Lindahl, K. *Conservation for Survival.* Morrow, New York, 1972.

One of the most recent books in the area of conservation. Quite broad and up to date, it provides a synopsis of many man-environment interactions. The section on animals is quite good.

Dasmann, R. F. *Environmental Conservation, 3rd edition.* Wiley, New York, 1972.

The most recent revision of a classic text. Its treatment is comprehensive, its author well known and respected. Includes a section on urban environments; otherwise, it is comparable to Curry-Lindahl's book.

Duffey, E., and A. S. Watt (eds.). *The Scientific Management of Animal and Plant Communities for Conservation.* Blackwell, Oxford, 1971. (Sold in the United States by Lippincott, Philadelphia.)

This volume, the proceedings of a symposium of the British Ecological Society, is rather technical and is meant for professional biologists. If you have the background, it is the best volume available, bar none, for information concerning the contents of this chapter and for the most recent and exciting developments in the field. See Naveh (1971), below.

Ehrenfeld, D. W. *Biological Conservation.* Holt, New York, 1970.

This book contains numerous examples of biological conservation problems of current interest and is a short and inexpensive volume. It is recommended for another overview on the subject of conservation. Ehrenfeld presents another view of the functions of zoos than that found in this chapter.

Elton, C. S. *The Ecology of Invasions by Animals and Plants.* Meuthen Monographs, Wiley, New York, 1958.

A classic book by one of the world's most famous ecologists. Descriptive and exhaustive treatment of one aspect of ecology, one that becomes more important as human activities disturb more of the biosphere.

Feder, H. M. "Cleaning Symbiosis in the Ocean." In *Symbiosis,* vol. 2 (S. M. Henry, ed.). Academic, New York, 1966.

Fisher, J., N. Simon, and J. Vincent. *Wildlife in Danger.* Viking, New York, 1969.

The illustrated and expanded version of the famous "Red Data Book" of endangered species (see below). This volume is easy reading, but it is listed here because the commentaries are extensive enough to permit corroboration of the ecological and evolutionary basis for species extinction.

Fittkau, E. J. "Role of Caimans in the Nutrient Regime of Mouth-lakes of Amazon Affluents (An Hypothesis)." *Biotropica*, 2 (1970), 138–142.

Fittkau's work has not received much attention but deserves more because it illustrates the systemwide importance of the caiman (crocodile) in the Amazon. Fittkau's evidence indicates that the abundance of fish is linked to the abundance of caiman; hence when caimans were slaughtered and transformed into "alligator hide," fish populations also declined. This work offers striking evidence of unanticipated systems effects.

Hediger, H. *The Psychology and Behavior of Animals in Zoos and Circuses.* Dover, New York, 1968.

The first monograph on the subject was published by Dover under the title *Wild Animals in Captivity* (1964). The original publications were written in German. Hediger's works still stand as classics and serve to deflate some of the optimistic (if noble) roles ascribed to present and future zoos.

International Union for the Conservation of Nature and Natural Resources. *The Red Data Book*, 5 vols. (all classes of vertebrates and the angiosperm plants). Morges, Switzerland, 1966–1975.

The Red Data Book is the definitive catalogue of endangered species, continually being updated by IUCN and popularized in Fisher et al. (1969). Not recommended reading, but a source of data for further analyses.

Krieger, M. H. "What's Wrong with Plastic Trees?" *Science*, 179 (1973), 446–455.

Lack, D. *Darwin's Finches.* Harper Torchbooks, New York, 1961.

Lack's book is a classic description of a famous case of evolution of a species of finch that reached the Galapagos Islands long ago. It is slightly out of place here, but is cited because it emphasizes the importance evolution can have for biological systems.

MacArthur, R. H. *Geographical Ecology.* Harper & Row, New York, 1973.

A technical work but also an excellent review of new directions in community ecology and quantitative biogeography.

Mech, L. D. "The Wolves of Isle Royale." *Fauna of the National Parks of the United States, Fauna Series, No. 7.* Washington, 1966.

Mech, L. D. *The Wolf.* Doubleday, Natural History Press, Garden City, N.Y., 1970.

These two volumes contain the bulk of Mech's work referred to so often as "the wolves and moose of Isle Royale" in this book. The 1970 book is the popular supplemental version of the 1966 monograph, and it has an extensive bibliography that will take students into the wolf literature.

Meggars, B., E. S. Ayensu, and W. D. Duckworth (eds.). *Tropical Forest Ecosystems in Africa and South America: A Comparative Review.* Smithsonian Institution, Washington, 1973.

Extremely technical, but for the serious student interested in the rapid change in the ecology of formerly untouched tropical forest, this book provides a valuable sourcebook. See also H. T. Odum and R. F. Pidgeon (eds., 1970), *A Tropical Rain Forest.* U.S. Atomic Energy Commission document TID-24270. Clearinghouse for Federal Scientific and Technical Information, Springfield, Virginia.

Naveh, Z. "The Conservation of Ecological Diversity of Mediterranean Ecosystems through Ecological Management." In Duffey and Watt (1971), above.

Rubinoff, I. "Central American Sea-level Canal: Possible Biological Effects." *Science*, 161 (1968), 857–861.

Rubinoff, I. "The Sea-level Canal Controversy." *Biological Conservation*, 3 (1970), 33–36.

> The first paper is quite well known, but the later one is more thought-provoking. See problem 8.

U.S. Department of Interior, Bureau of Sport Fisheries and Wildlife. *Rare and Endangered Fish and Wildlife of the United States.* Resource Publication 34. Washington, 1966, 1968.

> The U.S. equivalent of the *Red Data Book*.

van Dyne, G. M. (ed.). *The Ecosystem Concept in Natural Resource Management.* Academic, New York, 1969.

> Similar to Duffey and Watt. See especially the contributions by Coupland et al., Lewis, and Wagner.

Watt, K. E. F. *Ecology and Resource Management.* McGraw-Hill, New York, 1968.

> Plentiful examples of management efforts for animal populations and an extensive argument for computer simulation as a management tool.

16. Quiet Corners and Recreation for a Diversity of Needs

In Chapter 15 we were concerned with saving enough of the planet to conserve representative samples of the diversity of ecosystems that currently exist. Aside from whatever benefits such actions have for human interests, the thrust of this effort is true conservation in the sense that nonhuman creatures are the prime beneficiaries. In this chapter we shall consider the reverse side of our relationship with nature, namely the preservation and management of more or less natural environments to suit particular human needs. These needs all can be considered as recreational in the broad sense given by standard dictionaries, where "recreation" is defined as "refreshment of mind and body." Obviously we can create complex and wholly artificial diversions that serve recreational needs, but many people require more natural environments to fulfill all their recreational needs.

Hence there is a well-defined public interest in having a selection of natural environments designed to meet broadly defined recreational demands. We shall refer to such natural environments as "open space" in this chapter.

In urban areas open space can have a variety of meanings, but in rural areas open space tends to carry a more uniform connotation of natural, usually scenic, landscapes in public parks. In both cases there are many questions about the selection and protection of such areas, the political mechanisms to accomplish these ends, and the purposes they serve. Perhaps the most important questions, however, are concerned first with the meaning of "open space" with respect to its recreational aspects and then with the function of overt management of those parks containing natural assemblages of species. As we shall see, we are directly at the center of both questions.

The Imprecise Meaning of "Open Space"

The term "open space" is used so imprecisely that it is imperative first to analyze the different perceptions about what it can mean and then to consider why provisions must be made for them in the formulation of various recreational and developmental plans. Within our broad definition we may distinguish at least four separate interpretations of open space:

1. Open space explicitly set aside for the type of recreational purposes that require spacious parcels of land, such as gardens, baseball diamonds, playgrounds, picnic sites, zoos, and other developments associated with parks.

2. Open space as the planned restriction in development intended to accomplish certain objectives in urban design; as such it is one of a number of tools for developmental control (see "Zoning" in Chapter 8). One special example of open space is based on its use to ameliorate the effects of skyscrapers and their deeply shaded canyons, which demand the contrast of tree-lined plazas, miniparks, or even nothing more than shorter buildings or the development of such ingenious solutions as the roof-top gardens and recreational areas so popular with the Japanese.

3. Open space also can be the collection of vacant lots that result from contemporary land development practices. In typical American cities development is characterized by unbuilt areas that are only slowly, if ever, built up. In some cities, 50 percent or more of the land is this kind of open space; it is a characteristic of urban sprawl. This is open space by default, insofar as it involves no explicit planning, but it would be an error to assume that such open space is useless.

4. Open space as a general description for the undeveloped land at the periphery of any city, especially with respect to the control of urban growth. In this sense, open space is a resource to be used in whatever way seems fitting. It can be used to preserve existing conservational or agricultural uses, for urban design (Figure 16-1), or for a stock of future urbanized land.

There is obviously no real difficulty in defining the various meanings of open space. We run into difficulties only when we must translate these definitions into action and attempt to answer questions such as, "Do people really need open space?" and, more important, if they do, "How much do they require and what kinds?" As for so many questions in social research, the answers are extremely difficult to obtain.

The problem is that people vary widely in their needs and perceptions, depending both on their backgrounds and on their social milieu, as well as on all the forces of conditioning and circumstance that may influence particular individuals. Some people may feel they need no natural landscapes of any sort, while others may desire large parks nearby to retreat to at any time. The lack of clearly defined goals puts great pressure upon those responsible for securing and planning open space.

Open Space: Needs and Planning

One important consequence of the diverse meanings of "open space" is that the policies which people propose to deal with their perceptions of open-space needs also can vary widely. As examples of two different approaches to open-space needs, we will consider the ideas of William Whyte and Jane Jacobs. William Whyte's book *The Last Landscape* (1970) will represent the viewpoint that vast parcels of natural open space are an indispensable human requirement. In his book, Whyte displayed a strong sense of urgency in going about the task of open-space preservation. We must act, he argued, as if we were defending the last barricades and treat each loss of open space as irreplaceable. Whyte appealed for greater public participation in saving open space because he realized that his own commitment to open-space protection was not necessarily shared by a broad cross-section of American society. Once open space is acquired, Whyte favored its preservation in the form of large, natural open-space areas. He knew, however, that the acquisition of open space was only the beginning of the process, and he acknowledged that constant vigilance would be necessary to prevent the ultimate development of these areas. Therefore, though he preferred large, natural open spaces, Whyte realized that if these areas remained too wild, they might experience limited use and, as a result, be exposed to greater pressure to develop. As a compromise, he felt that the land should be used as intensively as possible without seriously modifying the environment, and he even favored multiple use of the land in order to attain this objective. The pressure for development that Whyte feared most is the one resulting from the American institution of de facto land-use planning (see Chapter 8). He believed that this constant pressure could be dangerous, since it potentially can erode gains already made as well as prevent further gains.

In *The Death and Life of Great American Cities* (1961), Jane Jacobs advanced the idea that urban parks—especially the large, more natural forest preserves cherished by large cities—were useless because they do not meet social needs. In a blunt attack on most large urban parks, Jane Jacobs pinpointed what she believed to be the factor of critical importance in determining the use of a particular park. She identified the opportunities that the park affords users for socializing in an atmosphere of confidence regarding personal safety. The presence or absence of these factors creates a situation of self-acceleration wherein the presence of some users stimulates the arrival of others; conversely, fewer users rapidly leads to desertion of the park. Jacobs argued that large, densely wooded parks create situations which do not inspire confidence regarding personal safety; thus they are fated to be used less.

412 QUIET CORNERS AND RECREATION FOR A DIVERSITY OF NEEDS

(a)

(b)

OPEN SPACE: NEEDS AND PLANNING 413

FIGURE 16-1 Examples of urban open space. (a) A garden in Central Park, New York City. (b) A mini-park on a cleared parcel of land on East 5th Street, New York City. (c) Urban and farm land abutting on the outskirts of Madison, Wisconsin. (d) Parking lots in downtown Los Angeles. (e) The beginning of a development in Florida. ((a) Courtesy of Jon Rawle; (b) courtesy of Katrina Thomas from Photo Researchers, Inc.; (c) courtesy of Daniel S. Brody from Stock, Boston; (d) courtesy of Diane Rawson from Photo Researchers, Inc.; (e) courtesy of Max and Kit Hunn from National Audubon Society.)

The same factors are evident in Jacobs's analysis of cohesive neighborhoods. She argued that the key element in tightly knit neighborhoods is social cohesiveness, fostered by continuous watch over the activities in the area by all the residents. This discourages intruders with harmful intentions because they are detected long before they can commit a crime. Because large parks lack this intense vigilance, they must be expected to simultaneously encourage fear and discourage use. The inevitable result is that they become "dispirited city vacuums . . . eaten around with decay, little used, unloved."

Jacobs argued that small neighborhood "pocket" parks can be successful where large parks must fail because smaller parks can constitute part of the cohesive neighborhood. She believed that these parks do not deny the presence of the city around them—as she believed large, more natural parks do—but are successful precisely because they fit into the urban environment and still provide maximum opportunities for the use of leisure time, for interpersonal communication, and for observation of the urban scene. Such parks are well integrated with the surrounding city—they are close enough to potential users for easy access; located in areas that promote use throughout the day; and amply supplied with playgrounds, benches, and other facilities that promote their use. All these things Jacobs combined in the term "diversity": diversity in users, diversity in the kinds of activities encouraged, diversity in the periods of the day when use occurs. These parks succeed because they do not make communion with a natural setting their primary objective.

Jacobs's position emphasizes the social aspect of human interaction in the park setting at the expense of man-nature communication. Certainly not all authorities accept her analysis: There has been considerable doubt that her faith in a closely knit urban society (and that is the basis of her kind of useful open space) is the all-inclusive answer to urban problem-solving. In other words, does the Greenwich Village, New York, society that Jacobs used as her typical neighborhood really represent all urban societies, and do the principles she derived from the study of this neighborhood apply to the specific open-space needs of all urban dwellers?

At first glance, Jacobs and Whyte would appear to hold completely contradictory positions, but there are some important common features. Whyte was searching for a viable strategy for the preservation of open space, and to this end he supported more intensive use of land that already had been urbanized in order to reduce the rate of conversion from agricultural and wild lands to urban use. He even supported the multiple usage of conserved open space in order to improve its resistance to development. Jacobs, on the other hand, did not deny that open space of the proper sort was useful, and she even admitted that certain large parks—such as Golden Gate Park in San Francisco—were exceptions to the rule that large parks were social failures. Though Jacobs obviously is obsessed with crime and social cohesiveness, her main goal is to have parks that meet social needs, a requirement which Whyte does not deny.

If urban open space is to serve as a center of social activity, then Jacobs's point is well taken; there can be no question that social purposes should dominate the planning of such spaces. On the other hand, Golden Gate Park and Central Park may be essential parts of urban living to some persons who depend upon them to counterbalance the urban experience. For urban planners who attempt to meet such diverse needs, there would appear to be many opportunities for innovative solutions; but for social science, there would seem to be even larger opportunities for productive research in defining the role of open space for human society. Until we have a better understanding, we shall have to be content with a variety of different interpretations and interim planning solutions. A better understanding of the importance of open space in the human social system might allow us to address Jacobs's anxieties, while at the same time it could help us to determine exactly what our open-space requirements are and thus aid us in evaluating the urgency of Whyte's arguments.

Rural Open Spaces

In urban areas, conservation of living systems is not very important compared with recreation, accessibility, and facilities. Urban open spaces are meant to be used frequently and intensively. As we leave the urban area, a transition usually measured in travel times exceeding two hours, the nature of open space changes. Most frequently such rural open spaces have conservation as an important part of their mission, if not always as the primary purpose, and the recreational experience is normally of a different kind. Depending upon the region, rural open spaces usually are not under so much pressure for urbanization that it has been necessary to preserve them by establishing formal parks, although this has been done for the most striking geological formations, for unique and rare species, or for particularly beautiful areas.

The remainder of the chapter is devoted to an analysis of such rural parks. No distinction will be made between regional, state, and federal parks, since the governing authority is not as important as the goals of the park itself. The U.S. National Park System (which will serve as our model) has placed increasing emphasis upon parks having different main goals: Some are devoted to recreation, some to the preservation of natural features, and some to the preservation of manmade structures. This differentiation within the National Park System seems far more important than mere jurisdictional authority would foster.

National Parks: Their Creation and Roles

The first national park (if not formally in name) was established in the Yellowstone region of Wyoming, Montana, and Idaho in 1872. The section of the U.S. code that established this park directed that Yellowstone National Park should be "dedicated and set apart as a public park or pleasuring ground for the benefit and enjoyment of the people" at the same time that the natural resources of the park—namely, its timber, minerals, "natural curio sites," fish, and game—were to be preserved. When the National Park Service was finally established in 1916, these objectives were formalized in its statement of purpose. The National Park Service was directed to "conserve the scenery and the natural and historic objects and the wildlife therein and to provide for the enjoyment of the same in such a manner and by such means as will leave them unimpaired for the enjoyment of future generations." These statements clearly differentiate between the aims of national parks and those of urban open space, where any preservation of the natural state is quite open to question.

FIGURE 16-2 *Exposure to people has altered the natural state of black bears in Yellowstone National Park.* (Courtesy of Bill Gabriel from Photo Researchers, Inc.)

objects has consequently been impaired, the Park Service has become concerned increasingly with the dilemma of incompatible priorities. National parks have been accepted so enthusiastically by the public that it has become necessary to manage (if not actually to limit) the volume of visitors in the most popular parks. Even though the number of national parks has grown steadily since Yellowstone (Table 16-1, Figure 16-3), visitation has more than kept pace (Table 16-2). Consequently, the National Park Service is seeking better methods to meet both preservation and recreation roles.

Preservation Objectives in National Parks

The conservation role of national parks is clear even if the means to do the job are not. The spectacular elements for which most of the parks were created—for example, the areas of thermal activity in Yellowstone, the glacial canyon of the Yosemite Valley, the rugged ranges of the Grand Tetons, the magnificent forests of the Great Smoky Mountains, and the giant trees of Sequoia National Park—are of primary importance. Protection of such features as well as of other park elements that could enhance the public's enjoyment should and does have high priority in park management policy. For example, the single most important feature of Yosemite National Park is the view of the valley walls and granite domes. Spectacular views of these features are available from the meadows on the floor of the valley. However, these views can be ephemeral, because the meadows do not persist indefinitely; they are invaded by trees and sooner or later become part of the forest. Policies of burning and cutting to favor the continued existence of meadows suggest themselves as possible park policy and

But the nobly phrased, multipurpose aims of national parks have brought conflict; it is not always easy or even possible to reconcile heavy use with preservation of the natural state of the area (Figure 16-2). In earlier years, park policy was directed primarily toward the provision of adequate facilities for visitors and the encouragement of visitor use; but as the number of visitors has increased and evidence has accumulated that preservation of natural and historic

TABLE 16-1a *Summary of areas administered by the National Park Service, by type.*

Classification	Number	Lands within exterior boundaries[a]		
		Federal (acres)	Nonfederal (acres)	Gross (acres)
Historic areas not federally owned or administered	1		7.00	7.00
International parks not federally owned or administered	1		2,721.50	2,721.50
National battlefield parks	3	6,295.49	856.25	7,151.74
National battlefields	6	4,467.95	1,478.55	5,946.50
National battlefield sites	3	781.28	4.33	785.61
National capital parks[b]	1	7,054.05		7,054.05
National cemeteries	10	220.13		220.13
National historical parks	14	43,946.57	21,093.89	65,040.46
National historic sites	44	9,088.98	1,164.78	10,253.76
National historic sites not federally owned or administered	9	2.74	212.39	215.13
National lakeshores	4	24,404.10	165,210.75	189,614.85
National memorial parks	1	69,528.31	907.69	70,436.00
National memorials	19	5,624.15	123.27	5,747.42
National military parks	11	30,384.70	1,403.13	31,787.83
National monuments	82	9,665,531.72	214,284.30	9,879,816.02
National parks	38	14,770,856.98	309,059.74	15,079,916.72
National parkways	5	130,410.86	19,939.33	150,350.19
National recreation areas	13	3,630,811.43	179,497.57	3,810,309.00
National seashores	8	256,491.74	223,513.49	480,005.23
National scenic riverways	3	48,635.72	96,728.50	145,364.22
National scenic trails	1	17,000.00	33,000.00	50,000.00
National scientific reserves not federally owned or administered	1		32,500.00	32,500.00
Parks (other)[c]	5	24,356.61	1,248.86	25,605.47
White House	1	18.07		18.07
Total	284	28,745,911.58	1,304,955.32	30,050,866.90

SOURCE: National Park Service. *Areas Administered by the National Park Service and Related Areas*, 1972.

NOTE: Areas identified as Living Farms conduct appropriate farming operations all year. Areas identified as Living History areas conduct appropriate historical and cultural activities. Depending on the area, these demonstrate Indian, Hawaiian, colonial, pioneer, frontier, and military life, crafts, and transportation methods, generally using costumes and techniques of the historical periods.

[a]Acreage as of June 30, 1971, except for Lincoln Home National Historic Site, added Aug. 18, 1971, Arches and Canyonlands National Parks whose acreages are of Nov. 12, and Capitol Reef National Park, as of Dec. 18, 1971.
[b]Comprises 720 reservations within the District of Columbia, Virginia, and Maryland.
[c]Without national designation.

TABLE 16-1b *The National Park System.*

Park, state	Established	Authorized	Gross area (acres)
Acadia, Maine	1919	1919	41,642
Arches, Utah	1971	1971	73,234
Big Bend, Texas	1944	1935	708,118
Bryce Canyon, Utah	1928	1924	36,010
Canyonlands, Utah	1964	1964	337,258
Capitol Reef, Utah	1971	1971	241,671
Carlsbad Caverns, New Mexico	1930	1930	46,754
Crater Lake, Oregon	1902	1902	160,290
Everglades, Florida	1947	1934	1,400,533
Glacier, Montana	1910	1910	1,013,101
Grand Canyon, Arizona	1919	1919	673,575
Grand Teton, Wyoming	1929	1929	310,443
Great Smoky Mountains, Tennessee-North Carolina	1930	1926	516,626
Guadalupe Mountains, Texas[a]		1966	81,077
Haleakala, Hawaii	1961	1960	27,283
Hawaii Volcanoes, Hawaii	1916	1916	229,616
Hot Springs, Arkansas	1921	1921	3,535
Isle Royale, Michigan	1940	1931	539,341
Kings Canyon, California	1940	1940	460,331
Lassen Volcanic, California	1916	1916	106,934
Mammoth Cave, Kentucky	1941	1926	51,354
Mesa Verde, Colorado	1906	1906	52,074
Mount McKinley, Alaska	1917	1917	1,939,493
Mount Rainier, Washington	1899	1899	235,404
North Cascades, Washington	1968	1968	505,000
Olympic, Washington	1938	1938	896,599
Petrified Forest, Arizona	1962	1958	94,189
Platt, Oklahoma	1906	1906	912
Redwood, California	1968	1968	56,201
Rocky Mountain, Colorado	1915	1915	262,191
Sequoia, California	1890	1890	386,863
Shenandoah, Virginia	1935	1926	193,537
Virgin Islands, V.I.	1956	1956	14,419
Voyageurs, Minnesota[a]		1971	219,431
Wind Cave, South Dakota	1903	1903	28,059
Yellowstone, Wyoming-Montana-Idaho	1872	1872	2,221,773
Yosemite, California	1890	1890	761,320
Zion, Utah	1919	1919	147,035
Total (38)			15,073,226

SOURCE: U.S. Council on Environmental Quality, Third Annual Report, 1972.
[a]To be established after certain conditions of the authorizing acts are met.

FIGURE 16-3 *The National Park System.* (Redrawn from 3rd Annual Report, U.S. Council on Environmental Quality, 1970.)

Park	Location	1968			1969			1970		
Acadia	Maine	2303.3	1477.8	64.2	2489.8	1541.1	61.9	2772.3	1400.6	50.5
Big Bend	Texas	191.8	180.6	94.2	199.8	172.2	86.2	172.6	180.7	104.7
Everglades	Florida	1251.5	221.4	17.7	1187.2	150.4	12.7	1273.5	137.7	10.8
Grand Canyon	Arizona	1986.3	803.9	40.5	2192.6	767.6	35.0	2258.2	769.0	34.1
Grand Tetons	Wyoming	2970.3	547.1	18.4	3134.4	607.4	19.4	3352.5	665.9	19.9
Great Smokies	North Carolina–Tennessee	6667.1	620.2	9.3	6331.1	593.7	9.4	6778.5	690.7	10.2
Hawaii Volcanoes	Hawaii	918.0	89.7	9.8	719.9	72.4	10.1	822.3	86.4	10.5
Mount McKinley	Alaska	33.3	35.1	105.4	45.4	29.5	65.0	46.0	35.5	77.2
Olympic	Washington	2013.8	313.9	15.6	2135.9	315.1	14.8	2283.1	385.0	16.9
Rocky Mountain	Colorado	2187.6	273.6	12.5	2217.2	278.2	12.5	2357.9	294.9	12.5
Shenandoah	Virginia	2273.2	403.9	17.8	2400.9	419.2	17.5	2411.5	450.2	18.7
Yellowstone	Wyoming–Montana–Idaho	2229.7	1387.7	62.2	2193.7	1411.3	64.3	2257.3	1300.4	57.6
Yosemite	California	2281.1	1908.7	83.7	2291.3	1774.1	77.4	2277.2	1535.5	67.4

TABLE 16-2 *Recent figures showing trends in usage of selected national parks. The data, for the years 1968–1972 inclusive, are in thousands of visitors. "Total" means the total number of visitors for the calendar year: This figure includes the total number of persons entering a park and using some part of the facilities provided. "Overnight" means the number who camp or use lodging provided in the park for an overnight stay. For locations of the parks, see Figure 16-1. For complete data on all parks, museums, historical areas, etc., see the annual summaries published by the National Park Service,* Public Use of the National Parks.

The data given in this table are worth examining. For example, notice that Acadia National Park has had only a small increase in the total number of visitors, many less of whom are staying overnight in campgrounds. Contrast this with the trend in Shenandoah National Park, or in Yellowstone.

Key: For each year, the first figure is total usage; the second is overnight visitors; and the third is the ratio (100 overnight per total) = percent staying overnight.

Park	Location	1971			1972			Net percent change in total visits (1968-1972)
Acadia	Maine	2455.7	239.4	9.7	2645.4	224.5	8.5	15
Big Bend	Texas	247.4	192.0	77.6	290.2	215.3	74.2	51
Everglades	Florida	1293.5	131.4	10.2	1773.3	166.6	9.4	42
Grand Canyon	Arizona	2402.1	803.9	33.5	2698.3	895.2	33.2	36
Grand Tetons	Wyoming	3284.5	593.2	18.1	3002.2	618.6	20.6	1
Great Smokies	North Carolina–Tennessee	7179.0	559.8	7.8	8040.6	626.1	7.8	21
Hawaii Volcanoes	Hawaii	980.7	88.7	9.0	1389.1	93.5	6.7	51
Mount McKinley	Alaska	51.4	58.3	113.4	306.0	91.0	29.7	919
Olympic	Washington	1859.7	402.5	21.6	3031.7	553.0	18.2	51
Rocky Mountain	Colorado	2457.3	275.6	11.2	2519.6	281.1	11.2	15
Shenandoah	Virginia	2406.5	450.5	18.7	2304.1	446.2	19.4	1
Yellowstone	Wyoming–Montana–Idaho	2126.3	1345.6	63.3	2251.7	1464.6	65.0	0
Yosemite	California	2416.4	1596.3	66.1	2266.6	1609.4	71.0	−1

have in fact been used to preserve the view from the valley floor.

In recent years the mission of the Park Service has been reaffirmed and expanded. New parks have been acquired or planned for the coastal redwoods of California (Redwoods National Park), the border lakes and rivers of Minnesota (Voyageurs National Park), the Guadalupe Mountains in Texas and the northern Cascades in Washington (Figure 16-4). Furthermore, in an excellent demonstration of foresight, the National Park Service has several areas of Alaska under consideration for possible inclusion in the park system. When these areas are selected, the Alaska acreage alone is expected to at least equal all the acreage now in the entire park system.

Besides national parks themselves, other sectors receiving increasing emphasis from the National Park Service include more historical monuments, which are seen as an increasingly important part of the preservation objective of the Park Service; the establishment and management of a rapidly expanding system of national recreation areas, gateways, and national seashores to channel much of the recreation demand away from the parks proper; and, most important, the growing use of a widening, ecosystem-oriented approach to the management of parks having a significant proportion of wilderness. Traditionally, the Park Service has been forced to set policy on the basis of general ecological information that might or might not be applicable to the particular situation, or in response to political pressures, or perhaps from whatever serendipity might arise from small individual research projects carried out within the borders of

FIGURE 16-4 *Capitan Reef, an ancient limestone reef complex in the Guadalupe Mountains National Park, Texas.* (Courtesy of Grant Heilman.)

various parks. The Park Service has notably lacked much capability to initiate research directed specifically at determining the impact of visitors and management practices on park resources. Because of this lack of information, the Park Service has elected to meet their preservation objective simply by limiting use within parks in the belief that natural forces would operate if unimpaired by human interference. However, this assumes that the goals of human beings and nature are synonymous—a situation that is seldom true, which is being recognized slowly.

As an example of the broadening ecological perspective of park management, consider the reversal of policy with respect to forest fires. In past years, immediate fire suppression was the common practice in national parks. However, evidence has accumulated indicating the beneficial aspects of ground fires. For example, the importance of periodic fires to restore deer and moose browse, to keep the Everglades from burning out in one enormous conflagration, to maintain meadows, and to suppress the buildup of litter and stunted young trees that could feed forest fires in western woodlands have all been well documented. As a result, the National Park Service now maintains that fires will be allowed to burn freely wherever there is a history of naturally occurring fires. Since there is practically no place where fire is not a natural event, this shift in policy heralds an important first step in comprehensive ecological management and has set a precedent for a greater role for enlightened intervention in park management policy. Sometime in the not too distant future, perhaps, we shall see a situation in which the effects of such management can be displayed to the public in national parks, for this would be the most effective kind of environmental education imaginable. Furthermore, fire management is a tentative first step toward admitting that national parks cannot be insulated from human intervention and that we might as well apply the best management practices we can.

The Recreational Role of National Parks

In the Third Annual Report of the U.S. Council on Environmental Quality, Robert Cahn noted that national parks have an outstanding image in the eyes of the public. He substantiated this statement by pointing out that 20 percent of the visitor use on federal lands occurred in national parks, despite the fact that these lands constituted only 7 percent of the total federal acreage. Furthermore, the bulk of visitor use was concentrated on the 5 percent of the park lands that have been developed specifically for mass use. This information reveals still another conflict built into Park Service objectives, because the flow of visitors is not to the wilderness areas that conservationists adore. In many ways, the parks have been most popular because they permit more or less conventional activities within magnificent (or at least unusual) settings. However, the development required for this kind of park use results from very different attitudes and policies (the parks cater to the needs of all potential visitors) from a wilderness experience (the visitors must accept all prior conditions and constraints upon use). Of course, in the real world neither of these attitudes and usages exists in a pure form, and so the question is one of relative effort expended in the two basic directions.

The two adversary positions regarding emphasis have not been developed equally, but each alternative has powerful arguments. Suppose, for example, that the predominant policy was to keep parks mostly as wilderness. This certainly would facilitate the preservation mission of the Park Service, because user impacts would be more limited in extent and there would be few user facilities to get in the way of periodic fires or any activities relative to Park Service management programs. Large tracts of wilderness would also mean that if current wilderness-use policy prevailed, fewer visitors would be allowed into the parks, although in these areas the visitors who were allowed in would have the highest possible quality recreational experience in terms of communion with nature. On the other hand, one might argue that a wilderness-oriented policy is highly exclusionary, with high-quality experience for a few being favored over exposure to the park environment for many more. The underlying concept for those who oppose the wilderness-oriented policy is the belief that mass recreation could be as fine an experience to many as isolation in wilderness might be to others. To answer those who would protect the parks from the detrimental impact which they feel is associated with a large volume of visitors, the proponents of mass recreation would argue that the direct linkage between visitor use and environmental impact can only be presumed at present and will remain unproved (except in general terms) without sufficient data. They further might maintain that the more important impact of visitor use is on the nature of the recreational experience and that, after all, the determination of a "desirable" experience is an individual decision. Therefore, though the wilderness experience in its purest form may well be what some people perceive as the recreational objective of national parks, it does not follow that this is what all or even most of the people who visit national parks want.

The problem is that while both positions can be sustained, they hardly are compatible. What then? Clearly we need more and better information, but it usually is not a simple matter to get it. Information is costly to obtain, usually far too expensive for available funds. Even with enough money, some information will be difficult to obtain for these reasons: (1) It may take a long time to get sufficient information to permit meaningful conclusions to be drawn, perhaps longer than is practically available; (2) natural variation may necessitate extensive sampling in order to obtain reliable data; and (3) there may be complex chains of mutual cause and effect that are nearly impossible to unravel. To make the problem-solving situation even more difficult, national parks have their problems imposed upon them—there is no choice about which problems are easiest or most interesting to solve; they all must be dealt with if adequate information is to be obtained.

Ecology and Recreation Planning in National Parks

Ecology is one science relevant to the planning of national parks, but for the reasons outlined above there is considerable doubt that pure ecological research can provide definitive answers when they are needed and at a reasonable cost. Consequently, the Park Service has relied little on this type of original research in the parks. Given continuously increasing use of the parks, the National Park Service has adopted the masterplan process, which, among other things, reflects a conservative attitude toward activities and facilities permitted in the parks.

A typical example of the master plan and its relation to protecting environmental quality has been formulated for Yosemite National Park, one of the most popular and heavily used parks in the National Park System. The plan involves an acceptance of present facilities and use patterns and concentrates on planning around these in an effort to deal with the variability in the ecological fragility of different areas within the park and in order to rectify problems long recognized. This plan was completed in 1971, and as policy has continued to change, it already has become obsolete in part. For immediate implementation, the plan called for the closure of part of the Yosemite Valley road network, the conversion of the remaining roads to one-way use, the institution of a bus shuttle system to move people about the valley, restriction in campground use, and the adoption of new rules for the use of the back country (see Figure 16-5). Many other specific recommendations too numerous to mention here, when taken all together, attempt to maintain the quality of the recreational experience for visitors while limiting use of fragile back-country areas; they also aim to establish zones of potential use as defined by the best available information about what uses these areas can withstand without degradation.

In recent years the National Park Service has gravitated toward the ideal of the "wilderness experience." In Yosemite this ideal is expressed as the long-term objective to remove as many of the support facilities from the Yosemite Valley as possible and to completely ban private automobiles. Visitors then would ride in buses to enjoy a view free from traffic and the clutter of buildings. But is this an improvement? Is the valley being degraded by visitors now? Moreover, do most visitors go to see the geological wonders of Yosemite in the context of a wilderness experience? Is the Park Service, in fact, merely reacting to political pressure from well-organized conservationists, mountain climbers, and back-packers? One might argue that the valley walls and granite domes themselves are virtually indestructible; if they are the most important attraction in Yosemite, the valley itself could be developed intensively while the remainder of the park's use is controlled rigorously. Of course, the situation is far more complex than this, but the fact remains that most of the questions posed above remain unanswered. Nor is the type of information that the science of ecology can supply of much help here. Therefore, given such information gaps and limited financial resources, it is not surprising that the Park Service should elect prudence when dealing with the maintenance and care of lands under its jurisdiction.

How well the National Park Service will be able to attain its objectives concerning use and preservation in the face of increasing public demand for recreational lands remains to be seen. Many lightly developed, little-used parks can serve as wilderness preserves, but this is probably not true of the more popular and accessible ones. In any case, the National Park System contains too little acreage to let us preserve all the wilderness that we ought to; fortunately, many other wilderness programs are in progress.

The Wilderness Act of 1964

After 15 years of study and several abortive attempts to produce viable legislation, the Wilderness Act was finally signed into law in 1964. It declared that not all public lands should be subject to development. Each federal agency administering a part of the public lands was required to declare that some part of its holdings would remain wilderness, to be administered in such a manner that the lands retain their primeval character "unimpaired for future use and enjoyment."

The Wilderness Act defines a "wilderness area" as one untrammeled by human beings; that is, there is no permanent human habitation within these areas, only visitors who pass through. The act mandated a review by the end of 1974 of all primitive lands (those having no permanent improvements) by the following government agencies: the National Park Service, the U.S. Forest Service, the Department of Agriculture, and the Bureau of Sport Fisheries and Wildlife in the U.S. Department of Interior. Actually, the idea of establishing wilderness areas in the United States is not new; as early as 1930, the Secretary of Agriculture had begun to declare wilderness areas in the public forests. These lands—which had been classified as "wilderness," "canoe," or "wild"—were included automatically in the wilderness system as long as they were so classified for 30 days before the law took effect and were either 5000 acres or more in area or large enough to contain systems that would be viable even if all surrounding land were eventually developed.

The Wilderness Act was finally passed over opposition strong enough to force limiting provisions and amendments. Thus the inclusion of new lands into the wilderness system is not an easy process; there is ample opportunity for many to lodge objections to a declaration of wilderness for any parcel of land. The law requires that inclusion of new wilderness lands shall be made only by an act of Congress, following review both by all the above agencies and by the President. The process must be accompanied by published notice of these actions followed by public hearings. Furthermore, despite the fact that the development of roads or any other permanent facilities is barred from wilderness areas, the use of aircraft and motor boats can be permitted if there is a past history of their use in the area; fire fighting, pest control, and disease treatment can also be undertaken; mineral workings established before 1983 are not affected; and livestock grazing, dam building, and even commercial development (in support of recreation) may be permitted in some instances.

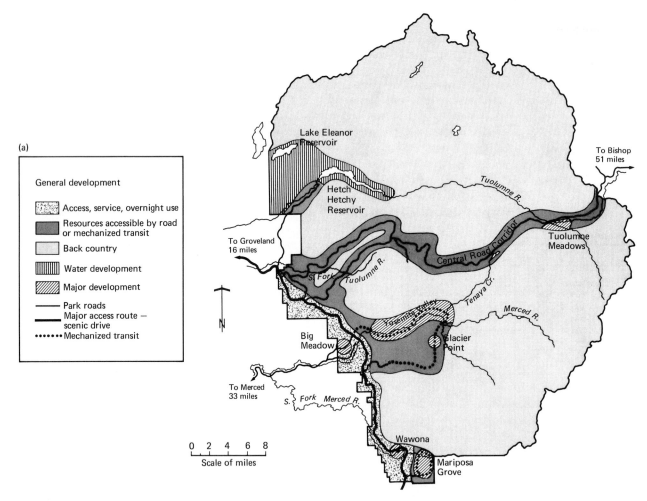

FIGURE 16-5 *A rendering of (a) the master plan and (b) the backcountry use proposals for Yosemite National Park, California. The master plan calls for the removal of the automobile from the Yosemite Valley and transportation by bus, bicycle, and foot only. The backcountry use plan, as mandated by the Wilderness Act of 1964, reflects existing uses of the Yosemite backcountry. This plan was ultimately rejected and a new one is being formulated in the master-planning process to include a broader spectrum of public opinion. (Redrawn from Preliminary Working Draft, Yosemite Master Plan, Yosemite National Park, California.)*

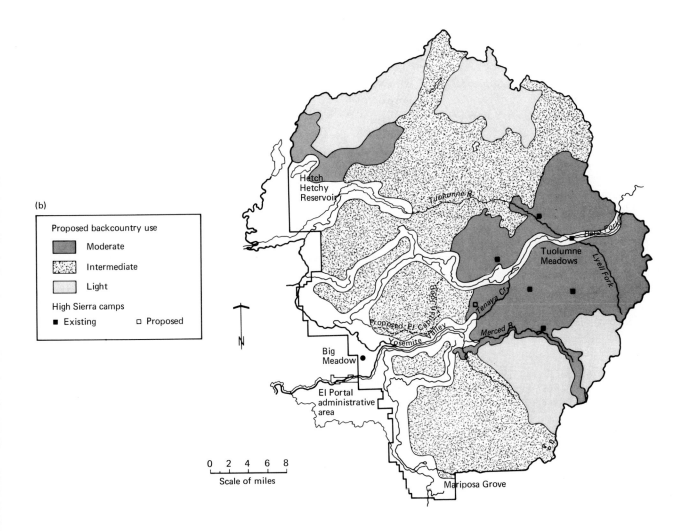

The loopholes and administrative flexibility of the Wilderness Act all add up to a guarantee that a large share of federal lands will not be preserved as wilderness (with the exception of national parks, which still will have large wilderness areas; see Figure 16-3 for one example). To illustrate the effect of all these limitations, consider the U.S. Forest Service's approach to the conversion of any of its lands to wilderness areas. The Forest Service has been the frequent target of conservation groups that have claimed the agency has adopted an unrealistically "purist" approach to wilderness because it essentially requires that there be no evidence of human presence before the land can be considered as potential wilderness. Given the agency's multiple-use philosophy, it is obvious that such an approach eliminates the vast majority of lands for consideration as wilderness areas. Another element of the law that has been criticized heavily is that it fails to require a public review if an agency decides a parcel of land is *not* suitable as possible wilderness, when possible inclusion always requires a public hearing. This provision can effectively remove excellent potential wilderness from the agenda for discussion, with no more justification than the whim of a few administrators.

Perhaps the largest oversight of the Wilderness Act is that the Bureau of Land Management (BLM), which controls nearly all desert lands, was not included in it. As a consequence, the BLM has been acting independently to seek out and include wilderness areas under its jurisdiction; its proposals are acted upon by the President without congressional involvement, but otherwise the procedure will follow the Wilderness Act.

Despite all the loopholes and provisional flexibility that could weaken the protection given to wilderness areas, the Wilderness Act represents the first systematic attempt to preserve the best wilderness remaining in this country. As such, it is a precedent that could lead to improved legislation. The Sierra Club already has been successful in using the Wilderness Act to force potential developers who wish to use wilderness areas for any purpose other than incidental recreation to first submit a statement of environmental impact under the provisions of the National Environmental Policy Act (NEPA). The Wilderness Act indicates a public mood that favors such developments as these not only in national parks but elsewhere. There even have been suggestions that some Forest Service lands in the wilderness-poor eastern United States be included in the wilderness system and then be allowed to revert to the pristine conditions that existed before we modified the landscape.

When the Wilderness Act was enacted, 9.1 million acres of land were included. Since then, Congress has added 1.3 million acres, and presidentially approved areas (if also approved by Congress) would bring the total to 15.2 million acres. In the same spirit as that associated with the Wilderness Act of 1964, Congress established a "wild rivers" system with the passage of the Wild and Scenic Rivers Act in 1968. This act recognized three kinds of rivers as suitable for protection:

1. Wild: rivers that are primitive and inaccessible (except by trail)
2. Scenic: accessible rivers that still are largely undeveloped
3. Recreational: developed rivers that are accessible

The act included parts of 8 river systems as an initial reserve, with 27 more under study for possible inclusion. Unlike wilderness areas, the wild and scenic rivers included under this act are scattered across the country (Figure 16-6). Moreover, this is not the only way to protect a river, since there are several other legal mechanisms to accomplish this aim. At the federal level, the National Park System is developing national scenic riverways such as the Ozark River system in Missouri. This system includes a long strip of land comprising 85,000 acres along the Current River and Jack's Fork in southeastern Missouri. In addition, the states and regions also may nominate candidates for the federal river systems, or they may erect systems of their own. Legislation affecting wild rivers has been passed in California because the extensive hydroelectric-irrigation development that already has occurred there has stimulated strong reaction against further development.

Despite its promise, however, the Wild and Scenic Rivers Act is a fairly weak piece of legislation. Nearly any river can be included in the system, but there are few guarantees that inclusion is much protection against change or degradation. Nevertheless, the Wild and Scenic Rivers Act, like the Wilderness Act, is a first step in a systematic assessment of primitive resources and their survival as viable entities.

In the past, the strategy of multiple use of protected lands has had an honored place in the management of this country's public holdings. As we noted above, the Forest Service in particular has been in the forefront in the multiple use of its lands for recreation, mining, grazing, timber harvest, game management, and wilderness preservation. However, some multiple uses can be incompatible—for example, strip mining and wilderness-oriented recreation—and incompatibilities can be difficult to resolve when economic pressures and historical precedents constrain freedom of choice regarding management alternatives. Therefore, the establishment of a wilderness system for lands and for flowing waters is a means for avoiding the difficulties involved in multiple-use conflicts. An immediate benefit from the establishment of such protected lands is the conservation of sufficiently large areas to allow the preservation of entire

FIGURE 16-6 *White water canoeing on the Skykomish River in Washington.* (Courtesy of Keith Gunnar from National Audubon Society.)

ecosystems, thus allowing the wilderness system to serve the desired purposes set forth in Chapter 15. However, in addition to these preservation objectives, there are also incalculable benefits for recreation, for the wilderness that is set aside wisely today may prove invaluable for future recreational activities. None of these actions, however, excludes these lands from development, for if a real need arises, the land then can be developed in whatever ways and to whatever levels are necessary. The caution underlying current National Park Service philosophy seems wise indeed when our present ignorance is considered.

Struggles Between Development and Preservation

Now we shall examine two cases that incorporate many ideas presented in this chapter.

WHAT SHOULD BE DONE WITH MINERAL KING?

The Mineral King area of the Sierra Nevada Mountains in California is part of the Sequoia National Forest and as such is administered by the U.S. Forest Service (Figure 16-7). It is much like other areas of the Sierra Nevada managed by the Forest Service, except for three pertinent facts:

1. The area was declared a National Game Refuge in 1926.
2. The area is surrounded on three sides by Sequoia National Park.
3. The Forest Service would like to see the area developed for intensive recreational use as part of their multiple-use philosophy.

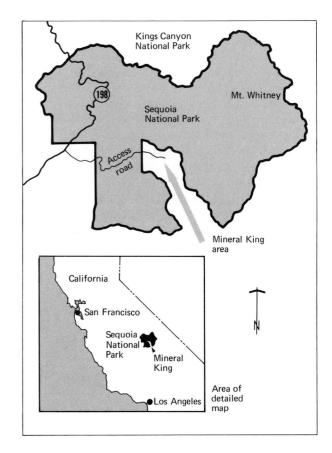

FIGURE 16-7 *Map of the Mineral King area. The inset shows its location in California.* (Redrawn with permission from *Natural History*, November, 1968. Copyright© The American Museum of Natural History, 1968.)

These three facts form the basis for a national dispute, at the heart of which is the chasm between the philosophies of the Forest Service and the Sierra Club.

As early as 1949 the Forest Service started looking for a private developer to build a recreational complex in the Mineral King area. Finally, in 1965, the Forest Service accepted a bid from Walt Disney Enterprises to build a complex centered upon the recreational activities of skiing during the winter and sightseeing in the summer. The Disney organization planned to spend up to 35 million dollars on the project. As plans developed, it became clear that they would involve what was essentially a permanent lease on the land. Furthermore, the Forest Service would have to accept all the risks involved in a projected volume of visitors that would *exceed* that of Yosemite Valley. Finally, and most important, the Forest Service would have to make a commitment to the construction of a road to the proposed recreational area, and the only logical route was directly through Sequoia National Park.

One might wonder why the U.S. Forest Service would agree to such commitments. At least three reasons seem plausible. First, the Forest Service had not made any commitments to the Disney Corporation that could not be modified if it became necessary. (The 1965 agreement was only for a three-year permit for further feasibility studies.) Second, the Forest Service had a substantial financial interest in the development of a partnership or working agreement (in reality, if not formally) with the Disney Corporation. Finally, this kind of project fits neatly into the long-established Forest Service policy of multiple use. In the Mineral King area, standing timber cannot be harvested (it is unthinkable that the National Park Service would allow the construction of a logging road across Sequoia National Park), the mineral resources of the area have been exhausted already, and only the recreational potential remains to be exploited. When the Forest Service became convinced that Mineral King could be a primary ski area for southern California, it was only a matter of time until the planning process moved forward. Concerning the fact that Mineral King would have to serve simultaneously as a recreational area and a wildlife refuge, it appears that the Forest Service finds no conflict between the two objectives. Laws governing the Wildlife Refuge System specify that recreational multiple use is permissible as long as it does not impair the primary objective of its function as a wildlife refuge.

The Sierra Club, on the other hand, has a different philosophical orientation. It is an organization with old, well-entrenched roots in conservation; it is one of the leaders—if not *the* leader—in this respect. The Sierra Club has accomplished much good for society by vigorously supporting certain partisan positions in the courts. One position most dearly held concerns any issue involving the relationship between wilderness and recreation—specifically, the belief that the highest form of recreation is the wilderness experience and that we cannot afford to lose even one more wilderness area to mass recreation. This position is highly reminiscent of William Whyte's philosophical stand on the urgency of defending open space. In the case of Mineral King, the Sierra Club's position as given by McCloskey and Hill (1971) focuses upon the risk that must be accepted if development of the resort is allowed to proceed, a risk involving not only the Mineral King area, but also Sequoia National Park. The immediate concern of the Sierra Club is the impact of the road and the volume of traffic—both people and automobiles—upon the biological nature of the area. A second concern is the impact of heavy visitor use upon the Mineral King area as a wilderness whose prime recreational resource is the solitude it can provide. Together, these things mean that to the Sierra Club the aesthetics of Mineral King are a function of the present landscape, and to disrupt this by heavy use would severely reduce the area's recreational value. Last, in looking to the future, the Sierra Club fears the precedent that completion of the project might set both for future Forest Service policy and for future erosion of other wilderness areas.

Given the great philosophical gulf between the Forest Service and the Sierra Club, it was inevitable that the Sierra Club would file a suit to stop this development. After attempts at administrative compromise failed, the Sierra Club filed suit in Federal Court in San Francisco, the result of which was a temporary injunction against the issuance of the final permit to begin work on the development. However, the Appeals Court overturned the injunction, and to date the U.S. Supreme Court has refused to recognize the standing of the Sierra Club as a plaintiff in the suit. However, definitive ruling on the substance of the suit is expected in the future, because it seems likely that the Sierra Club, acting in the names of members who actually use the Mineral King area, will qualify for standing.

Meanwhile, even while legal adjudication grinds slowly onward, the Forest Service and the Disney Corporation have not been idle. The Disney Corporation has announced that their original proposal has been scaled down so that the project now will cost no more than 18 to 20 million dollars, rather than the original 11.7 million dollars.[1] Moreover, the projected number of visitors has been revised downward, and increased attention has been paid to a cog railway system for transporting people into the area, rather than the all-weather highway that has proved so controversial. For its part, the Forest Service has been forced by public opinion to reaffirm that it will have considerable veto power over whatever development, if any, ultimately results. All these changes underscore the importance of the legal delay as a valuable tactic. With changing economic conditions and rising construction costs, a delay may be as effective in blocking the project as a court decision favorable to the Sierra Club would be. To date the Mineral King project remains stalled, apparently demonstrating the value of legal delay (Table 16-3). Meanwhile, the Disney Corporation has been busy elsewhere constructing a recreational development at Independence Lake. It is not clear yet whether or not Mineral King will be developed ultimately. The Forest Service could decide that the attendant adverse publicity is not worth the effort and call a halt to the project. Strong public opposition also could convince the Disney Corporation that the economic cost for their facilities would be too high and that the project is not a worthwhile risk. Finally, the National Park Service could resist pressures from both organizations and refuse right-of-way for either a highway or a railway. Any of these actions would scuttle the project, regardless of the decision of the courts.

No matter what the decision of the U.S. Supreme Court and the eventual extent of development of the Mineral King area, the basic philosophical positions of the two organizations involved are likely not only

---

[1] The original proposal allowed for expenditure increases up to 35 million dollars. The increases were to accommodate inflation and the rising cost of environmental protection.

TABLE 16-3 *The chronology of major events in the proposed development of the Mineral King area.*

Year	Event
1872–1888	Mining community of Mineral King grows to a population of 500 on the prospects of discovery of precious metals. Strikes not very rich, and after 1888 Mineral King declines to a tiny village of summer residences.
1926	Sequoia National Park expanded; Mineral King excluded but declared a game refuge under the continuing administration of the Forest Service.
1949	Forest Service issues prospectus inviting proposals from private interests for the development of a year-round recreational complex in Mineral King. No takers.
1965	Second prospectus. Disney proposal submitted, accepted, and permit for further study issued.
1969	Forest Service announces proposed issuance of final permit, but Sierra Club files suit to delay it.
1973	Suit still in courts. No development of the Mineral King area yet started. Disney Enterprises shifts its funds to Independence Lake.

to remain unchanged, but even to be strengthened in future confrontations. The Sierra Club so disagrees with the management philosophy of the Forest Service that they would like to see another agency in control of the Mineral King area. Specifically, the Sierra Club prefers that the Mineral King area be transferred to the aegis of the National Park Service by inclusion in Sequoia National Park. The preservation orientation of the National Park Service probably would insure de facto accomplishment of the Sierra Club's goals for Mineral King if it were included. On the other hand, the Forest Service has forcefully declared its intention to see the Disney project through. As reasons for this stand, the Forest Service has emphasized the great economic potential of the Disney proposal because private capital would bear the risk and still yield dividends to the government. The Forest Service also has argued that the ecological impact of Mineral King recreational development would be minimal; in short, it is no reason to withhold the final permit. It seems likely, no matter what the final judicial decision, that the Forest Service sees no reason why other projects similar to Mineral King should not be built. Mineral King represents one battle in what promises to be a lengthy war.

PRESSURE ON TSAVO NATIONAL PARK, KENYA

International conservation organizations have shown considerably more interest in the problems of conservation in such places as Africa than in the United States and are likely to regard such controversies as Mineral King as rather minor events in a country where conservation has already become a relatively well-established tradition. For example, the main animal species of interest to the Sequoia (Mineral King) Game Refuge is the deer (*Odocoileus hemionus*); yet the deer problem in California is one of overabundance, for which a game refuge is scarcely necessary. Under these circumstances, to worry about the fate of Mineral King as a game refuge certainly appears trivial when compared with the fate of the great migratory vertebrates of the East African plains, which

FIGURE 16-8 *If the elephant wants the succulent upper foliage, down comes the tree.* (Photos courtesy of Ira Kirschenbaum.)

are threatened by many factors. Outside of national parks, competition for land is steadily reducing suitable habitat, and poaching is scarcely controlled. Within the parks, there are problems with setting the best boundaries for lasting biological conservation and the same kinds of overabundance in a protected environment that characterize California deer. Apart from these immediate biological problems, there are many multiple-use conflicts such as heavy poaching for meat and ivory, wandering tribes that use the land for traditional grazing and farming practices, and growing demand for rapid industrialization. The situation is compounded by meager economic resources, undermanned park staffs, and struggling, often transient governments.

To illustrate the problems facing large African parks, we shall consider Tsavo National Park in Kenya, placing particular emphasis on the elephant population in and around the park. Tsavo National Park is a typical example of the parks of the East African plains. It is large (over 20,000 square kilometers—twice the size of Yellowstone, the largest American park); within its borders live large herds of herbivores (zebras, wildebeests, gazelles, giraffes, and many others) and their predators (lions, leopards, etc.). Of special interest are large numbers of elephants and black rhinoceros, two of the park's most conspicuous animals. The elephants, in particular, are important to the ecology and future of Tsavo National Park because their impact on the landscape has implications for all the organisms sharing it. For this reason, the elephants of Tsavo deserve primary attention in any management program for the park.

Ecologists have known for many years (see Laws, 1970) that elephants can completely change their environment. For example, during periods of low food availability elephants will uproot trees to browse their crowns (Figure 16-8). They can strip a region of all its trees and shrubs and turn it into a grassland—to the detriment of species (such as the rhinoceros) that must compete for browse. Furthermore, the problems that elephants eventually would impose on the national parks were foreseen even by many range experts who could see the increasing density of the elephant population and its effect on the vegetation. Myers (1973) calculated the size of the *ecounit* (the area needed to support the biomass of the elephants of the park itself) and found (Figure 16-9) that it encompassed more than the area of the park. In fact, it did not even respect the border with Tanzania. The size of the ecounit makes it clear that the Tsavo elephant population far exceeds what the park can support.

Myers analyzed the causes and consequences of an overabundance of elephants. The crisis occurred during the 1970–1971 drought, when dry conditions exacerbated the already-critical supply of elephant browse. During the year, about 4700 elephants died in the park as the drought interacted with human activities outside the park to worsen the situation for the elephants. Myers suggested that two factors have caused the extraordinary death rate:

1. Increasing cultivation of the lands within the watershed of the Voi River, which flows through Tsavo National Park. This cultivation has deviated water from the Voi and consequently has reduced water available to the park. The reduced flow of the Voi River exacerbates whatever droughts may occur.

2. A rapidly growing human population (3.5 percent per year at present) that increasingly restricts the habitat of elephants both in and outside the park, which

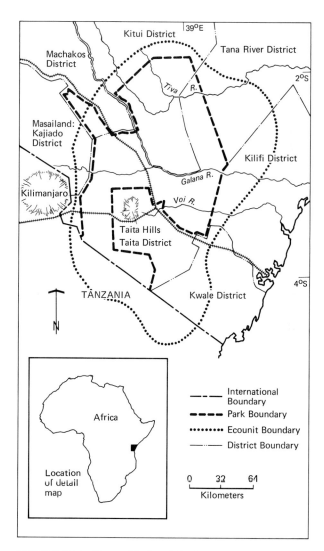

FIGURE 16-9 *Tsavo National Park, Kenya and the boundaries of the ecounit for elephants in relation to the park boundaries.* (Redrawn from N. Myers in *Biological Conservation* 5, p. 124, Applied Science Publishers, Ltd., 1973.)

frustrates the normal tendency of elephants to migrate from areas of short resources. There are fewer places to migrate to.

These changes trigger two specific responses in the elephant populations, namely:

1. Tsavo National Park is a refuge for elephants. Consequently, elephants are concentrated at abnormally high densities within the park.
2. The abnormally high densities of elephants make them susceptible to starvation whenever environmental conditions reduce growth of the browse they depend upon for food.

The size and activity of elephants are important to other populations in the park. As elephants continue to be concentrated in the park, the threat of their starvation and, more important, their capability to destroy vegetation both increase. Conversion of the tree-shrub-grasslands to pure grasslands may favor many grazing species, but it is detrimental to browsers like the lesser kudu and the black rhinoceros, many of which died with the elephants during the 1970–1971 drought because their forage had been eaten and destroyed by the elephants.

This brief sketch is a classic illustration of the effects on ecological systems that result when one population changes. In this case the elephants became very effective competitors when their numbers grew too large for the area to support. This example also illustrates the utility of some systems approach if management is to attain the goal of survival of all species in the park. In Tsavo National Park, sufficient evidence has accumulated to mandate an elephant management program, because there are just too many elephants, a situation that only can get worse as suitable elephant habitat outside the park continues to disappear.

Ironically, the elephant problem partly results from a simple fallacy. Park policy dictates that all biological events within the park should be free of human interference (for example, no shooting of elephants for any reason, even to prevent overpopulation). This dictum implicitly assumes, however, that park borders somehow will prevent the development of close interrelationships between the park and surrounding lands that lie within the elephants' ecounit (even though they fall outside park boundaries). It also assumes that encroaching development at the park's periphery will not affect the park, something we already know is incorrect. Hence it is simply wrong to regard Tsavo National Park as an ecologically isolated unit in which freedom from human interference is extended to mean refusal to interfere with the elephant herds within the park in any way. To take this approach merely means that park management, in fact, stems from the collection of forces that foster the high densities of elephants (human competition for water and land, poaching, drought conditions, etc.) that amount, in Myers's terms, to "management by default rather than design."

Myers argues that both the short- and long-term futures of the Tsavo elephant population are at stake if effective management is not instituted soon. The short-term future, specifically the dynamics of the elephant population and their consequences for other organisms, have already been shown to be entering a period of great instability as the ecosystem is forced to adjust to an environment undergoing continuous change as the area around the park continues to be developed. If nothing is done, one may confidently predict major fluctuations in elephant density in the

Tsavo area. If such a situation is allowed to persist and population growth is followed by repeated crashes, the elephants would be maladapted. Their long life spans and very low potential reproduction are characteristics developed by populations suffering low rates of natural mortality, not by populations suffering regular losses from harsh environments and insufficient food supplies. For the long term, there are two general alternatives: adaptation and selection processes operating on the elephants, perhaps to produce smaller elephants that reproduce more; or reduction and possible extinction of the species in the Tsavo area.

The Tsavo elephants represent a microcosm of the management problems of the entire region. The fate of that part of the ecosystem which hinges upon the well-being and survival of the elephants will depend finally on what happens in the entire region, not only on management activities within the park. Any proposal for regional planning with respect to the ultimate fate of Tsavo National Park must attempt to deal with at least two issues: (1) conflicts in multiple use of the land and its ecosystems in and around the park and (2) the source of funds to support management activities within the park. Activities included in the multiple-use issue are not only those within the park (tourism, research, ivory hunting, poaching, and fish farming) but also those outside the park but within the elephant ecounit (farming, urban development, lumbering, and charcoal production). The wide range of these activities, the lack of control over them, the fact that two countries are involved in the Tsavo elephant ecounit, and especially the poor economic condition of Kenya all make the problems of Mineral King look trivial. It is useless to pretend that answers will be simple and straightforward. Yet Myers sees hope as long as elephants are the principal problem, because he is confident that they can be dealt with by cropping their excess numbers and using them as a source of protein, with the ecological benefits that are derived from this as a valuable bonus.

The problem of a source of funds for an effective management program is dependent upon the political situation; but even given a favorable political climate, Kenya does not have sufficient capital to fund these programs completely when the people have so many more obvious needs. Myers sees no hope that Kenya can resolve the plight of Tsavo National Park alone; hence he sees Tsavo as an international problem requiring international resources. In a conclusion that applies well beyond the scope of Tsavo National Park, Myers writes:

Present funding is not remotely adequate for even minimal research into the Tsavo problem—research which was urgent already ten years ago. This is an opportunity for people far away from Africa, who frequently remind Africans that their wildlife is a heritage not of Africa alone, to show how far they are committed to conservation in Africa. Professional bodies could meanwhile reflect on their priorities during a decade when several hundred additional studies were added to the thousands on white-tailed deer (by no means an endangered species), while in Africa an episode of large mammal ecology underwent one spasm after another, and of a telescoped intensity that was virtually unprecedented since the emergence of modern biology. It has been a phenomenon on such a scale as would almost rival the appearance of a mammoth in Tsavo, yet a phenomenon largely unrecorded and unregretted by the world at large.[2]

---

[2]From N. Myers. "Tsavo National Park, Kenya, and Its Elephants: An Interim Appraisal." *Biological Conservation*, 5 (1973), Applied Science Publishers, Ltd.

At first glance Mineral King and Tsavo National Park would seem to have little in common, but in fact they and all other large natural reserves share many of the same problems, for they are all parts of larger areas that cannot be isolated as neatly preserved packages. They all require management to retain their distinctive character. They cannot simply be established and then ignored.

Questions

REVIEW QUESTIONS

1. The problem of definition of terms—the age-old problem of semantics—appears to be a particularly troublesome factor in the debates about open space planning, wilderness recreation, and ecosystem preservation. What do you think open space is? Can your definition be readily used by planners? Explain why or why not.

2. Is there a single kind of open space that would satisfy
   a. a city dweller
   b. a farmer
   c. a foreign tourist?

3. The Sierra Club believes that wilderness areas are incompatible with multiple-use practices. Yet—although the Mineral King area is permanently inhabited, was formerly mined for its mineral resources, and thus formally does not qualify for inclusion in the wilderness system—the Sierrra Club would like to see Mineral King preserved as wilderness in Sequoia National Park.
   a. Who would benefit from Mineral King as wilderness, and who would benefit from the Disney development?
   b. What parties would not benefit from each alternative?
   c. Can an area be wilderness after it has been exploited for other purposes? Why or why not?

4. Discuss the potential role of ecological research in helping the national parks meet the recreational needs of a public whose leisure time is rapidly expanding.

ADVANCED QUESTIONS

1. For many years the concept of "carrying capacity" has been one popular measure of the number of uses that an environment could support without permanent damage. Using your knowledge of the idea of carrying capacity (with further reference to elementary ecology texts as needed), specify how the concept may and may not be applied in the national parks.

2. You are in charge of the research program at Isle Royale National Park, Michigan. Design a program to encourage the coexistence of wolf, moose, and woodland caribou.

3. Official National Park Service policy states that the national parks should be restored, insofar as possible, to their original state before human entry. Under what circumstances would this be difficult to accomplish?

4. From time to time economists have suggested that visitation in the national parks be controlled by selling visitor rights to the highest bidders. Evaluate this proposal with respect to
   a. equity
   b. preservation
   c. in-park development

5. Read and criticize Diamond's hypotheses (*Biological Conservation*, 71 (1975), pp. 129–146) about the size and shape of nature reserves.

Further Readings

Note: This bibliography has undergone the same sort of selective treatment given to Chapter 15. An attempt has been made to delete unnecessary repetition in the ideas presented.

Brown, W. E. *Islands of Hope.* National Recreation and Park Association, Washington, 1971.

> A relatively recent book addressed to the problem of adequate recreational experience, the future options of natural preserves, and the role of environmental education in resolving conflicts between these two. Some may find the book simple and perhaps boring, but it is unique and should be read.

The Conservation Foundation. *National Parks for the Future: An Appraisal of the National Parks as They Begin Their Second Century in a Changing America.* Task force report. Washington, 1972.

> The most recent addition to the literature. Many prevalent ideas are summarized and evaluated here. This is undoubtedly the single most stimulating source currently available, even if it is doubtful that readers will agree with all its conclusions and recommendations.

Darling, F. Fraser, and N. F. Eichorn. *Man and Nature in the National Parks.* Conservation Foundation, Washington, 1967.

> A well-known and oft-cited forerunner to the 1972 evaluation. Summarized in short form by Eichorn in Darling and Milton, *Future Environments of North America.* Doubleday, Garden City, N.Y., 1966.

Hall, E. J. *The Hidden Dimension.* Doubleday, Garden City, N.Y., 1966.

> More discussion of Hall's work will be found in Chapter 18. See also his earlier work, *The Silent Language,* published by Doubleday in 1959.

Hart, W. J. "A Systems Approach to Park Planning." IUCN publications, new series supplemental paper no. 4. Morges, Switzerland, 1966.

> There is little in the systems literature on park planning, although some of the environmental planning literature cited in Chapter 8 approaches it.

Hartmann, F. (ed.). "The Metro Forest." *Natural History,* 82, No. 9 (1973), 45-84.

> A demonstration of the often exciting viability of forest patches in urban settings.

Hendee, J. C., and G. N. Stankey. "Biocentricity in Wilderness Management." *Bioscience,* 23 (1973), 535-538.

> Most recent paper; has an up-to-date bibliography useful for entries into the literature.

Houston, D. B. "Ecosystems of National Parks." *Science,* 172 (1971), 648-651.

> Houston presents the arguments for the promotion of scientific research in the national parks. See Cunningham (1970) *Canadian Field-Naturalist* for similar arguments based on Canadian experience.

Jacobs, J. *The Death and Life of Great American Cities.* Vintage, New York, 1961.

Laws, R. M. "Elephants as Agents of Habitat and Landscape Change in East Africa." *Oikos,* 21 (1970), 1-15.

> The literature on biological management and conservation is quite large and an excellent source for student projects. This paper is one of the keys into the literature, but unfortunately much of it is in hard-to-get journals. (For additional information in this area, see Duffey and Watt, especially the paper by Parker and Graham and the whole of part 4 of this book—cited in full in Chapter 15).

Lee, R. F. *Public Use of the National Park System, 1872–2000.* National Park Service, Washington, 1968.

A history and predictions on the future written by a veteran of many years of NPS duty. By far the best-detailed summary available and highly recommended.

Leopold, A. S., S. A. Cain, C. M. Cottam, I. N. Gabrielson, and T. L. Kimball. "Wildlife Management in the National Parks." Report of the Advisory Board on Wildlife Management. *Transactions of the North American Wildlife Conference,* 28 (1963), 28–45. Reprinted by U.S. National Park Service (1969).

The "Leopold Report" set off a chain of inquiries, the most recent being the 1972 Conservation Foundation report cited above.

McCloskey, M., and A. Hill. "Mineral King: Wilderness Versus Mass Recreation in the Sierra." In *Patient Earth,* (Harte and Socolow, eds.). Holt, New York, 1971.

For further developments in the Mineral King controversy, monitor the Sierra Club Bulletin.

Myers, N. "Tsavo National Park, Kenya, and Its Elephants: An Interim Appraisal." *Biological Conservation,* 5 (1973), 123–132.

Nash, R. *Wilderness and the American Mind,* revised edition. Yale, New Haven, Conn., 1973.

This book is, in my opinion, the best available on the functions and history of the establishment of wilderness. Comprehensive, well written, and now updated in this revised edition, *Wilderness and the American Mind* is the next logical step from this chapter on the subject of wilderness values.

Outdoor Recreation Resources Review Commission, 27 vols. of reports. Washington, 1962.

The 27 titles alone would fill this page. The ORRRC reports were the first systematic attempts to quantify the supply and demand of recreation. The volumes I have found most useful, particularly for independent analysis, are

3. *Wilderness and Recreation: A Report on Resources, Values, Problems*

19. *National Recreation Survey*

20. *Participation: Factors in Demand*

27. *Literature Survey*

Owen, J. S. *"Some Thoughts on Management in National Parks."* Biological Conservation, 4 (1972), 241–246.

A short and recent synopsis of the First World Conference on National Parks (1962) and on more recent developments in park management policy, blended with Dr. Owen's decade of experience as director of Tanzania's parks (including the famous Serengeti Plains National Park). The Second World Conference was held at Yellowstone National Park in 1972 in commemoration of the centennial of the U.S. National Park Service.

Sikes, S. *The Natural History of the African Elephant.* American Elsevier, New York, 1971.

A large, lavishly illustrated compendium of background information on the African elephant. In combination with the other elephant references, Sikes provides a starting point for detailed analysis of the elephant in Africa. The last decade has seen numerous books on large mammals that provide insight into vertebrate dynamics on the African and Asian continents. Prominent contributions have been made by van Lawick, van Lawick-Goodall, Schaller, Geist, and Kruuk.

Udall, S. *The Quiet Crisis.* Holt, New York, 1963.

U.S. Council on Environmental Quality. *Environmental Quality*, the 3rd annual report. Washington, 1972, chap. 9, national parks.

This section of the CEQ report was written by Robert Cahn, who served as a member of CEQ and has subsequently returned to the *Christian Science Monitor.*

U.S. Forest Service. *Prospectus for a Proposed Recreational Development at Mineral King in the Sequoia National Forest.* Washington, 1965.

U.S. Forest Service. *Mineral King: A Planned Recreation Development.* Washington, 1969.

Whyte, W. H. *The Last Landscape.* Anchor/Doubleday, Garden City, N.Y., 1970.

Section F
People and their
Creations

SECTION F / SELF-STUDY ORGANIZER

- Begin
- Read Chapter 17.
- If you wish to develop greater insight into the problems and dynamics of cities, you will require a firm base of knowledge in the social sciences and in the technological-ecological interface. This is such a large area that it is hard to know where to begin. Start by reading Rose and the Scientific American collection *Cities*.
- If you still feel inadequately prepared to progress any further, you will have to enter the social sciences in a depth beyond the scope of this text. This area is too large to be covered well by reading one or two texts, and reference to one or more authorities is needed. You should, however, know enough from reading this text to be able to consult intelligently with people who know these sources.
- Do you wish to examine some practical applications?
  - Yes → You should examine some of the solutions for planning and design of urban areas and regions. Review Chapter 8 and references therein, especially McHarg.
  - No → Read Chapter 18.
- If you feel that environmental perception is a valid field of study, you may want to enter the literature in depth. Read Saarinen carefully and references therein that interest you; then read *Image of the City*. If not, go on.

```
                                          ┌─────────────────────────────────┐
                                          │ Study Appleyard et al. and       │
                                          │ Sonnenfeld. Using these authors  │
                                          │ and the problems of Chapter 18   │
                                          │ as a guide, devise a simple      │
                                          │ project that you can do your-    │
                                          │ self.                            │
                                          └─────────────────────────────────┘
                                                         │
                                                         ▼
                                          ┌─────────────────────────────────┐
                                          │ If this awakens your interest,   │
                                          │ there is a very highly detailed  │
                                          │ literature in this area in the   │
                                          │ fields of psychology and         │
                                          │ anthropology. See the volumes    │
                                          │ by Esser and by Downs and        │
                                          │ Stea for leads.                  │
                                          └─────────────────────────────────┘
```

Interested in planning applications? — **Yes** → (above)

No ↓

Read Chapter 19. This chapter summarizes the mechanism that mediates human being-environment relationships and is exceedingly important.

To obtain the maximum benefit from this information you will require a firm base of knowledge in the fields of economics and policy analysis. Review Thompson (1970) again, and if your economics is rusty, reference to Samuelson or another basic text will be useful.

Review Section D, especially Chapter 12 on the economics of pollution control and pertinent references therein. Integrate these with Chapter 19.

You should be ready for applications. For the steady-state economy and its implications, compare Ayres and Kneese to Daly. Next, if you wish, examine the interface between science and policy. Compare the positions of Commoner and Ehrlich for a good example summarizing very different positions.

If you are still interested after all of this, there remain the political-legal institutions. Wish to explore these?

Yes →

There are a number of ongoing issues you could follow in the legal and governmental literature. Among these are:
1. The vicissitudes of DDT.
2. The implementation of the Clean Air Act.
3. The interpretation of section 102 of NEPA
4. Nuclear power plant siting.

To do these things requires considerable sophistication in searching the legal literature; few sources (besides newspaper accounts) are readily available to the public. One of these is *The Environmental Reporter.*

No ↓

You are near the end. Go on to Chapter 20.

Exit

In previous sections we considered specific subsystems of the human ecosystem. The emphasis was upon the various interactions between us and our environment. In this section, we shall devote our attention to some special systems that are human creations. Chapter 17 considers the urban areas that are simultaneously the source of our greatest pride and our greatest despair. We are gregarious animals that increasingly find our fate bound more closely with urban centers; we shall either live well within cities or live miserably within them. Idyllic rural existence is not in the future for most of us.

In Chapter 18, we shall consider a special kind of creation—a mental process in which images replace reality in importance. As the search continues for means to build new cities and improve older ones, the evidence leads us back again and again to that mental process. As environmental planning develops, the role of image formation once more proves to be an important influence. This area, usually called *environmental perception*, is an attempt to understand the process of image formation and its environmental applications in cities and elsewhere.

In Chapter 19, we turn to the mechanism by which all human activities are given a value that can be compared with others and the political system which serves to temper and modify economic processes used to make evaluations. In recent years the failures of free-market economies to deal adequately with environmental problems have drawn considerable attention in both the technical and the popular press.

Selected General Sources for Environments of Human Creation

Appleyard, D., K. Lynch, and J. R. Myer. *The View from the Road.* M.I.T. Press, Cambridge, Mass., 1964.

Ayers, R. U., and A. V. Kneese. "Economic and Ecological Effects of a Stationary Economy." *Annual Review of Ecology and Systematics,* 2 (1971), 1–22.

Commoner, B. "The Closing Circle." *Environment,* 14, No. 3 (1972) 25, 40–52.

Daly, H. E. (ed.). *Toward a Steady State Economy.* Freeman, San Francisco, 1973.

Downs, R. M., and D. Stea (eds.). *Image and Environment.* Aldine, Chicago, 1973.

Ehrlich, P. R., and J. P. Holdren. "One Dimensional Ecology." *Bulletin of the Atomic Scientists,* 28, No. 5 (1972), 16, 18–27.

Esser, A. H. (ed.). *Behavior and Environment.* Plenum, New York, 1971.

Holden, C. *Science,* 177 (1972), 245–247. Analysis of the Commoner-Ehrlich debate.

Lynch K. *The Image of the City.* M.I.T. Press, Cambridge, Mass., 1960.

McHarg, I. *Design with Nature.* Doubleday, Natural History Press, Garden City, N.Y., 1969.

Rose, H. M. *The Black Ghetto.* McGraw-Hill, New York, 1971.

Samuelson, P. A. *Economics,* 9th edition. McGraw-Hill, New York, 1973.

Saarinen, T. F. *Perception of Environment.* Commission on College Geography Resource Paper, No. 5. Association of American Geographers, Washington, 1969.

Scientific American. Cities. Knopf, New York, 1968.

Sonnenfeld, J. "Environmental Perception and Adaptation Level in the Arctic." In *Environmental Perception and Behavior* (D. Lowenthal, ed.). Resource Paper No. 109, Department of Geography, University of Chicago, 1967.

Thompson, D. L. *Politics, Policies, and Natural Resources.* Free Press/Macmillan, New York, 1972.
Thompson, D. N. *The Economics of Environmental Protection.* Winthrop, Cambridge, Mass., 1973.

Selected Periodicals

ABSTRACTS

Urban Affairs Abstracts

PERIODICALS

Note: The list that follows is by no means complete. Its breadth reflects the degree to which this particular literature is scattered.

The American Anthropologist
American Journal of Psychology
American Journal of Sociology
Annals of the Association of American Geographers
Ecology Law Quarterly
Environmental Law
Environmental Reporter
Environmental Science and Technology
Environment and Behavior
Geographical Review
Journal of the Air Pollution Control Association
Journal of the American Institute of Planners
Journal of Experimental Psychology
Journal of Law and Economics
Journal of Psychology
Journal of Regional Science
Journal of Social Issues
Journal of the Water Pollution Control Association
Land Economics
Natural Hazard Research
Natural Resources Journal
102 Monitor
Papers and Proceedings of the Regional Science Association
Psychological Review
Science and Society
Water Resources Research

17. Ecology and Urban Environments

In the previous section we examined environments not permanently inhabited by large concentrations of people. This chapter is concerned with just the reverse situation. As the U.S. population continues to be concentrated overwhelmingly in urban areas, the environments we shall be most familiar with (and perhaps best adapted to) are cities. Whatever their problems, modern cities are still the most notable achievements of a succession of civilizations; they are the spatial synthesis of a number of different social functions that somehow must be integrated thoroughly. They are an environment entirely created by and shaped for human beings, but in spite of its synthetic nature, this environment is characterized by functions that parallel the ecological functions found in other ecosystems.

The first question to evaluate is how exact the parallels between ecology and urban system dynamics are. Viewed from the perspective of the city as a system, there are certainly some striking similarities: For example, cities are as dependent upon energy resources from the outside as any population is upon its food; in this sense, the term "feeder route" that is applied to the pathways by which a city receives its energy supplies is remarkably appropriate. Nevertheless, though such analogies are found easily, they are superficial, they presume an erroneous set of causal pathways, and they do not contribute much in the long term toward urban environmental problem-solving. The evaluation of analogies between ecology and urban ecology forms one of the two primary questions dealt with in this chapter.

The other question concerns our future fate as a social, urbanized animal. As populations grow larger and resources become more limited (especially energy), it will be more difficult to retain a highly dispersed, highly mobile lifestyle; unless solar power or some other suitable technological development becomes operational, the choice may very well be between rural self-sufficiency or even more compact urban centers. Since it is likely that most of us will continue to live in cities, it would be desirable to know what we might expect in the future so that we would know how to structure cities to provide the best compromises among convenience, personal comfort, quality of life, energy efficiency, environmental quality, and impact on agriculture. In somewhat oversimplified terms, do we build future cities upward or outward? Is there some optimal internal organization or not? Is there an ecological imperative for city design?

A Synopsis of the Structure and Function of Urban Areas

In this section we examine the idea that a city may be viewed as an ecosystem which has structure and which evolves and functions in certain ways.

FUNCTIONS

In past years cities were the centers of civilization and human activity, and so they arose where it was natural to have such an aggregation of people. For example, St. Louis was founded at the confluence of the Missouri and Mississippi rivers (both of which were major trade and transportation routes), and San Francisco also was founded at a major transportation center (San Francisco Bay was an important harbor for Pacific Ocean shipping). Cities are, however, dynamic assemblages which are shaped by changing conditions. For example, in recent times, San Francisco has become more important for commerce and as an entertainment center than as a port for marine commerce. In this sense, then, because they are dynamic and able to adapt to changing conditions, cities may be visualized as "superorganisms" that integrate various functions much as organ systems are integrated in the body. In the same way, it is possible to view cities as creations that must combine a number of functions well enough to attract a population of inhabitants and promote economic activity. Cities that do not achieve a satisfactory balance of these functions do not survive as dynamic entities.

The functions that a city must possess can be sorted into six categories. The biological requirement for shelter means that *housing needs* must be met, with their attendant questions about quantity, location, type, and cost. There also must be provision for each

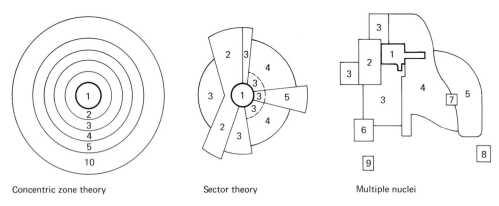

FIGURE 17-1 *Three theories of the structure of cities based on their growth patterns.* (Redrawn from *Internal Structure of the City: Readings on Space and Environment* by Larry Bourne. Copyright © 1971 by Oxford University Press, Inc. Used with permission.)

person to earn a living and a system for valuation of labor and goods: Thus a city must have an *industry-manufacturing base* and an adequate *commercial establishment* to produce and market goods and services needed within an interdependent society. This also includes the handling of foodstuffs and other goods produced outside urban areas. The *transportation-communication network* is the means of distribution, both within and between cities; it binds the other activities together and enables widely separated activities to function smoothly. Next, a *governmental system* is needed to adjudicate disputes and provide public services, protection, and welfare. Finally, the *recreation-entertainment* function includes the provision of open space, parks, cultural and sports events, and other activities that occupy time not needed for earning a living. Because of their concentrated populations, cities are the only places that can support various cultural and sporting events which require extensive public financial support.

These six functions describe a highly evolved, socially regulated network that allocates materials and energy, moves them from place to place, and provides essential services to members of the population. In this sense, at least, there is a clear parallel between the structure and function of any ecosystem and the urban counterpart. It is another question, however, whether these similarities reflect shared ecological properties or whether they are simply generalized systems properties. Urban specialists tend to view urban structural patterns as the product of the land-use requirements of the different urban functions. Three classic models of city structure are the concentric, centrally organized model; the discrete sector model; and the diffuse model (Figure 17-1). The concentric model hypothesizes that city development is orderly outward from the center and arranged such that heavy industry and commerce are inside and residential areas are more peripheral in even rings outward from the center. The discrete sector model hypothesizes a traditional center but has the different functions localized in discrete sectors or radii that project outward from the center. The diffuse model assumes that development subsequent to the center's formation will be highly dispersed; all activities will develop together in different neighborhoods and suburbs. The generalized model presented in the paragraph above is a hybrid of these more specific but less realistic models in that it recognizes the central location of some functions but also notes the tendency of other functions to disperse widely. Bourne (1971) has pointed to the weaknesses of the three models because of their static nature. In fact, recent work has shown that modern American cities rarely show any of the classic forms, because the factors that determine the structural arrangements of functions in a city interact in complex ways which actually can produce a wide range of different structural patterns.

Two major determinants of urban structure are the control of land use and the nature of the transportation system. A city will have its structure partly determined by the forces that govern land use, whether these forces are de facto land planning or the more explicit public planning discussed in Chapter 8. The results are to be seen in the internal structure of the city and at its growth on the fringe. Analysis of these will reveal (at least roughly) the extent and nature of structural growth controls.

The nature of the transportation network as well as its location is another important determinant of urban structure. The large eastern cities of the United States all developed before the advent of the automobile, and hence railroad and waterway networks were the only available modes of transportation. Since both these systems are limited in the area they can serve, cities developed compactly; the transportation network limited the extent to which they could spread. The invention of the automobile, however, revolutionized urban structure. The automobile made highly personalized transportation available to all segments of society, improved both the speed and the flexibility of transportation, and made it possible for urban functions to be more widely separated without sacrificing any distributional convenience. In the United States there is much evidence of the automobile's impact on urban structure. Most western cities experienced major growth and development after the automobile had been integrated thoroughly into urban society, and as a result, these cities show a more pronounced tendency to be spread over the landscape, with the role of the central business district being much weakened in favor of regional centers of commerce and government. This drastic change in city structure was promoted by the spiral of development of vast road networks and the further dependence upon the automobile that this produced. This, in turn, made it possible for an urban region to be spread over even greater distances.

Los Angeles is the prototype of this kind of city. It developed from an area with 2.9 million people before World War II to one with over 7 million people by 1970. Its dependence on the automobile and truck is total and has become famous throughout the world; Los Angeles's freeway system has been a model for other cities—some emulate it, others avoid it. The effects of this total dependence on road vehicles can be seen today: Los Angeles has spread from the San Fernando Valley in the north to Orange County in

FIGURE 17-2 (a) Downtown Los Angeles has wide streets and innumerable parking lots. Note how close the freeway is to the business district. (b) Chicago was designed before road vehicles were as important as they are today. As a consequence, Chicago's streets are narrow, its public transportation is more developed, and its center is much more compact. Suburban Chicago's later development evidences the same sprawl growth that all of Los Angeles does. (Photos courtesy of Tom Stack and Associates.)

(a)

(b)

the south (and even promises to merge with San Diego), and from San Bernardino in the east to Malibu in the west. The central area of Los Angeles is vastly less important than in any other American city of comparable size (Figure 17-2); it has practically no public transportation network and is characterized by the diffused localization of industry, commerce, government, and housing in a number of centers (Los Angeles, Long Beach, Orange County, San Gabriel Valley, San Bernardino, Riverside, West Los Angeles, and the San Fernando Valley). Interestingly enough, an examination of urban development in older American cities after 1940 reveals the same pattern. The range of the automobile has promoted suburbanization of these cities too, no matter what the earlier pattern had been.

EVOLUTION

The influence of the automobile on urban structure is just one example of how a city may change, and cities certainly have changed in response to various factors. In his monumental book, *The City in History* (1961), Lewis Mumford traced the evolution of cities from their function as relatively small centers for trade, commerce, and manufacturing in agricultural regions to their role as centers of regional aggregation of power (Greek city-states, Rome), and finally to their present roles as the predominant loci of human activity. San Jose, California (Chapter 8), is one excellent example of this evolutionary trend. Yet the trend seen in the history of metropolitan regions is not one of unmitigated blessings, nor is its study merely an esoteric exercise. In fact, it is to be hoped that through a better understanding of urban evolution we might develop greater insight into future urban trends. For several decades, our cities have exploded rapidly across the countryside, changing from compact urban centers to widespread megalopolises as populations have continued to grow and people have gravitated to cities for employment. Urban planner Constantine Doxiadis has predicted that if the world population continues to grow, cities will spread until they meet, resulting in a world that is mostly one large city separated only where physical barriers are unbridgeable. However, other planners are confident that the present growth of urban areas cannot continue unchecked. They point out that urban concentration already has created an enormous range of problems—disruption of family units, social segregation, frustration, crime, inefficient transportation, energy waste, and many others—problems which certainly are not going to be solved by further expansion.

The direction of further evolution of American cities is unknown. There are certainly strong feelings that we cannot simply continue blithely onward, but there is little agreement about what to do to improve the situation. Should we assume that American cities eventually will merge with one another in a manner reminiscent of Ecumenopolis? Should we instead attempt to reconcentrate our cities into the compact form they once had? Or do we attempt to decentralize the urban areas we already have and all go back to the simpler rural life?

In any event, cities will continue to evolve because the forces of internal change—the residents and their activities—cannot stop and wait for some final strategy to emerge; neither will external conditions remain unchanged. This uncertainty makes the evolution of cities a fascinating topic. To illustrate this uncertainty, consider the possible changes in cities should petroleum costs rise so high that conventional automobiles are impractical for most of us. Cities may have to become compact again because there may be no alternative. On the other hand, public transportation may reassume its preeminence and in part counteract this tendency, allowing urban growth to disperse along selected routes; or cities may fragment once again into the semiurban centers of rural activity they once were, should petroleum or some other resource shortage also limit industrial activity.

Alternative Futures for Urban Areas

Our discussion of urban structure, function, and change suggests that cities may indeed be viewed as ecosystems. In a city, economic activity is the equivalent of an ecosystem's energy resource and is itself dependent on resources that must be imported from outside, permitting the development of diverse subpopulations, other economic activities, and various cultural activities. Transportation and communication provide the transfer mechanisms that promote integration of all these things and that help determine the structure of the assemblage. Thus it is possible to view the metropolis as an urban ecosystem having an energy-flow network and whose character depends on the magnitude and flow of this energy. If we search for more analogies, they are not difficult to find. But the more important question is whether or not ecological science can provide insight into urban problem-solving. In this section we shall assume that our cities are not perfect and that there are probably better ways to do urban planning. Readers who doubt the validity of this assumption should read Mumford (1961, 1968), Cassel (1971), or any of the authors discussed below.

Some Solutions for Urban Ills

We now shall examine solutions that have been advocated by various authorities with different specialties—planners, architects, engineers, and sociologists. Readers should bear in mind that these proposals can overlap (sometimes quite strongly), but they also often embody different underlying beliefs about the causes of urban problems and involve different levels of comprehensiveness, as well as different scales of city organization at which a proposal is to be effective (neighborhood, city, or region).

REHUMANIZATION OF CITY ENVIRONMENTS

It is now common knowledge that the large urban renewal projects so fashionable in the 1950s—the severe concrete blockhouses (Figure 17-3) which builders and city financiers hoped would be the single-stroke solution to slum clearance and urban renewal—were colossal failures. They provided no magic answers and in fact proved to be serious problems in themselves. The failure has subsequently been laid to the scant attention these designs gave to human factors, particularly to the social impact of these structures upon their residents. This discovery stimulated many books and articles and was followed by the same experience with one after another of these public tenements until now they are being abandoned. In St. Louis, the city government has begun to demolish some of these buildings (notably the Pruitt-Igoe complex) and to replace them with other, smaller clusters of buildings.

One of the first to challenge the emphasis of the planning profession upon large high-rise public housing projects was Jane Jacobs, who, after having dismissed large urban renewal buildings as hopelessly inadequate, proposed that cities be built as a collection of small and socially cohesive neighborhoods. Jacobs believes that these cohesive social groups could provide the essential human element that would produce the paragon of the ideal city. Fostered by an arrangement consisting of small blocks, high population densities, and the resulting high diversity of people and activities, Jacobs's ideal city strongly resembles the parts of New York City that have so influenced her thinking. Jane Jacobs uses the eastern American city of the 1800s and early 1900s as her model of the ideal urban environment, largely untouched by the automobile and suburbanization. As such, her plans represent one attempt to rehumanize the city by emphasizing new social considerations within the physical planning process of existing cities. The major difficulty in her proposal is that she advocates intense crowding, for after all, her ideal city is compact. To advocate the encouragement of higher densities cannot be condemned in itself (especially since many people apparently benefit from high densities and since total social involvement is the heart of Jacobs's proposal), but its implementation requires far more consideration of the possible consequences than she has made.

FIGURE 17-3 (a) *The Pruitt-Igoe housing complex in 1956, three years after it was completed.* (b) *In June 1972, the city of St. Louis began bringing the complex down with explosive charges.* (Courtesy of Arteaga Photos, Inc., St. Louis, Missouri.)

The Jacobs plan for cities has been criticized because it is inconsistent in places and represents a call for a return to cities of the past—cities that might not have been faultless communities after all. In their thoughtful book *Communitas* (1960) Paul and Percival Goodman put forth a rationale that they feel will guide rehumanized urban development. They perceived that the physical planning element—the formulations developed by the typical planning staff—is a mere technical solution with specified goals which do not necessarily consider moral, aesthetic, and cultural integration. From this starting point they developed a philosophy of planning under the name *neofunctionalism*. They maintained that form should be specified by function, but only after the functions themselves have been evaluated and questioned thoroughly and systematically. They ask that each particular function be examined carefully to determine whether it is really needed, and if so, to determine what it is needed for. They ask that we question whether a plan serves the needs and desires of the people it affects most, or if it really requires that the people it affects adapt to it. In particular, the Goodmans note that planning often is guided not by human considerations but by the self-generating need to have full employment of capital at all times. That is, a society and its structures are dependent upon getting the maximum use from all existing industrial machinery, and under these circumstances economic health is prevalent only when goods are being produced as fast as possible—whether they are needed or not. The net result is a society dominated by full employment rather than by consideration of the quality of life for individuals. Thus they decry this sort of planning because its function is not for the betterment of individuals at all; they feel its function is meaningless and should not be served.

Having denounced present-day cities and the planning that goes into their creation, the Goodmans present three paradigms that could lead to different kinds of cities, including

1. Cities built for efficient consumption
2. Cities built to eliminate the separation between the sites of production and consumption
3. Cities based on planned security with minimum regulation

These three alternative scenarios hypothesize different relationships between production and consumption within a city.

Efficient consumption is the paradigm closest to the current operation of American society. It differs, however, in proposing that the efficiency of consumption by occupants of cities be drastically increased. A city would be structured so as to produce this effect. The Goodmans present one example in which a city built for efficient consumption is arranged into four concentric zones. The innermost zone is the marketplace, where manmade goods are consumed (and in their consumption are destroyed). This is the zone of industry and commerce. In zone II manmade goods are not destroyed in their use. This zone includes such things as education, arts, culture, and journalism. Zone III includes such things as social and sexual intercourse and domestic life—it is, in short, the residential zone. Zone IV is open country; it includes goods produced by nature that are enjoyed but not consumed. This is open space with recreational, aesthetic, and environmental offerings.

Such a city is the rational result of a neofunctionalist analysis because the arrangements represent a continuum from human activities in the center to relatively undisturbed natural landscapes at the periphery.

FIGURE 17-4 *Washington Street in downtown Boston is the shopping district bordering on the business district: (a) normal weekday traffic, and (b) Sunday afternoon. (Courtesy of Ira Kirschenbaum.)*

Presumably, this avoids inefficiency as well as conflicts between incompatible activities and at the same time minimizes transportation costs by putting related activities close together.

Elimination of the separation between the sites of production and consumption departs from the prevailing idea that the sites of production and consumption must be rigidly separated in a city (Figure 17-4). The Goodmans argue that such separation prevents work from being integrated into lifestyles, thereby also preventing it from gaining value for individuals and allowing them to develop pride in the work process. In essence, this paradigm sees the ideal solution as being small individual workshops scattered throughout a city. The Goodmans, however, consider this paradigm most useful in undeveloped regions where evolution of cities is still mostly in the future, because only under these conditions could such cities develop.

Planned security with minimum regulation is the paradigm preferred by the Goodmans. It decouples government from the economy by having the state take over only the subsistence aspects of the economy; this, they feel, preserves individual incentives while maximizing opportunities for self-betterment without endangering lives.

Whereas the first two paradigms suggest the ultimate structure that cities would attain—one fairly rigidly specified and the other highly dispersed—the third one is the pinnacle of rehumanization in that all the planning elements for a city are purely a matter of choice and are subject to the same neofunctionalist analysis used in earlier steps. Clearly, this would produce a social system and cities quite unlike any we know. The government would withdraw from most economic activities, leaving a society with a unifying force that is based on the city as a social commune committed to an individual's freedom; structural elements are definitely secondary. This paradigm is also utopian, for the new society would have to be freed from conventional ideals. Moreover, it would be necessary to locate the city in a relatively undeveloped area where it could develop slowly along with the economy.

Communitas is appealing—even if utopian—because the ideas upon which neofunctionalism is based are in the main simple, logical, and unified. The common thread that runs throughout *Communitas* is that overspecification and overplanning result from putting human needs second. Neofunctionalism is the Goodmans' answer to irrelevant plans and nonfunctional designs; it asks us to question what we think our needs are, eliminating the frivolous and discovering what necessities we have overlooked. Neofunctionalism provides the rationale, and the planning process follows automatically. The basic faith in humanity that glimmers through the Goodmans' message is unmistakably the same as Jane Jacobs's.

Our last example in this category is the analysis of planning presented by Robert Goodman in his book *After the Planners* (1971). His thesis is also quite simple: People who live in cities should direct the planners and architects, for the inhabitants are best able to determine what they need and want. Thus Goodman suggests that people do their own planning and design; by having such power, they could control their own destinies and then would be able to design cities that meet their needs. In Goodman's view, this is the essence of a rehumanized city.

Goodman's arguments rest upon three premises. First, he argues that planners either intentionally or unwittingly have promoted the centralization of political power into the hands of an elite few—never into the hands of the people, especially the poor and disadvantaged people who need the power most in present-day cities. Second, he believes that within the present system, the mere development of plans serves to reaffirm these political and economic institutions which perpetuate the stratification of society and, therefore, the repression of the poor—even while proclaiming the opposite. Finally, Goodman decries the effectiveness of "advocacy planning"—the donation of professional services by planners and architects to the poor—because even the advocates cannot be truly liberated from their biases. Given these premises, Goodman's conclusion follows logically. If planners and architects have been trained to share certain biases and are either aided and abetted in the exercise of these biases by stipulated government funds and statements of policy or are so constrained by higher authorities that they are mere technicians involved in the execution of policies set from above, then there is no alternative to withdrawing the design and planning power from the whole system and giving it back to the people—lock, stock, and barrel.

Goodman's call for "power to the people" represents the most direct way to achieve humanization of urban environments, as long as "the people" know their needs as well as Goodman believes, and as long as they have the wisdom and ability to implement these needs. However, an even larger problem must be addressed: the potential incompatibility between the humanized cities so designed by the authors we have just discussed and economically viable, highly integrated cities. Would the small, cohesive neighborhoods Jacobs favors impair economic activity in the city? Alternatively, can an economically dynamic city be developed from small neighborhoods? Does the concept of neofunctionalism necessarily guarantee a rehumanized environment? Or are neofunctional structures just another set of spatial patterns that really do little or nothing to modify the social environment of the city? Finally, we must ask whether or not Goodman's solution would actually work: Would the city be an integrated economic, political, social, and cultural entity if his solutions were accepted? Or would it be a disorganized hodge-podge, an anarchistic nightmare?

GRAND DESIGNS IN PHYSICAL PLANNING:
NEW TOWNS

While some urban thinkers have concentrated on the renewal and rehumanization of existing cities, others have espoused the development of entirely new cities in rural areas, quite distinct from existing metropolises (that is, something other than suburbs), either because they feel that existing cities are beyond salvation or because they believe the best answer in urban planning ultimately lies in the rational distribution of smaller cities—not just one giant megalopolis—throughout a region. The new cities arising in the countryside are called *new towns*.

The originator of the new town concept was Sir Ebenezer Howard, who founded the "garden city" movement in England at the turn of the century. Garden cities were the first new towns. Howard believed that cities had strict physical limitations to their expansion, beyond which they would disintegrate functionally. The means proposed by Howard to accommodate population increase was to found new towns (garden cities) when existing ones reached a population of about 25,000. The essence of the garden city concept lies in the foundation of economically independent cities unlike our present suburban ("bedroom") communities. They would be designed with quantities of open space. Eventually, all the towns would form a regional cluster, each free of the self-strangulation characteristic of a single giant megalopolis and each surrounded by accessible open spaces and greenbelts. Howard's plans were meant to salvage all cities in that their implementation would prevent the inevitable degeneration resulting from the congestion so characteristic of unlimited growth. Toward this end, he embarked on a series of experiments aimed toward finding the best means for deciding when a town should be allowed to grow and when a new one should be founded instead. Howard did not discover the answers, but two new towns he created survive in England today, albeit as bedroom communities (Figure 17-5). His work finally led to the New Town Act in England in 1946, and his imaginative approach served as a springboard for many efforts that followed the path he blazed.

Since Howard's initial work, new towns have sprung up in many countries, particularly in Europe and in North and South America. We shall consider one of them: Tapiola, Finland, near Helsinki. It is a logical successor to Howard's original ideas about garden cities, and its design is a tribute to his influence. Tapiola was begun in 1952 with the aim of being industrially and commercially independent of Helsinki only 7 miles away. The city was designed to provide as much individuality and as much intermingled natural landscape as possible and to include considerations of environmental quality and environmental impact in the planning scheme.

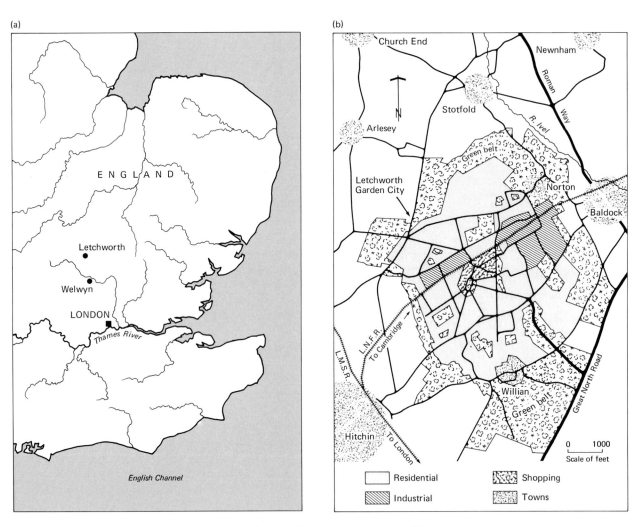

FIGURE 17-5 *Sir Ebenezer Howard's original Garden Cities in England.* (a) *Map of southeastern England showing the relationship of Letchworth and Welwyn to London.* (b) *The design of Letchworth with respect to greenbelts and open space.* ((b) Redrawn by permission of Faber and Faber Ltd., from *Garden Cities of Tomorrow*, E. Howard, 1965.)

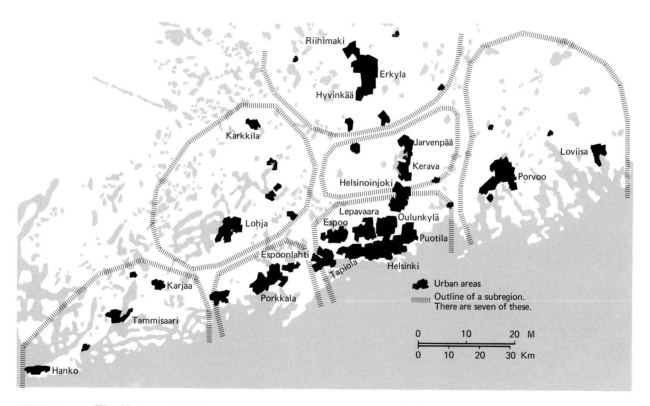

FIGURE 17-6 *The Uusimaa 2010 Plan, an outgrowth of the Seven Towns Plan for regional urban development in southern Finland. Each of the seven developmental regions of Uusimaa County is outlined by broken lines. The shaded areas are bodies of water.* (Redrawn from H. von Hertzen and P. D. Spreiregan, *Building a New Town*, 1971, M.I.T. Press.)

Tapiola was built and is administered by a private nonprofit consortium named *Asuntosaatio* (the "housing foundation") instead of by a public agency. As a result, the methods by which Tapiola was built were innovative. For example, the town was constructed in sections, each of which was designed by an architect who first was selected through competition. Without rigid rules other than their own standards and the determination to meet civic design objectives, the architects of Tapiola turned the new town into a humanized architectural showcase, but not one of uniform style; instead, one of the most refreshing aspects of the town is its great architectural diversity.

FIGURE 17-7 *Tapiola Garden City.* (a) *H-type semidetached houses separated by pedestrian paths only.* (b) *Evening view of the town center. The tower penthouse is illuminated as a beacon.* (c) *A walk-up apartment building.* (From H. von Hertzen and P. D. Speiregan, *Building a New Town,* 1971, M.I.T. Press.)

Tapiola ultimately will consist of seven sections totalling 80,000 residents. Because it is so close to Helsinki, Tapiola has not fully attained all its goals—a portion of the residents do commute into Helsinki to work. Nevertheless, the list of achievements in Tapiola is quite impressive. As of 1971, open space constituted 54 percent of the total area of Tapiola; roadways, so dominant in other cities, occupied only 9.5 percent. Tapiola's impact on Finnish city planning has been considerable: Plans have been drawn up already for six more new towns around Helsinki (the Seven Towns Plan, recalling Howard's regional cluster), and following this there are plans for a refined, regionally based version (the Uusimaa 2010 Plan). The Uusimaa 2010 Plan (Figure 17-6) is the logical successor to the successful completion of the Seven Towns Plan. Uusimaa County would be broken up into seven developmental regions, each having one of the seven towns as its nucleus for further development. Ultimately, the success of the Uusimaa Plan would serve as a model for development of the other counties in Finland.

It is difficult to assess the implications of Tapiola for the viability of new town planning elsewhere. In many ways Tapiola Garden City has not yet provided definitive answers. The amenities for residents are readily obvious (Figure 17-7), but Tapiola has not yet shown that it can be the independent regional center Asuntosaatio wants it to be and yet retain the combination of low density and high environmental amenity it presently has. Furthermore, the success of Tapiola may not guarantee similar successes in other countries, for Tapiola may be an ideal development only for Finland, whose population density is very low compared with most countries (4.7 million per 130,000 square miles, or 36 per square mile). Hence Finland is unusual in that it has the option (especially after the shattering impact of World War II) to develop at low densities since so much of the country is still wild land and there seems to be little antidevelopment sentiment. In England, the failure of Howard's garden cities to dominate the English countryside can be traced to the inability of the garden city concept to keep pace with the rapid population and economic growth of the time (for example, see Mumford's argument in *The Cities in History*); in Finland, this problem does not exist.

TABLE 17-1 *Summary of the statistics on Tapiola Garden City. Unlike conventional cities, housing and roads occupy relatively little space, leaving vast commons and parks. This is the result of the predominance of multiple family dwellings over single family homes.*

Portion	Acres	Hectares	Percent
Housing	161.6	65.4	24.2
Public buildings	34.0	13.8	5.1
Commercial buildings	13.8	5.6	2.3
Industrial buildings	24.7	10.0	3.7
Garages	7.2	2.9	1.0
Common green and parks	362.0	146.5	54.2
Traffic areas	63.7	25.8	9.5
Total	667.0	270.0	100.0

	Acres	Hectares	Percent
Single family	25.7	10.4	15.9
Row	51.6	20.9	32.0
Multistory	84.3	34.1	52.1
Total	161.6	65.4	100.0

SOURCE: Reprinted from *Building a New Town: Finland's New Garden City, Tapiola* by H. von Hertzen and P. D. Spreiregan by permission of The M.I.T. Press, Cambridge, Massachusetts. Copyright © 1971 by The M.I.T. Press.

All these unknowns raise a question for the future of new towns, wherever they may be built. Will it be universally possible for new towns to have all the necessary amenities and still be fully functional cities independent of all others? While the new town concept is intriguing and would seem to have great potential, there are still many questions to be answered.

Cities like Tapiola, Reston (Virginia), Ciudad Guayma (Venezuela), and Brasilia are in fact continuations of Sir Ebenezer Howard's pioneer experiments. It is hoped that they will provide some much needed answers.

GRAND DESIGNS IN PHYSICAL PLANNING: SUPERCITIES

The new town approach to urban design emphasizes environmental quality from the humanitarian, individual perspective. Other architects—who have reacted more strongly to the global, national, or regional explosions in population and have been alarmed at deteriorating environmental quality and resource supplies—have taken entirely different approaches to city design.

C. A. Doxiadis's (1969) warning of Ecumenopolis is one version of supercity by default. Ecumenopolis is a projection made from Doxiadis's calculations on population growth rates, resource supplies, and habitable areas of the earth. Doxiadis calculated the maximum world population and its distribution across the planet. In general, his projections extend present patterns of urbanization, with the only large low-density areas being either uninhabitable (for example, the Sahara Desert) or places in which the world's food supply is grown.

While it would be interesting to examine such projections on a finer scale, say your own town or state, Doxiadis restricted his predictions to the global scale on the grounds that his data were not good enough for projections at a lower scale. Nor could he specify how Ecumenopolis would function, how it would be governed, or what social impacts its existence would have. In general there are many reasons to doubt that Ecumenopolis would ever emerge, many reasons to feel it would collapse long before it grew into the megacity Doxiadis envisions. There is social and ecological support for believing that such a city is impossible—matters of resources, questions of controlling the megacity, and questions of social interactions. Ecumenopolis is the antithesis of Jacobs's cozy neighborhoods and of Tapiola; it is profoundly anti-human and embodies the worst of urban growth uncontrolled. For these reasons, Ecumenopolis is not likely to be any kind of solution, and we had better hope it does not come to pass. In the final analysis, Ecumenopolis is clearly not a uniquely planned supercity but only a projection of what might await us should the world population continue to increase.

R. Buckminster Fuller became widely known to the public when his geodesic dome housed the U.S. exhibition at the World's Fair of 1967 in Montreal, Canada. Actually, Fuller had been involved in a mixture of innovative architecture, technological projection, and metaphysics for some decades. He designed the geodesic dome (Figure 17-8) to emphasize a means for controlling the environment in combination with extreme lightness in weight. It proved to be the forerunner of an even grander design by Fuller for an entire city. He used the same principles of strong but light construction in this proposal, but this design would take the form of a gigantic equilateral tetrahedron composed of more than 300,000 dwelling units lining the exterior surfaces. Transportation corridors would not be numerous, but Fuller considered them adequate for the tetrahedral city, and he proposed further that the entire base would be covered by an enormous public park lighted by great windows at every fiftieth story. The city could grow at its own pace because the space allocated for dwelling units—that is, the faces of the tetrahedron—would be filled only as needed. The city could use its cavernous interior spaces to house essential machinery and a port and an airport; it also could house commercial and industrial enterprises. Built upon the same principles of strength and low weight used earlier in the geodesic domes, Fuller's tetrahedrons would have the structural strength to support a city of 1 million people, yet would be light enough in weight so that the mere addition of a hollow concrete foundation would allow it to float on water. Fuller claimed that the floating version of the tetrahedral city would be insulated from earthquakes on its liquid foundation and could even be anchored at sea.

Fuller's fascination with lightness and strength also lead him to return to the potential role of geodesic domes in future structures. Why not, he suggests, use geodesic domes to enclose entire cities (presumably including the tetrahedrons)? This would

1. Allow cities within the geodesic dome to control the weather and temperature conditions of their external environment.

2. Allow human beings to live in inhospitable environments, such as the polar regions, because they would not need to venture into the regions except in protected intercity vehicles.

3. Convert the cities into mobile units. Fuller calculated that the weight of a dome 1/2 mile in diameter would be only 1/1000th as much as the air it enclosed (he contended that the weight of the enclosed city and its inhabitants would be negligible). Therefore, when exposed to sunlight, the greenhouse effect produced by the dome would heat the interior air and reduce its density relative to the surrounding air, allowing the entire structure to become airborne.

FIGURE 17-8 *R. Buckminster Fuller's geodesic dome at Expo '67 in Montreal, Canada.* (Courtesy of Russ Kinne from Photo Researchers, Inc.)

FIGURE 17-9 *The life cycle of an archology. Each stage of the life cycle is shown from the elevation view and from above. Arrows indicate the direction of material flow during construction, and later, demolition.* (From P. Soleri, *Archology, The City in the Image of Man*, 1969, M.I.T. Press, p. 35.)

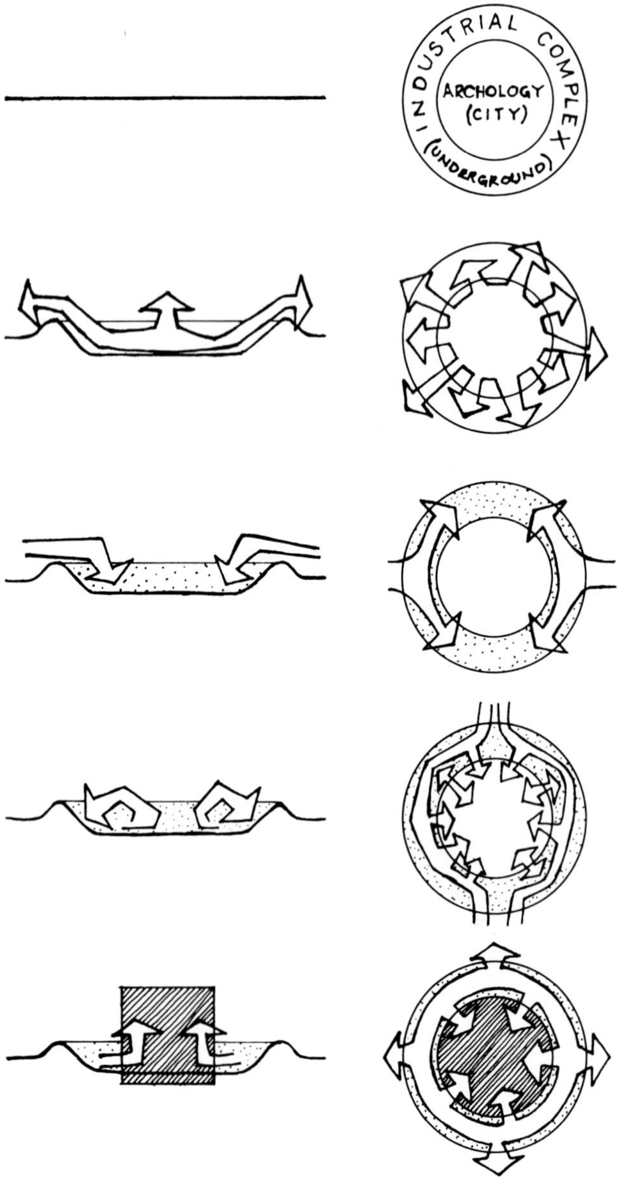

PHASE 1

Selection of site

Time: months–years

Ecological survey
Social and economical survey
Regional and state lineage via rail, road, air, radio, power station, water, gas . . . .

PHASE 2

Excavation and quarrying

Time: months

Production of building materials: sand, gravel, stone (cement)
Production of new topography and foundation platform.

PHASE 3

Instrumentation of the grounds

Time: months

Tooling for the production of standard and custom made structural and utility elements. The industrial base of the future city is established.

PHASE 4

Industrial production of city's parts.

Time: months–years

Stock pile of manufactured materials.
Establishment of inlets and outlets.

PHASE 5

Assemblage and construction of skeleton

Time: months–years

The 3-dimensional landscape is constructed and linked to the region.
Parts of the 3-D landscape go into use.

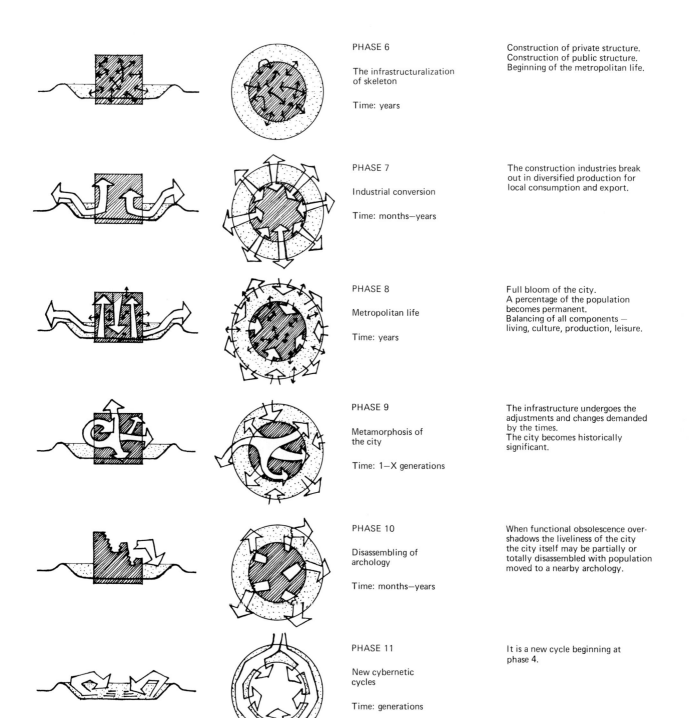

Fuller contended that if the heated air could be kept within the dome, the density difference would allow the dome and its contents to float overnight. (But he does not mention what might happen if there were a series of cloudy, overcast days.)

Fuller's visions are a utopian picture of future urban environments:

Passengers could come and go from cloud to cloud, or cloud to ground, as the clouds floated around the earth or were anchored to mountaintops [the floating clouds are his airborne cities]. While the building of such floating clouds is several decades hence, we may foresee that—along with the floating tetrahedral cities, domed-over cities [plus several other mobile cities and cities in unusual places]—man may well be able to converge and deploy at will around the earth, in great numbers, without further depletion of our planet's productive surface.[1]

Fuller's visions are idealized, highly technological, solutions through which we may temporarily escape the limitations of resources. They seem far-fetched and impractical, and they probably are.

The work of Paolo Soleri may seem superficially similar to Fuller's, but in reality it is vastly different. Both men have adopted the vertically expanded city with a high population density as their basic solution to the problem of housing vast human populations, but the similarity ends there. Soleri builds vertically in order to save wild lands; Fuller builds upward in order to demonstrate a compact and elegant future technology. Soleri pays attention to the functional relationships within the city; Fuller does not. Soleri's premise is that we must reconcile ourselves to nature;

---

[1]From R. B. Fuller. "City of the Future." *Playboy*, 15, No. 1 (1968), 230.

Fuller is confident that technology always will find answers that enable us to overcome the limitations of nature.

Soleri defines an "archology" as the fusion of ecology and architecture into a viable process-oriented city that will prevent human self-destruction through the senseless continuation of our two-dimensional, space-consuming megalopolises. By 1969 Soleri and his colleagues had designed 30 archologies to be built in particular kinds of environments. (The structure of each assumes that the topographic setting will be of certain kinds.) However, they all were designed on the basis of five ideas:

1. Soleri contends that human society is naturally three-dimensional, and so it is only natural to build cities three-dimensionally.
2. The functions Soleri identifies as essential to urban society (for example, commerce, industry, housing, government), and the transportation-distribution elements that link them, are given the highest priority in the design.
3. All archologies are provided with the means to regulate their internal physical environments.
4. All archologies provide specific physical elements that make it easier to organize private and public institutions.
5. All archologies are unabashedly of human design—there is no attempt to preserve nature within them. But in exchange, surrounding areas are left as completely natural as possible.

Soleri's concept of the evolution of a typical archology is shown in Figure 17-9. This figure depicts, both in overhead view and in profile, the entire process from site selection (phase 1) to construction (phases 2 to 7) to maturity (phases 8 and 9) to final dissolution and rebirth (phases 10 and 11). Soleri conceives of a typical archology as serving needs that change in time; hence the archology itself becomes outdated and must be replaced sooner or later.

Now let us look at the simplest of these archologies. Arcosanti (Figure 17-10) is also known as Mesa City; it will be the first archology to be built.[2] Perched upon an Arizonan mesa, it will be small (a total population not to exceed 1500) and is intended as (1) an archological prototype, (2) a school for Soleri and his students, and (3) an ongoing experiment in urban design. In his second book (1971) Soleri expanded the Mesa City project into a fully integrated and independent city of 20,000, but here we treat the smaller version of Arcosanti. In Figure 17-13 we see an elevation of Arcosanti in place on the mesa. Because Arcosanti is small and specialized, the design itself is fairly simple, emphasizing work, communication, and living quarters with much of these being located underground. In spite of its limited size and simplicity, however, Arcosanti is not dominated by any one type of structure; in fact, one of its virtues is its tremendous architectural variety.

Soleri's more advanced and speculative designs are so massive and awesome that further analysis can be carried out only by studying specific archologies. For students seriously interested in the subject, 30 archologies are detailed in Soleri's book *Archology: The City in the Image of Man* (1969).

The neatness of Soleri's archologies might make one suspect the practicality of his ideas. His solution is basically the accommodation of very high densities of people by building the ultimate in high-rise apartments. This suggests that one problem archologies might have is the same attendant to the high-rise slum-clearance buildings renowned for not meeting human needs. Is the archology concept an advance, an efficient and aesthetically appealing solution to minimize residential space needs? Or is it a throwback to ghettos of newer and grander design?

SYSTEMS ANALYSIS AND SIMULATION OF URBAN ENVIRONMENTS

Models of metropolitan regions began to appear in sizable numbers after the advent of high-speed digital computers. Most of these models were concerned with short-term predictions of economic importance, for example, housing demand, number and distribution of jobs, commuting travel times, or manufacturing output. Because of their concentration on more immediate problems, there has been a notable lack of concern in these models with such questions as the long-term cycle of urban deterioration, problems of urban renewal, and the environmental impacts of city growth. For example, the Bay Area Simulation Study (commonly called the BASS model) simply predicts trends in employment, housing, and population change (among other parameters) in the nine San Francisco Bay Area counties. Based largely on linear multiple regression methods (a widely used statistical technique), the BASS model is capable only of extrapolating known trends. While short-range predictions can be valuable and useful as an aid to local planners and government officials, the type of model that BASS represents is not the comprehensive model of urban environment that can have significant impact upon the urban planning process directed toward the solution of social ills.

---

[2]Under construction in 1975.

FIGURE 17-10 Arcosanti. (a) Elevation of Arcosanti completed on the mesa. (b) Existing Cosanti Foundation buildings. (From P. Soleri, *Arcology, The City in the Image of Man*, (1969, M.I.T. Press, (a) p. 122 and (b) p. 119.)

(a)

SOME SOLUTIONS FOR URBAN ILLS

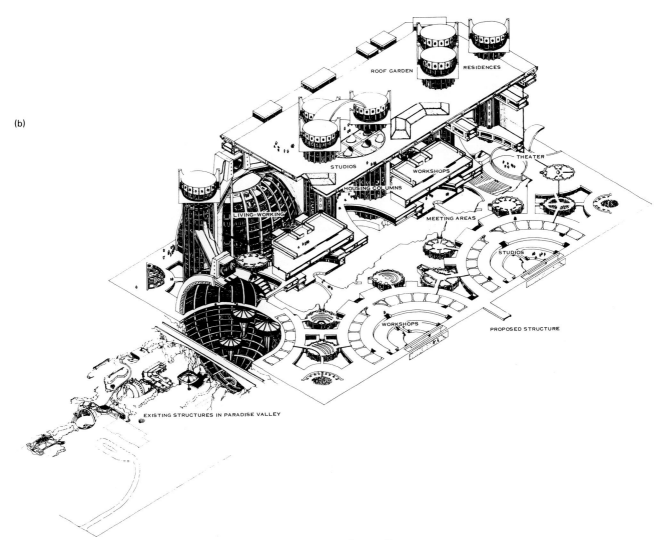

Arcosanti
(Mesa topography)

Population	1,500
Density	531/hectare; 215/acre
Height	50 meters
Surface covered	2.8 hectares; 7 acres

In 1969, J. W. Forrester of the Massachusetts Institute of Technology published his book *Urban Dynamics*. The urban model presented therein was a new beginning. Unlike previous models with intricate detail, extensive data, and limited scope, Forrester set out to build a comprehensive model of a hypothetical city, one that mimicked the essential characteristics of cities faithfully but that also would suggest policies to correct their ills. In so doing, Forrester injected mathematical modeling into areas traditionally reserved for sociologists, geographers, planners, and political scientists; as a result, he expanded the horizons of urban thinking.

Forrester's urban dynamics model created a sensation when it appeared. Using a new computer language (DYNAMO) and the advice of former Boston Mayor John Collins, Forrester applied his previous work in the modeling of industrial inventories to the modeling of his hypothetical city. Discussion of the actual model building process is beyond this text (see Forrester, 1968); it is important only to recognize that the methods are rational and lead to a specific model of a city. Forrester used the model to explore various policy alternatives. For example, a policy of providing low-cost housing for workers can be built directly into the model as an increased rate of low-cost housing construction. After specifying the rate of increase, Forrester could explore the effects of the changes upon various indicators of the state of the urban system.

After having evaluated a large number of policies in this way, Forrester drew a number of conclusions about the dynamics of cities:

1. Four policies widely believed to be cures for urban ills—job training for the underemployed, tax subsidies to cities for education and welfare, the construction of low-cost underemployed workers' housing, and direct financial aid to cities from state and federal governments—do not work. The model demonstrates that each succeeds only in worsening the very situation it was intended to correct. The city only decays further, and more workers are unemployed.

2. Forrester terms this kind of surprising behavior "counterintuitive behavior." He claims that counterintuitive behavior is widespread in systems because all real systems are so rife with various feedback loops that their behavior will be complex and generally unpredictable. To observers using simple reasoning to arrive at an expectation, the unpredictability is "counterintuitive."

3. Forrester suggests that the real remedies for reconstituting urban areas should consist of slum clearance (*not* renewal of more housing for underemployed workers), to be followed by replacement of underemployed housing by worker housing, favoring new industry at the expense of declining industry, and by discouraging new housing starts. His measure of success includes such things as reduced unemployment and less declining industry.

4. Forrester concludes that the key to a healthy city is the prevention of urban decay through the application of such policies as specified in item 3, either by use of tax powers or by direct government regulation. In the event that a city government refused to act, higher levels of government could impose the "proper policies" upon the city.

These conclusions were unconventional enough to raise a major controversy. One critical analysis of Forrester's work has been published by L. P. Kadanoff

(1971) and will serve as an example of a viewpoint very different from Forrester's. Kadanoff questioned the validity of some of the assumptions that Forrester made in constructing his model and then went on to examine the effects of different assumptions by constructing a similar model using them. For example, one basic assumption Forrester makes about the nature of his hypothetical city is that it exists in an unlimited environment which supplies the resources to support the city and acts as an infinite consumer to about everything the city produces (including excess people), yet which has no other effect upon the city. Kadanoff showed that even gross system behavior—such as general conclusions about the wisdom of constructing additional low-cost housing and the desirability of tax subsidies—would change dramatically simply by altering this single assumption. Kadanoff found that when he applied the same programs on a nationwide scale that Forrester had labeled as "counterintuitive failures," they worked and were not counterintuitive. He effectively showed that having an environment which affects the city as well as being affected by the city completely changes the results. The explanation is that the institution of a job-training program or a low-cost housing program by one city no longer serves to attract the disadvantaged and underemployed from cities without these programs; the application of the job and housing programs in all cities prevents the migration of the poor and jobless from canceling out the benefits in the one city that has the program. In Forrester's model, the relegation of all these other cities to a limitless environment insures that migration will be an important force. Furthermore, his excess poor can be shoved out into the environment with no thought to their fate—a good thing in his model. Kadanoff concluded that further analysis of the urban model actually suggests the importance of applying programs of urban renovation nationally more than the failure of current urban renewal programs.

There is wide agreement that Forrester's urban dynamics model is neither an adequate representation of city behavior nor (at this stage, at least) a suitable tool for the formulation of urban policy. However, Forrester deserves credit for opening up a new perspective on urban problems. He was responsible for an unconventional and fresh look at cities, and in this context he has stimulated a great deal of work. He has challenged a number of fundamental beliefs about cities, probably to their eventual benefit. Even if wrong or wrongly interpreted, Forrester's model is at least a testable, quantitative statement about the nature of cities.

APPLIED URBAN ECOLOGY

Following the widespread public realization that we must live within the limits of a global ecological system, a number of suggestions have been made relating to the fact that urban areas are ecosystems which also must obey certain ecological rules. One suggestion is that economic diversity and species diversity operate in the same way, so that if species diversity promotes stability of natural ecosystems, then economic diversity should promote the stability of urban ecosystems. Thus it might be argued that great industrial diversity would insulate a city from changing economic conditions in the same ways that a tropical forest is more resistant to change than an area of arctic tundra is. Consequently, the diversity-stability argument has attracted much attention because it at least offers a hope for a means to make cities function more smoothly and dependably.

If such rules of thumb as this are to be applied, it is critical to know what the potential role of ecological science is for urban problem-solving. Will ecology contribute as much to urban problem-solving as sociology or anthropology? How applicable is ecological theory to urban environments? Quite frankly, most answers are unknown. There are certainly similarities relating to the fact that both natural and urban systems are complex entities which must have some behaviors resembling one another; but it is equally true that many dissimilarities are just as pronounced. Chief among these dissimilarities are the unique biological and sociological properties of human beings and the potential impact of technology. It is impossible to overstress the role of technology in mediating the relationship between us and our environment, a reason economists have offered again and again to explain why economics is the most appropriate science for dealing with environmental problem-solving. Yet technology hardly exists in natural systems; consequently, it is one phenomenon with which ecologists have had little experience.

At this time, the best conclusion we can draw is that while ecological theory can be applied to urban problem-solving in a general way, such specific applications as the diversity-stability argument will require more careful analysis. One good reason for this cautious approach is that an uncritical application of ecological ideas to urban situations is not likely to prove either productive or useful. Perhaps the most fruitful application would be a mutually beneficial fusion of ecology with the social sciences: On one hand, social science might emerge strengthened by the best of ecological ideas; on the other, we would have a fresh approach to human ecology and perhaps a further commitment from ecologists to participate in research on human ecosystems.

Urban Cures and Panaceas: A Retrospective Conclusion

The foregoing examples raise an interesting question: Will it be possible to build cities that provide for human needs and are, at the same time, "ecologically sound"? The chapter concludes with an analysis of this question.

As long as the world's population continues to grow, the creation of new cities as well as the expansion of old ones are likely. However, this certainly does not relieve us of the need to solve the problem of stagnating or decaying cities. Many ideas have been proposed to correct urban ills, and some have been discussed in this chapter. Most of these ideas would fail to completely solve the problems we face; rather, they serve as partial solutions. The philosophy underlying *Communitas* and Tapiola, Finland, has a concern for the kind of urban design that will encourage human interactions, while Robert Goodman's solution goes even further and proposes a complete transfer of design authority to the urban residents who must live in the cities and who, Goodman believes, should have the right to determine the shape these cities will take. All these planners (and this applies still more to Jane Jacobs) fail to recognize whatever value there may be in natural landscapes apart from their direct benefits to human beings. Soleri's archologies, on the other hand, reflect his respect for the preservation of nature, since they are designed to have minimal impact on the biosphere. A possible cost of this is that the human interactions in Soleri's archologies are difficult to predict: We have only his sketches. No experience tells us whether the incredible densities he proposes are socially viable and whether materials and energy can be moved smoothly throughout such a complex system.

The suggestions we examined regarding the rehumanization of present urban areas (the ideas of Jacobs, Goodman and Goodman, and Robert Goodman) address themselves to issues that are well documented and that demand changes of the sort described. Yet if we examine these proposals from the point of view of attempting to reduce their destructive impact on natural systems, we find that the modifications offer no real solution; and to the extent that they introduce greater energy inefficiency, they actually may contribute to further decline in environmental quality. On the other hand, the new-town concept and supercities also raise questions, even if different ones. What will keep these creations from stagnating and deteriorating as our "old towns" have? Furthermore, do these new towns address themselves to the really difficult problems, or do they simply exclude them? For example, Soleri's archologies conspicuously lack heavy industries like steel making, which are isolated in special areas surrounding the city. In his discussion on the development of Mesa City, for example, Soleri plans to isolate heavy industry in canyons adjacent to the city; he cuts them off by the "river of waste," the outermost of three concentric rings of structures around the archology. This outer ring is dominated, in Soleri's words, by "second-hand stores and markets, the blight of the car dealers, the equipment dealers and renters, the junkyards, and all the various colorful and often distressing aspects of man's dependence on technological ephemeralities." Soleri proposes to put the people in lofty towers, but he cannot alter the existence of necessary but "distressing" elements (Figure 17-11). He can only put them as far away from people as possible.

(a)

(b)

FIGURE 17-11 (a) Suburban business district. (b) City junk yard. Both of these so-called blights would be relegated to the outermost ring of Soleri's city. ((a) Courtesy of Patricia Hollander Gross from Stock, Boston. (b) Courtesy of Jeff Albertson from Stock, Boston.)

We can dismiss the remaining proposals as not being very helpful. Fuller's tetrahedrons and airborne geodesic domes are too poorly specified even to assess their feasibility, although his tetrahedron appears to share some of the most distressing aspects of today's failure-ridden slum-clearance projects (high-rise buildings with central elevators and monotonous cell-like apartments). Ecumenopolis is also too poorly specified to assess and is more extrapolation than solution.

Finally, there are other proposals that require further consideration. The application of systems analysis and the use of ecological theory in urban problem-solving still remain as viable future hopes. Systems analysis has provided a useful tool for facilitating orderly thinking and has opened new vistas for investigation, but progress is hobbled by great gaps both in the understanding of relationships in social systems and in data quality. Ecological theory could be another way to explore still other vistas, but at present its potential remains unproved. Both the systems approach and the use of ecological theory may turn out to be most productive as links to other urban sciences. Planners have discovered the benefits of systems methods (see Steinitz and Rogers, 1970), and the next few years may show that ecologists and social scientists share independently developed but surprisingly similar theories. When fused, these theories may prove fruitful indeed.

Perhaps better than anyone else, ecologists recognize the nature of limitations on population growth rate and population size and how these limitations may operate. This idea applies not only to human populations, at least in general terms, but to cities as well, because it seems unlikely that we can have cities which are compressed and intensely active and at the same time which allow for individual expression and needs. Therefore, as the world's population continues to grow, we must understand that we cannot have the highest degree of individuality and still have cities which do not strangle themselves. It may seem strange perhaps, or even inappropriate, to end a chapter on urban places with a comment about population growth, but unless one accepts the belief that technology will allow us to escape all resource limitations, we shall sooner or later be confronted with the choice between living in an archology or some kind of ecumenopolis.

Questions

REVIEW QUESTIONS

1. Construct a table comparing urban solutions on the following characteristics:
    a. impact on residents and potential residents
    b. probability that the city will remain an exciting but peaceful place to live
    c. "economic vitality" (define in your own terms)
Draw conclusions you believe are appropriate.

2. Compare Fuller's tetrahedron and Soleri's archology as fully functional cities. Which would you expect to perform better (that is, to be a complex and dynamic entity), and why?

ADVANCED QUESTIONS

1. In the *Scientific American* collection *Cities*, Abel Wolman discusses the metabolism of cities as though a city were some sort of "superorganism." Do you think a city could become sufficiently integrated to be considered a superorganism? Support your argument.

2. Karp and Karp (*Landscape*, 13, 1963, 4-8) conceive of the "ecological city" as one that is part of, rather than a contrast to, its surrounding environment. Evaluate their ecological city. In what sense is it ecological? In what sense is it not?

3. Watt (*Principles of Environmental Science*, 1973, pp. 36-37) restated Margalef's idea that mature ecosystems exploit immature ones and keep them from maturing. One of the examples he used was that of a large city and its suburbs; the city draws energy, people, and materials from the suburbs and prevents these from growing into large cities themselves. Criticize this analogy.

4. Would you like to live in an archology like Arcosanti or like Babel IID (Soleri, 1969)? Why or why not? If you find you would not mind living in one but would mind the other, can you say what aspects of each affect you positively or negatively?

5. Sometimes an analysis of maps can reveal some interesting patterns without involving a lot of work. For example, compare a road map of Los Angeles in 1940 with one of New York, Boston, or Chicago in 1940. Then make the same comparison for 1970. What differences do you see? To what do you attribute these differences? Try to define some alternative explanations for the differences, then ask yourself how you might find out which explanation is true. (This question is ideal for a small group working together and sharing ideas.)

6. If the world's population continues to grow, it will become increasingly necessary to be able to construct cities that simultaneously meet human needs and have a minimal environmental impact. Will this be possible? Construct pro and con arguments.

Further Readings

Bourne, L. S. (ed.). *Internal Structure of the City*. Oxford, Fair Lawn, N.J., 1971.

> An excellent reader, systematically organized and with a high level of editorial comment. This volume provides a comprehensive background for the subjects of urban structure and function which were given only a cursory treatment in this chapter.

Calhoun, J. B. *The Ecology and Sociology of the Norway Rat*. U.S. Public Health Service Publication, No. 1008. Washington, 1962.

> This publication, along with Esser, Cassel, Selye, and Christian and Davis, summarizes information on the pathological effects of density and crowding.

Cassel, J. "Health Consequences of Population Density and Crowding." In *Rapid Population Growth* (National Academy of Sciences). Johns Hopkins, Baltimore, 1971.

Center for Real Estate and Urban Economics. *Jobs, People, and Land*. University of California, Berkeley, 1968.

> This publication is the least technical version of the BASS model. The details (save one part that has never been published) may be found in California State Water Quality Control Board publication *San Francisco Bay-Delta Water Quality Control Programs Task II-3* (mimeo.). Sacramento, 1968.

Christian, J. J., and D. E. Davis. "Endocrines, Behavior, and Population." *Science*, 146 (1964), 1550-1560.

> A summary of a number of years of work. This paper is most valuable for its bibliography. A valuable related paper, but unfortunately hard to get, is by D. Chitty: *Proceedings of the Ecological Society of Australia*, 2 (1967), 51-78.

Doxiadis, C. A. "The City (II): Ecumenopolis, World City of Tomorrow." *Impact of Science on Society,* UNESCO, 19 (1969), 179–193.

See also *Ekistics* (Oxford, 1968) and *Science,* 162 (1968), 326–334.

Esser, A. H. (ed.). *Behavior and Environment.* Plenum, New York, 1971.

This book is the proceedings of an international symposium that attracted many of the more prolific authors in the areas of social biology and the perception of space. Even though it is highly technical and not directly concerned with cities, the volume is still valuable.

Forrester, J. W. *Principles of Systems,* preliminary edition (10 chaps., paper). Wright-Allen, Cambridge, Mass., 1968.

Forrester, J. W. *Urban Dynamics.* M.I.T. Press, Cambridge, Mass., 1969.

Fuller, R. B. "City of the Future." *Playboy* 15, No. 1 (1968), 166–168, 228–230.

Goodman, P., and P. Goodman. *Communitas,* 2nd edition. Vintage, New York, 1960.

Goodman, R. *After the Planners.* Touchstone Books/Simon & Schuster, New York, 1971.

Holling, C. S., and G. Orians. "Toward an Urban Ecology." *Bulletin of the Ecological Society of America,* 52, No. 2 (1971), 2–6.

A thoughtful report written by two eminent ecologists. Anyone who wishes to apply ecological theory to urban dynamics would do well to read this short paper first.

Howard, E. *Garden Cities of Tomorrow,* new edition (F. J. Osborn, ed.). Faber, London, England, 1946.

Thoroughly reviewed by Mumford (1961), this citation is included for its historical importance.

Jacobs, J. *The Death and Life of Great American Cities.* Vintage, New York, 1961.

Jacobs, J. *The Economy of Cities.* Vintage, New York, 1970.

In her second volume Jacobs argues that high economic diversity is important to the well-being of a city. This argument deserves careful examination.

Kadanoff, L. P. "An Examination of Forrester's 'Urban Dynamics'." *Simulation,* 16 (1971); 261–268.

Kadanoff, L. P. "From Simulation Model to Public Policy." *American Scientist,* 60 (1972), 74–79.

These papers contain substantially the same arguments, although the results are presented in slightly different ways.

Lowrey, I. S. *A Model of Metropolis.* Rand Corporation, memorandum RM-4035-RC, 1964.

One of the best examples of a highly detailed model incorporating many data. This model was constructed for Pittsburgh, Penn.

Milgram, S. "The Experience of Living in Cities." *Science,* 167 (1971), 1461–1468.

Mumford, L. *The City in History.* Harcourt, Brace, Jovanovich, New York, 1961.

Mumford, L. *The Urban Prospect.* Harcourt, Brace, Jovanovich, New York, 1968.

Although his interests have changed slightly, Mumford remains one of the most respected and lucid observers of urban areas and urban change. *The Urban*

Prospect consists of a collection of Mumford's own writings. All are good, and surprisingly all fit together. His articles "Megalopolis as Anticity" and "Home Remedies for Urban Cancer" are especially good.

Rose, H. M. *The Black Ghetto.* McGraw-Hill Problems Series in Geography, New York, 1971.

This book explores a topic only skirted in this chapter. Despite this fact, the topic is so important and this particular treatment so lucid it is given here as a detailed extension of Chapter 17.

Scientific American. Cities. Knopf, New York, 1968.

A collection of papers that originally appeared in *Scientific American.*

Selye, H. "The Evolution of the Stress Concept." *American Scientist,* 61 (1973), 692–699.

Soleri, P. *Archology: The City in the Image of Man.* M.I.T. Press, Cambridge, Mass., 1969.

Soleri, P. *Sketchbooks of Paolo Soleri.* M.I.T. Press, Cambridge, Mass., 1971.

These volumes are full of technical jargon, and the second is strictly for professional planners and architects. Nevertheless, students who wish to derive the most benefit and make the critical analysis of Soleri's designs should study these volumes to get the flow of ideas and the underlying beliefs of Soleri's thinking.

Steinitz, C., and P. Rogers. *A Systems Analysis Model of Urbanization and Change.* MIT Report, No. 20. Cambridge, Mass., 1970.

This book is included because it presents the process of systems analysis and model building more clearly than most, even though the final product illustrates the various difficulties involved much better than demonstrating an elegant model.

von Hertzen, H., and P. D. Spreiregan. *Building a New Town,* M.I.T. Press, Cambridge, Mass., 1971.

This is the story of Tapiola, Finland.

Wallace, D. A. (ed.). *Metropolitan Open Space and Natural Process.* University of Pennsylvania, Philadelphia, 1970.

McHarg's ecological planning process (see Chapter 8) is reintroduced here because several papers are concerned with the physical impacts of town planning.

18. Environmental Perception: It's What You Think That Counts

For the purposes of this chapter we shall define *environmental perception* as those processes by which individuals interpret and react to their environment and to environmental change. We shall take the former set of perceptions to mean the reactions to spatial arrangements and physical objects that constitute human environments. The latter set, although not clearly separable from the former, is concerned more with reactions to changing conditions of environmental quality (and to the policy proposals designed to counter these changes) and hence involves social sciences more than the physical planning arts.

Even with the perspective limited in this manner, environmental perception would appear to be an exceedingly amorphous topic, and because of this the entire area, though very important, is a most difficult subject to study. Environmental perception is the indistinct interface between individuals and what is happening in their environment; hence it is a mediating influence that determines what actions are or are not taken to modify existing man-environment relationships. For example, environmental perception is relevant to pollution abatement—the rate of change of the pollution situation and citizen's feelings about what constitutes an acceptable amount of deterioration in environmental quality in trade for some gain in economic prosperity. The problem is that environmental perception is *not* a well defined, widely recognized concept with an agreed-upon place in decision-making processes. One reason that it is difficult to work with the concept of environmental perception in the decision-making process is that all individuals devise their own precepts and images of the environment, which makes categorizing and understanding this entire area difficult. A good example of this problem was examined in the last chapter in our consideration of city planning and the design of cities with the aim of achieving efficient and humanized environments. Various persons and groups are likely to hold widely divergent views on what constitutes such an environment. Architects, for example, are likely to have perceptions of optimal city structure conditioned by their training and the biases imposed upon them by building codes; builders are likely to have their perceptions "adjusted" by the market they attempt to reach and by the economics of construction and competition; different buyers are each likely to have preferences concerning residential environments. These differences can be influenced by the location, style, internal conveniences, and price of a house, as well as by neighborhood, schools, transportation, and many other factors. (And the builders are likely to know something about these too!) Besides all these groups, we also must consider the environmental perceptions of city planners, industrialists, shopkeepers, traffic departments, and tax collectors. This vast array of people, many of whom will disagree violently about what constitutes the optimal city, present a formidable problem for the person who would try to define the optimal urban structure. But, on the other hand, can a city be designed to meet human needs successfully without an understanding of the environmental factors that people desire?

It is frustrating (if also challenging) to realize how little is known about environmental perception when it seems so important. In his review of the subject Saarinen (1969) stressed that the field of investigation, being both interdisciplinary and recent, lacks well-developed methods of investigation; thus no coherent body of theory or set of general principles distinguishes the field. In fact, no single name for, or definition of, the topics of interest has emerged at this point; Saarinen lists 21 names that have been applied to the topic at one time or another!

We might ask why anyone would enter such a nebulous and difficult research field as environmental perception. One reason is that a firm understanding of the role of perception is valuable for explaining and predicting human responses. This would seem to be highly relevant unless we are willing to dismiss human behavior in the area of environmental problem-solving as unimportant. A second reason is that environmental perception is one research area where various social and health sciences could be united in an extremely useful synthesis that no single scientific discipline could cover. Despite these points, one could argue that environmental perception is still so broadly defined and empirical that it is only quasiscientific.

If one demands quantitative data and well-delineated procedures for study, the charge is true now; but this in no way reflects upon its potential or its future importance in environmental problem-solving.

Historical Analysis of Environmental Perception

Historians and geographers have written extensively about human attitudes toward nature in an effort to understand contemporary environmental problems as a product of closely held attitudes that have long been out of date. Insofar as the present state of cities and ecological systems is in part the result of important past events, historical analysis may make a significant contribution to understanding how perceptional systems have evolved or resisted change as environments themselves have been modified.

It was the publication of Lynn White's (1967) hypothesis that stimulated renewed interest in the evolution of environmental perception systems. White argued that early Judeo-Christian beliefs about nature were conditioned by a constant struggle to survive in harsh deserts, which reinforced attitudes that human beings must "struggle against and dominate Nature" by any means possible. All this might be merely an amusing and interesting footnote on local Mideastern history and of little significance had not these attitudes been incorporated into the Bible and accepted as part of a religious doctrine by peoples of other lands and other times—people who were only too willing to exploit natural landscapes under the command of deeply held religious beliefs.

White drew three important conclusions from his analysis. First, our callous treatment of natural systems is traceable directly to Christian teachings. Second, science and technology are the tools of this exploitation and as such cannot be depended upon to aid in the implementation of a change in our basic environmental attitudes. Third, a moral revolution that would permit us to adopt a patron saint like St. Francis of Assisi could lead to a lasting reversal to our doctrine of exploitation, because St. Francis's most important teaching—respect for all life—represents an antithesis in the Judeo-Christian attitude White feels is the root problem. Thus White's paper is, in essence, a thesis that religious teachings have influenced the environmental perception of Western cultures to the extent that only through a similar reindoctrination will an effective solution be achieved.

White's paper has attracted so much attention that it has been reprinted in countless environmental readers. However, other papers refute White's thesis from many different viewpoints. For example, if White's thesis is correct, then one might expect that cultures not strongly influenced by Christianity might have developed differently. Two authors who have made such comparisons between different cultures are Yi-fu Tuan (1970) and Daniel Guthrie (1971), who respectively have examined Eastern and Amerind attitudes toward and treatment of the environment. Tuan found that Asian attitudes toward nature are not substantively different from Western attitudes, despite some very different religious philosophies. In China, Taoism and Buddhism both taught respect for and harmony with nature, but they also encouraged some other practices that had exactly the opposite effect. For example, Tuan pointed out that the teachings of Buddhism were directly responsible for the preservation of trees surrounding temples in an otherwise deforested landscape—a landscape which was deforested because of fuel needs, needs exacerbated at least because Buddhism also introduced and encouraged cremation of the dead. The evidence presented by Tuan certainly does not support the idea that Western civilization is the supreme destroyer of environments; instead he showed that Chinese culture had at least a comparable history in its treatment of natural landscapes.

Guthrie questioned the idea that primitive human societies had any better relationship with nature than modern ones do. Using the American Indians as his documentation, Guthrie constructed a picture of opportunism and conflicting beliefs which casts considerable doubt upon widely held beliefs that Indian tribes were able to live in close harmony with nature, that they were well integrated into the ecosystems of their territories. Furthermore, Guthrie argued that there are no philosophical differences between Amerind and modern people—only that the Amerind had no choice about overexploiting resources which were necessary for survival. Compare this with the powerful modern technology and the doctrine of resource substitutability.

Lewis Moncrief (1970) has also presented arguments in rebuttal of White's hypothesis, but they are based more upon an analysis of the history of Western civilization than upon cross-cultural comparisons. In spite of his different approach, Moncrief's rebuttal of White's hypothesis supports the conclusions reached by Tuan and Guthrie. His argument, summarized in Figure 18-1, is that science and technology just happened to have developed earlier in Western civilizations, thus promoting the explosion in population, individual wealth, and environmental exploitation which were themselves the *direct* causes of environmental disruption. Therefore, he argued, Judeo-Christian beliefs are at best only an indirect cause of environmental disruption. Moncrief believes that the personal accumulation of wealth is the

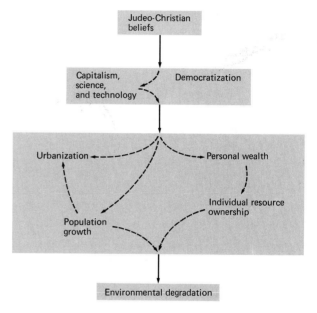

FIGURE 18-1 *Lewis Moncrief's model for the influence of Judeo-Christian beliefs on environmental degradation. One might argue that this model is itself imperfect, many other factors having been omitted from consideration, but from Moncrief's point of view, the important thing is that Judeo-Christian beliefs are not the direct cause of the "ecological crisis."* (Redrawn from L. W. Moncrief, 1970, *Science* 170, p. 511. Copyright © 1970 by the American Association for the Advancement of Science.)

prime reason for heightened environmental degradation. (It is evident that Moncrief accepts the inevitability of some level of environmental disruption in any society.)

The foregoing analysis suggests that it is our biological nature, aided by an array of scientific and technological tools, which is the source of the environmental attitudes these authors are concerned about.

The cross-cultural comparisons of Tuan and Guthrie certainly support this conclusion. In this context, then, science and technology are simply tools for implementing our perceived needs for food, shelter, and security; thus they are only indirectly the tools of environmental degradation. All this leads to a question: Is our perception of environmental disruption nothing more than a belated realization that environmental degradation has been occurring all along but now has become more obvious because it has worsened as our population and technology have expanded? It is difficult to separate real change from changes in perception, but it must be conceded that a change in this perception of environmental degradation is at least an important factor in our present increased environmental awareness.

Environmental Perception at the National and Global Levels

The historical analysis of environmental perception, traced as it must be through fragmentary written records, is so general in nature that its perspective must be at least regional and often global. Of course, other important aspects of environmental perception at the global level are not normally included in an historical analysis, and we shall examine some in this section. Unfortunately, the documentation for these is also scanty, and here too great gaps exist in our knowledge, a problem that will not be rectified easily when few dare to work with concepts so nebulous as the average citizen's global views. Yet further analysis of national and global levels of perception are well justified in order to expose the tangled webs of misconceptions and antiquated beliefs that make up our images of our country, other nations, and the world.

A most important characteristic of national- and global-level environmental perception is that human images seldom match reality. One reason for this is that most people have little time or interest in anything beyond their own immediate needs; furthermore, most have few opportunities for personal observation and must rely for information upon sources of widely variable reliability. Therefore, global and international perceptions frequently result from misconceptions based on national stereotypes and fragmentary impressions. In this situation, it is not surprising how far apart perceptions and the truth can be.

A second characteristic is the slowness with which perceptions are adjusted to changes in reality, a situation fostered by slow and incomplete communication of new information and the conservatism of human nature. For example, Tuan (1968) has shown that some religious and ethnic groups have been reluctant to accept the reality of deserts because this would imply, contrary to their beliefs, that some areas of the earth were not divinely blessed as much as others.

Lowenthal (1968) has examined some perceptions commonly held by various nationalities regarding the United States. These perceptions are very revealing to those who have grown up in the United States. Lowenthal showed that European visitors had four important images of the United States:

1. The United States is enormous.
2. The United States is characterized by wilderness and primitive landscapes.
3. The United States has no underlying order; it is an uncoordinated mass of individual structures.
4. The United States is notable for extremes in form and substance—such as natural landscapes being juxtapositioned with all sorts of manmade structures.

FIGURE 18-2 (a) This narrow street in Germany is characteristic of the intensive building practices of Europe. (b) Americans are used to open spaces like this wide avenue in a town in Kansas. ((a) Courtesy of Patricia Hollander Gross from Stock, Boston. (b) Courtesy of Grant Heilman.)

Upon visiting the United States, Europeans reinforce these impressions because they (consciously and unconsciously) tend to compare their impressions with their preconceived images, all in the light of their background of European history, tradition, and personal experience. In contrast, when questioned about these impressions, Americans are likely to react in entirely different ways because they are used to the open spaces and relaxed building practices surrounding them. Furthermore, European response is to wish for surroundings more like the intensively built landscapes of Europe, whereas Americans can hardly imagine being crowded more (Figure 18-2).

Lowenthal concluded that environmental planning and design cannot proceed without more intimate knowledge of how people perceive various environments. His analysis of cultural differences in perception certainly shows that people of different heritages would strongly disagree about how an environment should be changed or "improved" to suit human ends.

Perceptions at the Regional Level

Unlike the study of perception at higher levels, regional and local studies in perception can find immediate and direct application to planning efforts that are becoming increasingly regionwide in nature.

THE CONCEPTION OF WILDERNESS

Before we discuss planning at the regional level per se, let us consider some types of perceptional vagaries one is likely to encounter with regional planning. "Wilderness" is an excellent example of a relative concept—a perception of environment—that can have so many shades of meaning to different observers that it is difficult to agree even upon definitions, much less upon a course of action to be followed for its preservation. One group of people might consider wilderness to be only lands that are not used or entered by human beings, while another might include lands not permanently inhabited. Still others might consider wilderness to be any land that still retains significant natural features. Each position may be justified on the basis of the individual values involved. Since there can be no absolute definition of wilderness, the only realistic approach to wilderness planning must be related in some way to the perceptions people have about wilderness, which in turn requires some means to measure their ideas. Once again, we find ourselves confronted by the need to understand environmental perception and the interplay between perception and reality.

Similar comments may be applied to such concepts as "open space" or "quality recreational experience," for the problems involved with gauging perceptions of these items are essentially the same as those involved with the perception of wilderness. When we attempt to determine what "useful" or "quality" open space is, or when we attempt to derive a list of criteria that would constitute a "quality recreational experience," we inevitably must admit that historical, cultural, economic, philosophical, and environmental factors are all involved in this issue of environmental perception.

Once again, there is a clear need for perceptional analysis to untangle the web of conflicting opinions concerning not only the use of wilderness and open space but also the modes of recreational planning in making the "best" wilderness and open space available to all potential users. The only other choice is no choice at all, because to belittle attempts at understanding the nature and formation of perceptions leaves no alternative but the mute acceptance of the personal perceptions of those few who would end up making all the decisions about such questions as what wilderness is and what portions of it are to be saved, what open space is to be preserved, and how a quality recreational experience is to be obtained. We would not be escaping the perception problem at all.

THE ROLE OF PERCEPTIONS IN REGIONAL DEVELOPMENT

Modern urban areas have become so large and are such dominant forces in their environments that the only way to deal with them rationally is to consider them as centers for regional development (as in the Seven Towns Plan and the Uusimaa 2010 Plan described in Chapter 17). Under such circumstances, it is no surprise to learn that environmental perception is important for regional development, too, for several reasons. For one thing, regional development involves not only all the problems related to urban development but also such issues as the selection and preservation of open-space elements, land planning at the urban fringe, and total planning for the landscapes of the region; hence the same elements of perception required for all these facets of planning are also necessarily involved in regional development. Another reason why environmental perception is so important in regional development has to do with the need for comprehensive incorporation of so many factors into a single planning framework; this means that on an additional level of aggregation all the different perceptions and desires must be reconciled.

Geographers have been interested in one aspect of regional environmental perception that will serve as an example of the problems involved in regional planning and the kind of divergence that can occur between the planning process and the perceptions of citizens. This research, which has to do with the perception of potential natural disasters (such as hurricanes, tornadoes, floods, and earthquakes), has dealt with how people who live in areas of natural hazards adapt to the ever-present possibility of disaster. The most important conclusions (which probably can be generalized to many types of natural hazards) are summarized in Table 18-1a, which displays the perceptions of coastal residents of the chance of danger from natural hazards. It has been noted repeatedly that people subject to these hazards live with the danger by minimizing them in one of several ways (Table 18-1b). If a disaster does occur, the consequences often are accepted without any basic change in perception, although there may be a heightened interest in a potential solution for a short time afterward. A wide variation in perceptions has been noted among residents of natural-hazard areas, and this has been explained in terms of two factors: the personal experiences of different individuals with particular hazards (and this is well correlated with their future expectations concerning the hazards) and the threat of meteorological vagaries to the economic livelihoods of the residents so that those whose livelihood would be affected most by these hazards (such as those employed in the tourist industry) would have heightened perceptions of both the hazards and the probability of damage.

In contrast to the perceptions of residents who tend to minimize the dangers are the perceptions of the planners, engineers, and government officials who, if a disaster occurs, have some role in the crisis. Among these professionals, government officials are especially subject to the "crisis mentality"; when a disaster stirs them into action, a flurry of public-works projects results, followed closely by a quiet period in which all disaster plans and actions are again shelved. These government officials also are characterized by the belief that money is often the only factor limiting progress and that therefore the simple infusion of enough money into a project will solve any problem. For example, the U.S. National Science Foundation was ordered by Congress to spend *at least* 8 million dollars on earthquake research in fiscal 1973–1974, presumably in the belief that expenditures of such magnitude would promote satisfactory solutions to the many hazards of earthquakes (Figure 18-3).

TABLE 18-1a *Perceptions by the people of the risk from natural hazards as reported in a questionnaire survey conducted along the Atlantic Coast of the United States among coastal residents subject to storm hazard. The outstanding result is that although 89 percent of the respondents had experienced storms, only two-thirds of the total expected any more, and only one-third expected more damage from storms. Figures in parentheses are the percentage of each entry divided by the row total.*

Level of experience with coastal storm hazards	Neither storms nor damage expected	Both uncertain	Storms to be expected but not necessarily damage	Storms and damage both inevitable in future	Totals
No experience at all	0.8 (100%)	0.0 (0)	0.0 (0)	0.0 (0)	0.8
Heard about them	2.2 (22.7)	2.4 (24.7)	4.3 (44.3)	0.8 (8.2)	9.7
Experienced one storm	6.5 (21.7)	5.7 (19)	8.4 (28)	9.4 (31.3)	30.0
Experienced more than one	4.6 (7.8)	8.7 (14.7)	22.7 (38.5)	23.0 (39)	59.0
Totals	14.1	16.8	35.4	33.2	

Total respondents: 368

SOURCE: Kates (1967).

TABLE 18-1b *Popular perceptions of the risk from a wide range of natural hazards.*

Common responses to the uncertainty of natural hazards

1. Eliminate the hazard: deny its existence.
   "We have no floods here—only high water."
   "It can't happen here."
2. Eliminate the hazard: deny its occurrence.
   "Lightning never strikes twice in the same place."
   "It's a freak of nature."
3. Eliminate the uncertainty: make it determinate. "Seven years of plenty... seven years of famine."
   "Floods come every five years."
4. Eliminate the uncertainty: transfer to a higher power.
   "It's in the hands of God."
   "The government is taking care of it."

SOURCE: I. Burton and R. W. Kates, *National Resources Journal*, 3 (1964), 435. Reprinted with permission from *Natural Resources Journal*, published by the University of New Mexico School of Law.

Planners and engineers, like those in government, are also guilty of a tendency to rely upon technological answers rather than upon better planning or social-behavioral adjustments simply because technological answers are so much easier to implement. In recent years many towns with areas subject to flooding have finally instituted no-build zoning for these areas; but because these zones occupy a relatively small area, the implementation of this policy does not require extensive social-behavioral changes. For less localized environmental hazards such as earthquakes and hurricanes, it seems that reliance on technology will remain the primary approach to the problem.

A final characteristic of decision-makers in this area of environmental hazards again involves planners, engineers, and government officials. Government agen-

(a)

(b)

FIGURE 18-3 *The Los Angeles earthquake of 1972 may have had a great deal to do with the 1973 Congressional appropriations for earthquake research. Some of the most severe damage in the quake occurred (a) to the Golden State freeway and (b) at the Olive View Hospital.* (Courtesy of Tom McHugh from Photo Researchers, Inc.)

cies often depend upon the advice of these other groups to set policy. These consultants, with their scientific background, have a greater appreciation for statistical analysis and probability assessment and therefore are less likely to minimize the reality of natural hazards. Given these differences, the professionals are more likely to recommend greater expenditures to prevent loss of life and damage to property than the people actually at risk are willing to pay except perhaps immediately after an emergency.

In a study of the perceptions of coastal users, Kates (1967) showed that most interviewees were well educated and knowledgable about the risks of violent storms; nevertheless, their perceptions of the risks remain such that they preferred to accept these risks rather than to reduce the amenities of the area (Table 18-1a). Such things as sand dunes and sea walls would reduce the scenic value of beach-front property; hence despite the desirability of these barriers (from the professional point of view), the residents can be expected to resist efforts on their behalf. Kates suggested that the best procedure would be to attempt to modify the perceptions of these people so that they would be more amenable to changes which would reduce the hazards. He proposed that this be accomplished by promoting better scientific understanding of the physical processes involved in storms in order to assess the magnitude of the dangers involved, followed by dissemination of this information to coastal residents. If the perception of danger could be raised high enough, patterns of land use that minimized building in high-danger zones could result, but coastal amenities would be reduced only minimally.

Variation in public perception puts hazard planners in the unenviable position of having to make expenditures and push programs for which the current perception of risk may be diminished. This is particularly true of such hazards as air pollution, where the dangers are less demonstrable in terms of dramatic personal experience. Consequently, whatever control programs result will have to be strongly influenced by public perceptions of the magnitude and implications of the problem, the cost of possible solutions, and the political mechanisms that might be available. In Toronto, for example, Auliciems and Burton (1970) have pointed to the major role played by the mass media and two small anti-pollution groups in raising perceptions of a serious air-quality problem among Toronto citizens. The furor that accompanied these rising perceptions generated political support for pollution abatement, but as Auliciems and Burton are careful to point out, the success of the effort hangs on the tenuous enthusiasm generated by the mass media, a kind of enthusiasm with no promise of persistence. This example illustrates a kind of counterprinciple: While an appreciation of the processes of environmental perception is doubtlessly important, it is equally important that a particular set of perceptions be evaluated cautiously before it is given much credence in formulating future policies. If nothing else, this reemphasizes the need for the development of a rigorous research methodology to minimize such uncertainties that plague the study of environmental perception.

Perception of Local Environments

In Chapter 17 we considered various alternative patterns of city growth. In the broadest terms, all these ideas can be lumped into two categories so that our choice is between horizontal (if controlled) dispersion of urban areas or vertical development (with a great degree of internal integration) of high-density cities of distinctive architecture. Now suppose that we were to classify all the proposed alternatives with respect to the perceptions embodied in each toward the environmental impacts of their ideal urban form. We might emerge with something similar to Table 18-2.

Again we are confronted by the same perceptional issues. Each author proposes a means to build better cities (including the improvement of existing ones), yet their perceptions (real or implied) with respect

TABLE 18-2 *The role of environment as perceived by the architects and planners in the designs presented in Chapter 17. See bibliography in Chapter 17.*

Perceived role of environment	Author
1. No apparent concern, except for the human element	R. Goodman
2. Environment as a recreational resource	P. and P. Goodman
3. Environment as a life support system	Doxiadis; Fuller
4. City and environment integrated	von Hertzen and Spreiregan
5. City with minimal impact on environment	Soleri

to urban impact on environment are, in many instances, dramatically different. In fact, it should be obvious by now that environmental considerations compose only one of several criteria of design, albeit an important one.

THE IMAGE OF THE CITY

Kevin Lynch of the Massachusetts Institute of Technology is one pioneering author in the field of environmental perception. His two works, *The Image of the City* (1960) and *The View from the Road* (1964; with Appleyard and Myer), are highly enlightening and valuable because they reveal new insights into ordinary scenes that usually are buried deep in subconsciousness. *The Image of the City* is concerned with the visual quality of the city—the apparent cohesion of separate images to represent a single visual whole. To obtain information that would be most useful for planning, Lynch developed procedures which attempted to analyze perceptions which hardly ever were held consciously. Lynch introduced the idea of having each individual in his sample sketch a map showing the routes the user usually took in moving through the city and the distinctive features along the way. By requiring a written response, Lynch forced interviewees to structure their own thinking, freed respondents of any biases an interviewer might have imposed in an oral question, and provided a written record of responses that could be compared.

Lynch conducted his studies in Los Angeles, Jersey City, and Boston. Despite vast differences in the outward form of these cities (Figure 18-4), Lynch found striking commonalities in individual perceptions of the three cities, including:

1. *Visual space and the "breadth of view"* Residents remember broad vistas fondly because they provide a sense of order. Boston's Back Bay, the Palisades of Jersey City, and the Los Angeles Civic Center are prominent examples of dominant scenes remembered by the interviewees.

2. *Open space and structure* Formless, unstructured open space is not remembered well or is remembered unfavorably; but even the presence of a few buildings or sculptures can transform a highly unfavorable image to one that is favorable.

3. *Landscape features* Several interviewees in the various cities noted how much trouble they would take to pass ponds, parks, or even particular groups of plants because it was pleasurable to do so.

(a) Los Angeles

(b) Jersey City

(c) Boston

(d) Jersey City

(e) Los Angeles

FIGURE 18-4 *Images of three cities: (a) downtown Los Angeles, (b) Jersey City, view of Journal Square looking northeast, (c) downtown Boston from above the wharves, looking northwest, (d) a typical Jersey City street: Tuers Avenue going north toward Journal Square, (e) Pershing Square in downtown Los Angeles, (f) Commonwealth Avenue in Boston's Back Bay.* (Courtesy of (a) Rapho/Photo Researchers, Inc., (b) Jersey City Chamber of Commerce, (c) Ellis Herwig from Stock, Boston, (d) Dennis Simonetti, (e) Diane Rawson from Photo Researchers, Inc., (f) Ira Kirschenbaum.)

(f) Boston

4. *The image of economic well-being* Lynch noted stereotyped responses on the part of interviewees in which neighborhoods were regarded as having "high" or "low" socioeconomic status. Travelers noted that they would attempt to pass through high-status areas if possible.

5. *The evidence of contrasts in the age of structures along spatial gradients* Residents were quick to point out adjacent structures that represented the glorious past and present achievements of their cities. Apparently the contrasts were regarded as part of the excitement of the urban scene. Lynch also noted, however, that the contrast could be pushed to extremes: In Los Angeles, for instance, he received responses replete with the ghosts of past buildings, accompanied by bitterness, nostalgia, and a sense of loss. These attitudes were attributed to the rapidity of change in Los Angeles and the apparent inability of residents to adjust quickly enough.

6. *Critical features of urban environments* Lynch identified five critical elements in city imagery: paths, edges, nodes, districts, and landmarks. Of these, paths (including all transportation corridors) and their nodes (the junctions and branch points) predominate in importance. Edges (physical boundaries that differentiate various types of urban districts) are the main substance of the urban structure, while landmarks are the signposts for proper orientation.

Lynch used this study to begin other detailed analyses, confident that further work in urban environmental perception would prove fruitful for physical planning. In *The View from the Road*, Appleyard, Lynch, and Myer continued to develop the ideas introduced in *The Image of the City*, but they carried them a step further by devising a plan based upon their perceptional findings. They recorded everything that would be seen by a driver and passengers on repeated trips over various roadways in the Boston area. In summarizing their findings, the authors identified several factors they thought were critical perceptional elements in highway design as an art. These factors are

1. *The relation of roadway to surrounding elements* How a road enhances or diminishes attention to its surroundings is important. The attention of driver and passengers tends to be focused on the roadway ahead, on distant landmarks lying within this forward field of vision, and upon the objects (bridges, tunnels, roadway cuts) that the driver passes through. This is especially true as average speed increases, since observers have less opportunity to let their eyes wander peripherally; hence there are large differences in the images formed when traveling on a low-speed street and when using a restricted-access expressway.

2. *The sensation of continuous motion* This is an important component of perception, even if it is difficult to quantify. The road and the scenery must flow continuously and smoothly without confusing inconsistencies. For example, if a road should rise, it should do so because it is passing over something; or if it descends, it should be because the road descends into something. It should not curve or split unless the way is blocked by something tangible. In short, there must be a *reason* for any change in motion.

3. *The sensation of space* This refers to the impression a scene leaves on observers—that is, as they drive along the roadway whether they feel their surroundings are more "open" or more "closed." Clearly the sensation of openness or its lack is relative and is controlled by several factors such as vehicular velocity

and spacing as well as the shape of landmarks. Obviously observers will derive a sense of openness in surroundings if the landmarks are spaced far apart. But if they are traveling at a rapid velocity, they perceive the landmarks as closer together and, therefore, the surrounding space as less open. This, of course, can be compensated for by arranging landmarks so that they are further apart along roadways characterized by high-speed vehicular traffic. It also has been pointed out that the shape of landmarks (whether they are tall and thin or low and spread out) can contribute to the sense of whether surroundings are open or closed.

4. *Reestablishment of scale* The authors claim that the automobile is a tool for the regular reestablishment of our mastery and control over a city whose growth would make us utterly insignificant if we were on foot.

5. *Goal attainment and orientation* The impression of continuous motion is aided by an individual's establishment of intermediate objects (intermediate goals) along the roadway. These intermediate points simultaneously let the driver measure progress toward the ultimate objective of the journey.

6. *Meaning in perception* Observers also must be able to see enough of their surroundings as they travel on the roadway so that they can interpret impressions in terms of their ideas and previous experience or modify ideas on the basis of observations. One cost of high-speed, grade-separated highways is to deny the individual of these observations, thus reducing the road to a stultifying sameness that is meaningless and possibly dangerous as well.

7. *Relative motion of the surroundings* This refers to the smooth progression in size of roadside structures as one draws nearer to them until they are finally passed, and also to the progression of other vehicles on the road relative to the observer.

Having set forth the perceptual aspects of vehicular travel, Appleyard and his associates applied their results to the redesign of the proposed Central Artery (Figure 18-5), a ring expressway planned to roughly encircle Boston and Cambridge. They offered their plan from the viewpoint of optimal perceptual quality and not from the standpoint of efficiency, ease of land acquisition, capacity, cost, or safety factors, except to the extent that reasonable cost and traffic-flow considerations were kept in mind. Their design is considerably compressed from the original proposal and resembles a triangle more than a circle. Cambridge would be outside the ring, since the proposed road would skirt along the Charles River instead of passing near Harvard University.

For our purposes, the value of the proposal lies in the rationale behind its new design. The purpose of the Central Artery is to improve access to downtown. Using their criteria of a visually satisfying environment, Appleyard, Lynch, and Myer criticized the officially adopted plan on several grounds. Since all but the eastern (downtown) leg run far out into residential districts, the authors claim that a journey along most segments of the road would be indistinguishable from any other, incomprehensible, and just plain dull because the sameness results in a poor sense of goal orientation for travelers. The wide and indirect sweeps in this plan will be confusing because travelers going downtown will often be doing so indirectly, and as a consequence will sometimes be traveling in a direction other than what logically would take

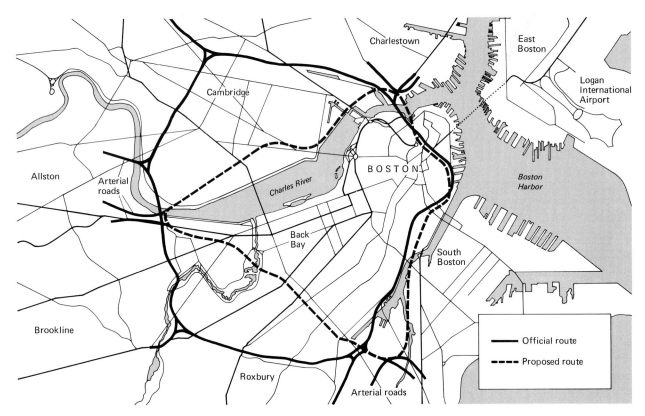

FIGURE 18-5 *The official Central Artery plan and the Appleyard et al. plan for metropolitan Boston.* (Modified from D. Appleyard, K. Lynch, and J. R. Myer, *The View from the Road*, M.I.T. Press, 1964.)

them downtown. Furthermore, such major visual features of Boston as its ocean, harbor, rivers, downtown–Boylston Street Center skylines—features that could aid travelers in orienting themselves—will be invisible from this road. Even the already-completed eastern leg runs in a maze of building tops that obscure the sea, the rivers, and, worst of all, even the downtown goal.

These criticisms lead directly to the proposed solution. Reference to Figure 18-5 will show that the northern leg as realigned by the authors exposes travelers to the expanse of the Charles River–Back Bay area as well as to the panorama of the Boston skyline. The southern leg is shortened to be a direct connection between the western and southern arterial roads;

as a result, it provides a better sense of goal orientation. This grouping of arterial roads completely eliminates the western leg of the Central Artery. Furthermore, each leg (no matter what the direction of travel) is aimed toward some landmark; the eastern leg would be built lower and would expose downtown landmarks to view; and finally, even the intersections of the arterials are placed so as to establish a sense of direction and motion without sacrificing visual quality.

THE VIEW FROM THE NEIGHBORHOOD

Many of the influences of physical environments upon perception are demonstrable at the neighborhood level also, the perceptual impact of the automobile upon American neighborhoods being one particularly instructive example. Jackson (1957a, b) has analyzed the role of the automobile in changing the nature of neighborhoods. The mobility provided by the automobile has permitted the functional neighborhood to expand enormously; as a result, businesses, recreational centers, commerce, and residences can be located farther apart and still be within convenient reach of each other. An excellent example of this expansion in the size of the functional neighborhood is seen in the process of suburbanization, the separation of work place and residence—all because of the mobility the automobile confers. Furthermore, the velocity of the automobile, Jackson contends, has promoted the existence of flashy, even gaudy buildings, signs, and landscapes of the kind we are all familiar with. Instead of a blanket condemnation of these phenomena, however, Jackson suggests instead that they are logical developments in the art of abstraction and

FIGURE 18-6 *Quiet, leisurely neighborhood interaction such as this can affect everyone's perception of a neighborhood.* (Courtesy of Owen Franken from Stock, Boston.)

generalization resulting from our fast-paced, superhighway lifestyle. The very blatantness of flashy advertisements is needed to catch the attention of passing motorists.

At the neighborhood scale, however, the interactions among people dominate environmental perception (Figure 18-6). Hence perceptions at the neighborhood level are primarily the concern of sociologists and psychologists, although their work is rarely called "environmental perception." To attempt to survey all the social science studies relevant to the understanding of social interactions would be presumptuous and far beyond our scope. Nevertheless, social factors cannot be ignored, and we shall examine the influences of just one—the threat of criminal attack—upon residents' perceptions of their neighborhood, especially as it relates to urban structural design.

Clearly the central focus in Jacobs's (1961; see Chapter 17) enclaves of socially cohesive neighborhoods is the elimination of neighborhood crime. Toward this end, her high-density buildings and pocket parks are designed to put people in contact so that they have the opportunity to develop mutual respect and a sense of neighborhood responsibility. Thus the essence of her plan is physical proximity leading to interpersonal involvement. Newman (1972) has elaborated the same theme by reaffirming the role of social interactions in suppressing crime in high-rise developments. His studies, directed at the suppression of crime through better planning designs, showed that the income, background, and social status of high-rise residents, as well as the actual mix of their backgrounds and social classes, were critical determinants of whether or not a residence building experienced low crime rates. Families that were relatively affluent, had the mother at home, and had lived in cities their entire lives could cope with this environment, whereas others could not. When the climate of unsuccessful adaptation prevailed, the tall elevators rapidly became playthings for children, who had nowhere else to go and nothing else to do. Newman reported that children became experts in disabling elevators, and because of this, the elevators were usually out of service. The isolation that threatened residents of upper floors because elevators usually were not available became a vital force in their abandonment. Furthermore, Newman showed that whether open space around these tall buildings was used or not was determined by the social understandings which were

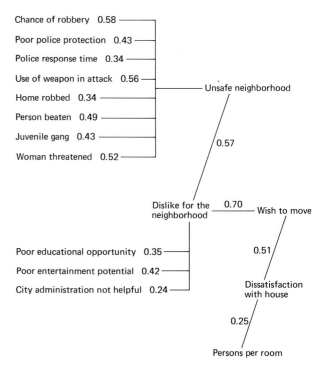

FIGURE 18-7 *Kasl and Harburg's model relating desire to move, neighborhood image, and criminal climate is based on correlation analysis (the numbers in the flow diagram). 1.0 indicates a perfect correlation. Remember that correlated variables are independent and do not necessarily total 1.0. (Redrawn with permission of S. V. Kasl and E. Harburg and the* Journal of the American Institute of Planners *from AIP Journal, 38, no. 5, September 1972.)*

worked out among potential users. The location of open space was critical in that it had to be situated so that residents had a sense of prior right to its use. If they did, they then could establish conventions by which conflicts with outsiders over use could be resolved peaceably. From his work, Newman concluded that buildings and their surrounding open space should be structured to maximize the sense of territorial protection (much like Jacobs's sense of neighborhood responsibility) and that this in turn would maximize their use and subsequently minimize crime.

On the subject of crime and neighborhood perception, Kasl and Harburg (1972) have presented a statistical model relating various factors, including the fear of criminal attack, to the desire to move out of a neighborhood (Figure 18-7). In this model, the perception of the neighborhood is affected most by the fear of criminal attack (correlation coefficient of .57), but the quality of education available to the children (.35), the availability of recreation and entertainment (.42), and the quality of city services to the neighborhood (.24) are also significant. The data were collected in Detroit, Michigan, and are perhaps specific to that city for the time of the survey conducted. But that is not a problem for our purposes, because the model can be tested in other cities and verified later. The important fact is that Kasl and Harburg were *able* to construct a model, for this shows that the quantification of perceptional images is possible, a hopeful sign for future research in social science and environmental design.

Proxemics: Perception of the Individual

At the lowest level of environmental perception we find our arena of interest reduced to the individual and his or her immediate environment—personal space. This is a topic of interest to anthropologists, behaviorists, and psychologists. Some of the basic questions here concern ideas about the variation in an individual's concept of personal space, reactions of people to others in restricted spaces, and the behavior of groups in different settings.

Two authors in particular have made prominent contributions to our knowledge of individuals and their perceptions of space. One of these, Edward T. Hall, has coined the term "proxemics" to describe the nature of the human use of space. Starting with the biological basis for space behavior, Hall (1966) used observation to determine the conceptualization of space by the average American (Figure 18-8). In his classification, Hall recognized four categories of space, each dependent upon one or more of the senses, and each considered by the individual to be appropriate for certain kinds of relationships. For example, intimate distance (from 0 to 1.5 feet) is defined by smell, touch, and thermal reception and normally is considered appropriate only for very personal relationships with family and sexual partners. Any other sort of relationship, such as with an employer, an acquaintance, or a stranger, would not be tolerated in this zone. The intimate zone shades imperceptibly into the personal zone, in which the voice and the eyes replace heat and smell as the primary senses.

Hall's diagram is a quantitative, and therefore testable, expression of the individual American's perception of space. One logical extension of these ideas is to use this conceptual schema, as Hall has done, to examine the differences in personal-space perception as a function of cultural differences. As one might expect, there are large differences between cultures. For example, Arabs depend on smell and touch in public where an American would not, because an Arab does not require a minimal personal space in this type of social interaction as an American does. As a consequence, in Arabic society cultural reinforcement has resulted in the tolerance of crowding

500 ENVIRONMENTAL PERCEPTION: IT'S WHAT YOU THINK THAT COUNTS

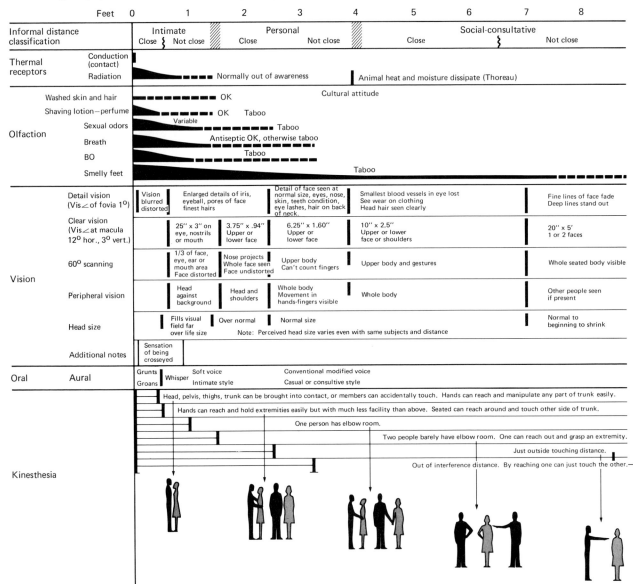

Chart showing interplay of the distant and immediate receptors in proxemic perception

Note: The boundaries associated with the transition from one voice level to the next have not been precisely determined

10	12	14	16	18	20	22		30	
←—Mandatory recognition distance begins here					Public			Not close begins at 30′ – 40′	

Slight eye wink, lip movement seen clearly	Entire central face included	Sharp features dissolve Eye color not discernable Smile-scowl visible Head bobbing more pronounced	Snellen's standard for distant vision employing angle of 1 min. Guild Opticians of America eye chart A person with 20-40 vision has trouble seeing eyes and expression around eyes though eye blink is visible.
	31″ x 7.5 Faces of two people	4′2″ x 1′6″ Torsos of two people	6′3″ x 1′7″ Torsos of 4 or 5 people
People often keep feet within other person's 60° angle of view		Whole body has space around it Postural communication begins to assume importance	
		Other people become important in peripheral vision	
		Very small	
People and objects seen as round up to 12–15 feet		Accommodative convergence ends after 15′ People and objects begin to flatten out	
	Loud voice when talking to a group Must raise voice to get attention Formal style	Full public speaking voice Frozen style	

Not so close as to result in accidental touching.

Two people whose heads are 8′–9′ apart can pass an object back and forth by both stretching.

FIGURE 18-8 *Proxemics in middle class Americans according to E. T. Hall.* (Redrawn from E. T. Hall, *Current Anthropology* 9, pp. 92-93, 1968.)

in public places, and there even is an implicit obligation for involvement (to the extent of smelling and touching) with people, even strangers, in public places. On the other hand, Arabs dislike closed views and small rooms, preferring high-ceiling, large spaces. However, this is not a contradiction, since these larger familial spaces allow family members to be regularly involved with one another rather than being more isolated, as they would be in several small rooms. In short, in the Arab concept of space, personal space takes second priority to social interaction, although it is not clear whether this is an adaptive response to a long history of crowding or a cultural characteristic that also happens to be adaptive.

In contrast to Arab proxemic behavior, the Germans treat personal space with considerable formality. Space should have precise structure (regarding such things as the arrangement of furniture in a room), and unauthorized modifications of the pattern are not tolerated. Furthermore, each person has a well-defined zone of personal space, similar to the American pattern but very different from the Arab habit. Hall discusses many examples of these contrasts in proxemic behavior between cultures, contrasts that are evidence of the influence of events and past history on attitudes of space use—strongly held if imperfectly understood attitudes, and consciously displayed if unconsciously motivated. The most immediate consequence of this maze of proxemic behaviors from culture to culture is the result of the great cosmopolitanism of people. There are few countries where cultures do not intermingle constantly, and Hall ascribes a large part of the misunderstandings between people to their interpretation of the behavior of members of other cultures in light of their own attitudes. This can be a serious matter, since misunderstandings frequently generate aggression and all of its consequences.

Hence thorough understanding of how people use space and interpret its use by others is important for the fields of environmental planning and architecture. There is a great difference in designing a marketplace, public park, or the interior arrangement of a home for Cairo (Egypt), Hamburg (Germany), or Wauwatosa (Wisconsin). This much is obvious, even if all the answers are not.

Psychologist Robert Sommer (1969) has attempted to define the impacts of proxemic behavior by combining experimentation with observation in small-group settings. One of his best-known contributions is concerned with the behavior of small groups under various seating arrangements using apparatus no more elaborate than a table and a group of chairs. In one experiment Sommer asked a group of students to choose a position from eight chairs arranged on either side of a rectangular table (Figure 18-9), assuming

FIGURE 18-9 *Aggression and defense around a table: the results of Sommer's preference experiment.* (Redrawn from Robert Sommer, *Personal Space: The Behavioral Basis of Design,* Copyright © 1969, p. 43, with permission of Prentice-Hall, Inc., Englewood Cliffs, N.J.)

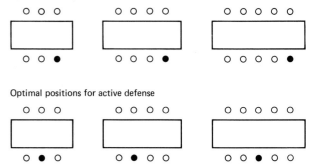

TABLE 18-3 *The results of Sommer's experiments in which students expressed preferences for different seating arrangements under four conditions of interaction with one other person. (Coaction: working at the same location but without involvement with one another.) Preferences also depended upon table shape: column (a), rectangular; column (b), circular. Figures show the percentage of subjects choosing each arrangement.*

(a)

Seating arrangement	Condition 1 (Conversing)	Condition 2 (Cooperating)	Condition 3 (Co-acting)	Condition 4 (Competing)
(x at top-left corner)	42	19	3	7
(x on both sides)	46	25	32	41
(x diagonal corners)	1	5	43	20
(x top-left and bottom)	0	0	3	5
(x both on left side)	11	51	7	8
(x top and bottom)	0	0	13	18
Total	100	100	100	99

SOURCE: Robert Sommer, *Personal Space: The Behavioral Basis of Design*, copyright 1969. Reprinted by permission of Prentice-Hall, Inc., Englewood Cliffs, N.J.

(b)

Seating arrangement	Condition 1 (Conversing)	Condition 2 (Cooperating)	Condition 3 (Co-acting)	Condition 4 (Competing)
(x adjacent)	63	83	13	12
(x across)	17	7	36	25
(x diagonal)	20	10	51	63
Total	100	100	100	100

that each was first to choose and (1) wanted to dominate the table or (2) wished to be as inconspicuous as possible. The results showed an invariate response: An aggressive position was signaled by taking a chair near the center of the table, a defensive one by taking a chair closer to the end. It is also important to note that these results are specific to the chair-table arrangement used, since a chair at the head of the table would be a signal for an aggressive, leading position; and the orientation of the door also would have an effect. (Sommer found that the defensive position was more likely to be facing *away* from the door.)

In a second experiment, Sommer asked students to indicate what positions they would take given variables of (1) the conditions of interaction and (2) the table shape. These results are tabulated in Table 18-3.

The face-to-face and side-by-side positions were favored for close conversation and cooperation; there was a less distinct tendency to assume more distant positions for separate work or active competition. The less distinct boundaries of the circular table did not change the results; they were substantially the same as those with the rectangular table. Once again, the students knew why they chose what they did; moreover, they usually could say why they did it.

These few examples underline an observation made for every level of perception we have discussed: There is demonstrable proof of the importance of perception for social relationships at many levels. Many authors have noted that such information has scarcely been used in the planning and design arts and that the oversight must be corrected. The question that remains is exactly how this is to be accomplished.

Environmental Perception and Normative Decision-making

We have seen that the threat of criminal attack could influence urban structure and perceptions of the neighborhood. If we were to examine such problems as air quality, "things in the public interest," and the quality of life, we would find that perceptional analysis could be applied to each. In fact, in some cases, perceptional analysis could be the dominant consideration in policy-making. While it would be ideal to base all environmental policy on accurate technical data, this situation will rarely be achieved for a number of reasons:

1. Decisions based on technical information require data on all relevant aspects of the problem; data on just some of the aspects of the problem are not very helpful. To obtain this sort of reliable data, we must have a realistic concept of the entire system involved, and this is not an easy order. Moreover, since this kind of complete data is expensive to collect, time and money must be practically unlimited, and there must be a strong motive to justify these expenses.

2. What data there are may contain several conflicting interpretations, each of which is backed by reputable authorities, and no one of which is clearly superior to the others. Furthermore, as so frequently happens, the perceived image of a problem does not necessarily correspond to absolute truth, even if the data are unequivocal.

3. Perceptional data, usually called *normative information,* can be obtained by survey and opinion sampling and are therefore easier and less expensive to gather than technical data. Since the sole reason something may be considered a problem is based on social perceptions of the situation, normative data can always be obtained by surveying these perceptions; whereas technical data, though more accurate, may not be nearly so easy to come by.

4. Except for a very general framework of technical information, normative data may be all that is needed to reach a decision, because decision-making is ultimately a political process in any case and therefore is based on the perceptions of the electorate.

Public perceptions of air quality are an excellent example of these ideas. On the technical side, the evidence of serious health hazard attributable to air pollution is not unequivocably clear, and it is not difficult to find authorities who disagree strongly about the interpretation of what evidence we have. This situation is not surprising, since getting better evidence requires time to run controlled experiments on human subjects. While the technical aspects are clearly important, it is more important that the majority of the public *believe* there is a "significant" health hazard and that, in general, air pollution is harmful to us in a variety of ways. No matter what the truth about air pollution effects may be, public perceptions of the seriousness of the problem overshadow them. Since opinions are easily shaped, easily measured, and cheaply obtained, normative information can be assumed to be very important in making decisions about environmental protection.

From the technical aspects of measuring perceptions about environmental problems, it is only a short step to the incorporation of opinions into political decisions and economic action, aspects that cannot be examined in depth here. This example suggests how important perceptional analysis could be once it attains an accepted role in social planning as well as in the design of structures and urban areas. In fact, it should be clear that there is a potentially fruitful triple interface between the "hard" sciences, the social sciences, and the design arts, with immediate implications of the broadest interdisciplinary type for environmental problem-solving.

Some Questions about the Future Applications of Environmental Perception

The literature, as dispersed as it is, suggests that there is an underlying process by which perceptions are formed, rationalized, and fitted into a pattern. Now we need to learn how to study the perception process in ways that not only are theoretically satisfying but are useful for applications in planning as well. The key element in this regard is the contribution a study makes to an understanding of the phenomena studied, because this understanding will eventually point the way toward the generality that is required for more practical applications of the information. We must strive toward the elaboration of better perceptional models that will, in turn, lead to testable hypotheses.

The prospects are not particularly bright at this time: There are difficulties even in reaching a satisfactory definition of the terms in the field, the resulting proliferation of names, and the multitude of disciplines involved in perception studies, each with its own approach to the problem. And one must consider the youth of the field and its consequent lack of a well-defined body of principles. Despite these difficulties there appears to be sufficient reason to believe that environmental perception will sooner or later make a substantial contribution to planning and architectural design.

The need for further studies is obvious, because the potential role of perception in the design of human environments is clear. However, there is still no consensus about what the completed studies imply for environmental planning. Nor is it clear that all we need is more studies. Sommer provides an excellent example of this by asking what room-design solution is to be adopted in a hypothetical situation. Suppose that French-Canadian students preferred closer contact in smaller rooms than did English-Canadian students. Should an architect in Quebec, where there is a French-Canadian majority, design smaller rooms?

Should the school administration increase class sizes or decrease interchair distances? Suppose now that the teacher is of English-Canadian heritage while the class is predominantly French-Canadian. What then? Sommer makes it clear that studies in perception cannot in themselves resolve conflicts such as these.

The wide range of studies in this embryonic field embodies some divergent views on what environmental perception is. The tone of Lynch's work is strongly aesthetic; Lynch measures perception as the succession of images encountered by individuals and these images themselves are based on aesthetic arrangements of structures. The studies of natural-hazard perception, on the other hand, are concerned with imagery that is linked only indirectly to aesthetics. In the case of urban structures, there must be a valid role for aesthetics in design, but there is presently no distinct limit to its relative influence. The case of the redesigned Central Artery in Boston is one example of primary aesthetic consideration in this area of urban design. Nevertheless, this does not mean that the redesigned artery is not equal in all other qualities as well as being far superior in the aesthetic respect; however, one must be careful when dealing with the concept of aesthetics to be sure that it is not used to justify a covert aim in a plan, such as the routing of a road through a slum which the local government would rather see gone anyway. Moreover, as our understanding of human perception grows, we must not let what is aesthetically pleasing or unpleasing dominate other considerations. Just such a plan, which fortunately failed in its attempt, is documented in a short paper, "The Old Men in Plaza Park," by Sommer and Becker (1969). In it they detail a case where a park was restructured specifically to drive out the old drifters thought to be a local form of blight. However, because the perceptions of the old men were not considered, the redesigned park ended up exacerbating the situation; in their effort to eliminate areas where the old men would gather, the city succeeded only in eliminating some of the areas, and the old men were just more crowded.

Thus, though the past history in this area has been one of insufficient attention to the needs of people, needs that only perception studies can adequately define, we must approach this endeavor with caution and wisdom, because studies in perception could also end up providing just one more tool to impose the will of planners and architects on the public. Given this dilemma we must strive toward research that is directed at better understanding the perception processes. In view of the needs of environmental planners, progress must be made as quickly as possible; for while environmental perception seems to be a diffuse, poorly defined, and even unscientific pursuit, a thorough understanding of it is vital. The problems merely serve to increase the challenge.

Questions

REVIEW QUESTIONS

1. Define "environmental perception," and draw up a list of at least four generalizations that characterize environmental perceptions.

2. Explain the failure of high-rise, low-cost public housing in terms of the environmental perception of the potential residents.

   Hint: You might find Newman (1972, chap. 17), and parts of Robert Goodman's book (particularly chaps. 3 and 4), helpful in answering this question.

3. Defend or refute Moncrief's theory and present arguments to support your case.

ADVANCED QUESTIONS

1. Goodman (1971, chap. 5) dislikes the work of E. T. Hall. Read his arguments and analyze them. Do you think (if Goodman is correct) that studies of culture and perception are socially dangerous?

2. Kates (*Natural Hazard Research*, Working Paper 14, 1970) contrasts the conflicting goals of perception models as generality versus detail. Given what you know about social systems and environmental perception, which goal seems more desirable, and which more easily attained?

The following two questions are group projects:

3. Read Lynch and Rivkin (1959) carefully. Then pick a block and repeat their experiment using a group of not less than 10 people and preferably more. Did you reproduce their results? If not, why do you think your results were different?

4. Experiments with personal space are lots of fun, if you dare. Try Sommer's first experiment on an audience that does not know your objectives for some relaxing practice. Then observe the behavior of a group of strangers in an elevator at different densities. (To really open your eyes, observe your own reactions; and no cheating by doing it with friends!)

Further Readings

American Institute of Planners. "Psychology and Urban Planning." *American Institute of Planners Journal*, 38, No. 2, 1972.

    Contains six papers of interest to readers of this chapter. The paper by Rozelle and Baxter is recommended particularly.

Appleyard, D., K. Lynch, and J. R. Myer. *The View from the Road.* M.I.T. Press, Cambridge, Mass., 1964.

Auliciems, A., and I. Burton. "Perception and Awareness of Air Pollution in Toronto." *Natural Hazard Research*, Working Paper No. 13, 1970. (Mimeo available from Department of Geography, University of Colorado, Boulder.)

Barbour, I. G. (ed.). *Western Man and Environmental Ethics.* Addison-Wesley, Reading, Mass., 1973.

    The first section includes White's 1967 paper, two rebuttals including Moncrief's, and an original answer to these rebuttals by White. The last section is concerned with the impact of technology in society (see also Mumford, 1970).

Burton, I., R. W. Kates, and G. F. White. "The Human Ecology of Extreme Geophysical Events." *Natural Hazard Research*, Working Paper No. 1, 1968.

Esser, A. H. (ed.). *Behavior and Environment.* Plenum, New York, 1971.

    A technical symposium with several papers of interest for personal-space perception.

Gans, H. J. *The Levittowners.* Pantheon, New York, 1967.

    Gans analyzes the dynamics of a particular suburban community (Levittown, Penn.). The viewpoint is sociological, but this book has much to say (if sometimes indirectly) about perceptions as well.

Glacken, C. J. *Traces on the Rhodian Shore.* University of California, Berkeley, 1967.

    This is an exhaustive account of human environmental attitudes to about 1800. Its value is largely historical.

Guthrie, D. C. "Primitive Man's Relationship to Nature." *Bioscience,* 21 (1971), 721-723.

Hall, E. T. *The Hidden Dimension.* Doubleday, Garden City, N.Y. 1966.

> See also the earlier book *The Silent Language* and his paper in Esser's (1971) volume.

Held, V. *The Public Interest and Individual Interests.* Basic Books, New York, 1970.

> "The public interest" is an excellent example of a normative judgment which is a perception. Held presents a full discussion of its meaning for those who wish to explore this particular aspect of perception in depth.

Jackson, J. B. "Other Directed Houses." *Landscape,* 6, No. 2 (1956), 29-35.

Jackson, J. B. "The Stranger's Path." *Landscape,* 7, No. 1 (1957a), 11-15.

Jackson, J. B. "The Abstract World of the Hot-rodder." *Landscape,* 7, No. 2 (1957b), 22-27.

Kasl, S. V., and E. Harburg. "Perceptions of the Neighborhood and the Desire to Move Out." *American Institute of Planners Journal,* 38 (1972), 318-324.

Kates, R. W. "The Perception of Storm-hazard on the Shores of Megalopolis." In Lowenthal, 1967.

Lowenthal, D. (ed.). *Environmental Perception and Behavior.* Research Paper No. 109, the Department of Geography, University of Chicago, 1967.

> This is a very useful volume.

Lowenthal, D. "The American Scene." *Geographical Review,* 58 (1968), 61-88.

Lynch, K. *The Image of the City.* M.I.T. Press, Cambridge, Mass., 1960.

Lynch, K., and M. Rivkin. "A Walk Around the Block." *Landscape,* 8, No. 3 (1959), 24-34.

Moncrief, L. W. "The Cultural Basis of Our Environmental Crisis." *Science,* 170 (1970), 508-512.

Mumford, L. *The Myth of the Machine.* Vol. I, *Technics and Human Development.* Vol. II, *The Pentagon of Power.* Harcourt, Brace, Jovanovich, New York, 1967, 1970.

> The impact of technology on human society is now Mumford's major interest.

Newman, O. *Defensible Space.* Macmillan, New York, 1972.

Saarinen, T. F. *Perception of Environment.* Commission on College Geography, Resource Paper No. 5, Association of American Geographers, Washington, 1969.

Schiff, M. R. "Some Theoretical Aspects of Attitudes and Perceptions." *Natural Hazard Research,* Working Paper No. 15, 1970.

> A brief and not overly technical report on the state-of-the-art on perception models.

Sommer, R. *Personal Space.* Prentice-Hall, Englewood Cliffs, N.J., 1969.

> Sommer's work is summarized here very well. He also has a contribution to Esser's volume.

Sommer, R., and F. D. Becker. "The Old Men in Plaza Park," *Landscape Architecture,* 59 (1969), 111-114.

Sonnenfeld, J. "Environmental Perception and Adaptation Level in the Arctic." In Lowenthal, 1967.

An example of the use of the semantic differential technique. See problem 3.

Tuan, Y. F. *The Hydrologic Cycle and the Wisdom of God: A Theme in Geoteleology.* Research Publication No. 1, Department of Geography, University of Toronto, 1968.

Tuan, Y. F. "Our Treatment of the Environment in Ideal and Actuality." *American Scientist,* 58 (1970), 244–249.

Tuan, Y. F. *Topophilia.* Prentice-Hall, Englewood Cliffs, N.J., 1974.

White, L. F. "The Historical Roots of Our Ecologic Crisis." *Science,* 155 (1967), 1203–1207.

19. Economic Systems, Government, and Ecological Systems: Coaction and Conflict

Human systems are dominated by conscious decisions, especially decisions about how resources are to be allocated among various alternatives. The distribution of food and energy within the human population is one well-known example, but the idea applies equally well to the protection of natural landscapes and air quality, because we have only to treat these as special kinds of resources to view them as allocation problems also. In the course of our evolutionary history, we discovered that money was a useful way to help us make these allocation decisions—and so developed the ideas upon which market economics is based. *Economics* may be defined as the science that seeks efficient and workable solutions to the problems of production, distribution, and consumption of commodities.

Despite the fact that economics is a human invention, it is defensible also to speak of natural economies based upon the allocation of energy. Since we also must participate in this energy economy, there can be differences between allocation priorities in natural and human economies. This possibility raises a series of questions that have serious implications for environmental policy:

1. What are the similarities and conflicts between human economies and natural economies?
2. What are the consequences if the conflicts are not resolved?
3. What policies have been suggested, and what policies have been followed, to promote conflict resolution?

These questions are the central topics of this chapter.

Comparisons of Human and Natural Economies

Natural economies are based upon solar energy and its derivatives; and to the extent that human beings have the same biological requirements, the human economy exhibits these same dependencies. Since all living systems require energy to survive, this is no surprise. As we have seen, however, nonhuman systems are highly empirical (trial-and-error) assemblages whose characteristics are shaped largely by selection and elements of chance. Thus nonhuman systems come to be dominated by species and combinations of species that are "lucky" and successful, within the limits set by the amount of solar energy and general environmental conditions. While in human societies many rules are imposed by the members upon themselves to help weaker members, there are few such rules in nonhuman ones. Thus even in advanced animal societies (Chapter 2) there is no regard for egalitarianism or for the protection of weaker members from rapid elimination. Moreover, nonhuman systems must operate only within natural limitations.

In contrast, human systems operate very differently. Whereas natural economies must stress adaptation to the environment and survival of the entire society over the fate of individuals, human priorities are reversed; much of our concern is over providing more equal opportunity for all people. Survival is accomplished less by adaptation to environmental conditions than by their modification. This modification of natural limitations, so much a characteristic of human systems, tends to mushroom and ramify in that our initial attempts at environmental modification have led to ever-increasing capabilities in this regard as our technological know-how has increased. As the magnitude of these modifications increases, so do their potential consequences. This point has been made in several earlier chapters, but it is worth restressing here, because economics is strongly rooted both in the problems that result from the kind of large-scale activities which usually are associated with human populations and in the activities that have the potential to ameliorate these problems.

Conflicts between Human and Natural Economies

The problems of production, distribution, and consumption have been addressed by two types of economic systems—the state-controlled, centralized economy and the free-market economy. The centralized economy requires central planning—explicit decisions about how much to produce and who is to get it. The free-market system, on the other hand, depends upon automatic adjustments of supply and demand. Each system has its advantages and disadvantages environmentally. In centralized economies there is no particular problem of resource allocation or unwanted production, as long as the citizenry agrees with the central planners. While the evidence does not show that central control of the economy has produced much environmental benefit, there is always the promise of it *if* the leadership is so inclined.

In free-market economies the system operates on the premise that allocation should occur automatically without the requirement of a centralized controlling agency. This system emphasizes individual freedom, but to do so requires two conditions: a means to establish the value of goods and services and a means to adjust the performance of the economic mechanism to changing conditions. "Value" is difficult to define, unless it can be related to monetary units. When this is done, economic theory begins to attain the precision of mathematics, because numbers are exact. The performance of the system is then controlled by feedback relationships among producers, consumers, and prices. Under ideal conditions, the system adjusts as follows: Producers set prices upon their products which potential consumers must decide whether they are willing to pay or not. If prices are set too low, too many consumers will decide to buy, demand will exceed supply, and prices will rise until supply and demand reach an equilibrium state. If, on the other hand, prices are initially set too high, demand will be less than supply, and prices will be depressed, again to the point of an equilibrium. The automatic, self-regulating control of negative feedback that characterizes the supply-demand-price system is a major reason for its widespread use. Systems of barter and trade are primitive examples of human economies based on this same system in which goods and services are physically exchanged, but they are basically no different from systems in which currency and the ledger sheet are the media of exchange. That some form of economic system would come to be the natural means to allocate energy and material may be seen by simply asking oneself how a society would operate without at least a barter system. As long as people cannot be completely self-sufficient, a rational system to run the exchanges of goods and services, to value labor, and to provide a livelihood is necessary. Indeed, it is improbable that such an economic system would not develop within a human social system.

Free-market economists freely admit that the market form of economic control is not perfect, especially with regard to environmental quality. Despite their apparent structural differences, it turns out that centralized economies have performed little better in this respect (see Chapter 11 and the following sections). The problems of environmental quality resulting from economic imperfections fall into three general categories:

1. Producers can impose part of their costs on consumers. For example, suppose that an electric power company can use very cheap coal and not be concerned with the air pollution it generates. If the company does not have to pay either for higher-quality coal or for air pollution control, it can sell electricity cheaper than the true cost of the power; hence it can produce more power for a given price, consumers can use more, and the nonmonetary costs are paid by the public in terms of deteriorating environmental quality, dirty laundry, and respiratory disease, rather than in elevated monetary costs. Costs such as these, which are not included in the production equation, are called *market externalities*. That the free-market system has no means of including market externalities in its framework is seen as one of its failures.

2. The second class of market failure is the dilemma of *free goods*, or as Garrett Hardin (1968) has called it, "the tragedy of the commons." The problem is that many resources are not owned by anyone in particular and therefore have no one to defend them from overexploitation. Market failure occurs with these free goods when no one group's economic interest is harmed by overexploitation to the point that it is willing to protect the resource from further abuse.

With free goods, there is, if anything, positive feedback from demand so that overexploitation frequently can lead to even greater excesses. And since the cost involved in the overuse of the resource is a free good, nothing controls this increasing supply and demand spiral, until the resource finally collapses. Throughout this book we have discussed many resources—air, water, biological resources—that are free goods and are overexploited for this reason.

3. The third type of problem is the assessment of value. If all values are included and can be expressed accurately in terms of price, the mathematical precision of economics is a virtue; but if, as is usually the case, all values cannot be expressed accurately in monetary terms, some things will frequently be undervalued. Examples include items normally thought to be aesthetic (such as building design, scenery, and wild animals) or free goods (such as whales or air quality). It seems that when the value of a resource is not easily accommodated in normal channels of market economics, the tendency is to assign it no value at all. This has been a common fault of benefit-cost analysis, wherein overall value is assessed as the net difference between all costs and benefits. Benefit-cost ratios have often involved variables that were difficult to quantify, and in these cases they were overvalued or undervalued. Recently, however, there has been some effort to improve this situation and to somehow properly fit aesthetic and free goods into normal systems of economic evaluation. For example, faced with the legal requirement to find some means to protect water quality, the Environmental Protection Agency is seeking ways to measure the value of any ecosystems covered by this mandate so that they can assign values to various ecosystems affected by a deterioration of water quality. Given the choice of making qualitative judgments (perhaps accurate but not precise) or placing monetary value on the ecosystems in question (precise but possibly not accurate or even valid), EPA has opted for the latter, because the need to assign quantitative values is so great.

All these problems are symptoms of a major difference in strategy between natural and human economies. The human economy is directed toward the greatest benefit to the human population and its individuals, whereas the natural economy is directed toward the greatest benefit to the entire ecosystem. When energy is dependably available, natural systems consistently respond by increasing their complexity; there are many more populations, but each is on the average small or even rare. In contrast, human economies could react to an increased and dependable resource with greater complexity (measured as the number of economic units using that resource), but these reactions could be expressed equally well as an enhancement of the competitive ability of a few large units or manufacturing firms to the point that they can dominate the entire market using the resource. Both outcomes can be expected to benefit the human population, but not the planetary ecosystem.

Whatever the full range of conflicts may be, the crunch is that our population-oriented strategy of development sooner or later must be reconciled with the ecosystem-oriented course of evolution. It is axiomatic that we cannot continue to make all our decisions completely on the basis of our economic inventions without taking into account some externalities which are likely to be important (such as a sustained yield of resources and minimal standards of environmental quality). Our biological limitations themselves demand the eventual resolution of the conflict between human and natural economies, and the actual shape of the means of resolution currently occupies much public attention, as it certainly should. It is wise, however, to remember that the means used to resolve this conflict will themselves have consequences which could shape society powerfully before the conflict is resolved.

"Spaceship Earth": The Economic Solution

Most contemporary economists believe in neoclassical Keynesian economic theory. This viewpoint holds that full employment of resources is required to insure full employment of people and therefore the maximum social good. It follows, then, that economic growth (measured by such indexes as the Gross National Product) is the most desirable state of affairs, because if the total economic output is larger, everyone will be better off. Viewed in reverse, if the economy did not grow, the chances for an individual's economic betterment would be vastly decreased, especially if the population continued to grow; unemployment would rise, and everyone would be worse off.

Thus Keynesian economists depend upon economic growth to avoid a social crisis. However, this premise assumes (1) that population growth will continue; (2) that social good means more material goods distributed more equitably; (3) that resources will not prove to be limiting; or, if they are, (4) that technological innovations will abrogate any such limits. Until recently, only a few economists challenged the validity of the Keynesian premise, one of the first being Kenneth E. Boulding of the University of Colorado. To illustrate his argument for transition to a steady-state (no-growth) economy, Boulding (1966b) constructed an analogy of the planet earth as a spaceship to denote the finite limitations upon the resources of the planet and the futility of using these resources in a "make-work" fashion simply to create full employment. Boulding saw the "spaceman's economy" as a closed system with minimal throughput. Only essential activities would be carried out, with the emphasis being on quality and not quantity of production and consumption. These things are clearly necessary in a spaceship. Boulding called the open economy the "cowboy economy" to signify its pioneer attitudes (operating at a frontier of unlimited resources, reckless about the consequences of its behavior, and obsessed with throughput). This analogy was immediately popular among environmentalists who appreciated the idea of absolute limits and who blamed "cowboy" economics as a root cause of environmental disruption.

Unless one believes in technological miracles, it is obviously true that steady-state economics must eventually predominate in a world increasingly crowded by human beings. It is a matter of controversy whether or not warning signs are evident already that such growth is no longer desirable even if it remains possible. However, even if a steady-state economy is inevitable, the means for achieving it are largely unknown, though extremely important, and this raises a series of troublesome questions. Specifically, what forces would promote steady-state economies? Could a steady-state economy be equitable? Would a no-growth economy really help preserve environmental quality? And finally, would a planned stationary economy be a better solution than a stationary but free-market economy? We shall consider each of these questions in turn.

TRANSITION TO STEADY-STATE ECONOMICS

In a world of increasing scarcity and political uncertainty, a smooth transition to some form of steady-state economy appears unlikely. Daly (1971, 1973) attempted to forecast what some problems of steady-state economies would be. Inequalities in income would have to be dealt with by some form of direct redistribution of income while, at the same time, population growth would have to end, and neither condition would be attained easily. Indeed, Daly was well aware that a revolution might be required to attain a steady-state economy, and that this attainment might only result in a violent counterreaction. Although he rejected both capitalism and Bolshevik socialism as the preferred steady-state economic system because both are growth-oriented, he could offer little else other than to reiterate that there is a great need to find a way to accomplish the sweeping redistributions of wealth and income he felt are required as part of the shift. Perhaps all this indicates that entirely new social institutions will have to accompany the economic transition.

One of the foremost radical economists, Robert L. Heilbroner, in his 1974 book also rejected both capitalism and socialism as modes to end economic growth. In fact, he argued that they are growing increasingly similar to one another because both are based upon technology and machines which emphasize consumption and production and ultimately result in a society based upon the expectation of continuing growth (in this regard, at least, Heilbroner's theme evokes the Goodmans' thesis of 27 years earlier). Thus he too feels that an entirely new tradition of economics would be required to implement a stationary economy. Nor does Heilbroner feel that the public would eagerly accept the repercussions of the new economies, for it would have to be accompanied by sacrifices of unknown magnitude. Heilbroner did not address the problems of implementation—in fact, he felt that the social upheavals which would result from such a change would be so massive that it would be impossible to predict even what sort of policies might be effective.

Ayres and Kneese (1971) rejected a social revolution and focused instead on more incremental changes leading toward a steady-state economy. They flatly rejected the possibility that a truly stationary economy would be attained, however; the best they could anticipate is a quasistationary economy, because they maintained not only that population growth would not end for several decades but also that there would not be a voluntary move toward the stationary state, and moreover, that resource shortages would not in themselves force a stationary economy.

Ayres and Kneese argued that technological innovation might be the one force which would permit a smooth transition to a quasistationary state (as characterized by slower rates of resource use). They saw many dangers in the transition process, however, most notably those associated with the possible failure of technology to keep pace with an increasingly desperate situation. The problem of maintaining social and economic stability would grow progressively difficult as increased population and dwindling resources in combination reduced the margin of error in decision-making. The possibility that the entire social system would collapse at the first minor crisis caused Ayres and Kneese to speculate that strong pressure for the development of a rigid and inhumane form of government would insure some form of economic and social stability. Furthermore, they speculated that the tensions created by inequities in resource use in different societies could lead to world war on a scale never before seen. Clearly, this is much worse than the pressure for redistribution of wealth that Daly forecasts, and it raises difficult questions about the transition process. Specifically, is such a transition even possible without world crisis and conflict?

EQUITABILITY IN THE STEADY-STATE ECONOMY

The Keynesian solution to the distributional inequities in the distribution of income and wealth is economic growth. This is because growth means more employment, more generation of capital, and thus more money to be distributed. Thus the Keynesian solution is based on the idea that if the economic "pie" is getting larger and larger, there will be increasingly more of it to devote to the less advantaged, and so everyone will be better off. Clearly this solution does not require any forced redistribution of income, but it generates expectations that the future will be brighter for all.

In contrast, the changeover to a steady-state economy would diminish rising expectations. No further growth means an unchanging pie, no bigger slices, and no hope that the future will automatically make the actual piece larger. Thus it seems certain, as Daly suggests, that the end of economic growth will create a socioeconomic crisis when everyone realizes that his own efforts will not mean a better life and that he is frozen into what distributional inequities might exist. The question is one of how inequities are to be rectified. The simplest answer is that the government would tax earnings regressively or would simply take from the richer and give to the poorer, but how feasible this kind of action is in a democracy is debatable. As Ayers and Kneese note, however, democracy could become a trivial matter compared to the abolition of income inequities.

The larger aspect of the equity issue is the means of implementation and regulation of a no-growth economy. These are questions of preventing growth, regulating the desired level of economic activity, and the identity of the authorities. For example, if the United States were to have a steady-state economy, would it be set at a level that meets only the basic needs of the population for food and shelter? Would industries be classified as "essential" or "irrelevant" and be encouraged or outlawed accordingly? What mechanism would be used to make all these decisions? Would an automatic self-adjusting market system be possible, or would a centralized economic authority be required? We must also consider the possible international ramifications of a changeover to a stationary economy. If the steady-state economy were restricted to the United States, one might expect dramatic effects upon the economies of other nations dependent on trade with the United States. On the other hand, a global steady-state economy brought on by world resource shortage would freeze countries in "have" and "have not" patterns that would generate immense pressures only too likely to end in armed conflict.

THE STEADY-STATE ECONOMY AND ENVIRONMENTAL QUALITY

Even if the steady-state economy is directed primarily toward reduction of resource usage rates, obvious benefits are to be gained in environmental quality. With less power consumption, less fuel consumption, and a reduced demand for raw materials, there are bound to be fewer power plants, automobiles, and factories to produce environmental pollution and to support the further expansion of resource-extraction industries. Thus we generally may expect the additional benefit of improved environmental quality from a steady-state economy. There is, however, much more uncertainty about the exact extent of the benefits, because we generally do not know the functional relationships between environmental quality and incremental changes toward a steady-state economy. Suppose, for example, that we are interested in putting only as much of the economy into steady-state as is needed to obtain a given amount of environmental improvement, or as much as possible for a limited investment. One obvious influence on these parameters is the exact parts of the economy to be restricted, for example, transportation versus hospital services. Information such as that given in Figure 19-1 (although hypothetical in this particular example) would be important. By comparing the four types of curves in Figure 19-1, we can see that the form of the actual relationship is a critical determinant of the cost per unit of improvement in environmental quality.

PLANNED ECONOMIES AND THE ENVIRONMENT

With all the kinds of market failure that we have reviewed already, why not simply resort to more perfect planning to eliminate the excesses of free markets? Why not, some have suggested, simply follow the socialist model of the planned economy?

Other authors have noted, however, that the USSR and other socialist societies are no less committed to economic growth than any other country. The low priority given to consumers in the USSR for many

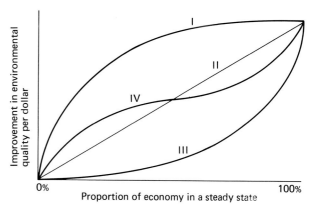

FIGURE 19-1 *Some hypothetical relationships between environmental quality and steady-state economy. No one really knows what the curve looks like, although we can hope that curve I is correct because it implies a great improvement in environmental quality with only small sacrifices in economic growth. Curve II implies a linear response, and curve III a very high initial cost (measured in terms of conversion of the economy and the sacrifice of growth) before substantial improvement results. Curve IV is an example of a compound curve in which there is a marked improvement in environmental quality with a relatively small drop in economic growth, i.e., only a small segment of the economy shifts to steady state. However, further improvement does not occur until a large proportion of the economy shifts to the steady state.*

years while funds were channeled into heavy industrial development suggests that the people were asked to sacrifice even more than in a capitalist society to insure maximum rates of economic development. Nevertheless, the fact remains that in these kinds of societies decision-making could be reversed quickly, and the growth emphasis could end. Goldman's (1972) analysis of environmental disruption in the USSR makes it clear, however, that such priorities have hitherto been nonexistent. For example, Lake Baikal is one of the world's finest examples of an isolated freshwater lake with a long and unique biological history. Lake Baikal is old, deep, and clear; with its highly endemic fauna, the lake is regarded as an international limnological resource. Unfortunately for the lake, it is surrounded by the rich northern coniferous forests of Siberia, an economic resource that could not remain unexploited for long (Figure 19-2).

FIGURE 19-2 *A view of Lake Baikal from a fir-covered ridge.* (Courtesy of Sofoto.)

Starting in 1966, the Soviet government authorized the construction of paper and pulp mills upon the shores of Lake Baikal. Although it was soon obvious that the lake was being degraded, the Ministry of Pulp and Paper Industries continued to plan and to construct yet more mills to handle a growing volume of timber cutting. In all of this there are numerous signs of the same underlying desires for economic growth that occur in free-market economies. One example that Goldman cites contrasts to the situation around Lake Tahoe (California-Nevada). On the basis of the cost—nearly 40 million dollars—the Ministry of Pulp and Paper Industries strongly opposed a sewage effluent line that would have exported all waterborne wastes out of the drainage basin of Lake Baikal. They preferred a less expensive sewage treatment plant on the shores of the lake, a treatment plant that by 1971 still had not performed to standards. The 55-dollar fines imposed for each water pollution conviction created pitifully little incentive for building the sewage diversion line. This socialist approach to the pollution of Lake Baikal might be contrasted to the decisions made in the United States concerning similar problems at Lake Tahoe, where sewage diversion from the Lake Tahoe basin has been adopted as the long-term solution for the prevention of pollution of that lake. Alternatively, it might be conceded, the difference could be attributed to the recreational importance of Lake Tahoe compared with Lake Baikal.

While the Baikal experience does not suggest that a planned economy will respect and protect its environment better than any other economic system, neither does it suggest that a planned economy does not have the potential to do better. For example, since the Soviet economy does not permit the existence of large consumer-oriented industries, it suffers little from such problems as planned obsolescence, the clutter of disposable bottles and other goods, and the like. This approach is the direct result of the policy of the authoritarian government which has the power to dictate policy and the power to change priorities rapidly. The USSR could decide that the purity and biological complexity of Lake Baikal are more important than paper—and pollution of the lake could be stopped with a simple telephone call or memorandum. The question, then, becomes one of whether or not there is any probability that priorities could be reversed so dramatically, especially since so much capital has been invested in the construction of the pulp and paper plants.

From this example, it should be fairly obvious that, to date, planned economies have certainly demonstrated no clear superiority to free-market economies in the realm of preserving and improving environmental quality. Goldman, at least, believes that industrialization has been the root cause of environmental disruption throughout the world, with the actual form of the economy being unimportant. This premise is certainly supported by Heilbroner, who maintains that the process of industrialization has served to greatly reduce the differences between the two economic systems.

Political Influences on Economic Processes

Although it can be useful to consider the differences between "planned" and "free" economies, in the real world the distinctions are blurred because most extant systems are some blend of pure-market forces and forces completely under political control. This evolution toward intermediate states is not surprising; experience has shown repeatedly that too much planning and control stifles individual incentives, while complete market freedom, in the process of maximizing profit and gain, can lead to all sorts of abuses.

It is becoming increasingly obvious that the world trend is directed toward greater governmental involvement in the economy. A number of arguments suggest why governmental intervention in free markets should be expected to increase:

1. One reason is that governmental institutions (judicial, legislative, and administrative systems) are the only ones with enough authority to arbitrate disputes among producers, distributors, and consumers. In addition, governments play a central role in regulating competition; antitrust laws and suits are examples of this important function. The phenomenon known as *clientele capture*, in which administrative agencies have come to be dominated by the very industries they were set up to regulate, indicates the important role of government in the economic sector. Because persons associated with the regulated interests have the most expertise to offer in that particular area of endeavor, and because they have the greatest interest and stake in the decisions of the regulatory agency, it will often happen that these representatives of the regulated interest are placed in the most important positions in the regulating agency. Ironically, when this occurs, the "regulated" interests end up all the more powerful, since the decisions they would have made anyway can be promulgated by a public agency with infinitely greater authority.

2. Governmental regulation of economic activity in the interests of industrial production is only one facet of a wide range of legislation that can be classified as being in the public interest. The government is the sole agency that can take actions to rectify situations deemed "contrary to the public interest." For example, times of war are deemed to be public emergencies when broad powers to set priorities and to ration resources are granted to government because the temporary suspension of the benefits of democracy are not considered important compared with the need for rapid and concerted action. The 1973 "energy crisis" has revealed that worsening resource shortages will precipitate the same kind of governmental responses: establishment of priorities, allocation and rationing, control of production processes and the rate of consumption, and also the use of public funds and tax powers to provide incentives to boost the production of the scarce resources. There is considerable disagreement (even among economists) whether or not free markets can cope with these shortages in ways that will prevent hardship upon some segments of the economy and the public. Nevertheless, in an atmosphere filled with public distrust both of corporate giants and of the quality of their products and services, governments at all levels have shown little faith in the ability of the free market to cope with either environmental quality or resource shortages—an attitude that seems likely to go unchanged. Therefore, in the face of continuously declining environmental quality and increasing shortages, it seems likely that government involvement will increase dramatically.

3. Through the collection of taxes, the federal government in particular controls enormous sums of public capital; how these funds are spent can produce powerful influences on the course of events in the economy. For example, the existence of many firms whose primary occupation is national defense is predicated upon the uninterrupted flow of federal funds. In more general terms, these kinds of public expenditures are but a part of the entire picture involving the allocation of public budgets, the so-called *political economy*. Aerospace, welfare economies, basic research, and defense contracts are prime examples of what have become de facto partnerships between government and industry. The vast sums involved create enormous incentives for the private economy to respond to, whether this response takes the form of bribes and kickbacks, or is manifested as clientele capture, or is manifested simply as increased incentives.

With all these forces actively working in society, it is no surprise that the economy should be increasingly dominated by governmental decision-making. If the steady-state economy is to become a reality, this economic control, along with a stable population, would seem to be necessary. Otherwise it is difficult to visualize what authority would have the power to dictate that less resources and less production could result from an economic system which has been geared to maximize them.

Political Responses to Environmental Quality Control

In the United States to date, environmental legislation has been shaped within the context of a growing economy. Here we shall examine the landmark piece of environmental legislation and contrast the legislative approach to the judicial approach advocated by others.

THE NATIONAL ENVIRONMENTAL POLICY ACT

The National Environmental Policy Act was signed into law in 1970 as a major step toward environmental protection. NEPA contained two major provisions, the first of which, Section 102 of Title I, directed all federal agencies to

1. Utilize a systematic and interdisciplinary approach toward all planning and decision-making activities that could affect environmental quality.
2. Consider variables that hitherto had been considered unquantifiable and that therefore had been ignored in planning and decision-making.
3. Submit statements detailing the environmental effects of any proposed federal activity, including all potential impacts, alternative courses of action, and comparisons of the long- and short-term effects. Furthermore, agencies are to enumerate the financial resources that, once committed, would not be recoverable should the project be canceled. This provision is the mandate for the so-called *environmental impact statement* (EIS).
4. Cooperate with all agencies at lower levels of government that may find any environmental information developed at the federal level useful.
5. Assist, in every possible way, the Council on Environmental Quality (CEQ). CEQ serves as the overall coordinator of federal activities related to the environment (see below).

The second major provision of NEPA, Title II, established the CEQ, which consists of three members as part of the executive branch of government. CEQ's duties include:

1. Providing assistance to the President in the preparation of the annual report on environmental quality as required by Title II of NEPA.
2. Serving as an advisor to the President in all matters relating to the environment.
3. Gathering all relevant data and formulating the appropriate indexes necessary to aid in the improvement of environmental quality in the United States.
4. Monitoring all statements required by NEPA Section 102 and making any necessary recommendations to the President related to the content of those statements.

GOVERNMENTAL REORGANIZATION BY EXECUTIVE ORDER

Immediately following the enactment of the National Environmental Policy Act, President Nixon reorganized the executive branch of the federal government and created two new superagencies with broad responsibilities from a number of existing departments having overlapping responsibility but little or no integration. These two agencies are the National Oceanic and Atmospheric Administration (NOAA) and the Environmental Protection Agency (EPA). A third, the Department of Natural Resources (DNR) was proposed, but the proposal has not been implemented. Of these agencies, EPA has been the most active and controversial. Given the responsibility for air, water, pesticides, solid wastes, and noise, EPA has had a turbulent history as it has tried to find workable solutions to meet the requirements of the Clean Air Act of 1970. Since the requirements for air quality are stringent, EPA was forced to propose all sorts of drastic measures to attain the standards by 1975, including gasoline rationing, banning motorcycles during summer seasons, drastic reduction of the total or cumulative mileages allowed in a given air basin (in the San Francisco air basin, for example, mileage would have to shrink to a "mere" 3 percent of 1972 figures to meet the Clean Air Act standards by 1975), and reducing the number of parking spaces. EPA even threatened to enter the land-use planning field in an effort to move air pollution sources outside of air basins with critically high pollution levels. Fortunately, with delays in the date of enforcement to at least 1977, the pressure on EPA has eased and allowed the agency to concentrate on more feasible measures.

EPA was created to be responsible for the enforcement of environmental quality laws in five areas—they are, in effect, environmental police. CEQ, on the other hand, is research oriented and has many advisory functions with no enforcement mandate. Nevertheless, in actual practice, the distinctions are not always clear, because since EPA has the manpower and funds, it has, in fact, been seeking programs of research relevant to its needs as well as engaging in its own research programs. In addition, EPA has issued preliminary guidelines relating to the form and substance of environmental impact statements (EISs) in their areas of concern. On the other hand, CEQ has neither the necessary staff nor the resources to meet its research objectives. However, in spite of these limitations, its role in receiving and advising on EISs is important because its recommendations are made directly to the federal agencies making the EIS.

THE IMPACT OF NEPA

NEPA has had a large impact on federal project planning because the law specifically requires EISs for all "major" projects having a "significant" environmental impact. As of March 1974, 5108 EISs had been filed either in completed or in draft form. There have been, however, a number of practical problems associated with filing EISs, both from the public point of view as well as from that of the agencies involved in filing them. For example, for the public there is a flaw in NEPA since the definition of what is "major" and "significant" is determined by the very agency filing the EIS. The agency could choose to avoid the whole problem of filing the statement by declaring that the project in question is not major and has no significant environmental impact. However, such a position can be challenged, and the courts can order an EIS if there is an appeal by individual citizens or by environmental groups; in fact, agencies have filed a report if there was any doubt about it. Because the Army Corps of Engineers and the Department of Transportation have filed EISs on nearly all their projects, no matter how minor, their reports form the majority of those filed. The difficulties encountered by the agencies filing the EISs are basically twofold. For one thing, since there is no single uniform set of guidelines or checklists, agencies involved in producing these reports can be in a difficult position when possible impacts are either known only in general terms or are completely unknown. Under these circumstances, the construction of viable project alternatives may be impossible. In practice, the problem of what to include in an EIS usually has been solved by having the statement prepared by one of the literally hundreds of firms specializing in EISs (an interesting employment benefit of NEPA); or by doing the report within the agency, using guidelines prepared by CEQ, EPA, or another agency; or by following the mandate of the courts.

The last alternative has been very important. Second, it must be remembered that even when the EIS has been prepared and filed, the agency may not be finished with it; the report is always subject to critical review and may be returned for additional work before its final acceptance.

NEPA and the governmental reorganization that followed it have established a new bureaucracy with an environmental mission beyond that of any previous time. However, NEPA itself mandates only that agencies (and private firms utilizing substantial federal funds in the development of environmentally significant projects) provide information about potential environmental impacts in their reports. There is no guarantee that the agencies will use this information in their decision-making, apart from the fact that they know outside interest groups may challenge them if they do not. Thus if the EIS is unfavorable, this does not necessarily mean either that the project will be canceled or that a more desirable approach will be selected. Despite the fact that NEPA does not mandate agency policy-making, through NEPA a structure has finally been established that will allow at least some public input into a governmental agency's decision-making process. The EIS is a specific document that demands special attention; if the requirements are neglected or slighted, there is a strong possibility that the agency will be publicly embarrassed and have its project delayed. Therefore, in spite of its apparent weaknesses, NEPA has proved itself surprisingly effective, and perhaps because of this, several attempts have been made recently to weaken it. NEPA requirements have often imposed project delays which opponents feel are intolerable, and because of this, there have been moves to circumvent NEPA in some cases. For example, some recent projects deemed by Congress to be "in the public interest," such as the Trans-Alaska Pipeline (Figure 19-3),

FIGURE 19-3 *One of the valve stations of the Trans-Alaska Pipeline. Construction is going on in the background.* (Courtesy of Ellis Herwig from Stock, Boston.)

have been exempted by an act of Congress from further public objections that are made on purely environmental grounds. In other cases Congress has bypassed NEPA simply by writing provisions into various bills that either exempt certain projects from NEPA requirements or state that any existing (favorable) EISs are accepted and not subject to further challenge. So far, however, attempts to overturn NEPA entirely (and specifically to nullify requirements relating to EISs) have not been successful, although complaints about the delays in completing public projects have increased.

The courts have made the most important decisions strengthening NEPA. Concern in district and circuit courts about procedural problems culminated in the "Calvert Cliffs" decision (*Calvert Cliffs Coordinating Committee v. U.S. Atomic Energy Commission*), in which the court ruled that the Atomic Energy Commission had not prepared an adequate EIS because it had neither considered the full range of alternative plans nor outlined its reasoning clearly enough to permit outsiders to evaluate its conclusions. Furthermore, the circuit court stipulated that, in order to be acceptable, EISs would have to be comprehensive, highly rational documents and contain sufficient technically detailed information to allow those reading it to evaluate the project. In doing this, the court put the burden of proof directly upon the agency—it now must prove that environmental impacts are spelled out, rather than opponents' having to prove detrimental impacts.

The second decision having a major impact was the Wallisville (Texas) decision, *Sierra Club v. Froehlke*. In this decision the South Texas District Court ruled that the EIS for the Wallisville Dam Project was inadequate because it was cast as having purely local impacts, whereas the court felt the project, which was part of the Trinity River Project, could only meet NEPA requirements if the EIS for the whole Trinity River Project were filed. Therefore they stayed construction of the Wallisville project until such an EIS could be filed or until the Corps of Engineers could show that the Wallisville project would indeed have purely local effects. This decision reaffirmed not only that the burden of proof lay upon the agency but also that the EIS would have to be technically detailed and rational. In addition, it broadened the meaning of comprehensiveness to the point where the courts now seem to be entering the policy-making arena. These decisions appear to put NEPA requirements into the center of agency decision-making and far beyond the mere gathering of information, because with all these stipulations to contend with, agencies are likely to eliminate dubious projects even before formal consideration is made and to abandon others if the prospects for long delays seem great.

THE JUDICIAL ALTERNATIVE

Many conservation organizations argue that there is little reason to have much faith in the legislative process. In their view, the process is slowed by its requirement for overwhelming majority support on any issue and weakened by its dependence on administrative discretion in the implementation of legislation once it has been passed. Legislation can thus be an exercise in futility, especially when an administrative agency has already been captured by its clientele and will do as little as possible to enforce the interest of the law. How, the disbelievers maintain, can these bankrupt solutions stop environmental degradation?

The ideas presented in Joseph Sax's book, *Defending the Environment* (1971), and the arguments of Victor Yannacone (1971) both champion the courts as the principal legal recourse for environmentalists. Both authors are extremely critical of administrators, and both document their beliefs. Sax concentrates upon what he calls the "insider's perspective," an attitude among administrators that results from a concern with their agency's image and a pronounced willingness to cooperate and compromise in order to achieve a favorable image—an attitude which Sax believes can destroy the power of any law. Yannacone's ideas are similar, differing only in that he chooses more colorful and more critical adjectives. He bluntly states that administrative agencies are "immune from criticism or public action, self-perpetuating, self-sufficient, self-serving, effectively insulated from the people and responsible to no one but themselves." In arguing for a primary role for the courts, Sax and Yannacone again follow similar reasoning. Both point to two major advantages of the courts: (1) direct public access by anyone with a grievance, and (2) a tradition of fairness and impartiality in handing down decisions. Sax argues that the insulation of judges from influence—both because they are concerned only with the law and because they usually are not trained in fields of substance in the case—protects them from the "insider's perspective." Yannacone emphasizes the advantages of the power of the courts to issue summonses and the court's respect for equity under the law. Under the rubric "use of one's property so as to not injure the property of another," Yannacone proposes the use of the judicial process to make common property resources, namely air and water quality, the de facto property of individuals and to use this precedent to prevent degradation of these resources. By his reasoning, since it is illegal to use one's property to injure that of another, it would also be illegal to use one's property to degrade air and water quality, since they also belong to other (indeed, to all) individuals. Yet it is difficult to see how Yannacone's guiding principle can be applied fairly. Since the use of any property probably will injure the property of some other individual, every action would be the basis for possible litigation by some disgruntled individual, and this would delay or possibly prevent some generally favorable activities with minimal environmental effects.

In general, however, Sax and Yannacone present some impressive arguments to support the judicial alternative, although several problems must be faced. First, there is the problem of expense. The judicial process is very expensive, and environmental litigation is no cheaper than any other. Second, the courts have demonstrated an unwillingness to set sweeping precedents in cases involving environmental litigation. For example, in cases involving NEPA, the courts have carefully pointed out that their judgments are specific to the case at hand. Third, litigation over controversial issues can be a long, drawn-out process; nevertheless, as the Sierra Club has found in the case of Mineral King, the slowness of the judicial system can be an important factor in itself. Since idle money often loses value because of inflation, any sizable delays may be as effective a deterrent to a project as a quick judicial decision against the project.

Solutions Beyond Present Mechanisms

Some authors see no hope for a workable solution to environmental protection short of a complete political revision. Let us consider a few such suggestions.

FEDERAL PLANNING: THE FOURTH BRANCH OF GOVERNMENT

The Center for the Study of Democratic Institutions in Santa Barbara, California, is a prestigious institution that concentrates upon sociopolitical innovations. From this institution has come a series of suggested drafts updating the Constitution of the United States. Authored by Rexford G. Tugwell, one of the principal advisors to President Franklin Roosevelt, these drafts are based upon the following premises:

1. That the three existing branches of our government have evolved to the point that they can no longer give open access to the people; that is, they are no longer democratic institutions.
2. The system of checks and balances works well *within* the governmental system, but it makes rapid responses to problems requiring the attention of all three branches nearly impossible.
3. The powers of the three branches of government were never defined well enough in the Constitution. Hence what powers each branch claims today result from free interpretations of the Constitution well beyond those originally granted.
4. The amendment process cannot be used to correct constitutional faults because it is part of the faulty system and therefore cannot be expected to reform it effectively.

Tugwell proposed the incorporation of a fourth branch of government into our present system. He argued that the fourth branch, the planning branch, would solve problems that result from the slow response and lack of interagency coordination in our present system. In his emerging constitution, version 36, Tugwell proposed that the planning branch be in charge of budget preparation pursuant to policies set up in six-year plans; that a chief administrator be appointed on the recommendation of professional planners; and that the six-year plan (with specific goals) and a twelve-year plan (a more general assessment of future possibilities) be constructed consistent with the needs of the republics (replacements for the states) and of international agencies. These proposals clearly envision a strong planning branch with dictatorial powers. Tugwell specifically mentioned its having control over activities of all planners at lower political levels and over the authority to plan land use, both public and private. It also would have the power to demand information from all public agencies and from private individuals and associations "as are affected with a public interest" related to expansion plans, members employed, and estimates of production and consumption. Furthermore, the planning branch would be the supreme coordinator with responsibility for advancing the quality of life, science, culture and art; for assimilating technological advances into society; and for ameliorating adverse environmental and social impacts.

Tugwell's constitutional reform shows a lot of faith in the concept of a powerful agency able to enforce its will upon government and society. The power to plan and to administer its own plans automatically guarantees the attainment of the goals of fast response and effective coordination, but not without social costs. Tugwell's constitution is, after all, little more than a glorification of the contemporary planning commission, but with an increase in its power. The suspicions that surround actions of local commissions would also be present with the federal planning branch, and the power of the branch is such that it could become the only important branch of government. It is doubtful that most people would concede the advantage of rapid response to emergencies to be worth having a single branch of government with unprecedented control over social and economic processes.

Despite its Orwellian possibilities, great authority may come to be vested in a planning branch of government when a series of resource crises reveal the full extent of interdependencies in social systems (Figure 19-4). The human desire for security could have precedent over the more esoteric benefits of individual freedom, as they do in several underdeveloped societies today. Moreover, the concept of a planning branch for the federal government has appeal for those who reject the thesis that free markets can deal with problems of environmental quality. Individuals who feel this way believe that a solution such as Tugwell's represents a chance to develop an ecological mandate within a branch of government powerful enough to enforce it.

Downs (1972) has reached conclusions that deny anything resembling an ecological mandate. He contrasts the systems analyst's definition of a system with a definition that could be applied to a nonhuman ecological system. He points out that a human system can be constructed only after a set of goals is established, whereas an ecological system does not need explicit goals. A nonhuman ecological system exists not because it accomplishes a set of goals but simply because it is able to survive under a particular set of conditions. Downs uses this difference to examine that contribution ecologists studying nonhuman systems can make to environmental problem-solving. He reasons that ecologists can see only natural solutions to ecological problems which are appropriate to survival of that system; they cannot see solutions for human problems that do not occur in nature. Downs sees this as a severe limitation. He then points out that the process of social development is based on aggressive pursuit of environmental modifications to attain social goals—in his words, by "expanding individual and social choices through transforming nature and society." Since this is at best a rare event in nature, it is obvious that ecology cannot provide the rationale for future choices of developmental alternatives, nor is it a suitable mandate for social goals, apart from the very general one of human survival by preventing environmental destruction.

If Downs is correct and no strategy of ecological systems is relevant for human policy analysis, what other systems can we use? We are forced to return, as Downs points out, to the "compromises and tradeoffs between alternatives—the basic stuff of economics and politics that operates in all social systems." Tugwell's planning branch is but one alternative proposal in this area and, like any other alternative, it ultimately must be measured by its ability to help determine the best trade-offs and compromises in comparison to the performance of any other proposal. Ecology cannot provide the critical yardstick in lieu of economics and politics.

THE ECOLOGICAL MORALITY

Even if we accept the idea that planning mandates must flow from socioeconomic forces rather than from ecology, we still might draw a distinction between the planning process and the impact of an ecological approach upon the establishment of human moral values. Therefore, even though ecology cannot provide the specifics, it can, through a change in attitudes toward the environment and its preservation, modify the priorities upon which the final planning process is based.

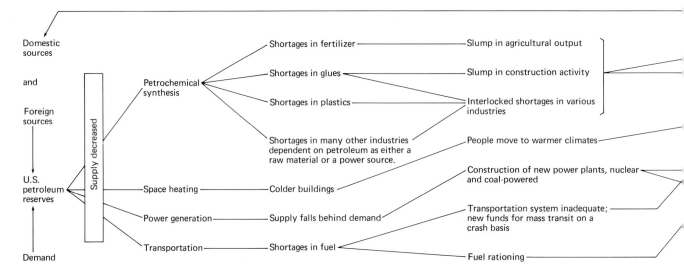

FIGURE 19-4 *An illustration of the interdependencies between social systems and their resources: the energy crisis. This particular system could easily be expanded to fill several more pages. The dependencies of the United States and World economic, political, and environmental systems upon petroleum for energy and as a raw material are so vast that any scarcity has worldwide implications.*

Charles Reich's book *The Greening of America* (1970) is concerned only indirectly with environmental quality, but its thesis is nevertheless applicable. Reich identified three stages of consciousness in American society:

1. *Consciousness I* The pioneer, free-enterprise attitude. This is the traditional, individualistic consciousness.
2. *Consciousness II* The corporate-liberal alliance. This is the dominant consciousness of middle-class America today.
3. *Consciousness III* Liberated youth. This is supposed to be the emerging new consciousness.

Reich sees in the youth-oriented rock and drug culture the future quiet revolution, a revolution involving the revision of values and the complete liberation of self that will end war, create social equity, and clean the environment. Clearly, Reich has no faith in present political and economic institutions (both of which are included in the arch-villain, consciousness II), because they are hopelessly enthralled with goals foreign to consciousness III; the only hope, as Reich sees it, is an entire new philosophy for life. In consciousness III, the people would be free of archaic institutions that cannot meet their needs. It

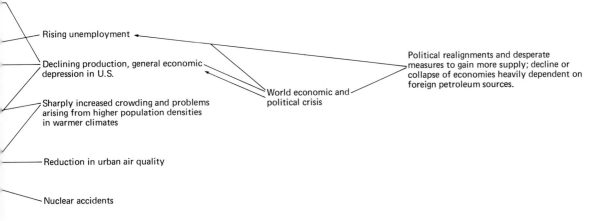

will be characterized by profound rejection of the now-dominant lifestyle and by a rich and expansive freedom of expression; in short, every individual will find satisfaction from inner desires and will not need to brook regimentation.

Reich's book has been criticized roundly in countless reviews for its sometimes muddled expression of ideas, its murky reasoning, and its highly idealized visions of a utopian future. Nor is *The Greening of America* a particularly novel appeal for moral reform. As usual, the problem is the mechanism of implementation, just as it is for the appeal of Garrett Hardin (1968) in his discussion of "the tragedy of the commons." Nevertheless, increasing frustration with the inability of market mechanisms and government to solve environmental problems will inevitably be accompanied by more radical proposals—including the inevitable appeals for moral reform.

In *Ecological Morality* (1972), Bruce Allsop has outlined an argument for a new ethic specifically applied to environmental problem-solving. His basic premise is simple: We compose but a single species that must fit into the global ecosystem—we cannot form an independent system of our own. Like Reich, Allsop displays little faith either in contemporary morality or in contemporary political and economic systems, whether free-market or socialist. His denunciations of the present are strongly worded and blunt, as Allsop feels they must be, because he contends the situation has been made so hopeless by our present morality that an entirely new morality is needed to compel us to adjust to environmental limits. Allsop disposes of all the problems of implementation with a pair of assertions: (1) He maintains that there is such a powerful imperative in the ecological morality that no barrier can be allowed to be insurmountable; (2) he believes that the political and economic mechanisms needed will naturally follow the adoption of the new ethic. Allsop simply says that such institutions will be created because they will have to be created.

Of course, the creation of new moralities is at best a difficult process and certainly cannot be instituted as easily as Allsop suggests; his prescriptions are no more satisfactory than the visionary speculations of Reich. Nevertheless, both authors' writings contain a valuable lesson, albeit a negative one: The frustrations with the imperfections and inadequacies of existing social systems, and the proposed solutions (involving such actions as revolution, calls for moral reformation, or just dropping out), do not automatically suggest a viable means for dealing with problems if we refuse to recognize the political realities relating to the distribution and use of power. The facility with which such solutions may be suggested is matched by the difficulty with which they are implemented.

SOCIAL REFORM THROUGH BEHAVIORAL CONTROL

B. F. Skinner has proposed a new program with many radical implications. In his book *Beyond Freedom and Dignity* (1971) Skinner started with observations similar to those made by Allsop—namely, that the survival of human society is at stake—but he did not stop with appeals for a new morality; instead, he provided a means for implementing this new order. He perceived that the mechanisms for a new morality and for new and better adaptations to a world of declining capacity were to be found in behavioral changes.

To justify his position on behavioral adjustments, Skinner first disposes of the concepts of "freedom" and "dignity." He claims that individuals cannot be said to have either freedom or individual dignity as absolute entities, since both imply predominant internal motivation, and that the social environment is the primary determinant of what these concepts mean through its influence on individual behavior. If freedom and dignity are relative terms, there is no reason why we cannot apply the best of behavioral science to recondition people so that their responses are adaptive to a world requiring less material and energy use and less reproduction, without sacrificing anything real in terms of true individual freedom or dignity. In fact, it would be just the reverse, because behavior which leads to continued excess reproduction will insure eventually that there will be little freedom or dignity left to enjoy; whereas restraint now will preserve at least the freedom and dignity which would be constrained by crowding.

Skinner's plan for shaping the future by behavioral means is built upon his previous work in the training of nonhuman subjects. His method, known as *operant conditioning*, involves the positive reinforcement (by using a reward of some kind) of desirable behavior. The method depends upon immediate application of reinforcement after desirable behavior in order to allow a subject to properly associate them. For example, Skinner has shown that small food rewards can be used to train responses in animals and in human babies; subjects soon learn which are "proper" responses (that is, which will result in a reward). The crux of operant conditioning is that these rather simple responses can be extended into complex behaviors, where one behavior serves to reinforce another, and that new stimuli can be introduced along the way and reinforcement can be used to shape the behavioral response. With the proper social conditions, it is obvious that operant conditioning can be used on the full range of human subjects and their social behaviors.

The powerful implications of Skinner's proposal are obvious and have been explored thoroughly by legions of critics and supporters. The controversy has been so intense that there is even substantial disagreement on exactly what Skinner wishes to accomplish, because specific meanings of his proposal are interpreted differently. Detailed examination of the arguments for and against Skinner's thesis are out of place here (see *The Center Magazine,* 1972, and countless reviews of his book). Nevertheless, Skinner has raised a valid proposal complete with a mechanism for implementation and arguments to justify the need for its adoption; his ideas have served as a platform for debate, and as *The Center Magazine* notes, Skinner "may be the most controversial scholar in the world." In the end, each person must decide whether or not freedom and dignity are really as relative as Skinner suggests, whether or not extensive behavioral conditioning is worth the social cost (whether a world of controlled behavior is better than a more crowded, chaotic one), and finally who would serve as the number 1 conditioner. For each of these questions Skinner proposes an answer—and the answers are themselves controversial.

A Footnote, an Opinion, and a Question or Two

The appearance of a literature in which the human role in nature is questioned heralds the approach of a new epoch in our history, an epoch that will see an end to our continual unquestioned expansion. Inevitably, this must mean new social institutions based on the steady-state. The difficulties we can anticipate in the transition period are formidable, and we certainly can expect disruptions and upheavals in all societies as a result. As we gain more knowledge and more experience in bringing about changes in economic and political systems, and as the importance of environmental quality is defined more clearly, to all people, it remains to be seen how important "environment" will be when the sacrifices are obvious. As the conflicts between human social systems and natural systems intensify (as they must), we can predict only more uncertainty. How will Skinner's proposal look then? Will entirely new social systems, as Reich suggests, be needed? Will NEPA be strengthened, or reduced to impotence?

If we indeed are entering the epoch of an end to growth, all we can be sure of is change, change which occurs so quickly that the darkest portrayals of Allsop or Reich or Skinner will prove true, to our despair. Behind all this remains the question of our role in the world. This is the question that ultimately will determine our future course, and we shall examine it in the final chapter.

Questions

REVIEW QUESTIONS

1. Contrast the goals of the natural economy and the human economy.

2. List the advantages of a judicial solution and of an administrative solution to an environmental problem.

3. For what reasons is the transition into a steady-state economy likely to be difficult for most of us?

4. What is the National Environmental Policy Act likely to accomplish? What is it unlikely to accomplish?

ADVANCED QUESTIONS

1. Skinner refers to the literature of social regulation of population size and its effect upon individuals. (Individuals often suffer for the good of the population—see Watson and Moss in *Behavior and Environment*, Esser, ed., 1971 for one example pertaining to red grouse.)

Try to construct arguments first to support Skinner and then to refute his argument that social regulation in animals is a model for human beings.

2. Compare the literature of social regulation with Skinner's proposal. Is his operant conditioning similar to the mechanisms of social regulation known for animals?
Here are some starters:
Watson and Moss, 1971
Wynne-Edwards, *Science*, 147 (1965), 1543
McLaren, *Ecology*, 48 (1967), 104
Williams, *Group Selection*. Atherton-Aldine, Chicago, 1966

3. Contrast the views of Downs (1972) and Allsop (1972) concerning ecology as a policy science. How do these views differ? Who do you think is more nearly correct?

Further Readings

Note: New developments in social, political, and economic mechanisms applicable to environmental problems have been occurring at such a rapid pace that it is impossible to cover them in this chapter. Readers (who are not economists or political scientists already familiar with the sources of information) are urged to consult various journals listed in the introduction to this section, particularly The 102 Monitor, The Environmental Reporter, *and various organization publications.*

Allsop, B. *Ecological Morality*. Muller, London, 1972.

Anderson, F. R. *NEPA in the Courts*. Resources for the Future, Johns Hopkins, Baltimore, 1973.

Even though this book bears a 1973 copyright, it is already out of date since the courts have been taking a very aggressive stance toward the requirements of NEPA. Nevertheless, it is useful for readers who are inclined toward detailed legal analysis.

Ayres, R. U., and A. V. Kneese. "Economic and Ecological Effects of a Stationary Economy." *Annual Review of Ecology and Systematics*, 2 (1971), 1-22.

One of the best, most thoughtful contributions on the impact of steady-state economics.

Barkley, P. W., and D. W. Seckler. *Economic Growth and Environmental Decay*. Harcourt, Brace, Jovanovich, New York, 1972.

Excellent introduction to economic theory relevant to environment.

Boulding, K. E. "Economics and Ecology." In *Future Environments of North America* (Darling and Milton, eds.), pp. 225-234, 1966a.

This paper is good for stimulating critical thinking. Boulding offers a number of ideas that will have variable merit to different readers. Be sure to read the discussions on pp. 289-314 and 390-409.

Boulding, K. E. "The Economics of the Coming Spaceship Earth." In *Environmental Quality in a Growing Economy* (H. Jarrett, ed.). Johns Hopkins, Baltimore, 1966b.

Caldwell, L. K. "Environment and Administration: The Politics of Ecology." In *Environment* (W. W. Murdoch, ed.). Sinauer Associates, Stamford, Conn., 1971.

A recent, fairly detailed paper by a famous author on his specialty, Caldwell provides an effective counterargument to Sax (1971) and Yannacone (1971). His bibliography is extensive.

Calvert Cliffs Coordinating Committee v. Atomic Energy Commission, 449 Federal 2nd District 1109, 1118-1119. District of Columbia Circuit, 1971.

Center for the Study of Democratic Institutions. "Beyond Freedom and Dignity: The Skinnerian Challenge." *The Center Magazine*, 5, No. 2 (1972), 33-65.

Provided that you *also* read the original book, this collection of pro and con views by Platt, Black, Toynbee, and Skinner himself will prove to be a valuable supplement.

Crowe, B. "The Tragedy of the Commons Revisited." *Science*, 166 (1969), 1103-1107.

An answer to Hardin (1968), but Crowe's thinking is also applicable to Allsop (1972) and to Reich (1970).

Daly, H. E. "Toward a Stationary State Economy." Chapter 14 in *Patient Earth* (Harte and Socolow, eds.). Holt, Rinehart, Winston, 1971.

Daly, H. E. (ed.). *Toward a Steady-state Economy.* Freeman, San Francisco, 1973.

Many ideas in this chapter were drawn from Daly's writings in his long introduction to this volume and his paper "The Steady-state Economy," also here. Boulding's "spaceship" paper is also reprinted in this volume.

Downs, A. "The Political Economy of Improving Our Environment." In *The Political Economy of Environmental Control* (Downs, Kneese, Ogden, and Perloff, eds.). Institute of Business and Economic Research, University of California, Berkeley, 1972.

Ehrlich, P. R., and J. P. Holdren. "Impact of Population Growth." *Science*, 171 (1971), 1212-1217.

Goldman, M. I. *The Spoils of Progress.* M.I.T. Press, Cambridge, Mass., 1972.

A short precis of this book was published in *Science*, 170 (1970) 37-42.

Hardin, G. "The Tragedy of the Commons." *Science*, 162 (1968), 1243-1248.

Heilbroner, R. L. *An Inquiry into the Human Prospect.* Norton, New York, 1974.

Holling, C. S., and M. A. Goldberg. "Ecology and Planning." *Journal of the American Institute of Planners*, 37 (1971), 221-230.

It is probably impossible for anyone to write a single definitive treatise on what applied ecology should be, but this paper and Holling and Orians (see Chapter 17) form an acceptable core.

Olsen, M., and H. W. Landsberg (eds.). *The No-growth Society.* Norton, New York, 1973. (Reprinted from *Daedalus*, Fall 1973.)

See particularly the paper by W. R. Johnson, "Should the Poor Buy No-growth?"

Reich, C. A. *The Greening of America.* Random House, New York, 1970.

Sax, J. L. *Defending the Environment.* Knopf, New York, 1971.

Sierra Club v. Froehlke, Decision No. 71-H-983 (Southern District of Texas, 16 February 1973). Reprinted in *The Environmental Law Reporter,* 3: 20249–20286.

Skinner, B. F. *Beyond Freedom and Dignity.* Knopf, New York, 1971.

Tugwell, R. G. "The Emerging Constitution, Version 36." Memoranda of the Center for the Study of Democratic Institutions. (Mimeo.; P.O. Box 4068, Santa Barbara, Calif. 93103). 1970.

This paper may be difficult to get, since it was intended for internal circulation only.

U.S. Council of Environmental Quality. *Environmental Quality: 1972.* Washington, 1972.

Appendix B contains the language of NEPA.

Yannacone, V. J. "Environment and the Law." In *Environment* (W. W. Murdoch, ed.). Sinauer Associates, Stamford, Conn., 1971.

20. Ecological Systems and Environmental Strategies

We at last have reached the point of final synthesis. Having considered various aspects of human society and environmental quality, we are ready to consider the total system and the kinds of problems we, as a part of this system, are likely to encounter in the future. We are hindered by a dearth of data in many areas; little is known about the behavior of the total system. Consequently, any investigations into possible future events must involve speculation. Thus I must begin this chapter by admitting that its arguments constitute a personal viewpoint based substantially upon fact but also, in places, drawing upon speculation and personal values. But this is not to be interpreted as an apology; rather, it is a challenge and an invitation to think independently. Although you may disagree with the personal conclusions here and may label this chapter as controversial or even as totally wrong, if it stimulates you to create something better, my aims will be fulfilled.

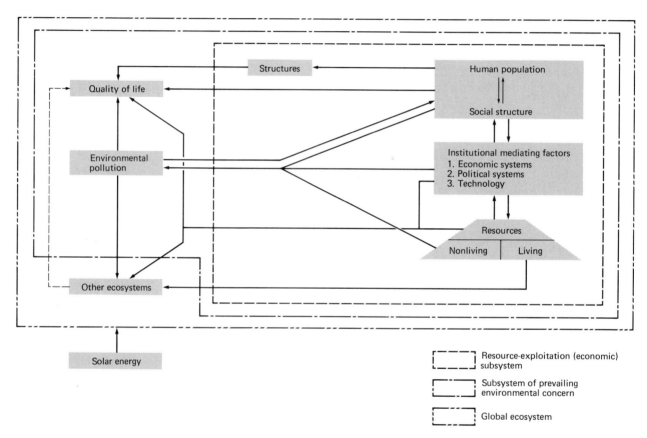

FIGURE 20-1 A highly aggregated view of the structure of the global ecosystem. Note the major subsystems keyed in the lower right corner. The organization of this world model roughly follows the organization of the Forrester-Meadows efforts, but departs from them in that it adds "other ecosystems" which normally fall outside the arena of human decision-making. The flows (arrows) connecting various components are conventions which mean that one element affects another without specifying whether the flow is information, energy, material, or some combination of the three. Furthermore, the flows are intended to represent only the major pathways. Certainly not all the known or probable interconnections that could exist have been included; it would be impossible to do this without totally destroying the readability of this diagram. It must be stressed, however, that this does not mean that such relationships are lacking. It would be a useful exercise to disaggregate this model, construct trial flows between selected components, identify those specific examples which are well known, and hypothesize what others might exist.

The Human Being–Environment System: A Synopsis of Earlier Chapters

Perhaps the best way to synthesize the man-environment system is to combine all previous discussions into one block diagram that shows their interrelationships. To do this meaningfully, we shall start with one of Barry Commoner's environmental axioms from *The Closing Circle* (1971): "Everything is connected to everything else." While we recognize the truth of this axiom, it is impossible to operationalize it in a diagram having more than a few components without totally obscuring our educational goals. Hence Commoner's axiom is most useful in a conceptual way, and we must restrict the interactions portrayed in a diagram of the global ecosystem to a few well-known clusters of effects. In addition, we shall treat the man-environment system as though the two components were truly separate, even though one might argue that such a distinction is unrealistic. Nevertheless, for our purposes, it is a good working assumption with which to begin.

In Chapters 4 and 5 (Section B), we examined the phenomenon of human population growth, its roots in cultural history and human psychology, its ramifying consequences for a wide range of environmental problems, and some alternative means to end the long history of expansion. There can be no doubt that the dynamics of the world population cause problems in the man-environment system, and so we start with this aspect as a fundamental block in our system (Figure 20-1). There are direct effects of population numbers on various elements of the system (for example, human wastes on water pollution, land conversion to urban and agricultural usages); but more frequently, instead of being direct effects, the population-related problems that arise result from a set of mediating factors—mechanisms which govern the use of resources and the degradation of environmental quality by the population (Chapters 11 to 14, 16 and 17). The model at this point is the sort of human-oriented system labeled in Figure 20-1 as the "subsystem of prevailing environmental concern," lacking only the addition of other living systems (Chapters 9, 15, and 16) to provide the outlines of a more comprehensive global model. Merely specifying the major parts of the global ecosystem is not very helpful, however, because this is the easier part of system development. The clusters of rates still must be identified and quantified, and in Figure 20-1 this process is barely begun. In addition, we must remember that any highly aggregated model can be broken down into further detail, which in turn requires respecification of the appropriate rate functions. This process is the major challenge, and it is here that extensive research is needed most.

This is the "rational" or scientific approach toward the optimal use of a model, but more often than not there are significant limitations to how extensively we can utilize this method. Let us assume that within the time and budgetary limitations we face, the purely scientific approach is impractical. How do we build upon the first step and work toward a comprehensive and useful model when such limitations have been imposed upon us? To elaborate upon the "everything is connected" theme is useful to a point, but that point is the relatively low-order one of a general awareness of this fact. We must question the utility of pursuing the general model further without replacing all the speculative and tentative relationships with something more concrete, for when the number of alternative possibilities is large, we can have little faith in the model until we improve our understanding of it.

Although data gathering requires time, we still can use the model immediately if we pursue goals consistent with the information we have. Depending upon how much we know, we can devise a strategy to obtain maximum benefit from what we have at the same time we pay attention to the risks of misuse (Figure 20-2). At one extreme—where a great deal of information is known—there would be essentially no limitations on how we could use the model, because we would have enough data to understand the system's structure and function; thus we could evaluate the impacts of various changes on the system and make confident predictions about the future. Unfortunately, almost no subsystems are known in such detail, and so we must scale our objectives accordingly. At the other extreme—where little information is available—any model would be largely speculative, would not be supported strongly by empirical evidence, and accordingly would deserve less confidence. Consequently, the use of a model such as this would emphasize improved data collection, general behavior of the system, and perhaps some very limited policy evaluation. However, these are not firm rules that dictate certain responses determined solely by the information level available in the model; in fact, each model must be judged individually on its own merits. Even the worst possible case—where a model is structurally inaccurate but correctly predicts the behavior of the system under certain conditions—might very well be useful as long as conditions do not change enough to expose the structural inadequacies. Thus, although potential problems arise from the misuse of fairly general models, they still have the potential to contribute to environmental problem-solving as long as we recognize their limitations.

Some Problems Raised by an Analysis of the Model

Throughout this book we examined some known environmental consequences associated with such facets of the human system as kinds of resource utilization, human population growth, and certain human institutions. Earlier in this chapter, we began to relate these elements, thus starting a post hoc synthesis of many diverse relationships. While there is clear benefit in doing this, we should not be satisfied only with looking backward, because we also must be concerned with anticipating problems before they are manifest, while we still have an opportunity to ameliorate them. When we examine our model with this concern in mind, certain potential problems are relatively obvious; for example, we may cite those associated with continuing world population growth, declining resource supplies, conflicts between economic welfare and ecological welfare, growing volumes of wastes that must be disposed of (especially radioactive substances), and unsatisfactory urban environment. You may find it profitable to make a systematic reexamination of earlier chapters concerned with these issues in light of the global model.

Perhaps the most important observation we can make here is that most of the above problems have reached the point where they at least are perceived to be problems, although solutions may not yet be evident. However, one problem has *not* penetrated human consciousness, and it has to do with our attitude toward nature and our perception of how we and nature should relate. This, in turn, stems partially from our perception of the meaning of "environment." The vagueness of this term has permitted a large variety of problems to be addressed under a single all-inclusive label. Everyone, for example, is in

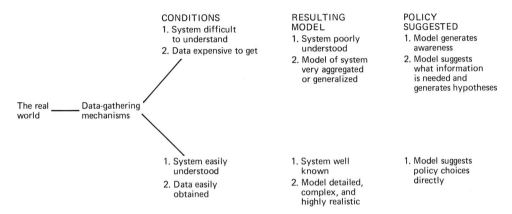

FIGURE 20-2 *The two extremes in what is actually a gradient related to the degree of knowledge available for a given system of interest, and the policy emphasis that results from the decisions based on the strengths and limitations of each extreme. This diagram actually represents a kind of rational decision-making model. It is important to remember that in the real world the absence of specific information does not always prevent highly specific decision-making, though this is perhaps not the preferable state of affairs. Thus, in reality, decisions are frequently based on some combination of generalized technical information and normative data.*

favor of a "better environment"; beyond this assertion, however, there is little agreement either on how to go about obtaining a better environment or on what constitutes a worthwhile cost toward this end. We have seen that widely different views about the meaning of "environment" lead to divergent meanings for the term "environmental pollution" and thus for the policies appropriate to the solution of environmental pollution problems. Four prevailing views of the meaning of environment are set out below.

1. *Environment as a tool of economic imperialism* The leaders of many less developed countries and of minorities in the United States view environmental concern as an excuse for suppressing their economic development; they see it, therefore, as a tool to promote the continuation of social inequities. This position implies to these persons, at least, that present environmental policies either are too costly or are unneeded compared with the toll they extract from social welfare. In recent months, anti-environment groups have been organized in the United States based upon some variant of the theme "sense about environment." And as the economic impact of steps taken to improve environmental quality continues to

increase, the increase in these kinds of anti-environmental groups results from the same kinds of suspicions that cause birth control programs to be labeled "racist," or international programs concerned with environmental quality "imperialistic." For a larger exposition of this meaning, and of the next one as well, see Neuhaus (1971).

2. *Environmental protection as an additional justification for other goals* In the relatively narrow context of middle-class America, environmental objectives are regarded as genuine in their own right. But as we have seen, there can still be much variation in the meaning of and in the support for environmental protection. Indeed, since it is difficult to conceive of any human activity that does not have some environmental effect, it is likewise difficult to conceive of a policy change that would not have some impact on the environment; perhaps because of this, the environmental issue has become a convenient vehicle for advocates of such diverse policies as those covering land-use control, large public works, the need for state-level planning, advertising control, and building-design reviews. Whereas some of these issues actually have a marked impact on the environment, it must be remembered that others fall into the category of aesthetics, another poorly defined subject. The danger is that the meaning of "environment" can be so diluted by diverse approaches that it no longer will be regarded as a legitimate concern for any issue.

3. *Environment as a resource: classical conservation* For many decades a conservationist was a person concerned with having adequate resources for future generations. Simply by defining "resource" as anything needed for human existence, one can automatically obtain a conservation philosophy that is recast in more comprehensive terms. It follows that the strong human orientation that is characteristic of the philosophy of conservation will also manifest itself when the environment is included as a resource. For example, EPA proposals in pursuit of Clean Air Act standards are based on the concept of the air basin; the air of any given basin is viewed as a high-quality resource to be protected by setting limits on the maximum concentration of various chemical substances therein. The ambient air standards themselves, however, were established on the basis of medical evidence of human health effects from air pollution. The direct threat of air pollution to human health and happiness was the only evidence deemed important enough to stimulate the tough air-quality standards of the Clean Air Act with all their powerful economic implications.

4. *Environment as an integral part of the system: the new conservation* This position is basically similar to the last, except that human interests are not accorded such overwhelming importance. In this context, "environment" connotes the performance and the structure of all of nature; there is an implied optimal state of harmony between human beings and nature. Insofar as this raises the question of our place in nature, it also implies a different strategy for development. For example, the continuous-growth economy is predicated upon improvement of the economic lot of all people, but it practically guarantees that short-term economic benefits will receive the highest priority and that natural processes with poorly known, long-term economic benefits will get the lowest. But if there were real concern for greater harmony between human beings and nature, then there would be greater interest in a shift to a steady-state economy

that, by stabilizing or reducing the level of economic activity, would limit our impact on the planet, depending upon the level of activity and the restrictions of the steady-state and its accompanying political institutions.

Abundant evidence shows that the first two meanings of environment are held widely, and the third is gaining recognition. There is not much evidence, however, that the fourth definition—"the new conservation"—has won much public acceptance. Nevertheless, this fresh approach is germane to the problem of environmental quality because the underlying question we must answer is what the future relationship between us and our environment will be, and the answer largely rests with us. This basic question will be evident throughout discussion of the six issues identified below.

THE POPULATION-RESOURCE SPIRAL

Little further detail is needed here. The simple facts are (1) that population growth has inertia, which means it is difficult to stop in a short time, and (2) that each increment of growth requires an accompanying increment in the growth of our resource supply. A major consequence of these two well-documented facts is that we can find ourselves simultaneously with resource shortages and with a population which continues to grow despite our best efforts to prevent it. If our response is not timely, we can expect a long period during which we shall have to adjust rapidly dwindling resources per capita. For instance, how will the remaining limited resources be allocated among the human populations? What criteria will determine this? The political and social ramifications alone are mind-boggling. A second example might concern the impact of the increasing demand for resources upon the environment. These familiar issues were treated in Chapters 4 and 5 and peripherally in later chapters, but they are sufficiently important to revive here because the population-resource spiral, if it continues, will force us to make many difficult choices including those concerned with such elemental human values as individual freedom.

ENVIRONMENTAL PROTECTION COSTS AND THE PROBLEM OF EQUITY

All policies, environmental and otherwise, result in some mixture of benefits and costs, and any policy change will alter the distribution of these. Construction of benefit-cost ratios, preparation of environmental impact statements, and analysis of economic impacts are all examples of attempts to accurately express the relative costs and benefits of any given project or policy. Elements of several earlier chapters were aimed at the question of balancing the relative costs and benefits of possible actions directed at solving problems with environmental effects. Examples include discussions of the costs and benefits of halting population growth (Chapter 5), combatting air and water pollution (Chapters 11 and 12), designing more effective urban environments (Chapter 17), and evolving toward a steady-state economic system (Chapter 19).

Obviously no one has all the answers, and few expect the answers to miraculously or instantly reveal themselves. However, recognition that environmental protection is a legitimate and important concern has become sufficiently well established already so that some people feel their economic interests are being subverted to those of environmental protection. As environmental and economic interests enter increasing conflict, it is reasonable to expect increasing attention for questions of equity—specifically for the question of who pays the costs and who gets the benefits from policies aimed at environmental protection. Already signs of such attention have been signaled by suggestions that proposals directed toward environmental protection be subject to "economic impact reports," an interesting situation, since environmental protection originally arose in response to economically motivated misuse of the environment.

ECONOMIC AND ENVIRONMENTAL
TRADE-OFFS

When resources are plentiful and the majority of the populace feels economically secure, it is easy to hold a generous view toward environmental protection, especially since most of the environmental protection that is implemented also benefits us (that is, improved air quality results in reduced respiratory problems for the human population). But as resources become more limited, there are no assurances that our systems of evaluation will be unaffected. In Sections C and D and in Chapter 19, the question of environmental and economic trade-offs was examined in many forms because the entire question of economic development necessarily involves processes that can result in environmental degradation. Much of environmental economics is concerned with finding suitable ways to balance developmental and preservationist objectives. As an extension of the consideration of economic and environmental trade-offs, we also must consider the problem of population stabilization because it is unlikely that a growing population will be able to stabilize its economic growth. The environmental consequences are likely to be severe unless we can both agree on and achieve population stabilization goals. In this regard, it is profitable to reflect on the biological precedents for something less than maximum population size. Wynne-Edwards's ideas (Chapter 2) come to mind particularly.

INTERNATIONAL ENVIRONMENTAL PROBLEMS

Although a large number of environmental problems are international in scope (population, air pollution, radiation, international resources), we are only beginning to search for effective international mechanisms to deal with them. Certainly the United Nations is a logical first alternative. In June 1972, the United Nations World Conference on the Human Environment convened in Stockholm, Sweden (for a report on the conference, see *Bulletin of the Atomic Scientists*, 28, No. 7, 16–56). This conference was the first attempt to define a global strategy for environmental problem-solving. After more than a week of discussion, the participating nations agreed to a final communique with the following eight points:

1. Establishment of a new Environmental Secretariat to coordinate international environmental activities
2. Establishment of an environmental fund to be financed with voluntary contributions
3. A call for an effective international pact to control offshore dumping of materials of shoreward origin
4. A moratorium on commercial whaling for 10 years
5. Establishment of a network of pollution monitoring stations throughout the world under the auspices of the Earthwatch Program
6. Statements that unique species and areas in the world must be conserved

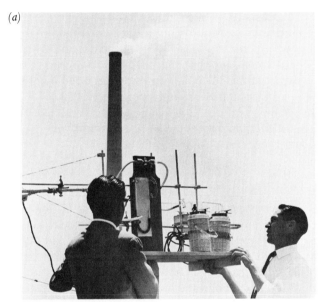 

FIGURE 20-3 *The United Nations Conference on the Human Environment in Stockholm in 1972 developed proposals for international cooperation in environmental control and protection. (a) Under one proposal the Institute of Applied Sciences at the University of Mexico under the direction of E.J.P. Stretta, a UNESCO expert, developed the apparatus show here, which, if installed in factory chimneys, will purify polluted air. (b) Wildlife experts doing an autopsy on an impala at the Moremi Game Reserve in Botswana. FAO experts and others from the Federal Republic of Germany are assisting Botswana in improving wildlife management. ((a) Courtesy of the United Nations. (b) Courtesy of the United Nations/FAO.)*

7. A list of 26 "environmental principles" outlining the responsibility of all member countries to promote international environmental problem-solving
8. A recommendation calling for developed countries to compensate less developed ones for any trade damages that might result from environmental improvement

One might obtain an optimistic view of the proceedings simply by emphasizing that it is a significant step forward even to convene such a meeting, much less to secure an initial platform of guiding principles. However, greater cynicism seems justified, especially if widespread international coordination on environmental policy is to be regarded as a prerequisite to any effective move toward solving worldwide environmental problems (Figure 20-3). The undisguised suspicion of the smaller and less developed nations toward the intentions of the larger and richer ones dominated the spirit of the conference. Many representatives felt the potential costs of environmental improvement would be a threat to their achievement of the highest possible rate of industrial growth, and these fears were responsible for the resolution on indemnity for trade damages (item 8). In more general terms, the Stockholm conference dramatically revealed the magnitude of the problem of getting close international coordination in environmental planning. The surrender of national power to an international agency is a problem that also characterizes much of international law. The apparent paradox is between a country's desire for national sovereignty and national economic self-sufficiency on one hand, and on the other, the requirement for a high degree of authoritarian control in order to meet environmental objectives that often conflict with national economic goals. Suppose, for example, that two countries secured an agreement to cut down a forest located in one of the countries, but that world opinion wanted the forest to stand because it contained a rare tree species and supported the world's only population of a rare animal species which world opinion favored preserving. One country stands to gain income from exporting the timber and from employment for its woodcutters; the other country gets the wood and employment for woodworking industries. Would the two countries be likely to sacrifice their economic benefits in response to world opinion? Would they take the next step and agree to empower an international organization with the authority to dictate that the forest would not be cut down?

If economic growth continues to be important, and if the major measure of successful human endeavor remains at the national level or below, anything beyond limited international cooperation appears out of the question. This observation is supported by the events of even more recent conferences (Bucharest and Rome, both in 1974) where, despite crisis problems of overpopulation and world starvation, international cooperation continues to be weak.

CHALLENGES TO HUMAN MORAL STANDARDS

As we work our way toward a new system of slower economic growth and greater concern for environmental protection, the traditional economic and political institutions that built the system in the first place are increasingly being challenged. These institutions favor economic and social development and are based upon the moral concept that we should pursue whatever policy leads to the greatest human benefit. Consequently, we think of steady-state economies and increased conservation of resources as less desirable than unrestricted development but necessary adjustments to reality and, therefore, ultimately the best alternative for our society.

Suppose, however, that we eventually encounter a situation where the "necessary adjustment" is not obviously our best ultimate alternative, though it is the preferable one for the environment. We then have reached the point where our choice involves the moral foundation upon which society and its institutions are based. One major precept that is likely to be challenged is whether the right of *all* people to greater health and improved safety is more important than environmental protection. To date, this concept has remained essentially unchallenged for several good reasons. Most of us, after all, are fond of living and fear death, and it is perfectly normal to want to live as long as possible, even to the point that we will condone rather extreme measures toward this end. Moreover, if we dare argue that to have better health and greater safety is not a right granted equally to all people, we raise a series of disturbing moral issues (for example, the sticky question of who is to survive and who is not if health and safety must be distributed unequally among the populace—even more important, who will make these decisions). Finally, one basic concept of human society during its long history has been that as individuals we are neither humane nor compassionate if we ignore the plight of others. This is a strong social commitment, long reinforced, that would be difficult to reverse.

These are powerful arguments that, once rejected, could threaten the nature of society. Despite this, it seems improbable that the concept of improved health for all at any cost can remain unchallenged indefinitely. In the environmental context, for example, the methods used to preserve human health and safety often can be extreme and detrimental to other species. Rethinking our present health and safety attitudes could lead to a revolution in our treatment of some natural ecosystems. For example, what if streams were no longer dammed for flood control or if marshes were not drained for mosquito abatement? Even without a reversal of our present human-centered attitude, I think that at least a general movement in this direction is both desirable and inevitable, even if we must face difficult moral and ethical decisions.

Policy Choices

The major choice confronting us involves whether future emphasis will remain upon socioeconomic development or whether it will shift toward environmental protection. This strategy choice emerges from the four meanings of environment that were presented above, because the first three can be derived from emphasis upon development, and the last can be obtained by stressing environmental protection. Although we have only approached the six issues just discussed as problems emerging from a shift to a greater emphasis upon conservation, these problems can only become more critical in the future as the situation is exacerbated by a succession of resource shortages. The 1973 energy crisis has been revealing in this respect (Figure 20-4). The threat of a petroleum shortage absolutely destroyed congressional opposition to the construction of the Trans-Alaska Pipeline. Before the stimulus of the energy crisis, the

FIGURE 20-4 *Scenes like this were the rule during the fuel shortage. Was it a harbinger of the future or will there be alternatives?* (Courtesy of Patricia Hollander Gross from Stock, Boston.)

construction of the pipeline had been stalled for several years. In the same vein, future energy crises may provide the impetus for a wide range of policies with deep environmental implications, such as allowing electric power plants to make greater use of high-sulfur coals by lowering air-quality standards, an increase in the strip mining of coal, more extensive offshore oil exploration and production, and more permissive siting procedures for nuclear power plants. The speed with which these proposals have been made in response to a minor crisis (compared with the specter of much worse) suggests a pessimistic future for the environment. After all, the energy crisis is probably only the beginning of what will be a series of crises in energy, metals, wood, plastics, and food. It will be interesting to see whether environmental quality will receive much attention then.

On the other hand, environmentally beneficial policies might emerge if resource shortages are perceived by society to be so severe and so inescapable that some form of responsible long-term accommodation is necessary; this approach would reject the attitude that the sacrifice of environmental protection is not to be regarded as a wise long-term solution to resource shortages. Thus in spite of limits in energy supply, recent laws have been passed to regulate strip mining more closely than ever, including a complete ban when the surface topography cannot be restored after the coal has been removed. Moreover, strong resistance to lowering air-quality standards and resuming subtidal oil exploration remains, and the prospects of serious attempts to cut energy usage are stronger than ever. Thus, on one hand, we see various indicators that perhaps society is not ready to sacrifice environmental quality at the smallest excuse. Yet in the previous paragraph, several indicators were cited that seemed to suggest just the opposite. That such contradictions should be commonplace is obvious from our study of complex systems. We may construct a series of different policies from what is already a much larger spectrum of possibilities.

PRESERVE PRESENT INSTITUTIONS, ADAPTING WHEN NECESSARY

This strategy is essentially a statement of faith that the power of technology and the strength of our long-evolved socioeconomic and political institutions will permit us to overcome all resource problems with due regard to environmental quality. This strategy values human society above all else and hence is basically accommodative to our society, since our traditions favor growth at the expense of natural environments where necessary and since value is measured only in terms of human good. Because all nations currently employ the same basic tactics, as many different substrategies of this approach may be identified as there are countries; but because of this basic similarity, in the end the effects from any substrategies are likely to be much the same. The human-centered aspects of our society that result in our attitudes about health and safety also favor development, and so it is not surprising that this strategy is so universal at present. This approach is viable only as long as our traditional institutions are able to adapt; but when this is no longer possible, unfortunately there will be an automatic switch to the last choice we shall discuss.

PLANNED TRANSITION INTO A NO-GROWTH SOCIETY

This is the theme of the steady-state economists. The differences between this and the first alternative are actually small (in particular, the basic human-centered orientation is the same), but there are two vital differences:

1. The emphasis upon a planned transition to a no-growth situation in this policy alternative is based upon the belief that our traditional institutions will not be able to deal with resource shortages indefinitely and that eventually our present system will be overwhelmed. Therefore, the wise move would be to plan and begin instituting the necessary revisions before a crisis occurs.

2. The requirement for transitional institutions, then, essentially involves an exchange of short-term social change for long-term system stability.

As in the first strategy, the emphasis is upon resource adequacy. Its weakness is that it is difficult to know exactly when the shift to the planned transition should be started except from a post facto analysis, and this will be of little use to what remains of society then.

REPLACEMENT OF TRADITIONAL INSTITUTIONS

Ecological moralists and advocates of revolution in social institutions insist that present institutions at least assist and, perhaps even worse, cause environmental disruption; hence, they maintain, a radical change in the institutions responsible for these problems is the best lasting solution for environmental protection. In one sense, this strategy differs from the second one only in degree (instead of a gradual change, this strategy is based on the premise that only radical change will work). But in addition, this strategy reflects the idea that our basic orientation to the environment is reversed. Thus the replacement of old, unworkable institutions with new ones could well favor a whole new policy toward the environment. The questions we have raised about the status of human health and safety are examples of this very basic change in attitudes; this strategy assumes that such dramatic institutional changes are possible.

AN INSENSITIVE SOCIETY OUT OF CONTROL

Those who believe significant attitudinal change is difficult to attain think that these responses will probably come too late and that a crisis undoubtedly will be necessary to stimulate appropriate responses. If this occurs, neither we nor the environment would benefit because this "strategy" portends a society out of control—exploitative, and ultimately destructive both of itself and of the global ecosystem. If this is really a strategy, it is a very short-sighted one, one that tells us what might happen if we are so crassly insensitive that we would permit such a situation to develop. We should fervently hope that our personal sense of dignity and our ability to reason will never make it a serious possibility.

SOME PERSONAL PREDICTIONS

The development-conservation dilemma involves essentially everything on this planet, giving us a bewildering variety of alternative choices and different consequences to consider. While it is too much to expect a clear consensus to emerge, it would be unwise simply to default to current standards of behavior without examining the consequences in as much detail as possible, as we have in some previous discussions.

It must be conceded that current patterns of morality will probably not change much. Thus the good of humanity will probably continue to be perceived as the primary objective of any actions that might be taken. For this reason, I believe that strategy choice 1 will dominate as long as there is any choice. Like Robert Heilbroner, I think that the next few decades will be difficult for us all, as a burgeoning world population simultaneously degrades the environment and strains its resources, then painfully adjusts to the realities of its alteration. The end to growth will not come easily—although come it eventually must—either in the population or in economic spheres (or both), and more than one government will commit foolish acts from sheer desperation. This scenario is much like that pictured in *The Limits to Growth,* and it is difficult to deny that such a possibility exists.

On the other hand, I do not believe that we will be foolish enough to let matters reach the point that we will destroy ourselves. After all, human societies are not quiescent entities completely incapable of change and inflexibly bound to tradition. Here I recall the strong faith in humanity that emerges from *Communitas,* and I put forth the hope that, when need be, our essential spirit of self-sacrifice and humanity will permit us to do whatever we must to meet each crisis. The marvelous strength of many patients, who, though terminally ill, are possessed with a calm acceptance of death without bitterness, is one example of the heights to which the human spirit can ascend. There is also ample evidence of the public's willingness to sacrifice for the good of the society. England during the early days of World War II comes immediately to mind. Even the United States has perhaps now begun to diminish its unbridled consumption, planned obsolescence, and rampant waste. In essence, I cannot believe that we are incapable of behaving as we think best for each situation, and thus I remain hopeful that we will seek to adapt to each resource crisis, to maintain a clean environment—and even to accept the consequences with dignity if we should fail.

The one thing I regret as an ecologist is that our preservationist responses will probably remain human-oriented, which certainly does not guarantee that, in the end, much will remain of the natural ecosystems of this world. As more and more coveting eyes are turned toward the Amazon basin, as tropical forests are felled at an increasing rate to supply us with lumber, as species go to extinction at rates unprecedented in history, and as natural areas shrink into smaller and more isolated enclaves, I may not have overwhelming evidence that it is a fatal mistake for us to do this, but at least I have a feeling of strong regret at seeing the earth become a more simplified life-support system directed entirely toward our desires. This thought is just the more depressing because it seems so inevitable.

A Final Perspective on the Tasks that Face Us

I shall end this narrative on a positive note. While many readers will disagree, sometimes violently, with my conclusions and perspectives, we cannot fail to agree that there is a lot more to learn before we can even begin to approach the development of a comprehensive environmental strategy. At the beginning of this book, we set out to examine the problem-solving role of ecology and systems theory in environmental applications, and now at the end we shall draw some tentative conclusions about these roles.

ENVIRONMENTAL PROBLEM-SOLVING

For some of our problems, science and technology seem closer to solutions than for others. Discoveries in atmospheric and hydrological dynamics and in air and water pollution control have been aided by such recent developments as economic mechanisms for encouraging pollution control, the development of new engines and anti-pollution devices, and of cleaner fuels. In contrast, biological systems are more complex, less predictable, and inherently more variable; they consequently are more difficult to manage either as whole systems (for example, national parks) or as resources. The most complex of all biological systems is complicated further by an intricate and highly evolved social structure—this system is, of course, human society. Because it is in the cities that the characteristics of our society are culminated, these centers present the most difficult environmental problems. This suggests a hypothesis: The smaller the scale of the problem and the smaller its biological complications, the better our chance to handle it adequately.

All these matters are relative, and even the most difficult problems still must be tackled. In an absolute sense, then, there are many opportunities for persons with the proper training to engage in useful work in the field of environmental problem-solving, not only in research but also in areas involving the more straightforward application of information. The preparation of environmental impact statements, both in governmental agencies and for private consultants, continues to demand people skilled in some area of environmental analysis yet with a broad perspective and the ability to analyze volumes of published literature. There is great need for environmental economic research of all sorts, ranging from research to provide us with answers to pollution control to less expensive means of utilizing resources. New governmental agencies and laboratories continue to appear, and even private companies are finding environmental analysts useful. Even the field of environmental engineering now involves something more than designing means for better sewage treatment. As our problems become more severe, it seems highly unlikely that demand for environmental skills can decline.

In all areas of research, governmental agencies have been the primary sources of funds. In recent years, such agencies as the National Science Foundations Research Applied to National Needs (NSF-RANN) and Sea Grant have sought research directly applicable to environmental problem-solving. Again, the needs are present and seem unlikely to diminish. Relative to the information we need to have, the information we now have is woefully small. In ecology, I have seen extremely speculative and untested ideas used as though they were proven facts, and generalities set into specific policies because there simply was nothing better. In the area of comprehensive land planning, in particular, there is a crying need for interdisciplinary research. We could easily name a hundred more areas where the same comments would apply.

THE ROLE OF ECOLOGY IN ENVIRONMENTAL PROBLEM-SOLVING

Is ecological science a useful tool for environmental problem-solving? No simple answer has yet emerged. Certainly from what we know of the purview of ecology, it must be a principal science in the management of biological systems. In this area of biological management, there are enough problems of research, both basic and applied, to keep future ecologists busy for generations just studying the principal disease, food, and resource management problems. Ecology remains a fairly diffuse, highly empirical science, and there is much to learn about how particular biological systems operate.

Once we go beyond applications of ecological theory in biological management and on to its application in human ecology, the usefulness of ecological theory becomes less certain. Apparently similar phenomena in human and other animals may not necessarily share the same systems of cause and effect, so that predictions in human systems based only upon ecological theory could actually prove to be wrong and misleading. This is not necessarily true of all ecological theory. Certain ideas—for example, the concepts of energy flow, mineral cycling, limited growth, natural selection—though highly developed in the field of ecology, are actually properties common to all general systems. Clearly, these kinds of general principles are applicable to human ecology just as to other biological systems. These sorts of general systems properties, however, are limited in their predictive abilities and frequently can only increase our awareness of the complexity of real systems and set only gross limitations on our policy choices. Nevertheless, it is important to remember that fairly general information may be useful for planning, especially when the area of concern is large and when normative information is considered important.

At the other end of the spectrum of potential applicability, we have highly specific ecological concepts. Since these frequently have involved discoveries based on characteristics specific to the system being studied, it is tenuous to attempt to extend the concept to other systems (including the human system) without corroborating evidence that the two systems are similar enough to make it likely that the concept would apply to both. Certainly the most prudent course in these situations would be the position taken by Downs (Chapter 19).

SYSTEMS THEORY IN ENVIRONMENTAL PROBLEM-SOLVING

The construction of comprehensive models continues to be a promising technique for environmental problem-solving. All the advantages of models—prediction, insight into processes, quantitative precision, and hypothesis formation—remain promising, as long as we keep their limitations in mind. As in the case of ecology, systems ideas can be used at several different levels. Even if the level is just one of awareness, it is still useful, whether this awareness is simply the realization that "everything is connected to everything else" or is a catalogue of specific examples of interactions. The elaboration of useful models for various environmental areas is uneven, and here the same physical and biological influences seen in the opening of this discussion are evident: air and water pollution models, solid waste handling networks, and some population-based resource models are well advanced, while complex ecosystems and human social dynamics continue to be too poorly known to permit highly specific modeling. At the highest level, global models aggregating a variety of submodels together are still quite crude, but in spite of this, they remain highly promising for future utility.

Certainly the greatest hurdle to the widespread pursuit and use of systems analysis in environmental problem-solving remains the scarcity of the kind of data needed to enable us to understand how real systems operate. Moreover, given the cost of obtaining such information, it seems unlikely that we ever shall have all the high-quality data we would like to have. Hence it is probably too much to expect most future models to yield highly reliable predictions. Nevertheless, no alternative approach will provide us with the information we need at a comparable or lower cost. Therefore, even if we cannot hope to obtain the highest possible level of precision in our models, at least we can anticipate using them to reveal more of the complex interactions that determine how a system reacts to changing conditions. Although having more information may actually make decision-making more difficult (if only because greater skill in balancing alternatives and making compromises will be needed), it is indefensible to argue against having this information. This is especially true if we resolve to adopt a comprehensive system for development and conservation involving reevaluation of traditional moral standards and human goals. The task of forging an environmentally oriented society will be difficult enough even with the organizing boost we can derive from systematic methods of analysis, and without it the complexity of the endeavor might overwhelm us before we start.

Further Readings

Note: This selected list emphasizes philosophical issues and some future problems. It has been kept short in order to give you a chance to read all of them.

Dubos, R. *Man Adapting.* Yale, New Haven, Conn., 1959.

Leopold, A. *A Sand Country Almanac.* Oxford, 1949.
 Reprinted along with *Essays on Conservation from Round River* by Sierra Club/Ballentine, New York, 1970.

McPhee, J. *Encounters with the Archdruid.* Farrar, Straus, and Giroux, New York, 1971.

Neuhaus, R. *In Defense of People.* Macmillan, New York, 1971.

Stebbins, G. L. "The Natural History and Evolutionary Future of Mankind." *American Naturalist,* 104 (1970), 111–126.

Stone, C. D. *Should Trees Have Standing?* William Kaufmann, Los Altos, Calif., 1974.
 Originally published in *The Southern California Law Review,* 1972.

Glossary

Acculturation The process of external change imposed on a population, with loss of traditional social and cultural institutions.

Acid rain The popular name for sulfuric acid aerosols heavy enough to be washed out of the atmosphere in rainstorms. The sulfuric acid enters the atmosphere as sulfate ions derived from combustion of sulfur-rich fuels. See also *Aerosol*.

Acute Critical or concentrated, as in an acute illness. Characterized by sudden onset and quick termination. See also *Chronic*.

Adaptive inferiority As used by Rubinoff, selection will operate to eliminate the hybrid in question.

Ad valorum tax Literally, a "value-added" tax. On forest lands, the trees have value as well as the land itself and consequently are taxable. Since the value is in the timber, and presumably this grows incrementally every year, the tax load grows with the timber.

Aerosol A particulate substrate to which various pollutant compounds are attached. Also called condensation nuclei. See also *Particulate pollutants*.

Aggregation and Disaggregation Literally, the degree of clustering of objects. In systems modeling, the degree to which variables are lumped together. A highly aggregated model is one which is simplified by lumping compartments and reducing the number of flows needed to interconnect them.

Air basin Most easily defined in rough or hilly country where valleys are obvious air basins. More specifically, a region where a metropolitan area affects air quality.

Ambient air standards Ambient air standards are formulated on maximum permissable concentrations in the air of the basin or region rather than on the emissions from various sources. See also *Emission control standards*.

Anadromous fish Any species which lives as adults in the sea but spawns and spends the younger stages of life in fresh water. The premiere examples are various species of Pacific salmon.

Anoxic A situation in which all oxygen is absent. Anoxic situations are most common in deep water in heavily polluted lakes. Typically only specialized decomposers survive in such situations.

Aquaculture High intensity production of aquatic foodstuffs under controlled or semicontrolled conditions. See also *Mariculture*.

Archology Soleri's concept of architecture and ecology combined, i.e., the goals and imperatives stemming from ecology input into the architectural design process.

Baleen whale The group of whales which have special filtering plates (baleen) for straining plankton from the water.

Benefit-cost ratio An economic summary in which the monetary benefits and monetary costs are summarized and compared.

Bilharzia A disease caused by parasitic flukes which enter the circulatory system. Severe infestations can be fatal and the disease is dehabilitating in any case.

Biocentric philosophy Biocentric: living systems as a prime concern. The biocentric philosophy extends human morality to all creatures. See also Chapter 20.

Bioconcentration; trophic chain concentration By feeding selectively, organisms can selectively concentrate certain materials (bioconcentration). Additional concentration will occur with each energy-material transfer to a higher trophic level (trophic chain concentration).

Biodegradable A substance having a biochemical or chemical structure which may be reduced to smaller elements by biological action.

Biological activity Refers to the role of living systems as energy-material filters. The broadest interpretation would include any biological function; in environmental pollution, biological activity usually means decomposition of organic waste.

Biological control Control of a pest achieved by releasing an appropriate parasite or predator.

Biomass The wet or dry weight of living organisms.

Biostimulation The result of eutrophication; increase in productivity as a result of nutrient inputs.

Biota All the living elements of an ecosystem or a given area.

Biotic vs. abiotic Biotic refers to any living or formerly living creature, or their products; abiotic is the remainder.

Block diagram; flow chart Synonyms referring to a technique used to portray conceptual models, i.e., the use of compartments and flows to show how various of the former are interrelated.

Causation, see *Correlation and Causation.*

Carrying capacity The level of maximum sustainable use or maximum size.

Centrally-planned economy One in which a centralized decision-making body makes explicit decisions about prices, allocation of resources, and production goals. See also *Market economy.*

Clientele capture The process by which a regulatory agency is first infiltrated by and then dominated by those interests it was originally designed to regulate.

Chronic Extended over time and usually at a fairly low, steady level, such as terminal cancer lasting twenty years. See also *Acute.*

Chronic bronchitis A respiratory disease characterized by swelling and inflammation of the upper respiratory tract. Like most respiratory diseases, chronic bronchitis is often difficult to distinguish from other upper respiratory diseases and can be related to a multitude of causes.

Closed system A system whose behavior is largely generated within its own structure.

Coevolution A special case of evolution in which two or more species develop interdependencies. Examples include various forms of symbiosis and stable predator-prey relationships.

Community In ecology the biota of a given area. The boundaries can be either arbitrary or based on natural physiographic boundaries. See also *Ecosystem.*

Community organization In ecology and in the way it is used in this book, community organization is analogous to population structure and close to the meaning of diversity, except that it is also concerned with the mechanisms by which characteristic organizational patterns are maintained.

Community stability A measure of the ability of an ecosystem to remain at, or restore, a steady state. Most definitions generally imply that the greater the duration of the steady state the greater the stability. There remains, however, considerable disagreement about how stability is to be defined precisely.

Compartment The components of a system that describe the system's condition at any particular time. Also called level, state, and state variable. See also *Rate variable.*

Competition In ecology, competition is generally interpreted as the interaction in which two individuals or species potentially have a negative impact upon one another. Organisms compete for food, space, and mates, among other things; in human ecosystems competition for money, social status, and personal prestige are also important.

Composting Aerobic decomposition of organic materials to form soil amendments with some limited fertilizer value.

Conceptual model Any model which attempts to portray the main causal pathways linking state variables without specifying the relationship qualitatively. See also *Compartment*.

Correlation and causation Correlation is a statistical technique which tests the association between two variables without necessarily implying cause and effect relationships. Causation, the condition when one variable is responsible for at least part of the behavior of the other, must usually be established in other ways.

Crustacean A member of the arthropod group characterized by an external skeleton. Most crustaceans occur in aquatic environments, although a few exist in terrestrial situations. Crabs, lobsters, and shrimp are well-known examples.

Crustal blocks and faults Adjoining areas of the earth's crust moving in different directions break along their margins. The breaks are called faults or more precisely surface faults. The sections between faults (which tend to be numerous and parallel but also accompanied by many transverse ones) are crustal blocks.

Defoliation The act of removing all the leaves from a tree or woody shrub. Depending upon the species and the defoliating agent, the event can be temporary or permanent.

Demographic transition The human population phenomenon in which birth rates decline after death rates. The cause is usually attributed to industrialization, higher standards of living, and less risk of death for children.

Development rights transfers A proposed mechanism which gives parcels of land certain numbers of development rights in accordance with physical and social constraints. Developments of various sizes would require some number of rights. Landowners who do not wish to develop could sell their rights to those who need them.

Disaggregation, see *Aggregation and Disaggregation*.

Disease vector Any organism which is part of the life cycle of a disease or parasite or which acts as a transmitting agent to the affected organism. Example: malaria and encephalitis transmitted by mosquitoes.

Dynamic vs. static The simplest interpretation of dynamic is the property of change, and of static, is immovable stability. However, a more useful distinction is that a dynamic system has mutually interconnected feedback loops (such as $A \leftrightarrows B$) whereas a static one does not ($A \rightarrow B$ or $B \rightarrow A$).

Easement A right reserved to the seller, government agency, or utility for continuing access or use of the land after sale. In a conservation easement the development right may be withheld to some degree.

Ecological determinism McHarg's name for a planning scheme in which the underlying principle is that ecological systems evolve from physical constraints to ecosystems.

Ecological engineering This term does not have a singular meaning, but it generally means (1) the application of biological systems to waste-disposal and similar problems, and (2) ecosystems management.

Ecological succession The process in which recognizable groups of species replace one another for a variety of reasons. The classic explanation is progressive habitat modification, but other explanations may be more important.

Ecology Suffers from a multitude of definitions and requires precise declaration to avoid confusion. Ecology basically is the description and subsequent quantification of interrelationships and ideally is causally based. See also Chapters 1, 2, 19, and 20.

Ecosystem The biotic and abiotic elements (that is, everything) in an area, usually defined as naturally as possible (as nearly a closed system as possible). See also *Community*.

Ecounit Myer's term for the area required to support a given population in steady state. Clearly it is highly situation and population specific. It may be regarded as the obverse of carrying capacity.

Emission control standards Emission control standards are those set for particular kinds of sources (e.g., automobiles) and occasionally for individual sources regardless of atmospheric pollutant concentration. See also *Ambient air standards*.

Emphysema A degenerative lung disease characterized by permanent loss of the small air sacs (alveoli) in the lungs.

Endemic Restricted to and characteristic of a certain area or environment. In disease, characteristic of an area and a possible epidemic threat without the restriction of rarity.

Energy flow The transfer of energy from source to sink (dissipation as heat). In ecological systems organisms act as filters to trap energy and slow down its ultimate loss as heat. See also *Heat sink*.

Environment Any environment is defined relative to the observer: it is everything outside the observer. In current usage environment usually means relative to human beings, and more restricted than that, it is often synonymous with environmental quality (pollution).

Estuary An area where fresh and salt water mix. Estuaries are often important biologically because they are characteristic of high productivity in some commercially important species.

Eutrophication Nutrient enrichment. While generally applied to aquatic environments, eutrophication is not environment-specific.

Evapotranspiration When plants photosynthesize, they draw water from the ground through their roots and evaporate much of it through their leaves. This is evapotranspiration as opposed to ordinary evaporation (a physical process that does not require plants).

Evolution The process of change in response to changing environmental conditions. Evolution is usually taken to mean systematic change without return to the original state, whereas the latter is normally treated as disturbance.

Exponential growth A special case of the J-shaped growth curve; the population growth curve can be described by the law of exponents.

Faults, see *Crustal blocks and faults*.

Fecundity Potential fertility. See also *Fertility*.

Feedback (loops) The process by which information (or information contained in energy or mineral flows) returns to the effector A from the reactions of the affected B in the relationship $A \to B$ such that the state of A is changed. If A changes in direct proportion to the change induced in B, it is positive feedback; if the change in A is inversely proportional, it is negative feedback. The process of feedback is vital to dynamic systems.

Ferrous vs. nonferrous metals Ferrous metals (the iron group) include iron, chromium, molybdenum, nickel, and others. Nonferrous metals include the precious metals, copper, tin, lead, zinc, and a few others chemically dissimilar to the ferrous metal group.

Fertility The number of children in the completed family. See also *Fecundity*.

Floodplain A part of a river valley which is inundated only when the river valley is in flood.

Flux Another name for flow. For example, energy flow = energy flux.

Free good A commodity of any sort for which there is nominally no charge for its use.

General systems theory The science which seeks to elaborate the features common to all systems.

Generalists and specialists Generalist species are able to successfully exploit a variety of environments at the cost of specializations which would improve the efficiency of exploitation of a subset of them. Specialist species have superior adaptations for a restricted number of conditions but the specializations often preclude switching to unpreferred resources in emergencies.

Granite dome A solid mass of granite rock derived from subterranean volcanic activity from which erosion has stripped away the sedimentary overburden. Halfdome (the Yosemite National Park symbol) is one example that was split by glaciation.

Greenbelts Open space zones used as public parks but intended to provide breaks between developed areas. Greenbelts are often used for aesthetic purposes.

Greenhouse effect The popular name for one form of differential heat transfer. The atmosphere is heated rapidly by short-wave solar radiation but loses heat at night more slowly, especially when CO_2 concentrations are high or there is an extensive cloud cover.

Heat balance; thermal balance It is possible to account for heat gain, heat loss, and heat distribution within any system, and thereby to predict what might happen with a perturbation of any of these. In this sense heat balance equations are very similar to energy flow models.

Heat island An anomaly in the heat balance of an area. The usual heat island in the landscape is a city.

Heat sink Any part of the environment large or cool enough to absorb indefinitely large quantities of heat. The earth is a heat sink for the sun; biological systems use water, air, and soil as heat sinks.

Homeostasis Physical and behavioral processes within individuals which promote the steady state. See also *Community stability* and *Population regulation*.

Hybridization The production of successful (viable) offspring from parents of two distinctly different populations. Usually the hybrid is itself different from the parents.

Indicator species Species whose presence, absence, distribution, or abundance measures the effect of some influence on the system (e.g., pollution).

Integrated control A method of pest control which combines chemical and biological control methods in some optimal fashion.

Land conversion Refers not to a change in physical properties but in use pattern.

Land subsidence Literally, settling (loss of elevation) of part of the landscape. Settling can be due to fluid withdrawal or compaction. See also Chapter 14.

Land tenure The concept that land is a commodity which can be owned and invests the owner with certain rights.

Malthusian solution Increase in population mortality to force adjustment of population numbers to food (or more generally resource) supplies.

Mariculture Marine ecosystems aquaculture.

Market economy Sometimes called the free-market economy or just the supply-demand mechanism. In free-market economies supply and demand are controlled automatically through price changes with central decision-making unneeded. See also *Centrally-planned economy*.

Market externalities Situations in which a commodity is under-valued and subsequently is used to overstimulate production.

Market failure Any situation in which the free-market mechanism fails to bring about a social goal or in which it creates a problem. The pollution of environment due to free goods and/or market externalities is an example of market failure.

Mathematical model Any model which is structured using quantitative statements (mathematics, statistics, and computational languages). It is expected to produce behavior characteristic of or even equivalent to the real system.

Microcosm In one sense, a working model of a system. A microcosm usually is a simplified version of a more complex system constructed from a few of the elements of the system.

Mineral cycles A descriptive picture of the flow of mineral elements throughout a system. It is much like an energy flow or trophic diagram, except that the system is closed (looped).

Miscible Mixable.

Model A mimic of any real object or phenomenon. See also *Mathematical model, Physical model, Simulation,* and *Conceptual model.*

Multiple use Designing a program or directing its execution such that several purposes can be carried out simultaneously. The opposite is exclusive use. The U.S. Forest Service and water development projects generally emphasize the multiple use concept.

Natality; natalist Births and pertaining to birth.

Natural selection The process by which some organisms are eliminated and others favored in particular environmental conditions. Special cases include sexual selection and group selection. The opposite to natural selection is artificial or human selection. See also Chapters 2 and 3.

Niche In ecological theory the niche may be defined as the place of an organism or species in the ecosystem—its food, habitat, activity periods, competitiveness, etc. The fundamental niche is all possible sets of "place"; the realized niche is that subset of the fundamental niche which is actually occupied.

Nonferrous metals, see *Ferrous vs. nonferrous metals.*

Normative information As opposed to technical information, normative information is derived from value judgments and decisions. In no way should this be taken as of lesser value.

Nuclear fission The splitting of an atom (usually uranium) to release potentially useful energy.

Nuclear fusion The amalgamation of two smaller atoms (hydrogen) to a larger one (helium), with the release of potentially useful energy.

Nutrient regeneration Nutrient release due to decomposition or burning.

Open space Another user-defined term referring to an unbuilt or lightly built area. Most often used for wilderness and parks.

Open system One whose behavior depends to a large extent on external forces.

Overexploitation Rates of harvest beyond sustainable yield that would eventually lead to the collapse of the exploited population.

Particulate pollutants Pollutants that are literally particles, usually 5 μm or more in diameter. These particles are complex chemical mixtures, not single compounds.

Pelagic Occurring in the open sea, either floating with the currents or actively swimming.

Perennial herb Botanically, herb refers both to broad-leaf plants and grasses (that is, plants which are neither shrubs or trees). Perennial growth is survival after seeding (reproduction), generally year to year or indefinitely.

Pesticide Any chemical compound used to kill target organisms. Sometimes called biocide to emphasize nonselective properties.

Pest resistance The development of resistance to biocides on the part of a target organism. The biocides are a form of selecting agent. The biological analogue of pest resistance is to be found in coevolution.

Photochemical smog A complex mixture of emissions and secondary reactants usually containing a variety of irritating chemicals. Photochemical smog is common where there is abundant sunlight to begin the secondary conversions, and may be detected as a yellow-brown band below the inversion layer.

Photosynthesis The process by which plants convert sunlight, carbon dioxide, and water into simple sugars and ultimately to starch and tissues. The essential catalyst is chlorophyll.

Physical model Any model which attempts to reproduce the major features of the real system. Examples include plaster and concrete basins to study wave action and tidal flow, and wind tunnels.

Plankton The free-floating organisms that move about suspended in the water column largely because of current forces rather than their own motive power.

Political economy The budget of the governmental structure and its subsequent allocation.

Pollution; pollutant Pollution is the process in which extraneous materials are incorporated into an ecosystem. Although there is no presumption of impact, pollution generally connotes a net negative impact on the ecosystem. The extraneous materials are called pollutants.

Pollution fees Pollution fee proposals are based on the concept that fees sufficient to mitigate environmental damage and/or discourage emissions above desired standards can be scaled to emission levels.

Pollution rights A system for pollution control based on the concept that desired pollution standards are best enforced by market mechanisms in which a maximum number of rights are subject to supply and demand.

Polyculture The simultaneous cultivation of a number of crops as opposed to stands composed of a single type (monoculture).

Population A group of individuals recognizable as members of the same species and potentially capable of breeding with one another.

Population regulation Analogous to some definitions of community stability, population regulation refers to the processes by which population size is kept under control and fluctuations in density are minimized. The forces which tend to increase fluctuations are collectively referred to as disturbance. See also *Feedback*.

Population structure The attributes of a population that distinguish one individual from another in the same population are said to give a population a structure. For example, a population is said to be young or old, and thus has a particular age structure. Other attributes: sex, race, profession, income, etc.

Productivity The rate of fixation of energy into tissue. Primary productivity is energy fixation by plants; secondary productivity is at higher trophic levels.

Proved petroleum reserves, see *Recoverable and proved petroleum reserves*.

Put-and-take fishery A fishery in which long term population equilibria are less important than throughput of catchable individuals. The "put" is stocking; the "take" is the catch. The object is to stock at rates sufficient to stay abreast with demand.

Rate variable The interrelationships between compartments. See also Figure 3-1 and *Compartment*.

Recoverable and proved petroleum reserves Recoverable reserves as known to be in the ground and extractable at some future higher price. Proved reserves are those which can be extracted at current prices.

Recycling As a general term, recycling refers to any secondary use of a virgin material. As an industrial process, recycling means the reuse of secondary materials to make new products.

Renewable and nonrenewable resources A simple classification scheme providing a rough indicator of the resources' regeneration time relative to use rate.

Reproductive isolation The condition in which two or more formerly contiguous populations are separated such that the probability of reproduction between members of different populations is very small.

Reverse osmosis A process used in desalinating water. Reverse osmosis acts by forcing water from a salt solution by placing pressure on it against a semipermeable membrane (one which allows water molecules to pass but not salt).

Sewage effluent The solution suspension of organic materials that is released into the environment.

Shortfall Shortage.

Simulation A form of mathematical modeling that usually emphasizes accurate predictions over accurate conceptual models.

Soil salinization The concentration of salts dissolved in irrigation water plus salts drawn surfaceward by capillary action due to evaporation in surface soils. High salt concentrations require salt-resistant plants.

Specialists, see *Generalists and specialists.*

Species diversity A measure of the variety in a set of organisms (sometimes referred to as uncertainty). A diverse assemblage is characterized by a wide variety of species; alternatively, there is great uncertainty because the probability that one individual belongs to the ith species is low.

Static, see *Dynamic vs. static.*

Steady-state economics and stationary-state economics The developing area of economic theory in which continuous growth is not an assumption. Steady-state economics is concerned with the allocation of wealth and the transition to indefinite periods when there is no net economic growth.

Sustained yield Sustained yield means that on the average an exploited population will produce enough biomass or individuals to recover the losses of exploitation. A population at sustained yield is in steady state with maximized production. See also *Biomass.*

Swidden agriculture A swidden is a temporary garden plot developed by burning off the forest in slash and burn agricultural practices. Swidden agriculture is the same as slash and burn agriculture.

Symbiosis A general term covering many types of cooperative species interrelationships. In some examples of symbiosis both species benefit; in others, only one.

Systems ecology That section of ecology which deals with (1) models and (2) entire functional assemblages (usually ecosystems).

Target system A system impacted by any activity, whether intentional or not.

Technology; technological innovation In this text technology and technological innovation are taken to mean scientific and engineering approaches to problem solving, as opposed to normative decision-making.

Thermal balance, see *Heat balance.*

Thermal inversion A layer in the atmosphere in which the top of the layer is warmer than the bottom of that layer. The collision of subsiding cool air and rising warmer air prevents diffusion through the boundary.

Trophic web A descriptive network showing what eats what in an ecosystem.

Turbulence Turbulence in fluids is unsmooth flow, i.e., the motion of the fluid is the net result of many focal flows and eddies of different directions rather than a single flow at a given velocity and a given direction.

Upwelling The phenomenon in which masses of colder water are forced to the surface to replace water masses blown out of the area by winds. Hence the name upwelling.

Watershed The area in which surface runoff eventually collects.

Wetting agent A detergent or other chemical solution which increases the effectiveness of a biocide by increasing surface penetration.

Wilderness This is another term that can only be defined relative to the observer. Absolute wilderness (very rare today) consists of areas unaffected by human beings or at least by human presence. At the other extreme, wilderness can be an area that only lacks many buildings and otherwise is heavily impacted by humans.

Wild rivers By statute rivers which are not modified by human engineering systems.

ZPG Zero population growth; a population in steady state.

Zoning variance A permit to allow a landowner or developer to escape a clause in the zoning ordinance.

Credits

Illustrated by James Loates

Cover photo by Marc Riboud from Magnum, Inc.

FRONT MATTER
i: Marc Riboud from Magnum, Inc.
iii: *top*, Mary Thatcher from Photo Researchers, Inc.; *bottom*, Ylla from Rapho/Photo Researchers, Inc.
v: Bill Byrne
vi: United Nations
vii: Martin Litton from Photo Researchers, Inc.
viii: J. Pavlovsky from Rapho/Photo Researchers, Inc.
x: Mark Boulton from National Audubon Society
xi: Russ Kinne from Photo Researchers, Inc.
xiii: Dombierer from Rapho/Photo Researchers, Inc.

SECTION OPENERS AND CHAPTER OPENERS
1: *background*, Ellis Herwig from Stock, Boston; *insert*, Bill Byrne
6: *top*, Ira Kirschenbaum; *bottom*, Frederick Ayer from Photo Researchers, Inc.
23: *top*, George Laycock from Photo Researchers, Inc.; *bottom*, J. Berndt from Stock, Boston
40: *top, left and right*, Grant Heilman; *bottom*, Ira Kirschenbaum
57: *background*, Inger McCabe from Rapho/Photo Researchers, Inc.; *insert*, United Nations
64: *top left*, Al Lowry from Photo Researchers, Inc.; *top right*, Jacques Jangoux; *bottom*, Joe Portogallo from Photo Researchers, Inc.
88: *top left*, FAO photo by P. Pittet; *top right*, U.S. Army photo; *bottom*, Derrick Te Paske
111: *background*, Jacques Jangoux; *insert*, Martin Litton from Photo Researchers, Inc.
117: *top*, Grant Heilman; *bottom*, Dombierer from Photo Researchers, Inc.
146: *left*, Georg Gerster from Rapho/Photo Researchers, Inc.; *right*, Joe Munroe from Photo Researchers, Inc.
173: *left*, Daniel S. Brody from Stock, Boston; *right*, Grant Heilman
210: *top left*, Jacques Jangoux; *top right*, American Forest Products Industries; *bottom*, Bill Byrne

234: *top left*, John Messineo from Tom Stack & Associates; *top right*, Grant Heilman; *bottom*, Owen Franken from Stock, Boston
255: *background and insert*, J. Pavlovsky from Rapho/Photo Researchers, Inc.
264: *top left*, Gordon S. Smith from National Audubon Society; *top right*, Phyllis McCutcheon Mithassel from Photo Researchers, Inc.; *bottom*, Bob Munns, Bureau of Commercial Fisheries
293: *top left*, Laurence Pringle from Photo Researchers, Inc.; *top right*, Paul Pougnet from Rapho/Photo Researchers, Inc.; *bottom*, Ron Engh from Tom Stack & Associates
329: *left*, Laurence Pringle from National Audubon Society; *right*, Ellis Herwig from Stock, Boston
345: *left*, Brookhaven National Laboratory; *right*, Theodore Foin
371: *background*, Bert Stern from Black Star; *insert*, Mark Boulton from National Audubon Society
378: *top*, New York Zoological photo by Bill Meng from Bruce Coleman, Inc.; *bottom*, San Diego Zoo photo by F. D. Schmidt
409: *left*, Keith Gunnar from National Audubon Society; *right*, Philip Jon Bailey
443: *background*, American Stock Photos from Tom Stack & Associates; *insert*, Russ Kinne from Photo Researchers, Inc.
448: *left*, Christopher S. Johnson from Stock, Boston; *right*, Dick Hanley from Photo Researchers, Inc.
480: *left*, Ellis Herwig from Stock, Boston; *right*, S. Kessler from Stock, Boston
510: Gene Ahrens from Bruce Coleman, Inc.
535: *left*, T. D. Lovering from Stock, Boston; *right*, Daniel S. Brody from Stock, Boston

Index

AAAS Herbicide Assessment Panel, 352
Abiotic influences, on living systems, 9–10
Aborigines, Australian
   population control in, 92
   relation to environment, 65
Abortion, 102
Abu Dhabi
   closed agricultural production system at, 132
   desalinization at, 157
Acid rain, 304–305, 319
   from cooling towers, 351
Activated sludge, sewage treatment using, 289
Acts, legislative
   Air Quality Act of 1967, 321
   Clean Air Act, 321–322
   Solid Waste Disposal Act, 332
   Wild and Scenic River Act, 175–176, 428–429
   Wilderness Act, 213, 425, 428
Adaptation
   and genetic types, 25
   patterns of, 28
Ad valorem tax, 223
Aesthetics
   and air pollution control, 314–315
   and perception of water pollution, 280
African plains herbivores, harvesting for food, 141
Age structure, 13–14
   changes in population under exploitation, 216–217
   of Dominican Republic population, 69
   effect of on population growth, 69
   and effects on reproduction, 24
   of human population, 69
   of Swedish population, 69
   and zero population growth, 104
Agency behavior, in water development, 147
Aggression and retreat, as signaled in proxemics, 502–504
Agricultural produce, price of, and irrigated land, 131
Agricultural production systems, closed, 132

Agricultural subsidy, air quality as, 132
   protection of, through urban growth control, 178
Agricultural technology, effects on an economy, 136
Agriculture
   high-intensity, environmental impact of, 137
   irrigated
      and California State Water Project, 166
      and Texas Water Plan, 168
   swidden, 138–139
Air, compared to water, 293–294, 309
Air basin
   carrying capacity of, 316–317
   Los Angeles, air quality control in, 315–318, 320
Aircraft, high-flying, climatic change due to, 306
Air pollutants, *see* Pollutants, air
Air pollution, *see* Pollution, air
Air quality
   as agricultural subsidy, 132
   control of
      in Australia, 321
      and Environmental Protection Agency, 321
      in Europe, 320–321
      in Los Angeles air basin, 315–318, 320
      social inequity problems, 322, 324
      and technology, 319–320
      in United States, 321
      wind pattern and air pollution distribution, 315–316
   economic tradeoffs in, 325
   and Environmental Protection Agency, 321, 540
   and internal combustion engine, 323–325
   regulation
      incentives for, 322
      through pollution rights, 322
Air Quality Act of 1967, 321
Air standards, ambient, 321

Algae, and water purification, 268
Allsop, B., 529-530, 532
Amazon, nutrient regeneration in, 385-386
Amish, social population regulation in, 96
Anadromous fish, 228
Analysis, systems, *see* Systems analysis
Anchoveta, Peruvian, 217
    fishery practice in, 217-218
    and guano birds, 217
    and price of American beef, 131
Animals, effects of air pollution on, 310
Anthropocentricity, 390-391
    and conservation, 391
    and human welfare, 374
Apple codling moth, and biological control, 135-136
Appleyard, D., 494-497
Aquaculture, 129
Aquatic areas and floodplains, zoning of, 184-185
Arable land, 120-121
Archology
    Arcosanti (Mesa City) as example of, 469, 470-471
    definition of, 468
    life cycle of, 466-467
Arcosanti (Mesa City), as example of archology, 469, 470-471
Arctic
    ecosystems, and oil exploration, 245, 247
    water pollution in, 276
Artificial reefs, use of solid wastes for, 340
Asia, Southeast, use of biocides in, 352
Asuntosaatio, 461
Aswan High Dam
    compared to California State Water Project, 166
    and fisheries, 159-160
    and infectious disease inflictions, 160
    as multipurpose project, 159
    siltation effects from, 159
    and soil fertility, 160
    as systems consequence, 160
Athwal, D. S., 128

Atmospheric CO_2, and greenhouse effect, 305, 307
Attitude(s)
    biocentric, 374
    toward environment, historical roots of, 482
    pronatalist, 60
Australia, air quality control in, 321
Australian aborigines
    population control in, 92
    relation to environment, 65
Automobile, role of in urban structure, 451-453, 497
Ayres, R. U., 515

Balance
    changes in heat, in water, 276
    heat
        Cayuga Lake projections for change, 349
        global, 308
"Balance of nature," 389
Baldwin, N. S., 274
Baldwin Hills reservoir failure, 361-365
    and subsidence, 365
Balikci, A., 91
Baltimore, Md., area, ecological determinism used in, 195-197
BART (Bay Area Rapid Transit System), 323
BASS (Bay Area Simulation Study) model, 469
Becker, F. D., 506
Beef, price of American, and Peruvian anchoveta, 131
Beeton, A. M., 270-274
Behavior
    counterintuitive, 472
    population, and social population regulation, 105
    prediction of in systems, 48
Behavioral reform, and environmental quality, 530
Belser, K., 184, 186-187
Benefit-cost assessment, 513
Benefit-cost ratio, 170
Benjamin, S., 283
Ben-Tuvia, A., 405
Biocentric attitude, 374

Biocides
  DDT and chlorinated hydrocarbons, 352–355
  definition of, 351
  in Southeast Asia, 352
Biological control
  and apple codling moth, 135–136
  definition of, 134
Biological management, and ecology and environmental problem-solving, 550
Biological systems, environmental problem-solving in, 549
Biotic communities, effects of water pollution on, 277–278
Birds
  and DDT, 354
  guano, and Peruvian anchoveta, 217
  Hawaiian, extinction of, 192
Birdsell, J. B., 65, 91, 92
Birket-Smith, W. C., 91
Birth control
  mechanisms of, resources and, 99
  and social population regulation, 96
Black lung disease, 311
Bloom, J. R., 286
Boerma, A. H., 118
Bohn, H. L., 326, 348
Borlaug, N. E., 127
Bottle law, in Oregon, 330, 337–338
Boulding, K. E., 514
Bourne, L. S., 450–451
Boyles, R. H., 165
Brahmins, Nambudiri, population control in, 91–92
Brandywine basin, Pennsylvania, ecological determinism used at, 198
Breeder reactors, U^{238} in, 243
Brown, L. R., 128
Bryson, R. A., 306
Buffalo, New York, air pollution health effects in, 312–313
Bureau of Land Management, and the Wilderness Act, 428

Burton, J., 488
Bushland, R. C., 136, 368

Cahn, R., 423
Caiman, and nutrient regeneration in the Amazon, 385–386
California
  Department of Transportation, land-use control by, 176
  Land Conservation Act (Williamson Act), 182–183
  State Water Plan, 161, 163
  State Water Project, 161–163, 166
Calvert Cliffs decision, 524
Canal, Panama, 396–400
Canal, sea-level, *see* Panama Canal
Carbon dioxide, and greenhouse effect, 305, 307
Carnivores, and DDT, 354
Carr, J. F., 274
Carrying capacity, air basin application, 316–317
Cassel, J., 454
Catfish, aquaculture of, 129
Cause-condition-effect matrix
  in ecological planning, 198–202
  example (San Diego), 200–202
Cauthorn, R. C., 326, 348
Cayuga Lake
  and nuclear power, 348–351
  waste heat in ecology of, 349
Celibacy, as population control mechanism, 94
Central Artery, Boston, design of, 495–497
Central Valley Project, 161
Centralized economies, and environmental quality, 517–519
Cesspool mentality, 265
Change
  chemical, in Lake Erie, 273
  climatic, 305–307
Changon, S., 305–306
Chant, D. A., 134–135
Chass, R. L., 316–318
Chaucey, J., 342

Chemical corrosion and soiling, effects of air pollution on, 309–310
Chemical pest control, systems impact of, 139
Chemical reactions, between air pollutants, 299–300
Chicago, and automobile's role in urban structure, 452–453
Chitty-Krebs hypothesis, 34
Chlorinated hydrocarbons including DDT, 351–355
Cities
   air pollution in, 295, 297
   as ecosystems, 448–449, 450–451, 454
   functional aspects of, 449–453
   garden
      Letchworth, 460
      Tapiola Garden City, 459–463
      Uusimaa 2010 plan, 461, 463
      Welwyn, 460
   importance of transportation, 449, 451–452
   influence of on water quality, 270–271
   models of development of, 450–451
   nature of, 448
   stability and diversity in, 474
   tetrahedral, 464–465
City development, land-use control in, 451
Clausen, J., 25
Clean Air Act, 321–322
   and Environmental Protection Agency, 522
Cleaning symbiosis, 388
Clearcutting, 223
Cleary, G. J., 321
Climate
   effect of on air pollution, 301–302
   effects of air pollution on, 305, 306
Climatic change, 305–307
Closed agricultural production systems, 132
Closed system, vs. open system, 44
Cloud, P. E., 248, 249–250
Coal gasification, 240
   and MHD (magnetohydrodynamic) power, 240
Coale, A. J., 269

Coastal environments
   land use control in, 189
   and mariculture, 130
Coefficient of population increase, 68
Cole, H., 286
Commission on Population Growth and the American Future, 103
Commodity, land as, 174
Common property resources, overexploitation of, 123–124
Commoner, B., 142, 265, 266, 537
Communities, biotic, effects of water pollution on, 277–278
Community stability, *see* Stability, community
Compartment, model, 258
Compartments, as model components, 46
Competition, 32–33
Competitor destruction, in fish and game management, 220
Composting, 338–339
Computer simulation, in resource management, 219
Concentration
   of air pollutants, and topography, 296–300
   of pesticides in the trophic chain, 353
Condensation nucleus, 306
Conditioning, operant, 530
Conservation
   anthropocentric orientation and, 391
   energy, 250–251
   and environmental awareness, 540–541
   hindrances of, 375
   of nonrenewable resources vs. substitution as modes of management of, 235
   rationales for, 374
   water, 152–153
   and zoos, 392–393
Control
   air pollution, and aesthetics, 314–315
   biological
      and apple codling moth, 135–136
      definition of, 134

disease, 140
governmental, and pollution control, 281
pest, 134-136, 139, 140
See also Air quality, control of; Growth control, urban; Growth control mechanisms; Land-use control; Pollution control; Population control
Conway, G. R., 136
Cooling towers, 346-348, 351
  acid rains from, 351
Corrosion, chemical and soiling, effects of air pollution on, 309-310
Council on Environmental Quality, 521
Counterintuitive behavior, 472
Courts, and environmental protection, 525
Criminal harassment, and urban design, 498
Crisis, environmental, 482-483
Crown of thorns starfish, 385
Currents, deep ocean, energy from, 241
Cutbill, J. L., 380-381
Cycle, hydrological, 147-148
  and development, 155-156
  in Florida, 155

Dales, J. H., 283-284
Daly, H. E., 515-516
Danielsson, B., 91
Darmstadler, J. F., 237
Darwin, C., 16
Dasmann, R., 141, 221
Data
  cost of, 504-505
  requirements of, and systems, 48
Davis, E. W., 313-314
Davis, K., 79
DDT, 274, 275, 279
  and birds, 354
  and carnivores, 354
  and chlorinated hydrocarbons, 352-355
  ecological effects of, 355
  global use pattern of, 359

  in marine and polar ecosystems, 355
  and pesticide policy, 359
  properties of, 352-354
  system ramifications of, 355
  in Wisconsin hearings, 358-359
Dean, G. W., 131
Decision-making, economic, and renewable resources, 211
Decomposers, and water purification, 267
Deep-well fluid injection, 359-360
Deer hunting, 221
Deevey, E. S., 66-67
Delafons, J., 184
Demographic transition
  definition of, 60
  and economies, 98
  in Europe, 73
  as a means for population stabilization, 73
Density, population, on the earth, 71-72
Department of Transportation, California, land-use control by, 176
Desalinization, 156-157
Design elements, of Tapiola, 461-462
Destruction
  of coastal environments, and mariculture, 130
  of competitors, in fish and game management, 220
  habitat, 382, 385
  mechanisms, of water pollutants, 294, 305
Detection, of air pollution health effects, 311-312
Detergents, water pollution by, 276
Determinism, ecological, *see* Ecological determinism
Deuterium, 243
Development
  city, land-use control in, 451
  land, *see* Land development
  of national parks, 415
  recreation, at Mineral King, 430-433
  urban
    models of evolution of, 450-451
    three paradigms for, 456-458

Development (*continued*)
  water
    agency behavior in, 147
    and naturally adequate supplies, 155
  water resource, 169
Development easements, as growth control mechanism, 179
Development right transfers
  description of, 185
  as growth control mechanism, 185
  implementation advantages of, 188
DeVore, I., 91
Diamond, J. M., 438
Dilution factors, in air pollution, 305
Disease
  black lung, 311
  control of, 140
  infectious, and Aswan High Dam, 160
Disney Enterprises, and Mineral King development, 432
Disposal
  methods of, for solid waste, 330
  ocean waste, 359
Disruption, social, and population control, 92, 94
Distribution
  of air pollution, 294-295
  water, 158, 165, 168, 170
    global, 148-151
    by use category, 151
Diversity
  and parks, 414
  species
    meaning of, 26
    and stability, 26, 34, 386
  and stability in cities, 474
Dodos, 383
Dominican Republic population, age structure of, 69
Donora, Pennsylvania, air pollution episode, 296
Douglas, M., 65, 91
Douglas fir, 215
Downs, A., 527, 532

Doxiadis, C. A., 81, 453, 464, 476
Drought, Sahel, 170
Dryden, F. D., 286
Dubos, Rene, 22, 279
Duffey, E., 402-404
Duke, J., 405
Dumping, of solid wastes, 331
Duncan, D. C., 236
Dunn, F. L., 91
Dunstan, W. M., 276

Earthquakes
  in Denver, 361-362
  and faults, 361
Earth's crust, plasticity of, 361
Easement(s)
  definition of, 179
  development, as growth control mechanism, 179
Ecological consequences, of thermodynamics, 10
Ecological determinism
  definition of, 193
  outline of method, 193-194
  used for Baltimore area, 195-197
  used for Brandywine basin, Pennsylvania, 198
Ecological engineering, 215
Ecological interdependencies, extinction and, 385
Ecological limitations
  in global ecosystem, 4
  and human beings, 4
Ecological mandate, 527
Ecological morality, 527-530
Ecological planning
  Hills's method of, 193
  land capability rating system method of, 202
  Lewis's method of, 193
  McHarg's method of, 193-198
  and simulation modeling, 204-205
  underlying concepts and framework of, 192-193
  use of cause-condition-effect matrix in, 198-202
Ecological principles, summary of, 37

Ecological science, and environmental problem-solving, 549
Ecological succession, 400–401
  and causal mechanisms, 36–37
  definition of, 35
Ecological systems, 7
  hierarchical organization of, 20
  model of, 18
  properties of, 8
  similarities in, 23, 24
Ecology
  of Cayuga Lake, 349
  definition of, 7
  effects of water diversion on, 164
  as empirical science, 20
  and environmental problem-solving, 7, 21, 550
  limitations in application to human systems, 21
  and planning, 527
  systems
    definition of, 41
    and environmental policy, 42
    mathematical models in, 41
  urban, 473–474
    evaluation of analogies, 454
Economic decision-making, in renewable resources, 211
Economic objectives, and pollution policy, 261
Economics, definition of, 510
Economic systems, closed vs. open, 514
Economic tradeoffs, in air quality, 325
Economic value, establishment of, 513
Economies
  centralized, and environmental quality, 517–519
  central planning of, 511
  and demographic transition, 98
  free-market
    and pollution rights, 284
    principles of operation, 512
  human
    and resource limits, 513–514
    strategy for development, 513
  individual, social population regulation in, 98
  steady-state, 515–517
    effect on environmental quality, 516–517
    implementation problems, 515–516
  of urban growth control, 176
Economy
  free, governmental role in, 520
  political, 520
Ecosystem(s)
  Arctic, impact of oil exploration on, 245, 247
  cities as, 448–449, 450–451, 454
  energy flow through, 14–16
  global, ecological limitations in, 4
  management of, 402–404
    consequences of absence of, 404–405
    future options and opportunities of, 403–404
    in Israeli mountains, 402–403
  marine, DDT in, 355
  modifier, water pollution as, 289
  natural
    evolutionary strategy, 378–379, 390
    and human health, 545
    protection of, 178
    and water purification, 266–267
  polar, DDT in, 355
  stability of, 387
Ecosystem Level Model (ELM), 224–226
Ecotypes, or yarrow (*Achillea lanulosa*), 24
Ecounit, 435
Ecumenopolis, 81, 453, 464
Eels, lamprey, in Great Lakes, 394
Effects
  of air pollution, 309–310
  on health, 310–314
  of nuclear power on water temperature, 347
  of temperature on organisms, 347
Eggshells of birds, and DDT, 354
Ehrlich, A. H., 84
Ehrlich, P. R., 84
Eipper, A. H., 348

Electrolysis, as desalinization method, 157
Elements
　design, of Tapiola, 461–462
　radioactive, see Radioactive elements
Elephants, at Tsavo National Park, Kenya, 434, 436
Ellis, H. T., 308
Elson, P. F., 278
Emission standards, 321
Emphysema, and air pollution, 311
Empirical science, ecology as, 20
Energy
　from atoms, 242–243
　conservation of, 250–251
　from deep-ocean currents, 241
　flow of
　　through ecosystems, 14–16
　　model of, 18–19
　　and trophic webs, 14–16
　from fossil fuels, 239–240
　from geothermal sources, 244–245
　impacts from exploration for, 245
　impacts from production of, 245–248
　from lunar gravitational pull, 243–244
　solar, 238
　from solid wastes, 240–241, 340–341
　from tidal movements, 241
　use pattern of, 237
　from wind, 241–242
Energy crisis, 520, 545–546
Engine, internal combustion, see Internal combustion engine
Environment
　attitude toward, historical roots of, 482
　and Australian aborigines, 65
　contrasting meanings of, 539–541
　definition of in pollution context, 258
　and primitive societies, 65
　and Shoshoni Indians, 65
　and Texas Water Plan, 168
Environmental awareness
　as exploitation tool, 539–540
　as justification for aims, 540
　as motive for conservation, 540–541
　water pollution as indicator of, 289
Environmental crisis, 482–483
Environmental hazard, of radioactive elements, 366
Environmental impact
　of high-intensity agriculture, 137
　of various forms of energy, 246
Environmental impact statements, 521
Environmental perception
　of the city, 491–494
　cultural differences in, 484–486
　definition of, 480–481
　design and policy implications of, 505–506
　difficulties in studying, 481
　and environmental crisis, 483
　and environmental policy, 486, 490
　at neighborhood level, 497–499
　and reality, 484
Environmental policy, 42, 486, 490
Environmental pollution, 258–259
Environmental problems, and population growth, 81, 83
Environmental problem-solving, see Problem-solving, environmental
Environmental protection
　advantages and disadvantages of, 525
　by the courts, 525
　and governmental reorganization, 521
Environmental Protection Agency
　and air quality, 321, 540
　and benefit-cost assessment, 513
　and Clean Air Act, 522
　land-use control by, 176
　and pesticide regulation, 352, 358
　and solid waste management, 332
Environmental quality
　and behavioral reform, 530
　and centrally planned economies, 517–519
　definitions, 258
Environmental studies, and systems, 7

Environmental variation, and population growth, 29
Equation, for exponential population growth, 66, 68
Equipment investment, and whale fisheries, 133-134
Equity issues, of pollution taxes, 283
Estimates
  of petroleum supply, 235-236
  population, 73-78
Europe
  air quality control in, 320-321
  land-planning process in, 188, 205
Eutrophication, 267
Everglades Kite, 383
Evolution
  and community and community stability, 33
  of natural systems, strategy of, 511
  urban, 453, 497
Evolutionary strategy, of natural ecosystems, 378-379, 390
Exploitation
  changes in population age structure under, 216-217
  of renewable resources, 211-212, 230-231
Exploration, of oil, 213
  impact of on Arctic ecosystem, 245-247
Exponential population growth, equation for, 66, 68
Externalities, market, 512
Extinction
  current rate of, 379
  and ecological interdependencies, 385
  factors affecting, 383-386
  of Hawaiian birds, 192
  of large mammals in Pleistocene era, 379
  of passenger pigeons, 379
  and prevention by isolation, 383
  probability of, 386-387
  and specialization, 383-384
  trends of in geological time, 380-381
  when population size is small, 383

Factors
  affecting extinction, 38-386
  dilution, in air pollution, 305

Failure
  of Baldwin Hills Reservoir, 361-365
  market, 512
Family planning, 100
Faults and earthquakes, 361
Feder, H. M., 33, 388
Feedback
  negative, 47
    example of, 55
    mathematics of, 55
    population control and resources, 105
  and population growth, 60-61
  positive, 46
    example of, 55
    mathematics of, 55
    in population growth, 69
  in systems, 46
Fees, pollution, 283
Fee-simple purchase of land, 179
Fertility, soil, and Aswan High Dam, 160
Fish and game, stocking of, 220-221, 394-396
Fish and game management
  destruction of competitors, 220
  goals of, 219
  by habitat management, 219
  harvesting regulations of, 221
Fisheries
  and Aswan High Dam, 159-160
  decline of in Lake Erie, 271, 274
  marine, 122
  shrimp, and Texas Water Plan, 168
  sport, in New Zealand, 396
  whale, and investment in equipment for, 133-134
Fishery
  put-and-take, 395
  salmon, model of, 228-230
  tradeoffs in Lake Erie, 285
Fish kills, by pesticides, 269
Fish production, estimates of, 125-126
Fish species composition, of Lake Erie, 271, 274
Fission reaction, 242, 243

Fittkau, E. J., 385-386
Floodplains and aquatic areas, zoning of, 184-185, 189
Florida, hydrological cycle in, 155
Flow, energy, 14-16
Flows, as model components, 46
Fluid injection
　deep-well, 359-360
　and Denver earthquakes, 361-362
Foin, T. C., 124
Food production
　and Green Revolution, 127
　population subsystem, model of, 121-122
Forest practices, 223-224
Forests, redwood, 215
Forest Service
　and Mineral King controversy, 431
　and Wilderness Act, 428
Forrester, J. W., 48-52, 472
　urban model of, 472, 473
Fossil fuels, 239-240
Free economy, governmental role in, 520
Free goods, "tragedy of the commons," 512-513
Free-market allocation systems, 511
Free-market economies
　and pollution rights, 284
　principles of operation of, 512
Freeway design, 494-497
Frejka, T., 103, 104
Fuels, fossil, 239-240
Fuller, R. B., 231, 464-465, 468, 476
Funnell, B. M., 380-381
Fusion, nuclear, 243

Garden cities
　Letchworth, 460
　Tapiola, 459-463
　Uusimaa 2010 plan, 461, 463
　Welwyn, 460

Gasification, coal, 240
　and MHD (magnetohydrodynamic) power, 240
Geertz, C. B., 94-96
General systems, properties of, 44-48
General systems theory, 42
Genetic types, and adaptation, 25
Geodesic dome, 464-465
Geological time, trends of extinction in, 380-381
Geothermal energy, 244-245
Glaciers and icebergs, as sources of water, 153
Global ecosystem, ecological limitations in, 4
Global heat balance, 308
Global model, 536-538
Goals, of national parks, 416, 421-424
Goldman, M., 281, 517-519
Goldsmith, J. R., 282
Goodman, Paul, 456-458, 474
Goodman, Percival, 456-458, 474
Goodman, Robert, 458, 459, 474, 507
Gould, E. M., 52, 226-227
Government, role in free economy, 520
Government control of production, for pollution control, 281
Government ownership of property, as growth control mechanism, 188-189
Governmental reorganization, and environmental protection, 521
Grasslands, as systems, 42-44
Graves, J., 165, 171
Great Lakes, lamprey eels in, 394
Grebe, western, 354
Greenhouse effect, and atmospheric CO_2, 305, 307
Greenough, J. W., 228-230
Green Revolution
　definition of, 127
　feedback to population growth of, 128
　and increased food production, 127
　as produced by IR8 and IR22 (rice), 127
　social inequities of, 128
　socioeconomic consequences of, 128-129

and technological input demand, 127-128
and technology in India, 136
Green turtle, 383
Griggs, D. T., 361-362
Group selection, 89-90
Growth
  exponential population, equation for, 66, 68
  human population, 60-67
  population, see Population growth; Zero population growth
Growth control, urban
  and agriculture, 178
  economies of, 176
  and natural ecosystems, 178
  and open space, 177
  rationales for, 176-178
Growth control mechanisms
  constraints upon, 178-179
  development easements, 179
  development right transfers, 185
  fee-simple purchase, 179
  government ownership of property, 188-189
  property taxation, 182
  recapture taxes, 185
  zoning, 183-185
Guano birds, and Peruvian anchoveta, 217
Guthrie, D. C., 482, 483

Habitat
  destruction of, 382, 385
  management of, 219
Hall, E. T., 499, 500-501, 502
Hamming, W. J., 316-318
Handling methods
  for radioactive elements, 366
  for solid waste, 330
Harburg, E., 498-499
Hardin, G., 512
Harte, J., 149-150, 171
Harvard Forest, 226
  model of timber production, 226-227

Harvesting regulations, in fish and game management, 221
Hatchery operation, in salmon, 396
Hawaii
  land-use control plan of, 178
  land-use plan of, 189-192
Hawaiian birds, extinction of, 192
Hazards
  environmental, of radioactive elements, 366
  health, 295-296
  natural, 487-490
  transportation, of radioactive elements, 367
Health effects, of air pollution, 295-296, 310
  in Buffalo, New York, 312-313
  evidence for, 313-314
  predications of in air quality control, 317-318
  problems of detecting, 311-312
Health, human
  and "human first" approach, 544
  and natural ecosystems, 545
  and water pollution, 279
Healy, J. H., 361-362
Heat, waste
  in Cayuga Lake, 349
  and nuclear power plants, 247
  sewage treatment using, 286
Heat balance, see Balance
Heat island, urban, 177
Heatwole, H., 16
Hediger, H., 392-393
Heilbroner, R. L., 515, 548
Hendricks, T. A., 236
Hepatitis, and water pollution, 279
Herbivores, 141
Hierarchical organization
  of ecological systems, 20
  and subsystems, 45-46
Hierarchies, systems as, 7
Hiesey, W. M., 25
Hill, A., 431

Hills, G. A., 193
Hiltunen, J. K., 274
History
  of human population growth, 60-67
  of national parks, 415
  of recreation development at Mineral King, 430-433
Homeostasis
  definition of, 18
  in population control, 28
Howard, E., 459, 460
Hubbert, M. K., 235, 236
Hubbs, C., 290
Hubschmann, J., 285
Human beings
  and ecological limitations, 4
  and nature, 390
Human ecology, and ecology and environmental problem-solving, 550
Human economies, *see* Economies, human
Human health, *see* Health, human
Human population(s)
  age structure of, 69
  growth of, 60-67
  and thermodynamics, 11
Human societies, primitive, social population regulation in, 90-91
Human strategy, and evolution of natural systems, 511
Human system, land development as aspect of, 175
Human systems, and ecology, 21
Human welfare, anthropocentric position on, 374
Hunting
  deer, 221
  market, 214-215
Hutterites, social population regulation in, 96-97
Hydrological cycle, 147-148
  and development, 155-156
  in Florida, 155
Hynes, H. B. N., 278

Icebergs and glaciers, as sources of water, 153
Idyll, C. P., 125, 126, 142, 250
Impact
  of elephants, on Tsavo National Park, 436
  environmental
    of high-intensity agriculture, 137
    of various forms of energy compared, 246
  health, 317-318
  systems, of chemical pest control, 139
Implementation
  design, and environmental perception, 505-506
  problems of, in pollution control, 281
Impoundment
  public opposition to, 158, 165, 168, 170
  of water, 154
Increase, population, coefficient for, 68
India, technology in, and Green Revolution, 136
Indians, Shoshoni, and environment, 65
Indicator species, 267
Individuals, adaptation patterns in, 28
Industrial melanism, 310-311
Infectious disease, and Aswan High Dam, 160
Injection
  deep-well fluid, 359-360
  fluid, and Denver earthquakes, 361-362
  of radioactive elements, 367
Innis, G. S., 225
Innovation, technological, and solid waste management, 334
Inorganic nutrients, and water purification, 268
Institute of Ecology, University of California, Davis, 223-224
Interdependencies
  ecological, and extinction, 385
  national, for mineral resources, 248
Internal combustion engine
  and air quality, 323-325
  and public transportation alternatives, 323-325
  stratified, 325
  as a system, 42-44

International Biological Program (IBP), 219
Interspecific relationships, 32–33
Inversion, temperature, 302–303
Ions, heavy metal, as air pollutants, 299
IR8 (rice), 127
IR22 (rice), 127
Irrigated agriculture
  and California State Water Project, 166
  and Texas Water Plan, 168
Irrigated land, and prices of agricultural produce, 131
Irrigation water, and desalinization, 157
Island, urban heat, 177
Isle Royale, wolves and moose on, 400–402
Ivory-billed woodpecker, 382

Jackson, J., 497
Jacobs, J., 411, 414, 415, 454–456, 474, 498
  and open space, 414
  park function analysis by, 411
Janzen, D. H., 33
Java, population control in, 95
Jordan, C. F., 22
Jordan, P., 401

Kadanoff, L. P., 473
Kantor, S., 313–314
Karp, E., 477
Karp, M., 477
Kasl, S. V., 498–499
Kates, R. W., 488, 490, 507
Keck, D., 25
Keever, C., 38
Kirtland's warbler, 384
Kite, Everglades, 383
Kneese, A., 515–516
Komodo dragon, 383
Krebs, C. J., 34, 35
Krieger, M., 393
Kyllonen, R. L., 52

Laboratory for Experimental Design, 198–201
Ladejinsky, W., 128
Lake Baikal, pollution of, 518–519
Lake Cayuga
  heat balance projections for change in, 349
  and nuclear power, 348–351
  waste heat in ecology of, 349
Lake Erie, 277
  chemical changes in, 273
  composition of fish species in, 271, 274
  decline of fisheries of, 271, 274
  fishery tradeoffs in, 285
  influence of cities on water quality of, 270, 271
  use of sewer, 285
  water pollution in, 270
Lake Nasser, 159
Lake stratification, 349–350
Lake Tahoe, 202–204
  pollution of, 270–280, 518
Lake Washington, pollution of, 270
Lamprey eels, in the Great Lakes, 394
Land
  agricultural use of, 212
  arable, 120–121
  as commodity, 174
  irrigated, and prices of agricultural produce, 131
Land-capability rating systems
  as ecological planning method, 202
  and plan of Lake Tahoe, 202, 204
  of U.S. Soil Conservation Service, 202
Land development, 174–175
Landfill, and solid waste, 330–331
Land Management, Bureau of, and Wilderness Act, 428
Land planning, 192
  process of
    in Europe, 188, 205
    in United States, 205
Land use, 174, 192

Land-use control
  by California Department of Transportation, 176
  in city development, 451
  coastal, 189
  by Environmental Protection Agency, 176
  at national level, 176
  and Wild and Scenic Rivers Act, 176
Land-use plan, of Hawaii, 178, 189–192
LaPorte, Indiana, climatic change at, 305–306, 307
Large mammals, extinction of in Pleistocene era, 379
Lave, L. B., 313–314
Law enforcement, and pollution control, 282, 283
Laws, R., 434
Leakage, of radioactive elements, 367
Lee, R., 91
Letchworth, 460
Levels, as model components, 46
Levins, R., 16
Lewis, P. H., 193
Liebman, J. C., 342
Life cycle, of archology, 466–467
Likens, G. E., 304
Living systems, abiotic influences on, 9–10
London Greenbelt, 188
Los Angeles, and automobile's role in urban structure, 452–453
Los Angeles air basin, air quality control in, 315–318, 320
Lowenthal, P., 484–486
Lua, and swidden agriculture, 139
Lukezie, F. L., 286
Lunar gravitational pull, energy from, 243–244
Lund, H. F., 320
Lynch, K., 491, 494–497, 507

MacArthur, R. H., 192
MacBeth, W. G., 316–318
Malthus, T. R., 114–115, 117
Malthusian dilemma, 114–115, 117, 119
Mammals, large
  extinction of in Pleistocene era, 379

Management
  biological, and ecology and environmental problem-solving, 550
  ecosystem, *see* Ecosystem(s), management of
  fish and game, *see* Fish and game management
  habitat, 219
  Land, Bureau of, and Wilderness act, 428
  of renewable resources, 230–231
  resource
    by controlling supply, 214–215
    use of computer simulation in, 219
    use of remote sensing in, 218
    using fire in, 215–216
  solid waste
    and Environmental Protection Agency, 332
    systems analysis in, 334–335, 342
    and technological innovation, 334
  at Tsavo National Park, Kenya, 433–434
  use of models for, 230
  waste, plan of Oregon, 332–334
Mandate, ecological, 527
Maneri, C. S., 313–314
Margalef, R., 477
Mariculture, 130
Marine, G., 165
Marine ecosystems, DDT in, 355
Marine fisheries, 122, 123–124
Marine fish production, estimates of, 125–126
Market economies, free-, and pollution rights, 284
Market externalities, 512
Market failure, 512
Market hunting, 214–215
Marks, D. H., 342
Martin, P. S., 379
Marx, W., 329
Materials recycled, 337
Mathematical model, 41
Maximization of profit, in renewable resource use, 211–214
McCloskey, M., 431
McGauhey, P. H., 338–339

McKelvey, V. E., 236, 251
McLaren, I. A., 532
Meadows, D. H., 49
Meadows, D. L., 48-51
Mech, L. D., 401
Mechanisms of growth control, see Growth control mechanisms
Melanism, industrial, 310-311
Mercury poisoning, in Minamata Bay, Japan, 275
Merrill, W., 286
Mertz, D. B., 405
Mesa City (Arcosanti), as example of archology, 469, 470-471
Metal ions, heavy, as air pollutants, 299
Meyer, J., 494-497
MHD (magnetohydrodynamic) power, 240
  and coal gasification, 240
Minamata Bay, Japan, and mercury poisoning, 275
Mineral cycles
  phosphorus, 11
  and thermodynamics, 11
Mineral King area
  and Disney Enterprises, 432
  and Forest Service, 431
  history of recreation development, 430-433
  and Sierra Club, 431
Mineral resources
  estimated stocks, U.S. and world, 248-249
  national interdependencies upon, 248
  offshore, 250
Mines, salt, as storage sites for radioactive elements, 367-368
Mining, strip, 246
Mitchell, J. M., 298, 307-308
Moas, 383
Model
  compartment, 258
  of ecological systems, 18
  of energy flow, 18-19
  of environmental pollution, 258-259
  global, 536-538
  mathematical, 41
  of population-food production subsystem, 121-122
  of proxemics, 500-501
  of salmon fishery, 228-230
  systems, 42-44
  of timber production in Harvard Forest, 226-227
  urban, of J. W. Forrester, 472-473
  use for management, 230
  of water pollution, 268
  world, 48-51
Model comparisons, 230
Model components, 46
Model sectors, 228, 230
Modeling, systems, and environmental problem-solving, 550-551
Models
  of city development, 450-451
  of perception, 491, 499
  quantitative, data limitations on, 550-551
Modern and primitive societies, and environmental crisis, 482-483
Modern society, and population stabilization, 106
Moncrief, L. W., 483
Monoculture, 137, 140
Montague, K., 368
Montague, P., 368
Montas, H. P., 69
Moore, C. L., 236
Moore, N. W., 357-358
Moose, and wolves on Isle Royale, 400-402
Morality, ecological, 527-530
Mosher, W. E., 313-314
Moss, R., 31, 32, 107, 532
Moth, tussock, 359
Mumford, L., 453, 454
Murphy, G. I., 124
Myers, N., 435-437
Myrup, L. O., 177

Nambudiri Brahmins, and control of population, 91-92

National Environmental Policy Act (NEPA), 521-524
  and Council on Environmental Quality, 521
  impacts on decision-making, 522-524
  provisions of, 521
  requirement for environmental impact statements, 521
National parks, *see* Parks, national
Natural ecosystems, *see* Ecosystems, natural
Natural hazards, 487-490
Natural pollution control, 285
Natural selection, 16
  goals of, 45
  and interspecific relationships, 32
Nature
  balance of, 389
  definition of, 375
  and human beings, 390
Nature conservancy, 179
Naveh, Z., 403
Nearshore ocean pollution, 275
Negative feedback, *see* Feedback, negative
Neighborhood, social structure of and urban design, 498-499
Neighborhood level, of environmental perception, 497-499
Neofunctionalism, 456, 458
NEPA, *see* National Environmental Policy Act
Network, trophic, 12, 13
Neuhaus, R., 540
Newman, O., 498, 506
New towns, 459-464
New Zealand, species replacement in sport fisheries of, 396
Niche, 10
Nietschmann, B., 94
Nonrenewable resources, *see* Resources, nonrenewable
Normative decision-making, 504-505
Normative (value) judgments, and land planning, 192
Norms, reproductive, 101-103
North, W., 290
North American Water and Power Alliance (NAWAPA), 154

North Carolina solid waste plan, 334-335
Nuclear energy, 242-243
  and petroleum energy, 235
Nuclear fission, 242-243
Nuclear fusion, 243
Nuclear power
  and Cayuga Lake, 348-351
  and escape of radioactive material, 247
  and heat disposal, 247
  by nuclear fission, 242
  and potential for explosion of power plant, 247
  and waste disposal, 247-248
  and water temperature, 347
Nuclear power plants
  and mariculture, 130
  and photosynthetic productivity, 348
Nutrient regeneration, in the Amazon, 385-386
Nutrients, plant, air pollutants as, 326
Nutrition, plant, sewage treatment and, 286-287

Ocean currents, deep, energy from, 241
Ocean dumping, of solid waste, 331
Ocean pollution, *see* Pollution, ocean
Ocean waste disposal, 359
Odum, E. P., 21, 141, 386-387
Odum, H. T., 14-15
Oil
  ocean pollution by, 275
  water pollution by, 280
Oil exploration, 213
  impact of on Arctic ecosystem, 245-247
Oil spill, at Santa Barbara, 247
Open space
  definitions of, 410, 411
  meanings, 410
  perception of, 486
  philosophy of defense of, 411
  protection of through urban growth control, 177
Open system, vs. closed system, 44
Operant conditioning, 530
Opportunistic species, definition of, 384
O'Regan, W. G., 52, 226-227

Oregon
    bottle law of, 330, 337–338
    waste management plan of, 332–334
Organic wastes, and water purification, 268
Organisms, effect of temperature on, 347
Organization, hierarchical, of ecological systems, 20
Oysters-champagne hypothesis, 91

Pack, D. H., 301
Paddock, P., 119
Paddock, W., 119
Panama Canal, 396–400
    proposed new, 396–397
        alternatives for, 399
        biological consequences of, 398
Park function, Jane Jacobs' concept of, 411
Parks
    and diversity, 414
    national
        goals of, 416, 421–424
        history of development, 415
        suppression of fire in, 422
    and preserves, protection of, 376, 382
Particulate pollutants, 299
Passenger pigeons, 213–214
    extinction of, 379
Patten, B. C., 52
Patterns
    adaptation, 28
    of land development, 174–175
Paulic, G. J., 124, 142, 217, 218, 228–230
Perception
    environmental, 480–481
    of natural hazards, 487–490
    of open space, 486
    of seriousness of water pollution, 280
    of water pollution, 289
    of wilderness, 486
Peripheral Canal, 165
Peruvian anchoveta, see Anchoveta, Peruvian

Pest control, 140
    chemical, systems impact of, 139
    strategy of, 134–136
Pesticide policy, see Policy, pesticide
Pesticides
    concentration of in trophic chain, 353
    effects of excessive use of, 356, 357–358
    fish kills by, 269
    persistence of, 352
    regulation of, and Environmental Protection
        Agency, 352, 358
Pest resistance, 356–357
Peterson, M., 178
Peterson, R., 192
Petroleum energy, and nuclear energy, 235
Petroleum supply estimates, 235–236
Phosphorus cycle, mineral, 11
Photochemical smog, 300
Photosynthetic productivity, and nuclear power
    plants, 348
Physical systems, environmental problem-solving in,
    549
Pinnel, C., 342
Plan(s)
    Hawaiian land-use, 178, 189–192
    North Carolina solid waste, 334–335
    Oregon waste management, 332–334
    Uusimaa 2010, 461, 463
Planning
    ecological
        land capability rating system method of, 202
        Lewis's method of, 193
        McHarg's method of, 193–198
        and simulation modeling, 204–205
        underlying concepts and framework of, 192–193
        use of cause-condition-effect matrix in, 198–202
    government branch of, proposed, 526–527
    and ecology, 527
    family, 100
    land-, process of, 188, 205
    land-use, 174, 192
    urban, by residents, 458–459

Plant nutrients, air pollutants as, 326
Plants
   effects of air pollution on, 310
   nutrition of, sewage treatment and, 286-287
Plants, nuclear power, and photosynthetic productivity, 348
Poisoning, in Minamata Bay, Japan, 275
Polar ecosystems, DDT in, 355
Polgar, S., 96
Policy
   conflicts in land-use, 174
   environmental, 42, 486, 490
   implications of, and environmental perception, 505-506
   pesticide
      DDT example, 359
      rationale for, 355-356
      and the tussock moth, 359
   pollution, 260-261
   population
      factors contributing to inaction in, 107
      in the United States, 103
Political economy, 520
Pollutants
   air
      chemical nature of, 299
      chemical reactions between, 299-300
      heavy metal ions as, 299
      natural sources of, 303
      particulate, 299
      as plant nutrients, 326
   interactions between, 260
   particulate, 299
   secondary, 260
   water
      distribution and toxicity of, 269-275
      types of, 269
Pollution, 260, 518-519
   air
      acute health hazard from, 295-296
      in cities, 295, 297
      control, and aesthetics, 314-315
      dilution factors, 305
      distribution and wind pattern, 315-316
      Donora, Pennsylvania, mortality episode, 296
      effect of climate on, 301-302
      effect of topography on concentration of, 296, 300
      effect of weather on, 302
      effect on animals, 310
      effect on climate, 305, 306
      effects on plants, 310
      effects on soiling and chemical corrosion, 309-310
      and emphysema, 311
      global distribution of, 294-295
      health effects, 310-314
      implications of temperature inversion for, 303
      in London, England, 295-296
      natural destruction mechanisms of, 294, 305
      sources of, 294, 298
      in southeastern U.S., 296
      urban-rural contrasts, 296
   environmental, 258-259
   fees, 283
   ocean, 274-275
   policy, 260-261
   rights
      definition of, 283
      and free-market economies, 284
      limitations of, 284
      as means of regulating air quality, 322
      and pollution standards, 282-283
      target systems, characteristics, 260
      taxes, 283
   visual, 259
   water
      in the Arctic, 276
      definition of, 264-265
      by detergents, 276
      as ecosystem modifier, 289
      effects on biotic communities, 277-278
      effects on salmon, 278
      effects on streams, 277

extent of, 266
by heat balance changes (thermal pollution), 276
and hepatitis, 279
and human health, 279
importance of quantitative aspect of, 266, 277
in Lake Erie, 270
in Lake Tahoe, 270, 280
in Lake Washington, 270
model of process, 268
as a natural phenomenon, 289
natural sources, 266
by oil, 280
perception of, 280, 289
roots in technology, 265
roots in western tradition, 265
Pollution control
by establishment of standards, 282-283
by governmental control of production, 281
natural, 285
problems with enforcing, 282, 283
problems of implementing, 281
by statute, 282
in the U.S.S.R., 281
Polyculture
definition of, 140
monoculture and, 137, 140
and pest and disease control, 140
using different genetic strains in wheat, 140
Population
age structure, 216
changes in, under exploitation, 216-217
and zero population growth, 103, 104
behavior, and social population regulation, 105
density, Earth, 71-72
Dominican Republican, age structure of, 69
estimates, comparisons of, 78
estimation
by United Nations, 73-75
by von Foerster's model, 75-78
food production subsystem, model of, 121-122
human, age structure of, 69
increase, coefficient for, 68

pressure, and arable land, 212
resource spiral, 541
shifts, urban, 80
size, 383
social regulation, 96
Swedish, 69
top carnivore, 354
Population control
in Australian aborigines, 92
and celibacy, 94
homeostasis in, 28
in Java, 95
and the Nambudiri Brahmins, 91-92
natural, 285
possible mechanisms of, 30
in red grouse, 30-32
and resource limitation, 114-115
and resources, 105
and social disruption, 92, 94
in the Tikopia, 93, 94
Population growth
effect of age structure on, 69
and environmental problems, 81, 83
and environmental variation, 29
exponential, equation for, 66, 68
feedback in, 60-61
Green Revolution's feedback to, 128
human, 60-67
positive feedback in, 69
and taxation, 70
in the United States, 83-84
zero, see Zero population growth
Population policy
factors contributing to inaction in, 107
in the United States, 103
Population regulation, social
in the Amish, 96
by birth suppression, 96
and group selection, 89-90
in the Hutterites, 96-97
in individual economies, 98
and population behavior, 105

Population regulation, social (*continued*)
  in primitive human societies, 90–91
  and resources, 99
Populations, adaptation patterns in, 28
Population stabilization, *see* Stabilization, population
Positive feedback, *see* Feedback, positive
Post hoc wisdom, in use of DDT, 355
Power, MHD (magnetohydrodynamic), 240
  and coal gasification, 240
Power, nuclear, *see* Nuclear power
Power generation, and solar energy, 238
Power plant leakage, of radioactive elements, 367
Power plants, nuclear
  and mariculture, 130
  and photosynthetic productivity, 348
Predation, 32
Predictions, of health impacts due to air quality control, 317–318
Preservation-recreation conflicts, in Yosemite National Park, 424–425
Preserves and parks, protection of, 376, 382
Primitive and modern societies, and environmental crisis, 482–483
Primitive society
  and environment, 65
  and population stabilization, 106
Principles, ecological, summary of, 37
Problems
  environmental, and population growth, 81, 83
  social inequity, in air quality control, 322, 324
Problem-solving
  environmental
    in biological systems, 549
    and ecological science, 549
    ecology and, 7, 21, 550
    information requirements for, 549
    in physical systems, 549
    social equity in, 541–542
    and solution strategy, 6
    strategies for, 546–548, 551
    and systems modeling, 550–551
  environmental and ecology, 550

  environmental and systems theory, 550–551
  urban, systems analysis in, 469
Production
  of energy from solid wastes, 340–341
  governmental control of, in pollution control, 281
  marine fish, estimates of, 125–126
  timber, in Harvard Forest, model of, 226–227
Productivity, photosynthetic, and nuclear power plants, 348
Profit maximization motive, and renewable resources, 211–214
Pronatalist attitude, 60
Property ownership, by government, as growth control mechanism, 188–189
Property taxation, based on potential vs. actual use, 182
Protection
  of agriculture, through urban growth control, 178
  environmental, and governmental reorganization, 521
  of natural ecosystems, through urban growth control, 178
  of open space, through urban growth control, 177
  of parks and preserves, 376, 382
Proxemics
  cross-cultural comparisons, 499, 502
  defined, 499
  model of, 500–501
  social signals in, 502–504
Prudhoe Bay, Alaska, 236
Public Law (P.L.) 480, 132
Public transportation alternatives, and internal combustion engine, 323–325
Puerto Penasco, closed agricultural production system at, 132
Pueschel, R. F., 308
Purification, water
  and algae, 268
  and conversion of organic wastes into inorganic nutrients, 268
  and decomposers, 267

and natural ecosystems, 266-267
and trophic concentration, 268
Put-and-take fishery, 395

Quality
air, *see* Air quality
environmental, definitions of, 258
water, and cities on Lake Erie, 270-271
Quantitative models, 550-551

Radioactive elements
as environmental hazard, 366
future estimated volumes of, 366
injection of, 367
methods for handling, 366
and power plant leakage, 367
salt mines as storage sites for, 367-368
transportation hazards of, 367
Radioactive waste disposal, 247-248
Rain, acid, 304-305, 319
from cooling towers, 351
Rainmaking, 158
Rainwater, L., 107
Raleigh, C. B., 361-362
Ratio, sex, and effects on reproduction, 24-25
Reaction, fission, 242, 243
Reactions, chemical, between air pollutants, 299-300
Reactor(s)
breeder, U^{238} (Uranium^{238}) in, 243
conventional (burner), 242
Recapture taxes, as growth control mechanisms, 185
Recreation
definition of, 409
development of, at Mineral King, 430-433
goals of, of national parks, 416, 423-424
Recreation-preservation conflicts, in Yosemite National Park, 424-425
Recycling, 335-337
Red grouse, population control in, 30-32
Red tide, 265
Redwood forests, 215
Redwoods National Park, 179

Reefs, artificial, 340
Reform, behavioral, and environmental quality, 530
Regeneration, nutrient, in the Amazon, 385-386
Regulation(s)
air quality, incentives in, 322
harvesting, in fish and game management, 221
pesticide, and Environmental Protection Agency, 352, 358
social population, *see* Population regulation, social
Rehumanization, urban renewal by, 455-456
Reich, C., 528-529
Relationships, interspecific, 32-33
Religion, and environmental crisis, 482
Remote sensing methods, in resource management, 218
Renewable resources, *see* Resources, renewable
Renewal, urban, 454-455
by rehumanization, 455-456
Reorganization, governmental, and environmental protection, 521
Reproduction
effects of age structure on, 24, 26
effects of sex ratio on, 24-25
Reproductive norms, factors affecting, 101-103
Reservoir failure, Baldwin Hills, 361-365
and subsidence, 365
Residents, urban planning by, 458-459
Resistance, pest to pesticides, 356-357
Resource management
by controlling supply, 214-215
use of computer simulation in, 219
use of fire in, 215-216
use of remote sensing in, 218
Resources
and birth control mechanisms, 99
common property, overexploitation of, 123-124
definition of, 114
limitations of, and population control, 114-115
mineral
national interdependencies for, 248
offshore, 250
nonrenewable, 234-235

Resources (*continued*)
   and population control, and negative feedback, 105
   renewable,
      economic decision-making in, 211
      exploitation of, 211–212, 230–231
      management of, 230–231
      profit maximization motive of, 211–214
      of whales, 212
   and social population regulation, 99
   substitution of, 214
Ricker, W. E., 142
Rights
   pollution, 283
      definition of, 283
      and free-market economies, 284
      limitations of, 284
      as means of regulating air quality, 322
      and pollution standards, 284
Rivkin, M., 507
Rogers, P., 476
Rubey, W. W., 361–362
Rubinoff, I., 397, 398–399, 405
Rudd, R. L., 352–353, 354
Rural-urban contrasts, in air pollution, 296
Ryan, J. M., 236
Ryther, J. H., 125, 126, 275–276

Saalfeld, R. W., 274
Saarinen, T., 481
Sahel drought, 170
Salmon
   effects of water pollution on, 278
   hatchery operation in, 396
Salmon fishery model, 228–230
Salt mines, as storage sites for radioactive elements, 367–368
San Diego, use of cause-condition-effect matrix in planning of, 200–202
San Francisco Bay–Delta, and water diversion, 164
Santa Barbara, oil spill at, 247
Santa Clara County, California, history of zoning in, 184, 186–187

Saunders, R. C., 278
Sax, J., 525
Schorger, Q. W., 405
Screwworms, sterile male technique as applied to, 136
Sea-level canal, *see* Panama Canal
Secondary pollutants, 260
Sectors, model, 228, 230
Selection, natural, goals of, 45
Seskin, E. P., 313–314
Sewage, ocean pollution by, 275
Sewage treatment
   and plant nutrition, 286–287
   types of, 286
   using activated sludge, 289
   using trickling filters, 288
   using waste heat, 286
Sex ratio, and effects on reproduction, 24–25
Shoshoni Indians, and environment, 65
Shrimp fisheries, and Texas Water Plan, 168
Sierra Club v. Froehlke, 524
Sierra Club, and Mineral King controversy, 431
Siltation effects, from Aswan High Dam, 159
Simulation, computer, in resource management, 219
Simulation modeling, applied to ecological planning, 204–205
Skinner, B. F., 530
Sludge, activated, sewage treatment using, 289
Smith, J. E., 275
Smith, K., 91
Smith, M., 301
Smog, photochemical, 300
Social disruption, and population control, 92, 94
Social inequities
   in air quality control, 322, 324
   of Green Revolution, 128
Social population regulation, *see* Population regulation, social
Societies
   modern
      and environmental crisis, 482–483
      and population stabilization, 106

primitive
   and environment, 65, 482–483
   and population stabilization, 106
Socolow, R., 149–150, 171
Soil fertility, and Aswan High Dam, 160
Soiling and chemical corrosion, effects of air pollution on, 309–310
Solar energy, 238
Soleri, Paolo, 81, 466–467, 468–469, 470–471, 474–475, 477
Solid waste
   as artifical reefs for fish, 340
   energy from, 240, 340–341
   estimated volume of produced, 329
   handling, 330
   methods for disposing, 330
   ocean dumping of, 331
   structural uses of, 339–340
   use of in counteracting subsidence, 340
Solid Waste Disposal Act, 332
Solid waste management
   and Environmental Protection Agency, 332
   systems analysis in, 334–335, 342
   and technological innovation, 334
Sommer, R., 502, 503–504, 506, 507
Sorensen, J., 198–199, 252
Southeast Asia, use of biocides in, 352
Southeastern U.S., air pollution in, 296
"Spaceship Earth," 514
Specialization, and extinction, 383–384
Species
   of fish, in Lake Erie, 271, 274
   indicator, 267
   opportunistic, 384
Species diversity
   meaning of, 26
   and stability, 26, 34, 386
Species replacement
   New Zealand sport fisheries in, 396
Spencer, P., 91
Sprague, J. B., 278
Spreiregan, P., 461–463

Stability
   and diversity, 386, 474
   of ecosystems, 387
   community, 33–34
      and community evolution, 33
      and diversity, 34
      and species diversity, 26
Stabilization, population
   through demographic transition, 73
   and modern society, 106
   and primitive society, 106
   world attitudes toward, and zero population growth, 105
Standards
   ambient air, 321
   emission, 321
   establishment of, in pollution control, 282–283
   pollution, and pollution rights, 284
Starfish, crown of thorns, 385
States, as model components, 46
Statute, pollution control by, 282
Steady-state economies, 515–517
   effect of on environmental quality, 516–517
   implementation problems of, 515–516
Stein, G., 286
Steinitz, C., 204, 476
Sterile male technique, 136
Stocking of fish and game, 220–221, 394–396
Stout, P. R., 136
Stratification, lake, 349–350
Stratified internal combustion engine, 325
Streams, effect of water pollution on, 277
Strip mining, 246, 546
Structure
   age, see Age structure
   population age, and zero population growth, 103, 104
   urban, and automobiles, 451–453, 497
Subsidence
   and Baldwin Hills Reservoir failure, 365
   and solid wastes, 340

Subsidy, agricultural, air quality as, 132
Substitution
    of nonrenewable resources vs. conservation as modes of management, 235
    resource, 214
Subsystem, population-food production, model of, 121–122
Subsystems
    and hierarchical organizations, 45–46
    and systems, 43–45
Succession, ecological, 400–401
    and causal mechanism, 36–37
    definition of, 35
Sun, energy from, 238
"Superexponential growth," 77
Supply and demand, failure of, 213–214
Suppression of births, as social population regulation, 96
Sustained yield, definition of, 210–211
Svala, G., 69
Swedish population, age structure of, 69
Swidden agriculture, 138–139
Symbiosis, cleaning, 388
System
    closed, vs. open, 44
    definition of, 40–41
    ecological, 7
    grassland as example of, 42–44
    as hierarchies, 7
    human
        and land development, 175
        limitations of, 21
    internal combustion engine as example of, 42–44
    open, vs. closed, 44
Systems
    biological, environmental problem-solving in, 549
    and data requirements, 48
    economic, closed vs. open, 514
    and environmental studies, 7
    feedback in, 46
    general, properties of, 44–48
    living, abiotic influences on, 9–10
    prediction of behavior of, 48
    and subsystems, 43–45
    and subsystems and hierarchical organization, 45–46
Systems, ecological, see Ecological systems
Systems analysis
    and solid waste management, 334–335, 342
    and urban problem-solving, 469
Systems consequences, Aswan Dam as example, 160
Systems ecology, 41–42
Systems impact, of chemical pest control, 139
Systems model, defining a, 42–44
Systems point of view
    advantages and disadvantages of, 51–52
    Commoner's axiom (statement) of, 537
    definition of, 51
    vs. specific model building, 51–52
Systems theory, and environmental problem-solving
    alternatives to, 551
    conceptual models of, 550
    data limitations on quantitative models of, 550–551

Tapiola Garden City, 459–463
    design elements of, 461–462
Target systems, pollution, characteristics of, 260
Taxation
    on forests, 223–224
    and population growth, 70
    property
        as growth control mechanism, 182
        based on potential vs. actual use, 182
    use-value, on land, 182
Taxes
    pollution, 283
    recapture, as growth control mechanisms, 185
    timber, 212, 223–224
Tchobansoglous, G., 287
"Technological fix," 250
Technological innovation, and solid waste management, 334

Technological input requirement, in Green
    Revolution, 127-128
Technology
  agricultural, as scaled to economy, 136
  and air quality control, 319-320
  and Green Revolution in India, 136
Temperature
  effects of on organisms, 347
  water, effect of nuclear power on, 347
Temperature inversion, 302-303
Tetrahedral city, 464-465
Texas Water Plan, 167-168
Thermodynamics, 10-11
Thompson, P., 198
Tidal movements, energy from, 241
Tikopia, population control in, 93
Tilapia (mouthbreeders), 129-130
Timber
  as agricultural crop, 223
  model of production of, 226-227
  taxation on, 212, 223-224
Topography, and air pollutants, 296, 300
Towers, cooling, 346-348, 351
  acid rains from, 351
Town and County Planning Act, 185
Towns, new, 459-464
Tradeoffs, economic, in air quality, 325
"Tragedy of the commons," and free goods, 512-513
TransAlaska Pipeline, 523
Transportation
  California Department of, land use control by, 176
  and cities, 449, 451-452
  public, alternatives and internal combustion engine, 323-325
Transportation hazards, of radioactive elements, 367
Triage system, 119
Trickling filters, sewage treatment using, 288
Trophic chain concentration of pesticides, 353
Trophic pyramid, 11
Trophic web(s), 11, 12
  energy flow through, 14-16

Tsavo National Park, Kenya, 433-437
  elephants in, 434, 436
  management problem in, 433-437
Tuan, Y. F., 482, 484
Tugwell, R. G., 526
Turnbull, A. L., 134-135
Turtle, green, 383
Tussock moth, and changes in pesticide policy, 359

U^{235} (Uranium^{235}), 242
U^{238} (Uranium^{238}), 242-243
United Nations Environment Conference, 542-544
United Nations population estimation, 73-75
United States
  air quality control in, 321
  evolution of land planning process in, 205
  pattern of energy use in, 237
  population growth of, 83-84
  southeastern, air pollution in, 296
  and world mineral resources, 248-249
United States governmental organization, defects of, 526
United States Soil Conservation Service, land-capability rating system of, 202
Upwelling, 11
Uranium, in breeder reactors, 243
Urban areas, systems analysis in, 469
Urban design
  elements of, 491, 494
  and fear of criminal harassment, 498
  and social structure of neighborhood, 498-499
Urban development
  models of evolution, 450-451
  three paradigms for, 456-458
Urban diversity, and stability, 474
Urban ecology, 473-474
  evaluation of analogies, 454
Urban evolution, 453
Urban growth control, *see* Growth control, urban
Urban heat island, 177
Urban model, of J. W. Forrester, 472-473

Urban planning by residents, as visualized by Robert Goodman, 458–459
Urban population shifts, 80
Urban problem-solving, systems analysis in, 469
Urban renewal, 454–455
  by rehumanization, 455–456
Urban-rural contrasts, in air pollution, 296
Urban structure, and automobiles, 451–453, 497
Use-value taxation on land, 182
U.S.S.R. (Union of Soviet Socialist Republics), pollution control in, 281
Uusimaa 2010 plan, 461, 463

Value, economic, establishment of, 513
Value (normative) judgments, importance of in land planning, 192
van Hyclkama, T. E., 149–150, 151
Variation, environmental, and population growth, 29
Visual pollution, 259
von Foerster, H., 75–78
von Foerster's model, population estimation by means of, 75–78
von Hertzen, H., 461–463
von Wodtke, M., 316–317, 326

Wagenseil, H., 182
Warren, C. E., 270, 288–289
Waste disposal
  ocean, 359
  radioactive, 247–248
Waste heat
  in Cayuga Lake, 349
  and nuclear power plants, 247
  sewage treatment using, 286
Waste management plan, Oregon, 332–334
Water
  compared to air, 293–294, 309
  use as dump, 265
Water conservation, 152–153
Water development
  agency behavior in, 147
  and naturally adequate supplies, 155

Water distribution, 158, 165, 168, 170
  by use category, 151
  global, 148–151
Water diversion, and San Francisco Bay–Delta, 164
Water impoundment, for water supply, 154
Water pollutants
  distribution and toxicity of, 269–275
  types of, 269
Water pollution, *see* Pollution, water
Water purification, *see* Purification, water
Water quality, and cities on Lake Erie, 270–271
Water resource development, 169
Watkins, T. H., 165
Watson, A., 31, 32, 107, 532
Watt, A. S., 402–404
Watt, K. E. F., 140, 219–220, 395, 477
Weather, effect on air pollution, 302
Welwyn, 460
Western grebe, 354
Whale fisheries, and investment in equipment for, 133–134
Whales, 383, 384
  and imminent population collapse of, 217
  in renewable resource use, 212
Whaling practices, legal controls on, 213
Wheat, and polyculture, 140
White, G. F., 169, 170–171
White, L. F., 265, 482
Whittaker, R. H., 9
Whooping cranes, 8
Whyte, W. H., 177, 183, 212, 411, 414, 415
  and open space, 411, 414
Wild and Scenic Rivers Act, 175–176, 428–429
Wilderness, perception of, 486
Wilderness Act, 213, 425
  and Bureau of Land Management, 428
  and Forest service, 428
  provisions of, 425
Williams, G. C., 532
Williams, R. B., 52
Williamson Act (California Land Conservation Act), 182–183

Wilson, E. O., 192
Wind, energy from, 241–242
Wind pattern and air pollution distribution, and air quality control, 315–316
Winkelstein, W., 313–314
Wisconsin hearings, 358–359
Wolman, A., 476
Wolves, and moose on Isle Royale, 400–402
Woodpecker, ivory-billed, 382
Woodwell, G. M., 278
World model, 48–51
Wynne-Edwards, V. C., 89, 106, 107, 532

Yannacone, Y. V., 525
Yarrow (*Achillea lanulosa*), ecotypes of, 24
Yield, sustained, definition of, 210–211
Yosemite National Park, recreation-preservation conflicts in, 424–425

Zapp, A. D., 236
Zero population growth
  and age structure, 104
  conditions for, 103
  and population age structure, 103, 104
  and world attitudes toward population stabilization, 105
Zinke, P., 138–139
Zoning, 183
  of floodplains and aquatic areas, 184–185, 189
  history of, in Santa Clara County, 184, 186–187
  variances in, 184
Zoos, 392–393
  and conservation, 392–393